SUSTAINABLE FOREST RESOURCES MANAGEMENT

Issues and Implications

SUSTAINABLE FOREST RESOURCES MANAGEMENT

Issues and Implications

Edited by

Jahangeer A. Bhat, PhD

Gopal Shukla, PhD

Manmohan J. Dobriyal, PhD

Sumit Chakravarty, PhD

A. Arunachalam, DSc

Rainer W. Bussmann, PhD

AAP | APPLE ACADEMIC PRESS

First edition published 2025

Apple Academic Press Inc.
1265 Goldenrod Circle, NE,
Palm Bay, FL 32905 USA

760 Laurentian Drive, Unit 19,
Burlington, ON L7N 0A4, CANADA

CRC Press
2385 NW Executive Center Drive,
Suite 320, Boca Raton FL 33431

4 Park Square, Milton Park,
Abingdon, Oxon, OX14 4RN UK

Library and Archives Canada Cataloguing in Publication

Title: Sustainable forest resources management : issues and implications / edited by: Jahangeer A. Bhat, PhD, Gopal Shukla, PhD, Manmohan J. Dobriyal, PhD, Sumit Chakravarty, PhD, A. Arunachalam, DSc, Rainer W. Bussmann, PhD.
Names: Bhat, Jahangeer A., editor | Shukla, Gopal, editor | Dobriyal, Manmohan J., editor. | Chakravarty, Sumit, editor | Arunachalam, A. (Ayyanadar), 1970- editor | Bussmann, Rainer, editor.
Description: First edition. | Includes bibliographical references and index.
Identifiers: Canadiana (print) 20240421639 | Canadiana (ebook) 20240421663 | ISBN 9781774917060 (hardcover) | ISBN 9781774918630 (softcover) | ISBN 9781003539070 (ebook)
Subjects: LCSH: Forest management. | LCSH: Sustainability.
Classification: LCC SD431 .S87 2025 | DDC 634.9068—dc23

Library of Congress Cataloging-in-Publication Data

Names: Bhat, Jahangeer A., editor. | Shukla, Gopal, editor. | Dobriyal, Manmohan J., editor. | Chakravarty, Sumit, editor. | Arunachalam, A. (Ayyanadar), 1970- editor. | Bussmann, Rainer, editor.

Title: Sustainable forest resources management : issues and implications / edited by Jahangeer A. Bhat, Gopal Shukla, Manmohan J. Dobriyal, Sumit Chakravarty, A. Arunachalam, Rainer W. Bussmann

Description: First edition | Palm Bay, FL : Apple Academic Press, 2025 | Includes bibliographical references and index | Summary: "Forests are one of our most valuable resources, yet they are fast depleting with an impending ecological crisis in view. The striking concept of sustainable forest management involves the use of socially justifiable, economically viable, and ecologically sound strategies for sustainably managing forest resources for present and future generations without diminishing their value. This new book, Sustainable Forest Resources Management: Issues and Implications, presents ideas of sustainable forest management from different viewpoints. The book details the issues of implicating sustainable forest management practices and highlights the aspects of forest resources, the importance of traditional knowledge, the need for conservation, the impact of climate change, and ecosystem services. It covers the scientific, technical, and socio-economical aspects of forests to further ensure the conservation of forest resources. The book explains the value of forest resources by emphasizing livelihood security and traditional knowledge. Moreover, it discusses the climate change effect on forest health and reviews ecosystem services from the management perspective. Topics included in the book include common ones such as the importance of forests, its benefits to communities, climate change, forest fires, etc. At the same time, it also covers specialized topics such as traditional knowledge of conservation status of medicinal plants, traditional methods of conserving medicinal plants, ecosystem services of agroforestry, trees as fodder, and so on. With special focus on Indian conditions and forests, the book presents possible suggestions and recommendations for managing forest resources on not just local but also the global level. The broad topics covered in the book will be useful to researchers, scientists, conservation activists, foresters, policymakers, and students in environmental science/conservation/biology/ botany programs across the world"-- Provided by publisher

Identifiers: LCCN 2024029202 (print) | LCCN 2024029203 (ebook) | ISBN 9781774917060 (hardback) | ISBN 9781774918630 (paperback) | ISBN 9781774917060 (ebook)

Subjects: LCSH: Forest management--India. | Sustainable forestry--India | Sustainable development--India

Classification: LCC SD431 .S87 2025 (print) | LCC SD431 (ebook) | DDC 634.9/2--dc23/eng/20240711

LC record available at https://lccn.loc.gov/2024029202

LC ebook record available at https://lccn.loc.gov/2024029203

ISBN: 978-1-77491-706-0 (hbk)
ISBN: 978-1-77491-863-0 (pbk)
ISBN: 978-1-00353-907-0 (ebk)

About the Editors

Jahangeer A. Bhat, PhD
Piramal Foundation for Education Leadership, Piramal University, Mumbai, India

Jahangeer A. Bhat, PhD, is currently associated with the Piramal Foundation for Education Leadership at Piramal University, Mumbai, India. Before taking on his current position, he was associated with the College of Horticulture and Forestry at Rani Lakshmi Bai Central Agricultural University, India. Earlier, he served as Head of the Department of Forestry at the College of Agriculture, Fisheries and Forestry at Fiji National University, Republic of Fiji Islands. His research work in India and Fiji Islands spans 12 years and encompasses emerging issues in forestry, including conservation of biodiversity, traditional knowledge of plants, and sustainable management of forest resources, particularly in the light of climate change scenarios. He has worked as counselor, mentor, and coordinator for forestry academic programs and has been instrumental in developing higher education and technical and vocational education and training streams of forestry and allied programs. He worked closely for accreditation with the Fiji Higher Education Commission and forestry stakeholders. He is the first researcher to report the first time growing of Acacia dealbata Link. (silver wattle), an invasive species in the high altitudes of the Himalayas. His focus of research is on vegetation ecology, ethnobotany, and evaluation of ecosystem services, forest plant biodiversity, climate change, and sociocultural issues in forestry.

Gopal Shukla, PhD
Professor, Department of Forestry, North Eastern Hill University, Tura Campus, Tura, Meghalaya, India

Dr. Gopal Shukla is currently a Professor & Head in the Department of Forestry at North Eastern Hill University, Tura Campus, Tura, Meghalaya, India. Prior to his current position, he worked as a faculty member in the Department of Forestry at Uttar Banga Krishi Viswavidyalaya, Cooch Behar, West Bengal, India. Dr. Shukla has over 13 years of experience in forestry teaching, research, and extension. His work has included projects such as the National Agricultural Innovation Project, National Innovations in Climate Resilient Agriculture (ICAR), and serving as a SERB (DST) young scientist on the project entitled "carbon stock of different land uses of Terai forest ecology. His research and development work focuses on agroforestry, medicinal plants, forest ecology, and conservation. Dr. Shukla has published over 100 papers and articles in refereed journals, 30 book chapters, and eight books. Additionally, he has participated in various conferences, handled three adhoc/extramural projects as PI and six as Co-PI, organized two National Seminars as Organizing Secretary, one ten-day training program as course coordinator, and nine farmer's training/awareness programs. Dr. Shukla has also contributed as a Member of the Faculty Council, Faculty of Horticulture, UBKV, Cooch Behar, and the Board of Studies (UG) Forestry, Bankura University, Bankura, West Bengal. Moreover, he was involved in framing the UG and PG forestry syllabus for Bankura University, Bankura, WB, and Amity University, Amity, UP. Dr. Shukla has guided four doctoral and six master's students in the field of forestry. He has been recognized with the SERB (DST) Young Scientist award in 2012, the Young Scientist award-2018, and has received the best oral and poster presentation award in numerous seminars/conferences. Dr. Shukla has also delivered invited lectures in seminars/conferences/training programs and chaired technical sessions in conferences. His educational background includes graduating from Dr. B. R. Ambedkar University, Agra, and obtaining postgraduate and doctorate degrees from Uttar Banga Krishi Viswavidyalaya, Cooch Behar, West Bengal.

Manmohan J. Dobriyal, PhD
Professor and Head, Department of Forestry, College of Horticulture and Forestry at Rani Lakshmi Bai Central Agricultural University, Uttar Pradesh, India

Manmohan J. Dobriyal, PhD, is a Professor and Head of Forestry at the College of Horticulture and Forestry at Rani Lakshmi Bai Central Agricultural University, Uttar Pradesh, India. He was a faculty member in different institutes, such as Doon (P.G) College of Agriculture Science and Technology (DCAST), Dehradun; Dolphin PG Institute of Bio Medical & Natural Sciences DIBNS, Dehradun, Uttrakhand; and College of Horticulture and Forestry CoHFat, Maharana Pratap University of Agriculture and Technology, Udaipur MPUAT, Udaipur/AU, Kota, Rajasthan, as Assistant Professor (Forestry). He also served as Associate Professor and Head of the Department of Silviculture & and Agroforestry at the College of Horticulture and Forestry, Navsari Agricultural University, Navsari, Gujarat, India. He has more than 19 years of experience in forestry research, teaching, and extension. His specialization is forest ecology and environment/silviculture, non-wood forest products (medicinal and aromatic plants) agroforestry, biodiversity conservation, and ecosystem services. He has published more than 100 papers and articles in refereed journals, attended various conferences, and operated research projects, as well as guiding many postgraduate and PhD students. He graduated with a degree in Forestry, followed by postgraduate and doctorate degrees from institutions that include GB Pant University, Pant Nagar; TNAU, Coimbatore; and Forest Research Institute, Dehradun, India.

Sumit Chakravarty, PhD
Professor, Department of Forestry, Uttar Banga Krishi Viswavidyalaya, Pundibari, Cooch Behar, West Bengal, India

Sumit Chakravarty, PhD, currently working as a Professor and Head of the Department of Forestry at Uttar Banga Krishi Vishwavidyalaya, Pundibari, Cooch Behar, West Bengal, India. With over 20 years of extensive experience in forestry training, research, and development, he holds a master's degree in forestry and a doctoral degree in Agronomy from Punjab Agricultural University, Ludhiana. Dr. Chakravarty has conducted research on various aspects of forestry, agroforestry, medicinal plants, and climate change, resulting in 120 research papers, 48 book chapters, and eight books. He has successfully completed and is currently involved in 15 research projects focusing on forest ecology, medicinal plants, and agroforestry. Additionally, he has mentored over 50 master's and doctoral students in the field of forestry. Dr. Chakravarty has participated in more than 50 national and international conferences as a presenter and invited speaker.

A. Arunachalam, DSc
Director, ICAR-Central Agroforestry Research Institute, Jhansi, Uttar Pradesh, India

A. Arunachalam, DSc, is a tropical ecologist globally noted for his contributions to soil ecology, ecosystem ecology, shifting agriculture, and land use dynamics, particularly in the Indian Himalayan region. He has more than 26 years of experience in teaching, research, and extension in the fields of forestry, biodiversity, ecology, and natural resources management. He has guided 15 PhD students; written/edited over 21 books in English and Tamil, and published over 240 research articles in national/international journals of scientific repute as well as over 120 book chapters, and 35 popular articles. Dr. Arunachalam worked as an Assistant/Associate Professor for Forestry at North Eastern Regional Institute of Science and Technology, Arunachal Pradesh, India; as Principal Scientist at the ICAR Research Complex for NEH Region (2009–2010); as Principal Scientific Officer at the ICAR HQ in New Delhi (2010–2020); and as Assistant Director General (International Relations) at

ICAR HQ during 2018–2020. Presently working as a Director at the Central Agroforestry Research Institute, Jhansi, Uttar Pradesh, he is also an adjunct professor at over 12 national/international universities. Dr. Arunachalam is the Project Coordinator of the ICAR-All India Coordinated Research Project on Agroforestry and the Coordinator of the Task Force of Himalayan Agriculture under the National Mission on Sustaining Himalayan Ecosystems, supported by DST, Government of India. Academically, he is a Gold Medalist for both his BSc and MSc programs at Madurai Kamraj University, India. He earned his doctorate from North-Eastern Hill University, Shillong, India, and did his postdoctorate work in Restoration Ecology at the University of Massachusetts, Boston, USA, and Biodiversity Monitoring at the Jawaharlal Nehru Centre for Advanced Scientific Research, Bengaluru, India.

Rainer W. Bussmann, PhD
Head, Department of Botany, State Museum of Natural History, Karlsruhe, Germany; Professor, Department of Ethnobotany, Institute of Botany, Ilia State University, Tbilisi, Georgia

Rainer W. Bussmann, PhD, is an ethnobotanist and vegetation ecologist. He currently serves as the Head of the Department of Botany at the State Museum of Natural History in Karlsruhe, Germany, and is a Full Professor of Ethnobotany at the Department of Ethnobotany, Institute of Botany, Ilia State University, Tbilisi, Georgia. Prior to his current roles, Dr. Bussmann held various academic positions, including Director of the William L. Brown Center at Missouri Botanical Garden, William L. Brown Curator of Economic Botany, and Senior Curator. He has also held academic appointments at the University of Texas at Austin and the University of Hawaii, both in the USA, and the University of Bayreuth, Germany. Dr. Bussmann earned his MSc (Diploma) in Biology at Universität Tübingen, Germany, in 1993 and his doctorate at the University of Bayreuth in 1994. His research focuses on ethnobotanical research and the preservation of traditional knowledge of the Andes, Caucasus, and the Himalayas. Dr. Bussmann has an extensive publication record, with over 350 peer-reviewed papers, over 1,300 peer-reviewed book chapters, and 38 authored or edited books. He is recognized as one of the most cited ethnobotanists globally and serves as editor-in-chief of the *Ethnobotany of Mountain Regions* book series published by Springer Nature. Dr. Bussmann has held leadership positions in various scientific societies and serves on the editorial boards of several journals.

Contents

Contributors

T. R. Ahlawat
Navsari Agricultural University, Navsari, Gujarat, India

Mohammad Afsar Alam
Department of Social Sciences, Fiji National University, Natabua Campus, Lautoka, Fiji

Nahila Anjum
Department of Botany, Baba Ghulam Shah Badshah University, Rajouri, Jammu and Kashmir, India

Shubham Anurag
Department of Environmental Sciences, Central University of Jharkhand, Ranchi, Jharkhand, India

Showkat Aziem
Department of Forestry and Natural Resources, H.N.B. Garhwal University, Srinagar, Garhwal, Uttarakhand, India

Archibold G. Bakare
College of Agriculture, Fisheries and Forestry, Fiji National University, Nausori, Fiji

Surendra Singh Bargali
Department of Botany, DSB Campus, Kumaun University, Nainital, Uttarakhand, India

Abdul Basit
Department of Horticulture, Faculty of Crop Production Science, The University of Agriculture, Peshawar, Pakistan

Jahangeer A. Bhat
Piramal Foundation for Education Leadership, Piramal University, Mumbai, India

Sarthak Bhattacharya
Department of Agriculture, School of Agriculture and Allied Sciences, The Neotia University, Sarisha, Jhinga, South 24 Parganas, West Bengal, India

Vedangini Bisht
Amity Institute of Forestry and Wildlife, Amity University, Noida, Uttar Pradesh, India

Bisworanjita Biswal
Agronomy Section, National Dairy Research Institute, Karnal, Haryana, India

Agniswar Jha Chakraborty
Department of Agriculture, School of Agriculture and Allied Sciences, The Neotia University, Sarisha, Jhinga, South 24 Parganas, West Bengal, India

Sumit Chakravarty
Department of Forestry, Uttar Banga Krishi Vishwavidyalaya, Pundibari, Cooch Behar, West Bengal

Murali Krishna Chattakonda
Amity Institute of Forestry and Wildlife, Amity University, Noida, Uttar Pradesh, India

A. D. Chaudhary
CAAST Secretariat, Directorate of Research, College of Forestry, Navsari Agricultural University, Navsari, Gujarat, India

R. S. Chauhan
Department of Forest Biology and Tree Improvement, College of Forestry, Navsari Agricultural University, Navsari, Gujarat, India

Nation Chikumba
Marondera University of Agricultural Sciences and Technology, Marondera, Zimbabwe

Pijush Das
Department of Agriculture, School of Agriculture and Allied Sciences, The Neotia University, Sarisha, Jhinga, South 24 Parganas, West Bengal, India

Rajeswari Das
Department of Soil Science, School of Agriculture, Gandhi Institute of Agriculture and Technology University, Gunupur, Rayagada, Odisha, India

Ningthoujam Surmina Devi
Department of Plant Protection (Plant Pathology), College of Horticulture and Forestry, Pasighat, Arunachal Pradesh, India

Shivani Dobhal
Forest Biology and Tree Improvement Department, College of Horticulture and Forestry, Pasighat, Arunachal Pradesh, India

Ashok Gehlot
Biotechnology Division, CSIR–IHBT, Palampur, Himachal Pradesh, India

Abhishek Ghosh
Department of Agriculture, School of Agriculture and Allied Sciences, The Neotia University, Sarisha, Jhinga, South 24 Parganas, West Bengal, India

Avijit Ghosh
ICAR–Indian Grassland and Fodder Research Institute, Jhansi, Uttar Pradesh, India

Dipti Grover
Kurukshetra University, Kurukshetra, Haryana, India

R. P. Gunaga
Department of Forest Products and Utilization, College of Forestry, Navsari Agricultural University, Navsari, Gujarat, India

Mohd. Hanief
Department of Botany, Baba Ghulam Shah Badshah University, Rajouri, Jammu and Kashmir, India

Asieleavio John
Department of Forest Products Utilization, College of Horticulture and Forestry, Pasighat, Arunachal Pradesh, India

Mohammed Latif Khan
Department of Botany, Dr. Harisingh Gour Vishwavidyalaya, Sagar, Madhya Pradesh, India

Abhay Kumar
Krishi Vigyan Kendra, Jehanabad, Bihar Agricultural University, Sabour, Bhagalpur, Bihar, India

Amit Kumar
Department of Geoinformatics, Central University of Jharkhand, Ranchi, Jharkhand, India

Ashok Kumar
Genetics and Tree Improvement Department, Forest Research Institute, Dehradun, Uttarakhand, India

Satya Kumar
Center for Ecological Economics and Natural Resources (CEENR), Institute for Social and Economic Change, Nagarbhavi, Bangalore, Karnataka, India

Sakul Kundra
Department of Social Sciences, Fiji National University, Natabua Campus, Lautoka, Fiji

Rakesh Kumar Maikhuri
Department of Environmental Science, H.N.B. Garhwal University, Srinagar, Garhwal, Uttarakhand, India

Ajay Maletha
Amity Institute of Forestry and Wildlife, Amity University, Noida, Uttar Pradesh, India

Arif Yaqoob Malik
Department of Education, UT of Jammu and Kashmir, India

M. S. Malik
Krishi Vigyan Kendra, Jehanabad, Bihar Agricultural University, Sabour, Bhagalpur, Bihar, India

Zubair A. Malik
Department of Biology, Government HSS Hardutooru, Anantnag, Jammu and Kashmir, India

K. Abha Manohar
Centurion University, Parlakhamundi, Odisha, India

Vivek Manyapu
Department of Biotechnology, National Pingtung University of Science and Technology, Pingtung, Taiwan

Mita Meher
Department of Soil Science, School of Agriculture, Gandhi Institute of Agriculture and Technology University, Gunupur, Rayagada, Odisha, India

Ajay Kumar Mishra
International Rice Research Institute South Asia Regional Center, Varanasi, Uttar Pradesh, India

Sukanya Misra
Rani Lakshmi Bai Central Agriculture University, Jhansi, Uttar Pradesh, India

Kingshuk Modak
Forest Ecology and Climate Change Division, Rain Forest Research Institute, Jorhat, Assam, India

Kalkame Ch. Momin
Department of Horticulture, North Eastern Hill University Tura Campus, Tura, Meghalaya, India

Murad Muhammad
Phyto-Ecology Lab, Department of Botany, University of Peshawar, Pakistan

Mohd. Hussain Naik
Department of Geography, Central University of Punjab, Bathinda, Punjab, India

Mohammed Nasir
Department of Forest Products Utilization, College of Forestry, BUAT, Banda, Uttar Pradesh, India

Sunil Nautiyal
G.B. Pant National Institute of Himalayan Environment (NIHE), Kosi-Katarmal, Almora, Uttarakhand, India

Ajeet Kumar Negi
Department of Forestry and Natural Resources, H.N.B. Garhwal University, Srinagar, Garhwal, Uttarakhand, India

P. R. Oraon
Krishi Vigyan Kendra, Jehanabad, Bihar Agricultural University, Sabour, Bhagalpur, Bihar, India

Vrutti Patel
CAAST Secretariat, Directorate of Research, College of Forestry, Navsari Agricultural University, Navsari, Gujarat, India

Subhashree Patra
Department of Environmental Sciences, Central University of Jharkhand, Ranchi, Jharkhand, India

Shaista Rashid
Department of Botany, Baba Ghulam Shah Badshah University, Rajouri, Jammu and Kashmir, India

Lakhpat Singh Rawat
Amity Institute of Forestry and Wildlife, Amity University, Noida, Uttar Pradesh, India

Khushnood Ur Rehman
Department of Botany, Islamia Collage University Peshawar, Peshawar, Pakistan

Qamer Ridwan
Department of Botany, Baba Ghulam Shah Badshah University, Rajouri, Jammu and Kashmir, India

Purabi Saikia
Department of Environmental Sciences, Central University of Jharkhand, Ranchi, Jharkhand, India

M. S. Sankanur
Department of Forest Biology and Tree Improvement, College of Forestry, Navsari Agricultural University, Navsari, Gujarat, India

Janmejay Sethy
Amity Institute of Forestry and Wildlife, Amity University, Noida, Uttar Pradesh, India

Swati Shabnam
Department of Agronomy, Faculty of Agriculture, Birsa Agricultural University, Ranchi, Jharkhand, India

N. N. Shahina
Siviculture and Forest Management Division, ICFRE–Rain Forest Research Institute, Jorhat, Assam, India

Sheetal Sharma
International Rice Research Institute South Asia Regional Center, Varanasi, Uttar Pradesh, India

Gopal Shukla
Department of Forestry, North Eastern Hill University, Tura Campus, Tura-794002, Meghalaya, India

Kaushal Singh
Department of Silviculture and Agroforestry, College of Forestry, Banda University of Agriculture and Technology (BUAT), Banda, Uttar Pradesh, India

Manoj Singh
Department of Zoology, Kalinga University, Naya Raipur, Raipur, Chhattisgarh, India

S. K. Sinha
Department of Forest Products and Utilization, College of Forestry, Navsari Agricultural University, Navsari, Gujarat, India

Sajitha Siril
Department of Forestry, Uttar Banga Krishi Vishwavidyalaya, Pundibari, Cooch Behar, West Bengal

P. S. Srikanth
Chemistry and Bioprocessing Division, Rain Forest Research Institute, Jorhat, Assam, India

N. P. Todaria
Department of Forestry and Natural Resources, H.N.B. Garhwal University, Srinagar, Garhwal, Uttarakhand, India

Saqib Ullah
Department of Botany, Islamia Collage University Peshawar, Peshawar, Pakistan

Vivek Vaishnav
Laboratory for Conservation and Genetic Improvement of Forest Trees, Department of Forestry and Environmental Science, Manipur University, Imphal, Manipur, India

Abdul Wahab
Shanghai Center for Plant Stress Biology, Center for Excellence in Molecular Plant Sciences, Chinese Academy of Sciences, Shanghai, China

Yashmita-Ulman
Department of Silviculture and Agroforestry, College of Horticulture and Forestry, ANDUAT, Kumarganj, Ayodhya, Uttar Pradesh, India

Mudasir Youssouf
Department of Environmental Sciences, Government HSS Chittergul, Anantnag, Jammu and Kashmir, India

Aadil Yousuf
Department of Civil Engineering, National Institute of Technology, Patna, Bihar, India

Titus J. Zindove
College of Agriculture, Fisheries and Forestry, Fiji National University, Nausori, Fiji

Preface

Forests are the backbone of the existence of lifeforms on Earth. However, the over-dependency on forest resources has led to their exploitation and degradation. Due to natural and anthropogenic causes, forests suffer from biodiversity loss, fragmentation, fire, and disease outbreaks. Persistent forest degradation may disrupt the ecological balance while widening the poverty gap. Hence, the scientific community has introduced the concepts of sustainable forest management as a socially justifiable, economically viable, and ecologically sound strategy for managing forest resources for present and future generations without diminishing their value.

Sustainable forest management includes the conservation of ecosystem biodiversity, prevention of forest fragmentation and conversion, support for forest-dependent communities, and formulation of site-specific policies. Therefore, this book, *Sustainable Forest Resources Management: Issues and Implications,* presents the ideas of sustainable forest management from various viewpoints. Additionally, the book details the issues related to implementing sustainable forest management practices and highlights aspects of forest resources, the importance of traditional knowledge, the necessity for conservation, climate change impacts, and ecosystem services.

Forest management encompasses scientific, technical, and socio-economic aspects of forests, ensuring the conservation of forest resources. More than half of the world's population depends directly or indirectly on forests. If the current exploitation trend continues, forests will not be preserved for future generations. Sustainable forest resource management aims to introduce advanced technologies and innovations to prevent resource depletion and ensure better resource accessibility for future generations. This book emphasizes the value of forest resources by highlighting livelihood security and traditional knowledge. Furthermore, the book discusses the impact of climate change on forest health and reviews ecosystem services from a management perspective. Sustainable forest management can only be achieved with a well-structured policy framework and systematic valuation studies. The book provides possible suggestions and recommendations for managing forest resources, particularly in Indian conditions.

Introduction

Forest resources, including timber, food, fodder, medicinal plants, and other non-timber forest products, are vital sources of health, food security, and socio-economic growth of forest fringe communities. However, degrading forests, combined with deforestation and resource exploitation, hamper the resiliency and balance of the ecosystem. This necessitates improved studies on managing forests and has resulted in the concept of sustainable forest resource management. This book opens up new horizons on the sustainability of forest resources through proper management and implications. A brief highlight of the chapters follows:

Chapter 1 discusses the socio-economic and ecological importance of forest genetic resources. In the era of diverse anthropocentric and natural threats, the chapter showcases how conserving forest resources through sustainable management practices ensures continuous accessibility and availability of forest genetic resources. Chapter 2 highlights the value of ethnomedical plants in health, nutrition, and livelihood improvement and the urgent need for their conservation and sustainable use. Chapter 3 emphasizes the demand and potential of forests and discusses the managemental aspects of forests for maintaining resource production and securing livelihoods.

Chapter 4 depicts forests as cores of ecosystem services and pinpoints the necessity of designing sustainable conservation strategies. Chapter 5 demonstrates the effect of climate change on forest health, including discussion of forest fires, pest and disease infestations, and invasive alien species invasion. Moreover, it evaluates biodiversity and productivity loss of global forests in extreme climatic events. Chapter 6 highlights that plant species, especially those growing in hilly areas, are extremely vulnerable to climatic alterations, as their distribution is predominantly determined by climatic factors. Chapter 7 assesses the susceptibility of forest fires to climate change geospatially. Also, the chapter recommends self-sustaining and community-based techniques for controlling forest fire incidents to encourage the forest's resilience.

Chapter 8 documents the plant-based traditional health care system in the remote regions of Govind Pashuvihar Wildlife Sanctuary, India. The study highlights the vital role of traditional knowledge in forest management by listing ethnobotanical uses of 87 species from the forests that have potential to cure different ailments. Chapter 9 reviews the importance of traditional knowledge and the conservation of medicinal plants in the Uttarakhand region. The chapter also recommends expert suggestions for socioeconomically improving the forest fringe communities through the scientific and systematic use of forest resources that have medicinal value.

Chapter 10 describes the concepts of traditional ecological knowledge (TEK) among the people of Northeast India. Furthermore, the chapter envisages TEK as keystone information for designing forest resource management without compromising biodiversity conservation and ecosystem services. Chapter 11 explains the Himalayas as a treasure trove of medicinal plants and reveals the use, opportunity, and threats of Himalayan medicinal plants in the post-COVID scenario.

Chapter 12 studies the impact of atmospheric nitrogen deposition on forest soil organic carbon and plant productivity. Also, the investigation exposes the need for simulated N fertilization modeling for framing policy-level strategies to alleviate the negative impacts of nitrogen accumulation. Chapter 13 discusses the utilization of forest waste for the creation of value-added wooden products and its contribution to livelihood developments. Chapter 14 deals with the perspectives and constraints of energy generation from municipal solid wastes and represented it as a sustainable waste management solution. Chapter 15 details the ecosystem services of agroforestry, a traditional sustainable practice, through a literature review. The chapter highlights the provisional, regulatory, and supporting roles of agroforestry systems. In addition, it recommends agroforestry practice as a suitable land-use option for achieving sustainable development goals.

Chapter 16 discusses the various tree species traditionally grown or retained in agriculture fields for providing multipurpose products like fruit, fodder, fuel, timber, and medicine in different agroclimatic zones of India. Chapter 17 discusses the scope of fodder tree resources for sustaining livestock feed. And it shows the need for proper awareness about the species, nutrients, and management among the farmer community. Chapter 18 emphasizes prospective methods for transitioning from degenerative agriculture through carbon farming, agroforestry, and bio-inputs. Furthermore, nature-based solutions are intended to improve ecosystem services, which in turn recover damaged land, boost carbon sequestration, and increase crop resilience and yield.

CHAPTER 1

Conservation Status and Sustainable Management of Forest Genetic Resources

SHIVANI DOBHAL,[1] ASHOK KUMAR,[2] KALKAME CH. MOMIN,[3] NINGTHOUJAM SURMINA DEVI,[4] and ASIELEAVIO JOHN[5]

[1]*Forest Biology and Tree Improvement Department, College of Horticulture and Forestry, Pasighat, Arunachal Pradesh, India*

[2]*Genetics and Tree Improvement Department, Forest Research Institute, Dehradun, Uttarakhand, India*

[3]*Department of Horticulture, North Eastern Hill University Tura Campus, Tura, Meghalaya, India*

[4]*Department of Plant Protection (Plant Pathology), College of Horticulture and Forestry, Pasighat, Arunachal Pradesh, India*

[5]*Department of Forest Products Utilization, College of Horticulture and Forestry, Pasighat, Arunachal Pradesh, India*

ABSTRACT

The genetic diversity observed in forest tree species reflects the genetic resources of the forest, an intergenerational resource of enormous social, economic, and ecological importance. The most important task in this context is the protection of forest genetic resources. Conservation of forest genetic resources includes methods and laws that ensure the continuity, development, and accessibility of these resources for the present and the future. Conservation of genetic resources is essentially a dynamic process, as are genetic resources themselves. Consequently, conservation of these resources should be understood as an effort to preserve individual genotypes or populations and the various combinations of genes that comprise them. Thus, the goal

Sustainable Forest Resources Management: Issues and Implications. Jahangeer A. Bhat, Gopal Shukla, Manmohan J. Dobriyal, Sumit Chakravarty, A. Arunachalam, and Rainer W. Bussmann (Eds.)

of genetic resource management is to maintain the environment that makes possible the genetic diversity of a species. The conservation status of genetic resources refers to their current state and the urgent threats they face. In general, it is desirable to protect forest genetic resources by maintaining the ability to continue evolutionary processes in designated conservation populations, ideally through *in situ* methods. Overuse, forest fragmentation, degradation, overgrazing, invasive alien species, and climate change pose significant challenges to forest genetic resources. Most of the FGR collected has yet to be classified and evaluated, which is critical for improvement and use in reforestation, plantations, and agroforestry. Apart from the current FGRs at various organizations, there is a need to study, collect, record, conserve, and manage FGRs of commercially important tree species.

1.1 INTRODUCTION

It has been assumed that by 2050, the present world's 7.2 billion population will be reaching 9.6 billion. As the population is increasing, it results in an increasing need for wood and energy products. Forest-related goods for both domestic and industrial usage will be increasing by 40% in the coming 20 years (FAO, 2014). FGRs (Forest Genetic Resources) comprise a significant biodiversity subset as they are unique and irreplaceable resources of forest products and renewable resources. For fodder and food security, as well as sustainable forestry, the most significant component is FGR. There exists a varied use of FGR, as well as a component range, is enclosed to inter-specific from intraspecific genetic diversity from the ecologically or taxonomically related species' set to forest species' complete range which are significant as well as used potentially (Young et al., 1999). Generally, plant species are included by the FGR which provides medicine, fodder, and food along with acts as a source of environmental protection, energy, shelter as well as other livelihood supports. To meet the demand and supply of wood and wood products, there is tremendous pressure in conservation and management strategies for FGR in India. It is important for changing the environment's adaptation and sustainable development (Rawat & Ginwal, 2009). Thus, FGRs' sustainable management, as well as conservation, must guarantee that future as well as the present generation should have trees as well as forests' benefits. The FGR management's main objectives include practices related to forest management and their future selection, land use patterns, accounting different threats levels, routine consideration of gene conservation during management and planning process, encouraging planners, land, and forest managers which affects the FGR and breeding program. The major threat involves human practices and encroachment, changing physical and biological environments, indiscriminate burning, pests and diseases, invasive weeds, and extreme climatic events. Land tenure issues and land use policies, FGR's sustainable utilization and conservation, lack of importance awareness, lacking the capacity (finances, infrastructure, and trained human resources) for FGR projects' development, implementation, and monitoring (Bhatt et al., 2013). Hence, it is required that FGR related issues must be addressed with the help of a thorough strategy of conservation and development of FGR as well as implementation planning (Stedman-Edwards, 1998; Brown & Pearce, 1994; Bawa & Dayanandan, 1998).

1.2 FOREST GENETIC RESOURCES STATUS

There exist 16 major forest types in Indian forests, as well as a wide array of species diversity housed by these forests. In the world, India is among 17 mega-diverse countries, and in terms of mega-biodiversity, it is ranked 8th in the world. In terms of cultivated plant origins, India is among 8 centers (Vavilov, 1951), and among the world's 12 mega-genes which possess 14.57% of the world flora. Among the 34 global biodiversity hotspots, it houses 4 parts, including the Sundaland part in the Nicobar Islands, Indo-Burma, Sri Lanka/Western Ghats, and Eastern Himalayas. In Asia and globally, India is ranked 4th and 10th, respectively, in terms of plant diversity. There are 26 floristically important regions that exhibit an exceptional concentration of endemic species, and with the higher occurrence of threatened plant species, habitat loss is also experienced (Katwal, 2003). Around 38% of the country's recorded flora is regionally endemic, primarily found in the Andaman and Nicobar Islands, the northwest Himalayas, the Western Ghats, and the Northeast. Considering the data available for non-flowering plants, they contribute about 28% as endemic non-flowering plants (pteridophytes and bryophytes) (BSI, 2017). Taking into consideration the high level of endemism and exceptional universal values in the Western Ghats, UNESCO (United Nations Educational, Scientific, and Cultural Organization) listed 39 sites in its 2012 list (MOEF & CC, 2014). In an investigative report by the Botanical Survey of India (BSI), it has been discovered that there are 4,65,688 plant species, representing 14.57% of the world's flora, with nearly 18,465 species accounted for as flowering plants. The BSI has also identified that locally threatened species exist in large numbers, and regular assessments are made for threatened plant species, including FGR. Among these threatened species are hundreds of Fungi, Lichens, Bryophytes, and Pteridophytes, as well as 1,700 species of flowering plants.

According to the current knowledge state, higher plants' 18,465 species (79 Gymnosperms and 18,386 Angiosperms) are recorded in India (BSI, 2017). Forest habitats contain higher plant diversity of 80%, which is approximately 14,500 species. Half of the forest plant diversity comprises FGRs (≈7,250 species), while others include herbaceous flora such as grasses, herbs, twiners, and soft climbers. The increasing human population and livestock in the country, along with their health and food security, are potentially ensured by FGR. Due to several factors, both man-made and natural, our country's biodiversity faces significant threats. It has been predicted that worldwide threatened plant species are 26,106. Although the geographical area covering the country represents 2.4% of the total landmass of the world, it embraces 48,655 plant species (BSI, 2017). Nearly 38% of Indian plant occurrences are country endemic. A comparative species account in greater plant groups like lichens, fungi, algae, bacteria, and viruses noted in the world and India is represented in Table 1.1. Furthermore, approximate species numbers recognized as endemic to India are also presented in Table 1.1. Additionally, 1.8% of the world's forest area and 2.5% of the world's land surface area are accounted for by the country. Due to the vast variability of the biogeographic zones in India, floristic diversity is exceptionally rich. The country's botanical wealth comprises 15,000 higher plant species, among which nearly 33% or 4,900 species are endemic.

TABLE 1.1 Status of Different Plant Species (Including Lower Microorganism)

SL. No.	Type	Known Species Number		Occurrence Percentage in India	Endemic Species in India	Threatened Species in India
		World	India			
I.	**Flowering Plants**					
1.	Angiosperms	2,68,600	18,386	6.84%	4,303	1,700
2.	Gymnosperms	1,021	79	7.73%	12	7
II.	**Non-Flowering Plants**					
1.	Bryophytes	16,236	2,748	16.92%	629	c. 80
2.	Pteridophytes	12,000	1,289	10.74%	66	414
III.	**Others**					
1.	Virus and bacteria	11,813	1,170	9.90%	Not known	Not known
2.	Algae	40,000	7,357	18.39%	1,924	Not known
3.	Fungi	98,998	15,115	15.26%	c. 4100	ca. 580
4.	Lichen	17,000	2,511	14.77%	c. 520	Not known
Total		**4,65,688**	**48,655**	–	–	–

Source: BSI (2017).

1.2.1 FOREST COVER AND CLASSIFICATION

Table 1.2 presented the classification of land use in India. In the country, the most significant land uses are agriculture as well as forestry. It has been suggested by the present pattern of land use that there exists a possibility to cover areas under non-cultivation and fallow land that comprised up to 17% of the land area in the coming years. In India, there was only 40.48 million hectares (14.2%) area under forest during 1950–1951. But in 2017–2018, improvement has been seen to 80.20 million hectares (24.39%). This category includes the land which is used for marshes, deserts, mountains, quarries, canals, transport routes, human settlements, etc. It is accounted for as India's total land, 12.11%. Under this category, more area is possessed by Madhya Pradesh, Gujarat, Rajasthan, and Arunachal Pradesh states. Including fallow land, other uncultivated land consists of cultivable waste, groves, crops, land under miscellaneous trees, grazing area, and permanent pasture. The country's total reporting land, 8.6%, is covered by this category. India's total reporting land, 45.82%, is recognized as the net sown area (Table 1.2). During the last five decades, the net sown area has phenomenally increased. Such increase results from pastureland, uncultivable land, barren reclamation, etc. The maximum level has been reached by the net sown area in Haryana. Net sown area is less than 30% in Uttar Pradesh, Himachal Pradesh, Jammu and Kashmir, Meghalaya, Punjab, Nagaland, Mizoram, and Arunachal Pradesh.

TABLE 1.2 Landusepattern Study from 2009–2010 in India

Land Uses	2009–2010	Percentage Land Areas (%)
Reporting area for land utilization statistics	3,05,611	–
Forests	70,042	22.92
Not available for cultivation	42,954	14.06
Permanent pastures and other grazing lands	10,149	3.32
Land under miscellaneous tree crops and groves	3,351	1.10
Culturable waste land	12,857	4.21
Fallow lands other than current fallows	10,484	3.43
Current fallows	15,753	5.15
Net area sown	1,40,022	45.82
Total cropped area	1,92,197	62.89
Arca sown morc than oncc	52,175	17.07

Source: Rawat & Ginwal (2009).

In the world, India is the seventh-largest country with 7,08,273 sq. km forest cover that is the country's geographic area's 21.54% (Anon, 2017). According to the India Forest State Report's latest assessment, since 2015, in two years an increase of 8,021 sq. km or 1% has been registered by India's forest and tree cover. Against the decreasing forest global trend, an increasing trend has been shown by India in tree and forest coverage as covered as per the last decade's report, 7,08,273 sq. km forest cover that is the country's geographic area's 21.54%. The combined tree and forest coverage is total geographical area's 24.39% or 8,02,088 sq. km. Indian forests' 60% area is presented in the ecologically sensitive zones, for example, the arid and semi-arid tracts, the mountain areas, the Western Ghats, and the Himalayas. The total forest cover according to the latest assessment is 7,08,273 sq. km that is the country's geographic area's 21.54%. In terms of density classes, the VDF covered area is 98,158 sq. km, which with OF is 301,797 sq. km and MDF is 308,318 sq. km equivalent to the country's geographical area's 2.99, 9.18, and 9.38% respectively (Table 1.3).

TABLE 1.3 Class-based Details of Forest Cover of India

Class	Area	Geographic Area Percent (Percentage of Land Area)
Moderately dense forest	3,08,318	9.38
Non-forest	25,33,217	77.06
Open forest	3,01,797	9.18
Scrub	45,979	1.4
Total forest cover	7,08,273	21.54
Very dense forest	98,158	2.99
Total geographical area	3,287,469	100
Include 4,921 sq. km under mangrove cover percentage rounded off.		

Source: Forest Survey of India (2017).

In a specific analysis of forest cover status based on greenwash area in the states of North East region (Table 1.4) reveals that forests within green wash areas have shown a considerable decrease in the "Arunachal Pradesh, Meghalaya, Mizoram, Nagaland, and Tripura's North" Eastern States due to shifting cultivation as well as other development pressures. Much of the decrease in Sikkim has been observed outside the Green wash area.

As per the present assessment, the forest cover in the Northeastern region accounts for 65.34% of the geographical area, which is 171,306 sq. km, compared to the national forest cover of 21.54%. A reduction of 630 sq. km

TABLE 1.4 Forest Cover of the States in the Region NE for 2017

State	Geographical Area (as per census)	Green Wash Area	Forest Cover Within Green Wash 2015 (Updated)				Forest Cover Within Green Wash 2017		
			Total	OF	MDF	VDF	OF	MDF	VDF
Arunachal Pradesh	84743	64045	58782	11815	27799	19168	11876	27629	19095
Assam	78438	27560	19999	8922	9735	1342	8686	8775	2553
Manipur	22327	17542	15241	8969	5545	272	8568	5983	898
Meghalaya	22429	17562	14868	6606	7897	383	6587	7791	410
Mizoram	21081	20001	18019	12277	5607	135	11578	5592	126
Nagaland	16579	10672	9046	4469	3387	1190	4269	3323	1185
Sikkim	7096	2747	2360	343	1577	440	337	1069	952
Tripura	10486	7765	6600	2467	4033	100	1393	4510	546

in forest cover has been observed in the North Eastern region. This decrease is mainly attributed to shifting cultivation and other biotic pressures in the area. Table 1.5 presents the State-wise forest cover area, along with changes compared to the previous assessment.

According to news reports in India, there has been a 1.36% increase in dense forest compared to the 2015 results. The country's total forest and tree coverage have also increased by 8,021 sq. km compared to the 2015 assessment. It is encouraging to note a significant increase in the Very Dense Forest (VDF) category in forest cover. Arunachal Pradesh has the largest area in the country's northeastern region with 66,964 sq. km. Mizoram has the highest forest cover at 86.27% in relation to the total geographical area. Additionally, five states have experienced decreased forest cover: Meghalaya (116 sq. km), Tripura (164 sq. km), Arunachal Pradesh (190 sq. km), Nagaland (450 sq. km), and Mizoram (531 sq. km). These states, being part of the Northeast, each have more than 70% forest cover. The primary reasons for the reduction in forest cover include natural disasters, agricultural expansion leading to forest cover submergence, diversion of forest lands for developmental activities, rotational felling, other biotic pressures, and shifting cultivation.

In the world, India has the 10th rank with 24.4% forest and tree coverage land area, although its surface area accounts for only 2.4% of the world's total. India also sustains 18% of the livestock population and meets 17% of human needs. An analysis of observed changes in the country generates a change matrix that indicates the extent of change in forest cover classes. The change matrix (Table 1.6) shows changes in forest cover across three density classes, non-forest, and scrub areas. A growth of 12,183 sq. km in MDF density is observed, while there is a change from VDF to NF/Scrub/OF and MDF by 4,148 sq. km, respectively, indicating deforestation or degradation in these areas. Similarly, a decrease in density is shown by MDF of 17,520 sq. km as it converts into NF, Scrub, and OF. Additionally, Scrub of 6,360 sq. km and NF of 24,175 sq. km have been enhanced into classes of higher density, indicating improved plantation and conservation procedures. Forest land diversion for encroachment, non-forestry purposes, as well as other natural and biotic forest disturbances, are the reasons for degradation and deforestation.

1.2.2 STATUS OF NON-TIMBER FOREST PRODUCTS

Across the globe, the NTFPs have a significant part in millions of urban and rural people's livelihoods (Pandey et al., 2011, 2016; Malhotra & Bhattacharya, 2010; Shackleton et al., 2015). It has already been recognized that

TABLE 1.5 Northeastern States Forest Coverage

State	Geographical Area	2017 Assessment								Change	Scrub
		VDF	% VDF	MDF	% MDF	OF	% Of	Total	Percent Forest to GA		
Tripura	10846	656	8.49	5246	67.9	1824	23.6	7726	73.7	–164	15
Sikkim	7096	1081	32.3	1575	47.1	688	20.6	3344	47.1	–9	307
Nagaland	16579	1279	10.2	4587	36.7	6623	53	12489	75.3	–450	503
Mizoram	21081	131	0.72	5861	32.2	12194	67.1	18186	86.3	–531	0
Meghalaya	22429	453	2.64	9386	54.7	7307	42.6	17146	76.5	–116	505
Manipur	22327	908	5.23	6510	37.5	9928	57.2	17346	77.7	263	1111
Assam	78438	2797	9.95	10192	36.3	15116	53.8	28105	35.8	567	217
Arunachal Pradesh	83743	20721	30.9	30955	46.2	15288	22.8	66964	80	–190	247
Total	**262539**	**28026**	**16.4**	**74312**	**43.4**	**68968**	40.3	**171306**	**65.3**	**–630**	**2905**

TABLE 1.6 Change in Forest Cover of India between 2015 and 2017

Class	2017 Assessment					Total ISFR 2015 (Updated)
	NF	**Scrub**	**OF**	**MDF**	**VDF**	
Very dense Forest	404*	7*	357*	3380*	84485	88633
Moderately dense Forest	6003*	307*	11210*	283036	12183	312739
Open Forest	14647*	2389*	263822	7928	1337	300123
Scrub	4505	31437	5833	552	5	42302
Non-Forest	2507658	11839	20575	3452	148	2543672
Total ISFR 2015 (Updated)	2533217	45979	301797	308318	98158	3287469
Net Change	–10455	3677	1674	–4421	9525	

*Loss in comparative assessment from 2015 to 2017.

Source: Forest Survey of India (2017).

human wellbeing is supported by various functions by NTFPs. Various products for energy, fibers, medicines, shelter, food, and a considerable proportion of less poor people and the world's poorest people's cultural artifacts are provided by the NTFPs (Saha & Sundriyal, 2012; Vantomme, 2003; Saxena, 2003; Kaimowitz, 2003; Hegde & Enters, 2000). These daily net resources' contribution to livelihoods usually falls within the range of 10–60% of total household income (Babulo et al., 2009; Asfaw et al., 2013). Also, various households are provided by NTFPs which help in generating income; it can be the primary income that helps in the generation of cash or supplementary income that helps in other livelihood activities (Shackleton et al., 2008; Marshall et al., 2006; Mahapatra et al., 2005; Babulo et al., 2009; Areki & Cunningham, 2010). In developing countries, NTFPs' significance in rural livelihoods has been widely acknowledged. Furthermore, in India, a US$ 2.7 billion per year equivalent income is contributed by NTFPs, and in the forestry sector, 55% of total employment is absorbed by it. Additionally, these resources help in forest-based export income, accounting for 70%, as well as forest revenue, accounting for 50% (Chauhan et al., 2008; Shiva & Verma, 2002). According to an assessment, in India, there are 275 million poor rural people who are dependent on NTFPs for minimal livelihoods of cash as well as subsistence (Pandey et al., 2016). However, one of the primary concerns is NTFPs' resource depletion, which explains forest degradation, deforestation, and indiscriminate exploitation, affecting the economics and livelihoods based on NTFPs. In comparison to developed countries, the significance of NTFPs (Non-timber Forest Products) for rural livelihoods in developing countries like India is less appreciated. Furthermore, according to policy-makers, an assumption has been made that harvesting wild plant or animal products from any modified or natural ecosystem and forest will be harmful to the long-term viability of target species and populations (Shackleton et al., 2018).

NTFP harvest sustainability is based on the organs that are harvested as well as the harvested species' life cycle. Standardized good harvesting/collection practices have been established for a few significant medicinal plants such as the bark of arjuna (*Terminalia arjuna*), kalmegh (*Andrographis paniculata*), sarpagandha (*Rauvolfia serpentina*), gudmar (*Gymnema sylvestre*), baheda (*Terminalia bellerica*), baividang (*Embelia tsjeriam cottam*), and aonla (*Phyllanthus emblica*) (Pandey, 2009). The quality of produce and economic returns, the socio-economic status of the community, and resource conservation have been positively impacted when sustainable harvesting practices are adopted at the right time.

1.2.3 CONSERVATION AND CULTIVATION STRATEGIES FOR MEDICINAL PLANTS

It is of utmost importance to develop strategies for sustainable conservation of medicinal plant species. It has been estimated by the experts that for pharmacological activity for the world's 5–10% plants a systematic investigation has been performed. The tropical forest as the most threatened species. There is a need to develop a systematic approach in the direction of conservation with the help of legal matters as well as biotechnology. Medicinal and herbaria plant gardens are established by the ICFRE (Indian Council of Forest Research and Education) research institutes as well as for economically significant medicinal plants' cultivation various packages have been developed with the latest techniques which include genetic engineering and tissue culture. For addressing the need for native medicinal plant species' conservation, biogeographic regions' forest sites network must be established in India. Although, the most cost-efficient means of managing intra- as well as interspecific diversity is *in situ* (field) gene banks network, which is an actual forest habitat. Furthermore, various ICFRE institutes are researching a particular species for germplasm conservation.

1.2.4 RARE AND THREATENED MEDICINAL PLANTS

As per Indian Plants' Red Data Book (Nayar & Sastry, 1990), there exist various species of medicinal plants that come under threatened and rare plants category of IUCN. As below, a list of these species has been provided and in for their conservation, they require sufficient protection so that they may come out of threatened status (IUCN, 2015).

1.2.5 BAMBOO RESOURCES

Under 23 genera, there exist 124 exotic and indigenous species that are discovered either under cultivation or naturally or both (Naithani, 1993). Bamboos' total genetic resources of 50% have been supported by North-east India (Naithani, 2008). Bamboos are a very significant non-wood forest resource that is discovered in the country's non-forest as well as forest areas. They are natural resources and are low-cost, versatile, renewable, widely present, and fast-growing. They can also be referred to as poor man's timber and green gold. Eastern India has greater than 50% bamboo species, which occurs in West Bengal, Tripura, Sikkim, Nagaland, Mizoram, Meghalaya, Manipur, Assam, and Arunachal Pradesh. In the eastern region, like West Bengal, Assam

Assessment of Bamboo Resources of India 2, and Northeastern Himalayas, the commercially important bamboos are *B. tulda, D. hamiltonii*, and *Melocanna baccifera*. The information about the bamboo area's state-wise distribution in recorded forests by site quality-wise is given in Table 1.7. Pure bamboo's occurrence was observed maximum as 226 km^2 in Mizoram, then 217 km^2 in Arunachal Pradesh, with 192 km^2 in Manipur and 101 km^2 in Nagaland. The dense bamboos were discovered in maximum areas of 8,681 km^2 in Arunachal Pradesh, then by 6,116 km^2 in Mizoram and then 5,825 km^2 in Manipur.

1.3 NATIONAL PROGRAMS FOR FOREST GENETIC RESOURCES' CONSERVATION AND MANAGEMENT

In current trends of available data, many species are already extinct, and a few are threatened and on the verge of extinction if suitable conservation measures are not taken. Since time immemorial, practices related to the conservation of individual species and habitats have been carried out in India. The increasing focus on conservation reflects a growing concern for the overall productivity, long-term health, and alterations of forest ecosystems. The 1992 "Convention on Biological Diversity" affirmed that states have sovereign rights over biological resources, and it is the responsibility of states to utilize and conserve biological diversity sustainably. This focus can be attributed to the ethical, economic, social, and ecological values of biodiversity. National programs and policies related to Forest Genetic Resources (FGR) encompass a wide range of activities, including conservation procedures to protect endangered and rare populations and species, regulations governing seed collection, transfer of socio-economically significant tree species, and comprehensive methods for managing ecosystems and landscapes. In India, considerations related to FGR have been integrated into broader frameworks such as national biodiversity action plans and forest programs (Indian Forest Act, 1927; Forest Conservation Act, 1980; Seed Bill, 2004; Biodiversity Bill, 2002). The primary objective of the National Forest Policy (1988) was conservation, focusing on preserving the country's natural heritage. The most effective and secure conservation strategy combines two complementary procedures: *ex-situ* and *in-situ*.

1.3.1 IN-SITU CONSERVATION

In a conservation network, target species that are identified must have systematic and planned management as per *in situ* conservation which aims

TABLE 1.7 Bamboo Acreage in Each State of NE Region by Density Classes

States	Pure Bamboo	Dense	Scattered	Bamboo Present but clumps completely hacked	Regeneration Crop	Total	RFA
Arunachal Pradesh	217	8681	6953	144	88	16083	51540
Assam	105	4049	2878	166	40	7238	26832
Manipur	192	5825	3101	124	61	9303	17418
Meghalaya	63	2815	1830	68	17	4793	9496
Mizoram	226	6116	2757	104	42	9245	16717
Nagaland	101	3064	1644	65	28	4902	9222
Sikkim	0	481	684	8	8	1181	5841
Tripura	67	2039	1079	43	18	3246	6294
Total	971	33070	20926	722	302	55991	143360

(Source: Tripathi et al., 2015)

to maintain the intra-specific genetic variability. These conservation areas include community reserves, nature reserves, sanctuaries, national parks, and biosphere reserves. Furthermore, the main focus is on maintaining the target species' genetic variability within a socially and economically acceptable land-use option mosaic. It mainly includes protected areas along with agroecosystems, managed forests, and multiple-use reserves.

1.3.2 PROTECTED AREAS

Over the years in India, for protecting critical ecosystems as well as non-commercial and commercial values' unquantifiable genetic resources, conservation reserves, wildlife sanctuaries, and national parks are established. These areas' primary aim is that these must be preserved as genebank capital as well as ecological gene pools' interdependent combinations samples. In a comparative assessment of protected areas from 2002 to 2019, there is a substantial increase in the number and area covered under such conservation efforts in the country. It has been observed that the number of national parks increased from 89 to 104, and the total area expanded from 37,803 km^2 to 40,501 km^2 (Table 1.9). The number of wildlife sanctuaries increased from 485 to 550, and the area expanded from 108,862.5 km^2 to 119,775.6 km^2. There were no community reserves in existence during 2002, and in recent reports, the number is 127, which covers a total area of 525.22 km^2. Similarly, the number of conservation reserves in the country is 87 with a total area of 4,286.31 km^2 (Table 1.9).

Presently, in India, there are 104 national parks covering 40,501.13 km^2, which is 1.23% of the country's geographical area (Table 1.9). In the Northeast (NE) region, there are 16 national parks covering a total area of 6,749 km^2 (Table 1.10). India also has 550 wildlife sanctuaries covering 119,775.55 km^2, which is 3.64% of the country's geographical area (Table 1.10). In the NE region, there are 57 wildlife sanctuaries covering an area of 10,718 km^2 (Table 1.11). Additionally, India has 127 community reserves and 87 conservation reserves, along with wildlife sanctuaries and national parks (Table 1.12). Currently, a network of 868 protected areas covers 165,088.10 km^2, approximately 5.02% of the country's total land area. The protection of endangered species and prime habitats is significant, and there is a need to design and protect more natural sites. The assessment of the total number of national parks in the NE region reveals that nearly 4.2% of India's geographical area is allocated for broad *in situ* conservation of ecosystems and habitats. A visual representation of the current status of protected areas in the NE region is provided in Figure 1.1.

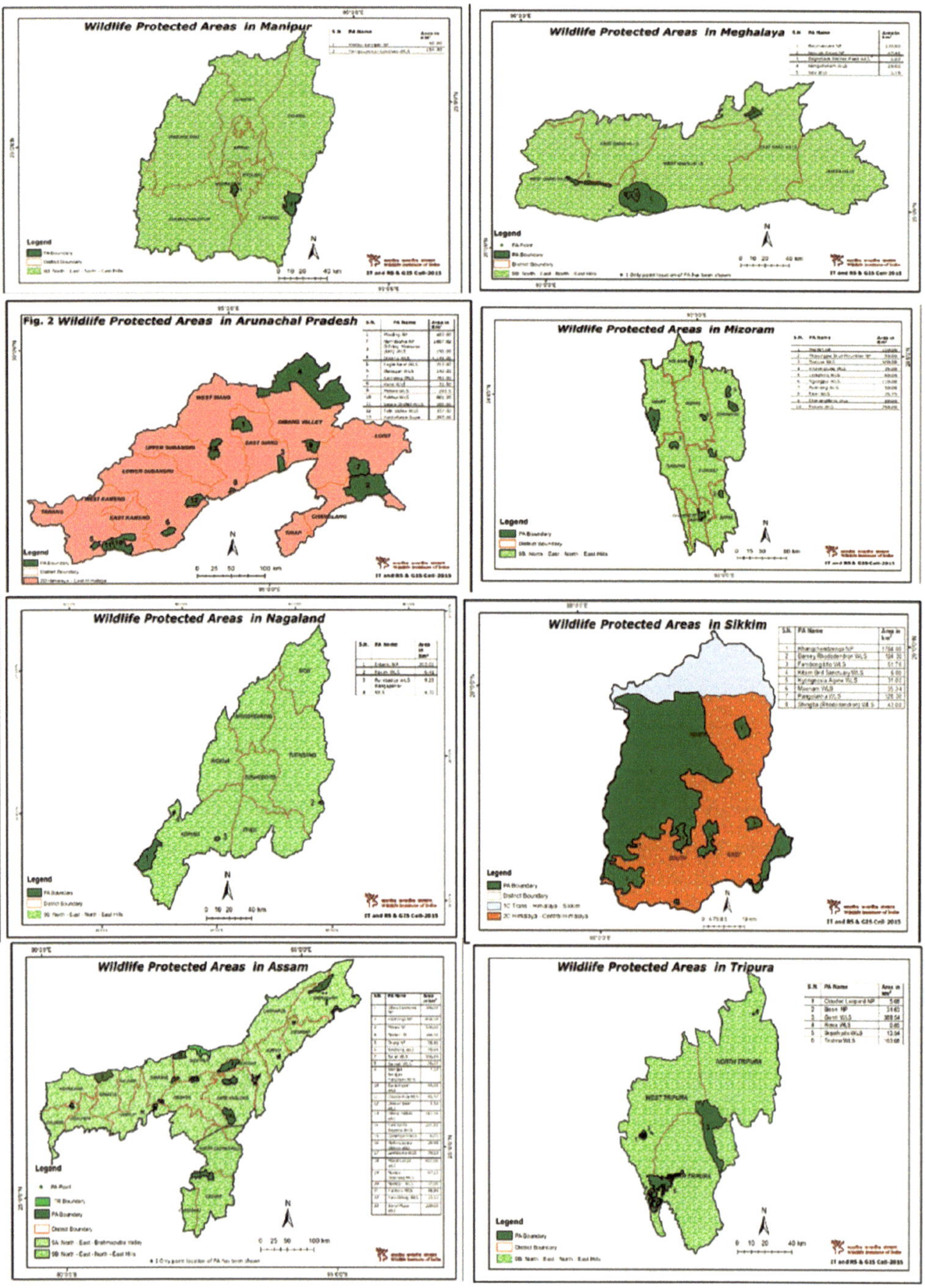

FIGURE 1.1 Status of wildlife protected areas of different states of NE regions.

Source: WII (2017).

1.3.3 EX-SITU CONSERVATION

The *ex-situ* genetic resources conservation is performed when, rather than an original, an outside growth environment is used for maintaining the reproductive material, individuals, or populations. Generally, well-defined situations are used for applying plants' *ex-situ* conservation (FAO, 1989). The biological diversity or genetic material's conservation outside the species' natural habitats is referred to as *ex-situ.* Usually, for the conservation to take place, some facilities are used that support storage or conditions' continuity that suits in maintaining the diversity of genetic material's genetic constitution and viability. For breeding and research, a required germplasm supply is provided by *ex-situ* conservation, which also provides virtual safeguards. The FGR's *ex-situ* conservation's several possible approaches are discussed in further sections.

1.3.4 BOTANICAL GARDENS

Under various management systems, there are 122 botanical gardens in India situated in various bio-geographical areas. An arboretum, Arboreta, usually specifies an area used for tree species' conservation. Furthermore, a 53 species bambusetum and 130 forest tree species arboretum are established in the Forest Research Institute in Dehradun. However, for tropical tree species, an additional arboretum is required. For seed storage, their behavior is either poorly known or unknown. For such species, the establishment of such arboreta will be considered as the first step for conservation (Singh et al., 2004).

1.3.5 CLONAL REPOSITORIES

In India, most of the research institutes like Forest Research Institute (FRI) in Dehradun deal with vegetatively propagated domesticated forest tree species that have field repositories for FGR conservation. Thus, such institutes deal with aromatic and medicinal plants as well as agroforestry species. In FRI, the Indian Council of Forestry Research and Education (ICFRE) has field repositories known as VMGs (Vegetative Multiplication Gardens) for species like *Dalbergia sissoo, Eucalyptus tereticornis, Acorus calamus, Asparagus racemosus,* and other region-specific species as germplasm banks and clonal banks at its regional institutes in Shimla, Ranchi, Bangalore, Jabalpur, Jorhat, Coimbatore, and Dehradun.

1.3.6 SEED ORCHARDS

"Seed orchards" term is mostly utilized in the context of breeding, which denotes a genetically improved seed mass-production unit. For fulfilling the particular objectives of genetic conservation, seed orchards are established, which are not part of breeding programs, and then conservation seed orchards are denoted. The primary concern is to increase or conserve the genetic variability of small populations or single trees, which have unwelcome pollination, insufficient fructification, or endangered status in their natural environment. Such plantations are mainly established for the production of seeds with proven genetic quality. Generally, there are two types of seed orchards: seedling seed orchards and clonal seed orchards. Under selective conservation, such orchards are characterized as the objective of *ex-situ* conservation. Such seed orchard establishments are considered as part of conservation management as well as a long-term breeding program.

1.3.7 CRYOPRESERVATION

In the past decades, there has been a rapid evolution in plant cryopreservation technologies which opens new options for various crop and forest species' valuable genetic resources' long-term storage. Originally, slow-cooling methods were used, but now due to various researches, more reproducible and easier techniques have been developed that allow intra- and extracellular liquids' complete vitrification with the help of explants' direct immersion in liquid nitrogen (Panis & Lambardi, 2005). Biological samples' storage at liquid nitrogen's ultra-low temperature (–196 to –150°C) in viable conditions is known as cryopreservation. At NBPGR (National Bureau of Plant Genetic Resources), forest tree species' pollens are being cryopreserved with the help of some alternative complementary approaches. Furthermore, there is a challenge in front of researchers when conservation is made at the DNA level. As this area is just being initiated in research, there is a great need for further research.

1.4 MANAGEMENT OF FOREST GENETIC RESOURCES

The following are the concept and components for forest genetic resources' management (Palmberglerch, 1998):

What	How?	Where?
Conservation	*In situ* conservation	• Protected areas. • Natural forest is managed for productive and protective purpose. • Plantation and planted trees.
	Ex-situ conservation	• Plantation and planted trees. • Clone bank, breeding populations. • Seed and pollen *in vitro* culture.
Enhancement	Selection	• Natural forests are managed for productive and protective purposes. • Breeding population.
	Improvement and breeding	• Plantation and planted trees. • Clone bank, breeding populations, *ex-situ* conservation stands.
Sustainable utilization	Sustainable forest management	• Natural forests are managed for productive and protective purposes. • Plantation and planted trees. • Protects trees only occasionally applicable.

1.5 CONCLUSION

The forest genetic resources' sustainability is threatened as forest areas are shrinking, and forest resource utilization is also increasing. The significance of sustainable and conservation management of these resources is highlighted. Many species are already extinct, and a few are threatened and on the verge of extinction if suitable conservation measures are not taken. There is great richness and diversity of flora and fauna in India. The nation's economic progress is substantially contributed to by the proper utilization of biodiversity. There is a need for the identification and conservation of areas that have richness and diversity of biological resources. On a priority basis, limited distribution areas with rare endemic species must be conserved so that they will not become extinct. There is great pressure on Indian forests to fulfill the increasing demands for non-timber forest products, pulpwood, timber, grazing, fodder, fuel, and alternative land uses due to constantly increasing needs for infrastructure, industrial development, and human as well as livestock populations. Multiple threats have also been faced by these resources regarding exotic species' invasion, climate change, forest fires, and habitat loss. For world security, its biological basis is the forest resources as well as on earth every person's livelihood is supported by them either

TABLE 1.8 Protected Areas of India as of January 2019

Year	Number of National Parks	Area Under National Parks (km^2)	Number of Wild Life Sanctuaries	Area Under Wild Life Sanctuaries (km^2)	Number of Community Reserves	Area Under Community Reserves (km^2)	Number of Conservation Reserves	Area Under Conservation Reserves (km^2)	Number of Protected Areas	Total Area under Protected Areas (km^2)
2000	89	37803.1	485	108862.5	–	–	–	–	574	146665.6
2006	96	38392.12	503	111229.5	1	0.31	4	42.87	604	149664.8
2007	98	38428.88	507	111529	5	21	7	94.82	617	150073.7
2008	99	39441.74	510	113123.4	5	21	45	1259.84	659	153845.9
2009	99	39441.74	512	113395.4	5	21	45	1259.84	661	154117.9
2010	102	40283.62	516	113842.9	5	21	47	1382.28	670	155529.8
2011	102	40283.62	518	113998.8	5	21	52	1801.29	677	156104.7
2012	103	40500.13	526	114933.4	5	21	59	2012.93	693	157467.5
2013	102	40500.13	532	117123.6	19	30.94	64	2232.61	717	159887.3
2014	103	40500.13	535	118290.7	43	58.22	64	2232.61	745	161081.6
2015	103	40500.13	541	118866.4	44	59.51	71	2548.82	759	161974.9
2016	103	40500.13	543	118917.7	45	59.66	72	2566.2	763	162043.7
2017	103	40500.13	544	118931.8	46	72.61	76	2587.95	769	162092.5
2018	104	40501.13	544	118931.8	46	72.61	77	2594.03	771	162099.5
2019	104	40501.13	550	119775.6	127	525.22	87	4286.31	868	165088.1

TABLE 1.9 Protected Areas of India

	No.	Total Area (km^2)	Coverage Percentage of Country
National Parks (NPs)	104	40501.13	1.23
Wildlife Sanctuaries (WLSs)	550	119775.6	3.64
Conservation Reserves (CRs)	87	4286.31	0.13
Community Reserves	127	525.22	0.02
Protected Areas (PAs)	868	165088.1	5.02

TABLE 1.10 Status of Protected Area (National Park) in NE Region

Name of State & Union Territory	State Area km^2	Number of NPs	Area km^2	% of State Area	Name of State & Union Territory	State Area km^2	Number of WLS	Area km^2	Percentage of State Area
Arunachal Pradesh	83,743	2	2291	2.7	Arunachal Pradesh	83,743	11	7,487.75	8.9
Assam	78,438	5	1978	2.5	Assam	78,438	18	1,840.14	2.4
Manipur	22,327	1	40	0.2	Manipur	22,327	2	184.81	0.8
Meghalaya	22,429	2	267	1.2	Meghalaya	22,429	4	94.1	0.4
Mizoram	21,081	2	150	0.7	Mizoram	21,081	8	1,090.75	5.2
Nagaland	16,579	1	202	1.2	Nagaland	16,579	3	20.34	0.1
Sikkim	7,096	1	1784	25.1	Sikkim	7,096	7	3,99.10	5.6
Tripura	10,486	2	37	0.4	Tripura	10,486	4	5,66.93	5.4
Total	2,62,179	16	6,749	3.4					

(Source: WII, 2017)

TABLE 1.11 Distribution of Wildlife Sanctuaries in Different States of NE Region

Name of State & Union Territory	State Area km^2	Number of WLS	Area km^2	% of State Area
Arunachal Pradesh	83,743	11	7,487.75	8.94
Assam	78,438	18	1,840.14	2.35
Manipur	22,327	2	184.81	0.83
Meghalaya	22,429	4	94.1	0.42
Mizoram	21,081	8	1,090.75	5.17
Nagaland	16,579	3	20.34	0.12
Sikkim	7,096	7	3,99.10	5.62
Tripura	10,486	4	5,66.93	5.41
Total	2,62,179	57	10,718	29

(*Source*: WII, 2017)

TABLE 1.12 Details of the Community Reserve of the NE Region (WII, 2017)

SL. No.	State	Name of Community Reserve	Year	Area
3	Meghalaya	KpohEijah Com R	2014	0.17
4	Meghalaya	Miewsyiar Com R	2014	0.87
5	Meghalaya	Umsum Pitcher Plant Com R	2014	0.40
6	Meghalaya	Lumkohkriah Com R	2014	6.11
7	Meghalaya	Ryngud Com R	2014	5.22
8	Meghalaya	Thangkharang Com R	2014	1.11
9	Meghalaya	Nongsangu Com R	2014	1.00
10	Meghalaya	Raid Nongbri Com R	2014	0.70
11	Meghalaya	LumJusong Com R	2014	0.70
12	Meghalaya	Jirang Com R	2014	2.00
13	Meghalaya	Raid Nonglyngdoh/PdahKyndeng Com R	2014	0.75
14	Meghalaya	Phudja-ud Com R	2014	1.20
15	Meghalaya	Lawbah Com R	2014	2.10
16	Meghalaya	Ryngibah Com R	2014	0.80
17	Meghalaya	Mongalgre Com R	2014	0.20
18	Meghalaya	Dangkipara Com R	2014	0.03
19	Meghalaya	Aruakgre Com R	2014	1.00
20	Meghalaya	ResuHaluapra Com R	2014	0.50
21	Meghalaya	Kitmadamgre Com R	2014	0.70
22	Meghalaya	KaKhlooThangbruUmsymphu Com R	2014	0.20
23	Meghalaya	KaKhlooPohblaiMooshutia Com R	2014	0.34
24	Meghalaya	KaKhlooLangdoh Kur Pyrtuh Com R	2014	0.15
25	Meghalaya	Baladingre Com R	2013	0.50
26	Meghalaya	Bandarigre Com R	2013	0.67
27	Meghalaya	Chandigre Com R	2013	0.37
28	Meghalaya	Daribokgre Com R	2013	1.73
29	Meghalaya	Dumitdikgre Com R	2013	0.70
30	Meghalaya	Dura Kalkgre Com R	2013	0.60
31	Meghalaya	EmanAsakgre Com R	2013	0.30
32	Meghalaya	KhlooAmrawan Com R	2015	1.29
33	Meghalaya	KhlooBlaiKaRaij U Landohlonglang Com R	2016	0.15
34	Meghalaya	KhlooBlaiKongwasanKhlooBlaiChyrmang Com R	2014	0.07
35	Meghalaya	KhlooBlai Sein Raij Tuber Com R	2014	0.97
36	Meghalaya	Mandalgre Com R	2013	0.50
37	Meghalaya	Mikadogre Com R	2013	0.01
38	Meghalaya	Nongumiang Com R	2003	0.31
39	Meghalaya	RongmaParomgre Com R	2013	0.62
40	Meghalaya	RongmaRekmangre Com R	2013	1.92
41	Meghalaya	Sakalgre Com R	2013	1.22

indirectly or directly as they provide energy, medicine, timber, wood, shelter, clothing, fibers, feed for domestic animals, food, and various other advantages. Also, they are used as raw material for new plant varieties production either through biotechnology or traditional plant breeding. In practices of silviculture, like superior genotypes' selective felling, may result in genetic diversity loss. Therefore, for biodiversity conservation, suitable methods for forest management should be developed and ensured and must be manifested in working plans or management. Thus, there exists a significant requirement for conserving and using genetic resources as a future safeguard (Tables 1.8–1.12).

KEYWORDS

- **conservation**
- **dynamic process**
- **genetic diversity**
- **genetic resources**
- **plantations**
- **reforestation**

REFERENCES

Areki, F., & Cunningham, A. B., (2010). Fiji: Commerce, carving and customary tenure. In: Laird, S. A., McLain, R. J., & Wynberg, R. P., (eds.), *Wild Product Governance: Finding Policies that Work for Non-Timber Forest Products* (pp. 229–242). Earthscan, London, UK.

Asfaw, A., Lemenih, M., Kassa, H., & Ewnetu, Z., (2013). Importance, determinants, and gender dimensions of forest income in Eastern highlands of Ethiopia: The case of communities around Jelo Afromontane Forest. *For. Policy Econ., 28*, 1–7.

Babulo, B., Muys, B., Nega, F., Tollens, E., Nyssen, J., Deckers, J., & Mathijs, E., (2009). The economic contribution of forest resource use to rural livelihoods in Tigray, Northern Ethiopia. *For. Policy Econ., 11*, 109–117.

Bawa, K. S., & Dayanandan, S., (1998). Climate change and tropical forest genetic resources. *J. Ecol. Environ., 23*, 449–466.

Bhat, S. J. A., Geelani, S. M., Bhat, H. A., Banyal, R., Shazmeen, Q., Shah, A. A., & Jeelani, M. I., (2013). Status, conservation and management of forest genetic resources in India. *International Journal of Environment and Bioenergy, 5*(1), 17–41.

Brown, K., & Pearce, D. W., (1994). *The Causes of Tropical Deforestation: The Economic and Statistical Analysis of Factors Giving Rise to the Loss of the Tropical Forests*. University College Press, London.

BSI, (2017). *Botanical Survey of India Database (BSI)*. Dehradun, India.

Chauhan, K. V. S., Sharma, A. K., & Kumar, R., (2008). Non-timber forest products subsistence and commercial uses: Trends and future demands. *Int. For. Rev., 10*, 201–216.

FAO, (1989). *Plant Genetic Resources: Their Conservation in Situ for Human Use*. FAO, Rome, Italy.

FAO, (2014). *Food and Agriculture Organization of the United Nations, Rome, Italy*. Web. http://faostat.fao.org/default.aspx (accessed on 22 April 2024).

Forest Survey of India (2017). *Indian State of Forest Report*. Forest Survey of India. Ministry of Environment and Forests, Dehradun.

Hegde, R., & Enters, T., (2000). Forest products and household economy: A case study from Mudumalai Wildlife Sanctuary, Southern India. *Environ. Conserv., 27*, 250–259.

Kaimowitz, D., (2003). Not by bread alone… forests and rural livelihoods in Sub-Saharan Africa. In: Oksanen, T., Pajari, B., & Tuomasjukka, T., (eds.), *Forests in Poverty Reduction Strategies: Capturing the Potential* (pp. 45–64). European Forest Institute, Joensuu.

Katwal, R. P. S., Srivastava, R. K., Kumar, S., & Jeeva, V., (2003). *Status of Forest Genetic Resources Conservation and Management in India.* Asia Pacific Forest Genetic Resources Program (APFORGEN). Forest Resources Development Service Working Paper FGR/65E, Forest Resources Division FAO, Rome, Italy.

Mahapatra, A. K., Albers, H. J., & Robinson, E. J. Z., (2005). The impact of NTFP sales on rural households' cash income in India's dry deciduous forest. *Environ. Manage., 35*, 258–265.

Malhotra, K. C., & Bhattacharya, P., (2010). *Forest and Livelihood* (p. 246). CESS Publisher, Hyderabad, India.

Marshall, E., Schreckenberg, K., & Newto, A. C., (2006). *Commercialization of Non-timber Forest Products: Factors Influencing Success: Lessons Learned from Mexico and Bolivia and Policy Implications for Decision-Makers.* UNEP World Conservation Monitoring Centre, Cambridge, UK. https://doi.org/10.1505/ifor.8.3.368 (accessed on 22 April 2024).

MoEF & CC, (2014). *Report of the National Forest Commission.* Ministry of Environment and Forests, Government of India, India.

Naithani, H. B., (1993). *Contributions to the Taxonomic Studies of Indian Bamboos* (Vol. I). PhD Thesis. H.N.B. Garhwal University, Srinagar, Garhwal.

Naithani, H. B., (2008). Diversity of Indian bamboos with special reference to North-East India. *Indian Forester, 134*(6), 765–788.

Nayar, M. P., & Sastry, A. R. K., (1990). *Red Data Book of Indian Plants* (Vol. II, p. 268). Botanical Survey of India.

Palmberg-Lerche, C., (1998). Management of forest genetic resources: Some thoughts on options and opportunities. *Forest Genetic Resources, 26*, 45, 46.

Pandey, A. K., & Kori, D. C., (2011). Non-destructive harvesting practices of *Terminalia arjuna* (Arjuna), *Phyllanthus emblica* (Aonla) and *Andrographis paniculate* (Kalmegh). *Indian Forester, 137,* 1269–1279.

Pandey, A. K., (2009). *Sustainable Harvesting Standards and Limits: Medicinal Plants in Central India* (pp. 4–10). Community Forestry, USA.

Pandey, A. K., Tripathi, Y. C., & Kumar, A., (2016). Non timber forest products (NTFPs) for sustained livelihood: Challenges and strategies. *Res. J. For.,* 1–7. doi: 10.3923/rjf.2016.1.7.

Panis, B., & Lambardi, M., (2005). Status of cryopreservation technologies in plants (crops and forest trees). In: *The Role of Biotechnology* (pp. 43–54). Villa Gualino, Turin, Italy.

Rawat, G. S., & Ginwal, H. S., (2009). Forest genetic resources conservation and management: Status in seven South and Southeast Asian countries. *Bioversity International Publication*, 23–46.

Saha, D., & Sundriyal, R. C., (2012). Utilization of non-timber forest products in humid tropics: Implications for management and livelihood. *For. Policy Econ., 14*, 28–40.

Saxena, N. C., (2003). *Livelihood Diversification and Non-timber Forest Products in Orissa: Wider Lessons on the Scope for Policy Change?* ODI Working Paper No. 223, Overseas Development Institute, London, UK.

Shackleton, C. M., Pandey, A. K., & Ticktin, T., (2015). *Ecological Sustainability for Non-timber Forest Products: Dynamics and Case Studies of Harvesting* (p. 280). Routledge, New York, USA.

Shackleton, S., Campbell, B., Lotz-Sisitka, H., & Shackleton, C., (2008). Links between the local trade in natural products, livelihoods and poverty alleviation in a semi-arid region of South Africa. W*orld Dev., 36*, 505–526.

Shiva, M. P., & Verma, S. K., (2002). *Approaches to Sustainable Forest Management and Biodiversity Conservation: With Pivotal Role of Non-timber Forest Products*. Centre for Minor Forest Products, Valley Offset Printers, Dehradun.

Singh, A. K., Tyagi, R. K., & Kameshwar, R. N., (2004). Ex-Situ approaches for conservation of plant genetic resources: Achievements and challenges. In: Dhillon, B. S., Tyagi, R. K., Arjun, L., & Saxena, S., (eds.), *Plant Genetic Resource Management.* Narosa Publishing House, Delhi, India.

Stedman-Edwards, P., (1998). *Root Causes of Biodiversity Loss: An Analytical Approach.* World Wide Fund for Ba Nature, Washington DC, US.

Tripathi, H. K., Somasundaram, P., Kumar, R., & Lakchaura, P., (2015). *Assessment of Bamboo Resources of India* (pp. 15–26). In Bamboo in India, ENVIS Centre on Forestry National Forest Library and Information Centre Forest Research Institute (Indian Council of Forestry Research), Dehradun.

Vantomme, P., (2003). *Forest Products Division.* Forestry Department, Food and Agriculture Organization, Rome, Italy.

Vavilov, N. I., (1951). Phyto-geographical basis of plant breeding. In The origin, variation, immunity and breeding of cultivated plants, Chester, K. S., transl. Selected Writings of Vavilov, N. I. *Chronica Botanica, 13*, 364, Waltham Mass., USA.

WII, (2017). *National Wildlife Database.* Wildlife Institute of India (WII), Dehradun, India.

WII, (2019). *National Wildlife Database*. Wildlife Institute of India (WII), Dehradun, India.

Young, A., Boshier, D., & Boyle, T., (1999). *Forest Conservation Genetics: Principles and Practice.* CSIRO Publishing, Australia & CABI Publishing, Wallingford, UK.

CHAPTER 2

Conservation and Management Approaches of Ethno-Medicinal Plants for Livelihood Improvements

PIJUSH DAS, SARTHAK BHATTACHARYA,
AGNISWAR JHA CHAKRABORTY, and ABHISHEK GHOSH

Department of Agriculture, School of Agriculture and Allied Sciences, The Neotia University, Sarisha, Jhinga, South 24 Parganas, West Bengal, India

ABSTRACT

Ethno-medicinal plants are recognized as a valuable source of herbal medicines. Medicines that are extracted from these wild plants are used by most of the population. Herbal products such as dietary food supplements and herbal teas, which can improve personal health, are made from ethno-medicinal plant species, enhancing their market demand globally. These plants are also major sources of revenue generation and livelihood options for forest neighboring communities since historic times. However, the number of these plant species is disappearing from nature at a much higher rate due to unauthorized harvesting, improper management, lack of knowledge for preservation, ongoing habitat eviction, and so on. Therefore, there is an immediate need to undertake appropriate measures to support the judicious use and conservation of these rich floras. Long-term sustainability in the management and preservation of these unique ethno-medicinal plant species should be accomplished through both *in situ* and *ex situ* conservation, combined with sustainable management approaches, so that rural indigenous communities can benefit.

Sustainable Forest Resources Management: Issues and Implications. Jahangeer A. Bhat, Gopal Shukla, Manmohan J. Dobriyal, Sumit Chakravarty, A. Arunachalam, and Rainer W. Bussmann (Eds.)

2.1 INTRODUCTION

Nature is inevitably the most dependable source of all the basic necessities of life. It supplies all the fundamental necessities like food, fiber, fuel, clothes, and herbal drugs, i.e., natural healing remedies primarily obtained from various plant species. Several plant species have traditionally served as a source of essential resources for humanity. Since ages, the ethnic community or tribal society has had a considerable reliance on the native flora for the natural healing of various diseases and other purposes for the existence of mankind (Thakur et al., 2015; Abu-Rabia, 2005). Because of this dependency, the indigenous people who lived in harmony with nature gathered all the knowledge of those plants, which serve a variety of vital roles in human existence (Nasab & Khosravi, 2014; Rashid et al., 2015). They use a unique system of remedies that dates back thousands of years, some of which are not mentioned in the literature (Asase et al., 2008). Understanding how various indigenous people interact with natural resources, particularly for medical and pharmaceutical uses, has benefited greatly from the systematic study of these plants (de Albuquerque & Hanazaki, 2009). The majority of the forest-dwelling indigenous populations are fully dependent on these natural herbal plants for curing common health problems because of poverty and a lack of access to modern medications (Shukla & Chakravarty, 2012). The consequence of human interaction with the most desired herb species existing in the present environment and their selection at a certain time is attributed to the local knowledge system related to those plant species (Venkataswami et al., 2010; Lulekal et al., 2013).

Various studies anticipated that about 70–80% of people globally solely rely on regional medicinal herbs as their primary form of healthcare (Mesfin et al., 2013; Calixto, 2005; WHO, 2002). Many herbal products are made available in the market for sale as dietary supplements, nutritional foods, herbal teas, and for a number of uses related to personal care and health in developed nations, enhancing the market demand for herbal medicines by 10–20% each year. Additionally, the market demand for ayurvedic medicines in India is rising at a rate of 20% annually (Subrat, 2002). Therefore, the demand for these herbal medicines is growing rapidly in tandem with the world's population growth and also due to the adequate effectiveness of allopathic medications in developing nations. In recent times, the demand for these traditional medicines is not only high on a global scale but also increasing more quickly every day. Consequently, these ethno-medicinal plants can be a sizable source of revenue generation and livelihood options for the indigenous population of developing as well as underdeveloped nations. Indigenous people from rural and tribal communities, as well as other economically oppressed groups like landless people and women, can make a living by extracting these priceless herbal drugs from their natural environments. In contrast, medicinal plant flora is gradually eroding due to the loss of these herbal plants from nature over time as a result of excessive harvesting, unauthorized collections, and ongoing habitat evictions. Therefore, sustainable utilization of these priceless resources might be a viable option for their conservation in respect of their rising demand in the primary healthcare system.

2.2 ETHNO-MEDICINAL PLANTS

Ethno-medicine is the study of indigenous knowledge, thoughts, faiths, or beliefs and practices followed by many ethnic groups of tribal and rural communities to prevent, treat, and cure disease by bioactive compounds obtained from natural herbs (Sharma et al., 2012). Ethno-medicinal plants are considered to be a tangible source of bioactive substances that may help treat ophidian envenomation immediately or, in the long run, act as complements to traditional serum therapy. Ethno-medicines are an effective tool for the preparation of herbal medicines that would benefit ethnic people (Negi & Subramani, 2002). One of the most reliable methods for discovering new drugs is ethno-medicinal plant research, and even now there are still a lot of undiscovered facts in this field (Fabricant & Farnsworth, 2001). In recent years, scientific society has given major consideration to research on the conventional use of ethno-medicinal plant species. Notably, knowledge regarding ethno-medicinal plants is typically conveyed verbally through family members to the next generation (Nadembega et al., 2011). Therefore, it is crucial to properly document such knowledge, especially traditional ethno-medical practices because people using ethno-medicines have a rich coalescence with herbal plants and their therapeutic properties (Kabir et al., 2014).

2.3 USES OF ETHNO-MEDICINAL PLANTS

The traditional system of medicine cannot provide a complete cure for all human pathological disorders (Redzic, 2007). In contrast, there has been a continuous retrenchment in these practices due to a lack of willingness among modern youth in traditional medical treatments. The swift decline in this rich expertise is attributed to rural depopulation, widespread deforestation, and the movement of traditional medical practitioners to new occupations (Kadir et al., 2013). However, rural people respect ethno-medical knowledge, and it has occasionally been demonstrated to be helpful in treating different health-related issues and the development of new medications. Plant-based traditional medicine has the potential to serve as the foundation for many pharmaceuticals that treat chronic diseases. Indeed, research on ethno-medicine has become a key contributor to the invention of both natural and artificial medications. The scientific community has become more interested in ethno-medical research in recent years (Heinrich, 2000).

Different ethno-medicinal plant components such as herbs, shrubs, and trees are employed to treat ailments including jaundice, neurological disorders, inflammatory diseases, anthelmintic diseases, diaphoretic diseases, diuretic diseases, and rheumatic pains (Table 2.1). Ethno-medicinal plants have contributed a lot to the traditional health care system for treating many of these ailments since ancient times. The therapeutic benefits of ethno-medicinal plants are due to a number of bioactive substances such as glycosides, terpenoids, alkaloids, steroids, essential oils, and tannins, etc., present within them (Figure 2.1). These substances are mostly used therapeutically and act as precursors for the semi-synthesis of chemo-pharmaceuticals (Suriyavathana et al., 2011). On average, 25% of medicines prescribed globally derive from these plant species (Rates, 2001; Jabbar et al., 2007), and the origins of approximately 50% of modern-day pharmaceutical drugs may be attributed to ethno-medicine (Van Wyk et al., 1997).

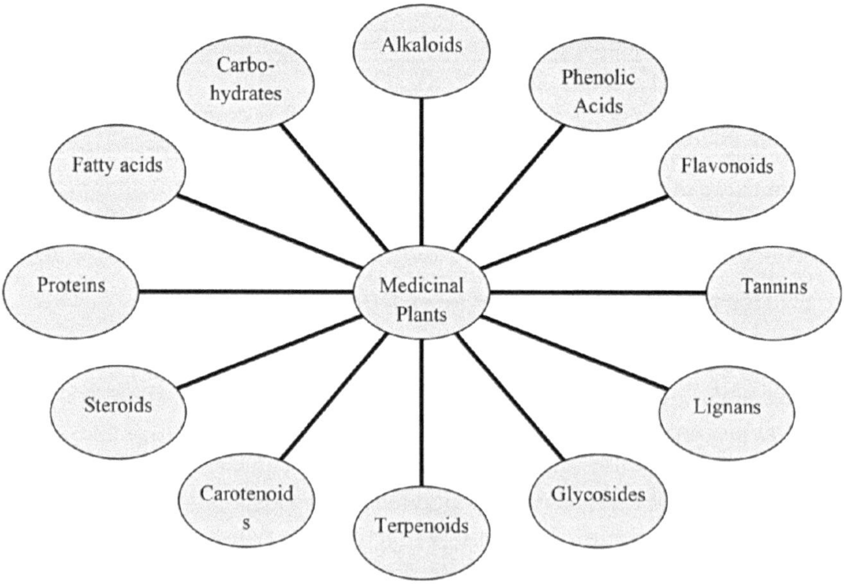

FIGURE 2.1 Bio-active substances present in different ethno-medicinal plants.

TABLE 2.1 Ethno-Medicinal Plants and Their Use by Ethnic Groups of West Bengal, India

Common Name	Botanical Name	Family	Parts Used	Ethno-Medicinal Importance
Vasaka	*Adathoda vasica* (L). Nees.	Acanthaceae	Leaf, root	Leaf and root extract cures asthma, diarrhea, and dysentery.
Akanda	*Calotropis procera* R. Br.	Apocynaceae	Root, flower	Paste prepared from root is useful against toothache and powdered flower against inflammation.
Mango	*Mangifera indica* (L).	Anarcardiaceae	Leaf, fruit	Mango leaf extract can cure diarrhea, cough, and dysentery effectively.
Amla	*Emblica officinalis* Gaerth.	Euphorbiaceae	Fruit	Powdered fruit is used to cure jaundice, anemia, diarrhea, and gastritis.
India Sarsaparilla	*Hemidesmus indicus* (L). R. Br.	Asclepiadaceae	Root	Helps to manage blood sugar levels and prevent damage to liver.
Arjun	*Terminalia arjuna* (Roxb.)	Combretaceae	Bark	Reduces the risk of heart diseases and useful against inflammation.
Ashok	*Saraca asoca* (Roxb.) Willd.	Leguminosae	Bark	Useful for treatment of skin related problems like eczema, dermatitis, acne, and psoriasis.
Ashwagandha	*Withania somnifera* Dun.	Solanaceae	Root, leaf	Root powder used in rheumatic swellings and leaf paste is taken as a remedy for anthelmintic.
Sugar apple	*Annonas squamosa* (L).	Annonaceae	Fruit	Fruits can be used to alleviate vomiting and treat tumors.
Atish	*Aconitum heterophyllum* Wall. ex Royle	Ranunculaceae	Rhizome	*Can* cure common cold, sore throat, cough, and flu symptoms.
Bach	*Acorus calamus* (L).	Acoraceae	Rhizome	Treats a wide range of health ailments such as neurological, gastrointestinal, respiratory, metabolic, kidney, and liver disorders.
Bael	*Aegle marmelos* (L.) Correa.	Rutaceae	Fruit	Used to cure constipation, diarrhea, diabetes, and other ailments.
Lantana	*Lantana camara* (L).	Verbinaceae	Leaf, flower	Juice of the leaves is effective in relieving snake and insect bite irritation.
Malabar nut	*Justicia adhatoda* (L).	Acanthaceae	Leaf	Extensively used for treatment of cough and other respiratory problems.
White goosefoot	*Chenopodium album* (L).	Chenopodiaceae	Leaf, seed	Used as a blood purifier, diuretic, sedative, and as an anthelmentic against round and hookworms.
Bhumi amla	*Phyllanthus amarus* (Schumach. and Thonn.)	Euphorbiaceae	Leaf, stem	Useful in curing liver problems and jaundice.

TABLE 2.1 *(Continued)*

Common Name	Botanical Name	Family	Parts Used	Ethno-Medicinal Importance
Pale persicaria	*Persicaria lapathifolia* (L).	Polygonaceae	Leaf	Crushed leaves are used against dental problems.
Bahera	*Terminalia bellirica* (Gaertn.) Roxb.	Combretaceae	Fruit	Acts against different health problems like bronchitis, asthma, hepatitis, diarrhea, dyspepsia, coughs.
Brahmi	*Bacopa monnieri* (L.)	Plantaginaceae	Leaf, stem	Reduces the symptoms of anxiety and mental fatigue while stimulating memory span.
Chirata	*Swertia chirayita* (L).	Gentianaceae	Leaf, fruit	Useful for treating numerous ailments viz. liver disorders, malaria, diabetes.
Camphor tree	*Cinnamomum camphora* (L.) J. Presl	Lamiaceae	Essential oil	Treats inflammation, muscle pain and infection in various regions.
Datura	*Datura metel* (L).	Solanaceae	Leaf, fruit	Leaf extract can cure chronic ulcers, asthma, inflammation, and wounds.
Vegetable fern	*Diplazium esculentum* (Retz.) Sw.	Athyriaceae	Leaf	Protects from liver infections contagious cold-cough and various types of pain.
Bermuda grass	*Cynodon dactylon* (L.) Pers.	Poaceae	Leaf, root	Plant extract can efficiently cure diabetes, ulcers, soothe body temperature.
Fig	*Ficus racemosa* (L).	Moraceae	Leaf, fruit	Cures liver disorders, diabetes, diarrhea, inflammation, respiratory diseases.
Barbary fig	*Opuntia ficus-indica* Haw.	Cactaceae	Stem	Promoted for treating diabetes, high cholesterol, obesity, and hangovers.
Marigold	*Tagetes erecta* (L).	Asteraceae	Leaf, flower	Help decrease inflammation and spasms, calm the nerves, and reduce blood pressure.
Aloe	*Aloe vera* (L). Burm. f.	Xanthorrheaceae	Leaf	Plant juice is useful in curing fever, jaundice, piles, and rheumatism.
African dream herb	*Entada rheedii* Spreng.	Fabaceae	Seed	Improves blood circulation to the brain and heal the after-effects of a stroke and heal jaundice.
Guduchi	*Tinospora cordifolia* Willd.	Menispermaceae	Stem	Used for treating jaundice, diarrhea, dysentery, skin disease.
Black pepper	*Piper nigrum* (L).	Piperaceae	Fruit	Efficient in curing diseases like cold, intermittent fever, throat diseases, stomachache, and dysentery.
Gymnema	*Gymnema sylvestre* (L).	*Apocynaceae*	Leaf	Helps to fight sugar cravings and lowers high blood sugar levels.

TABLE 2.1 *(Continued)*

Common Name	Botanical Name	Family	Parts Used	Ethno-Medicinal Importance
Guggal	*Commiphora wightii* (L).	Burseraceae	Resin	Useful for treatment of inflammation, arthritis, obesity, microbial infection, and gastrointestinal diseases.
Turmeric	*Curcuma longa* (L).	Zingiberaceae	Rhizome	Aids in the management of oxidative and inflammatory conditions, anxiety arthritis and hyper-lipidemia.
Haritaki	*Terminalia chebula* Retz.	Combretaceae	Fruit	Good to incite appetite, digestive aid, liver stimulant and mild laxative.
Veld grape	*Cissus quadrangularis* (L).	Vitaceae	Vine	Used as a pain reliever, and to heal broken bones.
Indian berberry	*Barberis aristata* DC.	Berberidaceae	Root, stem	Reduces cholesterol and sugar levels in blood.
Isabgol	*Plantago ovata* Forssk.	*Plantaginaceae*	Seed, husk	Seed husk is used as a laxative and stool softener and to help reduce cholesterol.
Flame lily	*Gloriosa superba* (L).	Colchicaceae	Rhizome	Cure parasitic skin diseases and used as a cataplasm in urological pains.
Kalmegh	*Andrographis paniculata* Burm. f.	Acanthaceae	Leaf	Used to treat diabetes, infectious diseases, skin diseases and fevers.
Pongamia	*Pongamia pinnata* (L.) Merr.	Fabaceae	Seed	Used as a liniment for skin diseases and rheumatic illnesses, treat stomachache and liver disease.
Saffron	*Crocus sativus* (L).	Iridaceae	Flower	Prevents nervous system disorders, boosting mood, promoting libido and weight loss.
Kulekhara	*Hygrophila auriculata* (L).	Acanthaceae	Leaf	Cure problems like anemia and diabetes.
Kuth	*Saussurea costus* (Falc.) Lipsch.	Asteraceae	Rhizome	Cures rheumatoid arthritis, asthma, bronchitis, chronic gastritis, ulcers.
Kutki	*Picrorhiza kurroa* Royle ex Benth.	Scrophulariaceae	Root	Useful in treating liver related problems and disorders of upper respiratory tract.
Long pepper	*Piper longum* (L).	Piperaceae	Fruit	Used to treat chronic bronchitis, asthma, respiratory infections, cough, diseases of the spleen and tumors.
Makoy	*Solanum nigrum* (L).	Solanaceae	Leaf, flower	Leaf and flower extracts cure skin diseases, piles, and ulcer.
Fenugreek	*Trigonella foenum* (L).	Fabaceae	Leaf, seed	Control sugar absorption in the stomach and stimulate insulin.
Mulethi	*Glycyrrhiza glabra* (L).	Fabaceae	Root, rhizome	Root extract is used for conditions such as digestive problems, cough, and bacterial and viral infections.

TABLE 2.1 *(Continued)*

Common Name	Botanical Name	Family	Parts Used	Ethno-Medicinal Importance
Nayantara	*Catharanthus roseus* (L).	Apocynaceae	Leaf	Leaf extract is useful in treating diabetes.
Neem	*Azadirachta indica* A. Juss.	Meliaceae	Leaf, bark	Bark and leaf extracts can efficiently control liver problems and skin diseases, respectively.
Chinese chastetree	*Vitex negundo* (L).	Verbenaceae	Leaf, flower	Used in curing cholera and liver problems.
Papaya	*Carica papaya* (L).	Caricaceae	Flower, fruit	Flower paste is used in anthelmintic and fruit extract regulates blood pressure.
Indian sandalwood	*Santalum album* (L).	Santalaceae	Heart wood	Extract of bark is used as sedative and also give relief from gonorrhea.
Sarpaganda	*Rauvolfia serpentine* (L).	Apocynaceae	Root	Improves immune system, lower down cholesterol levels in blood and cures skin diseases.
Satamuli	*Asparagus racemosus* Willd.	Asparagaceae	Rhizome	Treats constipation and stomach ulcers.
Sefaliful	*Nyctanthes arbor*-tristis (L).	Oleaceae	Leaf, flower	Can be used as a sedative, control rheumatism and skin ailments.
Tamarind	*Tamarindus indica* (L).	Fabaceae	Leaf, fruit	Soothes abdominal pain, also cures respiratory problems and diarrhea.
Indian pennywort	*Centella asiatica* (L).	Apiaceae	Leaf	Extract of leaves can efficiently improve memory and juvenility.
Tulsi	*Ocimum sanctum* (L).	Lamiaceae	Leaf, flower	Leaf extract is useful in curing cough and fever.
Ulatkambal	*Abroma augusta* (L).	Malvaceae	Leaf	Treat rheumatic, diabetes, inflammation, pains.

Many plant species possess and generate chemicals that defend them by eliminating or deterring insects feeding on them (Qin et al., 2010). Several ethno-medicinal plant species have also been thought to have some insect repellent qualities as well as insecticidal properties. Herbal products prepared from these ethno-medicinal plants contain a number of bioactive and therapeutic substances that can effectively be employed in agriculture (Farzaei et al., 2013). These plants are used for the preparation of many botanicals which act as a natural repellent against many insect pests and prevent infestation of harmful pathogens (Table 2.2). Several ways to use these plants as biological plant protection compounds include raw plant, extract of different plant parts, decoction, powder, etc. This bio-rational approach offers very little or no risk to other beneficial organisms or the environment. Studies have also demonstrated that many plant extracts work both as insecticides and repellents without posing any toxicity issues or adverse impact on the environment or non-targeted organisms (Oljira, 2015).

TABLE 2.2 Ethno-Medicinal Plants and Their Uses in Agriculture

Common Name	Botanical Name	Family	Importance
Nishinda	*Vitex negundo* (L).	Lamiaceae	Leaves incorporated in many pulse seeds for long time preservation.
Garlic	*Allium sativum* (L).	Amaryllidaceae	Spraying of garlic paste minimizes occurrence of many insect pests including caterpillars, armyworms, aphids, beetles, cutworms, and flies.
Neem	*Azadirachta indica* (L).	Meliaceae	Plant extract exhibits insecticidal properties; seed kernel extract is very much effective against leaf hopper.
Pongamia	*Pongamia glabra*	Fabaceae	*Essential oil* is used as insect repellent and act as natural fungicide.
Artemisia	*Artemisia monosperma*	Asteraceae	Leaf extract can effectively control cotton leaf worm.
Gliricidia	*Gliricidia sepium*	Fabaceae	Plant leaf paste is used as a rodenticide.
Chives	*Allium schoenoprasum*	Amaryllidaceae	Very much effective in controlling the infestation of aphids, beetles, and flies.
Marigold	*Tagetes erecta* (L).	Asteraceae	Flower extract is utilized by fruit and vegetable farmers to control aphids, plant lice, beetles.
Basil	*Ocimum basilicum* (L).	Lamiaceae	Can be used as mosquito repellent.
Mint	*Mentha spicata* (L).	Lamiaceae	Used as naturally repeling agent against mosquitoes.
Papaya	*Carica papaya*	Caricaceae	Leaf extract is effective to control brown spot disease of rice.
Cinnamon	*Cinnamomum verum*	Lauracea	Can check mycelial growth of fungus.
Arjun	*Terminalia arjuna* (Roxb.)	Combretaceae	Plant extract used against nematodes.

2.4 ETHNO-MEDICINAL PLANTS AND LIVELIHOOD OPPORTUNITIES

Livelihood is a term which is generally used for the services or work that people normally do for the purpose of surviving. The forest has historically provided millions of people with their primary source of occupation. People in forest surrounding areas normally depend on forests for their daily necessities including food, fodder, fuel, wood, etc., while urban people rely on forests indirectly for handicrafts, the wood industry, non-timber forest products (NTFPs) like medicinal plants, aromatic plants, honey, plant resins, etc. Not only do people rely heavily on the forest to meet their economic requirements, but they also interact with it on a cultural and social level. This is mostly done to acknowledge the close connection between the forest and its indigent residents (Sunderlin et al., 2008; World Bank, 2001a).

The economy of India is significantly influenced by its forest resources. According to a World Bank report, India's forest resources have enormous potential for reducing poverty and boosting the country's rural economy while simultaneously advancing important national conservation goals (World Bank, 2006). A variety of goods like fodder, lac, aromatic, and medicinal herbs, etc., also make significant contributions in their day-to-day livelihood activities. Medicinal (used in herbal drugs industry) and aromatic plants (used in cosmetic industry) are very lucrative options among the forest surrounding communities for their alternate livelihood. So, this medicinal and aromatic plant sector has enormous possibilities for the forest neighboring community to improve their standard of living. Many wild plants do have significant medicinal properties as pointed out by local ethnic communities as well as scholars worldwide. If properly harvested and processed, these can fetch significant income for the rural marginal population. As the processing chain of all these herbal products in the country continues to be underdeveloped, hence a great scope for establishing different processing industries of these products exists over there. It ensures the provision of alternate livelihood opportunities, generating employment for the rural population and ultimately preserving all these ethno-medicinal plant species.

Forest resources play a crucial part in the livelihoods of indigenous people through direct paid employment. Forest-neighboring communities can earn by selling these ethno-medicinal plants in the market. These forest products have significant value in the market. This also depends on an efficient marketing process and a lucrative marketing channel. Ethno-medicinal plants are considered a major way of earning a livelihood and generating employment. By following good cultivation practices of these ethno-medicinal plants, rural farmers can easily improve their livelihood (Joshi et al., 2014). The fringe marginalized populations should be encouraged to cultivate ethno-medicinal species that are commercially viable through capacity building, implementing timely legislative changes, and fostering strong market linkage (Raj et al., 2018). People generally find many plant species with ayurvedic importance that can be used as herbal medicines to treat various diseases. These ethno-medicinal plants, naturally grown in the forest, are organic in nature, increasing the demand for these products in the market. The collection of these valuable forest products, following proper management to process them, involves a significant portion of the rural community as well as forest dwellers, thereby generating more employment. On the other hand, cultivating these ethno-medicinal plants helps preserve biodiversity while increasing income (Kandari et al., 2012).

2.5 CONSERVATION STRATEGIES OF MEDICINAL PLANTS

2.5.1 RATIONALE FOR CONSERVATION

Importance of ethno-medicinal plant species in human life and livelihood is basically immeasurable. They make essential contributions to our health. Biologicals, natural goods, or entities that are derived from natural items made up 11 out of the 25 top-selling pharmaceutical medications in 1997, with a combined market value of US $17.5 billion (Laird & Ten Kate, 2002). These 11 pharmaceuticals accounted for 42% of the overall market. The global herbal medicine market, which stands at $165.66 billion during 2022, is projected to grow to $347.50 billion by 2029, with a CAGR of 11.16% in the foretold period. There is a lot of trading in medicinal plants at various levels, ranging from local to international. Large amounts of medicinal plants are traded regionally, globally, and from rural areas of poor countries into urban centers. During the unprecedented global COVID-19 pandemic, there has been a surge in demand for herbal remedies and ethnomedicine as people all over the globe were looking for ways to boost immunity and counter the common symptoms of the disease. Most of these go unreported in official figures or are documented inadequately, which is why decision-makers are usually unaware of the value of trade to economies and healthcare of citizens, as well as issues with sustainability and the frequent adverse effects of wild collection on natural habitats.

Medicinal plants, usually found in the wild, are recognized as a valuable source of traditional herbal medicines. In recent years, many of these valuable medicines have been developed from various therapeutic plants (Chen et al., 2010; Chacko et al., 2010). Medicines extracted from these plants are used in primary healthcare by a large population in developing countries, and 25% of prescribed drugs are derived from these wild plants in developed countries (Hamilton, 2004). While almost everyone on Earth benefits from medicinal plants, people from rural and ethnic societies, especially those with the lowest incomes, often rely on them culturally for remedies and as a source of income. Rural residents in poor nations can greatly benefit from the sale of medicinal plants, especially when collected from the wild. Herders, shepherds, and other economically marginalized groups in the community, including women and landless individuals, often work as collectors.

With the rising demand for herbal drugs extracted from wild plants with medicinal value, there is an increasing risk of uncontrolled and excessive cutting of forest plants. According to the report by UNEP/FAO, almost 16.8 million hectares of tropical moist forest area are being destroyed annually. Worldwide, there are almost 50,000 to 80,000 plant species that have some medicinal value as detailed by the International Union for Conservation of Nature (IUCN), but this number is gradually decreasing due to the loss of these herbal plants from nature over time. Due to excessive cutting and continuous destruction of wild habitats, about 15,000 plant species are under the threat of complete extinction (Bentley et al., 2010). Meanwhile, with the growth of the human population and increased consumption of plant products, about one-fifth of natural resources have been depleted (Ross et al., 2005). Several countries such as India (Heywood et al., 2003; Hamilton, 2008), China (Nalawade et al., 2003; Heywood et al., 2003), Nepal (Hamilton, 2008), Tanzania (Zerabruk et al., 2012), and Uganda (Zerabruk et al., 2012) are also facing the threat of complete loss of these valuable herbal plants due to the increased rate of destruction of wild habitats. The typical threats that endanger plant diversity in general also impact medicinal plants. In general, the main threats are habitat loss and degradation, including overharvesting of these valuable resources (Hamilton, 1997). Conversely, there is further reduction in the diversity of wild habitats due to natural and man-made disasters, leading to genetic erosion of medicinal plants (Hamilton, 2004).

Since the use of ethnomedicines among various rural and ethnic communities is essentially tied to the traditional knowledge carried on from generation to generation, documentation, and preservation of traditional knowledge will go a long way in not only creating awareness about ethnomedicinal plants but also helping formulate conservation strategies. In order to prevent overexploitation, destructive bioprospecting, and biopiracy, the Global Convention on Biological Diversity's Article 8 (j) on Traditional Knowledge, Innovations, and Practices recommends taking steps and creating certain procedures to ensure that forest fringe communities receive a good and reasonable share of benefits by applying their traditional knowledge and its uses; to standardize the manner in which impact assessments are conducted with regard to any future progress on holy areas or on land and water resources; to aid governments in creating laws or other enforcement measures to guarantee that traditional knowledge and its broader uses are appreciated, maintained, and protected.

2.5.2 CONSERVATION STRATEGIES

Conservation strategies in support of the sustainable development of wild resources must serve the purpose of saving biological diversity with the formation of a plant species inventorying system along with subsequent status monitoring and no further depletion of the natural ecosystem. Preservation of medicinal or herbal plants can be considered as the microcosm of plant conservation because major parts of plant species found in the wild habitat have some medicinal value. The use of wild resources in a sustainable way might be an effective alternative for medicinal plant conservation in respect of its increased demand in the healthcare system with limited supplies. Preservation of these medicinal plants through continuous protection of their genetic resources can be executed by both *in situ* and *ex situ* conservation. Now, the type of conservation activities (i.e., *in situ* or *ex situ* conservation) needs to be followed depending on the biological characteristics of a particular plant species or its geographical distribution.

By adopting these kinds of priorities, management practices, and philosophical frameworks, the acknowledgment of the significance of medicinal species and our remaining biological diversity, in general, must be included in our laws, as well as our land and resource management techniques. Without taking immediate action, we risk losing countless treasures that could save lives, frequently without even being aware that they existed.

2.5.3 MEDICINAL PLANTS CONSERVATION BY IN SITU METHOD

On-site or *in-situ* conservation refers to the preservation of any species of either plant or animal origin in their natural population or in surrounding areas from where their genetic development commenced. It mainly aims at the protection of endangered wild species in their natural habitat either by protecting the species or by defending them from predators. Therefore, it is of great importance to execute the rules and regulations as issued by the forest department and also follow potential scientific nurturing of medicinal plants within their natural environment for successful *in-situ* conservation (Volis & Blecher, 2010). The creation of preserved areas as well as employing an ecosystem-oriented strategy have been the main focuses of *in-situ* conservation efforts around the world (Ma et al., 2012). It is estimated that about 4% of the total geographical area of India is used for *in-situ* conservation currently. This type of conservation is generally achieved through protected areas namely Biosphere Reserves, Wildlife Sanctuaries, National Parks, and Biodiversity Hotspots.

2.5.3.1 BIOSPHERE RESERVES

A biosphere reserve, also known as a natural reserve, is a protected area that includes both terrestrial and coastal ecosystems and offers a mechanism for sustainable use of abundant flora and fauna occurring in natural habitats. Biosphere reserves are established to maintain and replenish biodiversity where the important wild resources are preserved and protected for assessing the impacts on different natural ecosystems (Rodriguez et al., 2007). They uphold the preservation of genetic diversity within the natural ecosystems by defending wild plants and animals, native ways of life, and genetic resources from domesticated plants and/or animals. It was first introduced as the "Man and Biosphere Program" by UNESCO in 1971 with the objective of conserving the entire ecosystem and its rich genetic resources. State governments generally manage these reserves thoroughly by legislation and allow tourists and researchers only to enter for carrying out various scientific studies.

Biosphere reserves often span areas larger than 5,000 square kilometers. They have traditionally been employed to safeguard various species. Around 18 Biosphere Reserves have been established in India as of now, 12 of which are included in the World Network of Biosphere Reserves (Ministry of Environment, Forest, and Climate Change, 2019). Some of these biosphere reserves are Nilgiri and Agasthyamalai Biosphere Reserve of Western Ghats, Nanda Devi Biosphere Reserve of Western Himalayas, Sunderbans of Gangetic Delta, Nokrek, and Manas of Eastern Hills, Simlipal of Deccan Peninsula, etc.

A biosphere reserve is comprised of three distinctly identified zones that serve specific objectives. These are:

1. **Core Zone:** The ecology that supports nature as such is called the core zone and it is legally protected by the state government. Human intervention is totally forbidden in this area and handled without

any disruption. Understanding the sustainability of operations or the preservation of environmental quality in the contiguous areas is made easier by keeping a careful watch on these zones.

2. **Buffer Zone:** The core zone is encompassed by a buffer zone that is employed to provide for a wider range of resource usage methods as well as to perform research and educational activities without interfering with the core zone. It is also possible for people to engage in activities like maintaining natural vegetation, practicing agriculture and fisheries, or harvesting forest products in the region. Campaigners for recreation and tourism are still active, although in a more constrained way.
3. **Transition Zone:** The exterior segment of a biosphere reserve where active interaction between the locals and reserve management can be seen is called the transition zone. The people who live there can perform activities like agriculture and forestry under the supervision of local governments and reserve management organizations. These activities contribute a lot to repairing the affected region through sustainable management and utilization of natural resources by ethnic communities, cultural groups, researchers, conservation organizations, and other stakeholders.

2.5.3.2 NATIONAL PARKS

National Park is a small natural reserve or developed territory that has been established as protected zones to preserve worthy resources treasured from the wild in order to maintain and replenish biodiversity (Chiarucci et al., 2001; Rodriguez et al., 2007). These areas are fully propagated or possessed by a sovereign state or country. There is a common objective, i.e., conservation of "wild nature" for future generations, despite the variations in how unlike nations designate their own national parks. It can cover an area of just between 100 and 500 square kilometers. The boundaries are clearly defined and encircled. In certain regions, it is forbidden to engage in activities like grazing, cultivation, and forestry. There are about 106 national parks listed in India including Gir National Park (Gujarat), Kanha National Park (Madhya Pradesh), Ranthambore National Park (Rajasthan), Kaziranga National Park (Assam), Corbett National Park (Uttarakhand).

2.5.3.3 WILDLIFE SANCTUARIES

Overexploitation, habitat destruction, and invasive plant species are heavily affecting the populations of herbal plants that are utilized for manufacturing medicines. Wild sanctuaries can provide a practical solution for the *in-situ* preservation of these herbal plant species that are rare, endangered, and in huge demand in the herbal industry (Soule et al., 2005). This is a place where natural habitats of plants, animals, and their surroundings are shielded from any kind of interference. These are protected areas where animal trapping, killing, and poaching are outright forbidden, and endangered species, in particular, are monitored. Activities like cutting down trees, gathering forest products, cultivating land, etc., are allowed until they don't interfere with the project's goals. In other words, sanctuaries are authorized to allow restricted biotic intrusion, which permits tourists to visit for leisure with requisite permission from the government. There are almost 553 wildlife sanctuaries in India currently, such as Bharatpur Bird Sanctuary (Rajasthan), Hazaribagh Sanctuary (Jharkhand), Sultanpur Bird Sanctuary (Haryana), Abohar Wildlife Sanctuary (Punjab), Nal Sarovar Bird Sanctuary (Gujarat), Mudumalai Wildlife Sanctuary (Tamil Nadu).

2.5.3.4 BIODIVERSITY HOTSPOTS

A bio-geographic area or region with high biodiversity levels is empirically referred to as a biodiversity hotspot. Any territory that needs to be considered a biodiversity hotspot must meet two basic requirements: (i) at least 70% of its primary vegetation must have been lost; and (ii) possess a minimum of 1,500 native vascular plant species (about 0.5% of the total plant species in the world). Many of the biodiversity hotspots have exceeded these two requirements. Some of these hotspots have more than 15,000 indigenous plant species, while in other hotspots, up to 95% of the original habitat has been lost. Approximately 60% of the world's plant, amphibian, reptile, bird, and mammal species are found in these areas, with a significant portion of those species being endemics. In biodiversity hotspots, only a fraction of the overall geographic area is currently protected. The World Wide Fund for Nature (WWF), Critical Ecosystem Partnership Fund (CEPF), and Compensatory Afforestation Management and Planning Authority (CAMPA) are just a few of the international organizations currently working to preserve biodiversity hotspots. India is home to four main biodiversity hotspots, namely The Himalayas, The Western Ghats, The Indo-Burma Region, and The Sundaland.

2.5.3.5 GENE SANCTUARY

Gene sanctuary refers to the area where preservation of germplasm can be achieved under natural habitation without any human interference. In this type of conservation, natural or semi-natural ecosystems and wild species are both conserved. Gene sanctuary generally contains a little part of entire diversity for a given crop species and multiple areas will need to be preserved for that particular species in this regard. India has established its first gene sanctuary for wild relatives of citrus in the Garo Hills of Meghalaya. Additionally, initiatives are also being made to establish gene sanctuaries for rice, mango, banana, and sugarcane.

2.5.3.6 COMMUNITY RESERVES

The term "community reserve" refers to land that cannot be recognized as national parks or wildlife sanctuaries and is totally owned privately by an individual or a community and preserves wildlife resources. The Wildlife Protection Amendment Act of 2002 established this form of protected area which typically acts as a transition zone between wildlife sanctuaries, natural parks, and reserve forests. In India, there are currently 219 Community Reserves, covering 0.04% of the nation's geographical land area (National Wildlife Database, 2022).

2.5.3.7 SACRED GROVES/FORESTS

Sacred groves are the forest reserves where all of the flora and fauna are revered and given complete protection by the local community owing to their religious beliefs and culture. These are the least disturbed forest sections serving as a natural biodiversity hotspot and shelter for a greater number of endangered, rare, and endemic wild species. It represents a long-standing practice of environmental protection based on local knowledge, noble culture, and religious faiths.

2.5.4 MEDICINAL PLANTS CONSERVATION BY EX SITU METHOD

The *ex-situ* conservation, which literally means "off-site" conservation, entails maintaining and rearing threatened plant and animal species in regions that are far from their natural place of origin in order to provide them with a much safer environment. *Ex-situ* conservation is a practice of protecting particular plant and animal species outside of their natural habitats. The primary objective of this method is to support conservation programs effectively by ensuring the continued existence of threatened and endangered species and associated genetic diversity. The idea of reintroducing species into their natural environment is further supported by the fact that threatened species are kept in secure custody until the incidental elements endangering their survival in the wild are addressed, making the avenue of reintroduction feasible. Botanical gardens, seed gene banks, field gene banks, germplasm banks, cryopreservation, tissue culture, DNA libraries, and plantation in school and temple premises are a few of the strategies followed in *ex-situ* conservation.

2.5.4.1 BOTANICAL GARDENS

Botanical gardens are generally referred to as public gardens that preserve collections of living plants primarily for scientific study, research, education, and conservation of wild populations. They are one of the primary organizations engaged in *ex situ* conservation of wild species. They can prevent the extinction of wild plant species through integrated conservation techniques. Botanical gardens preserve plant variety by cultivating sizable collections of threatened species and conserving them in cultivation or seed banks in case wild populations are wiped out. Botanical gardens, which contain a network of arboreta that keep endangered plants for research facilities, are home to around 105,634 species or 30% of the world's known plant variety (Mounce et al., 2017). They have developed into locations of multidimensional significance that encourage both the scientific community and the general public to become more environmentally conscious. Above all, the primary direct contributions of botanical gardens lie mainly in research and education for the *ex situ* preservation of endangered and rare wild species rather than in conservation.

2.5.4.2 SEED GENE BANKS

The simplest technique to save plant germplasm from both wild and domesticated species at a low temperature is in seed gene banks. These are the seed banks that store seeds to preserve the genetic diversity of plant species present in their natural habitat. They are considered one of the key methods of *ex situ* conservation techniques used to store, catalog, and preserve seeds, cuttings, or any significant genetic material from both domesticated and wild crops for future study. The main goal of seed gene banks is to maintain the genes that plant breeders require to improve the yield of crops, resistance against diseases, nutritional value, drought tolerance, and other traits.

2.5.4.3 FIELD GENE BANKS

Field gene banks are used to maintain crop diversity in fields that are vegetatively propagated, have long life cycles, or produce unusual seeds, such as recalcitrant seeds. Many significant crop species in the fields of agriculture, horticulture, and forestry are either difficult to grow because their seeds, when produced, are difficult to germinate or reproduce by vegetative means. When a crop species cannot reproduce through seed or produces only seeds with a short lifespan, its genetic resources can be preserved by planting it in a field gene bank. It conserves and maintains as much diversity as possible in a minimum sampling, which is more or less equivalent to diversity blocks. The field gene bank serves as a learning and entertainment facility in addition to being used for preservation, production, and distribution. Again, maintaining genetic resources permanently in a field gene bank is much more cost-effective and feasible compared to other gene banks.

2.5.4.4 GERMPLASM BANKS

The most effective way to preserve the genetic characteristics of threatened and economically valuable species is through germplasm banks. All of the genes that are present in a particular plant can be found in its germplasm, which can be preserved for a long time and reproduced anytime as needed in the future. Again, *in vitro* testing of germplasm for genetic modifications (transgenes) and the selection of the best germplasm among them are both made possible by this gene bank. It aims at the collection and preservation of plants that are deemed to be of the utmost value to our civilization, as well as to learn scientific information that will help to preserve and make use of genetic resources. Maintenance of biological integrity and provision of germplasm phenotypic and genetic descriptions are two of the key objectives underlying germplasm conservation. Germplasm conservation is crucial in preserving genetic data for threatened, extinct, or even any surviving species since genetic variation results in the loss of genetic information from earlier generations.

2.5.4.5 CRYOPRESERVATION

Many agricultural crops have been entirely reproduced from the tissues that are kept at extremely low temperatures for long months to several years (Ding et al., 2008). At present, the most economical and everlasting method to retain germplasm is cryopreservation, which involves preserving biological material at an extremely low temperature (about –196°C). In cryopreservation, any plant material can be stored without being altered or modified due to the complete cessation of metabolic activity of cells at this low temperature. It is the sole *ex situ* preservation technique for species that cannot be preserved in seed banks for a very long period. Plant tissues including meristems, seeds, endosperms, ovules, embryos, cultivated plant cells, etc., can easily be preserved via cryopreservation. It aids in overcoming the issues with conventional techniques of germplasm conservation such as seed dormancy, transient seeds, diseases transmitted through seeds, and high input costs. In addition to preserving plant samples for a long period, cryopreservation also significantly reduces the amount of pathogens present in the plant sample.

2.5.4.6 TISSUE CULTURE

In order to maintain rarely available, endangered plant species, there is a great deal of strain on natural habitats and populations. Tissue culture can be utilized to reduce this pressure and to regenerate these species through *in vitro* seedless culture of cells and tissues or complete plants which take place under carefully regulated environmental and nutritional conditions (Thorpe, 2007). It is more than just a different method of propagation. The practice of propagating and conserving endangered plant species is undoubtedly becoming more popular among environmentalists. Tissue culture is used to reproduce and save plant species whose numbers have declined owing to overexploitation in areas where regeneration is challenging using traditional means in order to keep them from going extinct. Tissue culture techniques have many advantages, including the ability to preserve endangered plants *in vitro* and the reduction of storage requirements for both space and time (Engelmann, 2011).

2.5.4.7 DNA LIBRARIES

Low genetic diversity among rare and threatened plant species is causing a decline in the natural survival rate of these species. Researchers now have access to genes of interest and a means of reviving these threatened species through the establishment of "DNA libraries." The preservation of plant species' genomic DNA at low temperatures might be thought of as a complementary technique for conservation of wild plant species. These libraries are made up of numerous DNA vectors, each carrying a unique DNA fragment that contains a particular genomic characteristic of a plant species. Numerous details describing the plant family, genus, and type of DNA product or marker made from DNA are available in the databases pertaining to DNA libraries. This method of biological data preservation gives researchers the chance to look up novel genes and their byproducts. In an effort to address the biodiversity issue, manage genetic resources globally, and realize their full potential, these libraries' DNA collections have emerged as significant global resources.

2.5.4.8 PLANTATION IN SCHOOL AND TEMPLE PREMISES

A mother stock or nursery is required in order to reintroduce medicinal plants that are rarely found in the wild and from which many offspring can be grown (Khan et al., 2005). Because of this, it is possible to plant trees on the premises of schools and temples to preserve and prevent the extinction of the rare flora that are connected to these educational institutions or places of worship. This natural flora and its habitat are nurtured and preserved in large part by temples and educational institutions. The aim of this nursery is to cultivate sparse and threatened medicinal plants that have been recognized by the community's "Boiddo," a rural health expert who could easily reach it and offer herbal treatments to the locals.

2.6 MANAGEMENT APPROACHES

Medicinal plants that are collected from the wild population or those cultivated commercially may become contaminated with other plant species or their different parts due to false identification, unintentional alteration or occasionally purposeful intrusion of impurities, which could have harmful consequences. This may lead to additional concerns for the preservation of endangered species. Therefore, it is urgently necessary to implement a number of appropriate management approaches for sustainable production of herbal products and to offer farmers or local collectors comprehensive technical direction on collecting high-quality medicinal plant materials. The conservation of these quality medicinal plant species, which are under threat of becoming extinct, can be made possible following various management approaches that include good cultivation practices, proper collection methods, documentation of indigenous traditional knowledge, training of nursery growers, herbal fairs, legislations, etc.

2.6.1 GOOD CULTIVATION PRACTICES

Preservation of medicinal plants is of utmost need in order to fulfill their growing demand. The appropriate cultivation techniques or good agricultural practices (GAP) must be promoted in order to preserve them. Through capacity training, timely policy intervention, and strong market links facility, the communities must be encouraged to cultivate commercially viable ethno-botanical species using improved cultivation methods. The likelihood of misidentification and adulteration also significantly decreases due to proper cultivation techniques. Good medicinal plant growing techniques can therefore contribute to the preservation of wild populations and meet the needs of global health.

Cultivation of medicinal plants involves rigorous supervision and care. Primarily, a proper cultivation method involves the selection of good quality seeds and other propagation materials in accordance with the required quality of herbal products to be obtained. The propagation materials must be of the highest caliber and uncontaminated by any potential contaminants or infections to encourage healthy plant growth. Depending on the desired herbal plant species and its parts; area of cultivation, optimal climatic and soil conditions need to be fixed. Fertilizer is very much inevitable to harness maximum production from cultivated medicinal plants. Therefore, precise use of fertilizers in respect of the right type and optimum quantities must be ensured for a particular plant species. Again, timely exercising the methods such as pruning, nipping, topping, and shading might be employed to regulate the growth of the plant and its development, thereby enhancing the production of desired medicinal plant products. Thus, good cultivation practice plays an important role in the preservation of those endangered plants, retaining the quality plant products of desired ethno-medicinal species.

2.6.2 COLLECTION OF MEDICINAL PLANTS FROM WILD HABITAT

The composition and quality of medicinal species taken from the wild typically vary according to a number of factors including plant growth stages, harvested plant parts, procurement time, environment, and genetic variations. However, all applicable national and/or regional quality standards should be met by raw medicinal plant materials. To secure the collection of correct plant material from the wild in a sustainable way, it is most needed for the collectors to have vivid knowledge, in particular, about the plant species, its vegetative cycle, population density, appropriate time and techniques of harvesting, etc. It is crucial for collectors to distinguish between real plants, consider conservation measures, and find strategies to protect the production of high-quality plant material suitable for use in pharmaceutical manufacturing. Again, collection methods of medicinal plants from wild habitats are still little known, underestimated, and often seen as unimportant, which must be addressed for sustainable procurement of those herbal plants and their conservation. A large number of plant species that are currently abundantly available will go extinct if irrational harvesting continues.

Conservation of these plant species also involves imposing certain regulations on their collection or harvest from the wild. Good harvesting practice plays the most significant role in sustainable utilization of wild plant resources. When it comes to their utilization, plants should be collected when they are at their peak in terms of the required medicinal quality and therapeutic efficacy. To minimize undesirable admixtures, it is crucial to distinguish properly between the medicinal plant and its relatives during collection. To ensure that the raw material harvested is of the highest quality, collectors should also possess sufficient knowledge of the best methods for harvesting, as well as the subsequent conservation process and storage conditions.

2.6.3 DOCUMENTATION OF INDIGENOUS TRADITIONAL KNOWLEDGE

Application of herbal remedies is significantly influenced by indigenous traditional knowledge (ITK) relating to medicinal plants. Traditional knowledge regarding the use of herbal remedies is crucial for the sustainability of wild plant resources, especially medicinal plants. Local knowledge specific to a given area and the transmission of knowledge, skills, and strategies between generations can be used to improve biodiversity protection. However, due to inadequate documentation and a lack of information transfer across generations, traditional knowledge is being depleted day by day. Hence, it is crucial to properly document and retrieve the venerable ITK system (Nautiyal et al., 2009; Kala, 2003). Documentation and stimulating practice by incorporating the local population are two of the finest strategies to preserve medicinal plants. Exploring the age-old traditional associated knowledge, recording it, and using those documented learnings to establish conservation plans are crucial steps in ensuring the long-term sustainability of medicinal plants. Therefore, gene banks, taxonomists, and conservation organizations would require documenting indigenous knowledge in order to properly maintain a record of medicinal plants. Doctrines subsisting within the ancient medicinal systems and found in various locations today are carried out generation after generation by elderly and aged people who bear profound knowledge of wild plants and their use that has to be transferred and documented for age-long preservation of these plants.

2.6.4 TRAINING OF NURSERY GROWERS

Technical information necessary for nursery preparation and propagation techniques of medicinal plant production was unknown to the local nursery growers. A training program on nursery establishment and plantation practices, for at least a few motivated local nursery growers, can help to make them more efficient and skilled. A resource person or another expert in the field must lead and conduct the training program. Training can be given to nursery growers on topics like techniques for gathering seeds or seedlings of particular species, choosing a nursery location, preparing the soil in the seed bed, pre-treating of seeds, planting seeds, vegetative propagation, uprooting, and transplantation, seedlings hardening, maintenance of nursery stock, seed storing, and continuous management practices like watering, weed control, protection against pests and diseases, application of fertilizers, etc.

2.6.5 HERBAL FAIR

Herbal fairs are landmark events that aim towards revitalizing relationships between people of diverse ethnic groups who attend the fairs to connect with nature, practices, and culture. The purpose of this fair is to educate the general public about the value and application of medicinal plants that thrive throughout the biosphere. Many stalls and shops display rare medicinal species, several herbal medicines, and explain their use and important phytoconstituents present in them, creating much enthusiasm among local ethnic communities. Through this innovative approach, hundreds of people from around the region can participate, exchange their traditional knowledge, and become aware of the preservation of medicinal plants along with their judicious use. In addition to promoting herbal products, herbal fairs offer a venue for direct communication between consumers and vendors of Ayurveda and herbal remedies.

2.6.6 ARRANGEMENT OF MEETING AND WORKSHOPS

Awareness among the people of different ethnic groups about the need to protect medicinal plants by identifying a comprehensive range of native wild herbs and their medicinal uses can be raised through meetings and workshops at the local, regional, and national levels. The main goal is to educate the target audience about the current situation and motivate them to manage and preserve medicinal plants efficiently. These workshops teach participants how to create their own safe, effective, and feasible herbal medicines using some lesser-known plants that naturally grow in the wild. Participants can acquire enough knowledge and confidence to start using herbal products immediately to help treat many common minor ailments, once they know the correct amounts and safe dosages for application. Each participant can learn to make tinctures, infused oils, and ointments after learning about the various qualities of medicinal herbs from these workshops. Additionally, these meetings and workshops aim to explore the potential for replicating and expanding this approach, as well as marketing opportunities for herbal products and seedlings in the region.

2.6.7 LEGISLATION

Numerous ethno-medicinal plant species are in danger of getting extinct due to indiscriminate harvesting from its natural habitat. The state and central governments, in this regard, must enact special Act to protect these priceless medicinal plant species before they become extinct and also provide training to forest department employees not only for today's prosperity but also for wholesome supply of these resources to our future generations. No plants or parts of plants on the endangered species list, whether local or global, should be harvested without any special permission granted by the concerned state and/or central authorities. Currently in India, there are three Acts viz. Indian Forest Act (1927), the Wildlife Protection Act (1972), and the Forest Conservation Act (1980) covering medicinal plant issues. Any violation of these Acts consequences in a minimum sentence of six months imprisonment, which may be increased to two years, or a minimum fine of 1,000 rupees, which may be increased to 5,000 rupees or a combination of both.

2.7 CONCLUSION

Ethno-medicinal plants found in natural ecosystems provide substantial economic benefits to indigenous people. These plants act as a remedy for several diseases like diarrhea, jaundice, diabetes, asthma, rheumatism, dysentery, gonorrhea, skin diseases, and livelihood options for rural communities in developing nations. Their increased market demand as food supplements or herbal tea has created a dependable source of income for rural indigenous people. Further pharmacological and clinical research on ethno-medicinal plants may help in treating many chronic diseases that are not curable with conventional allopathic medicines. These rich sources of herbal medicines should be managed properly and conserved following good cultivation practices so that rural indigenous communities benefit. It can be seen that overexploitation of wild resources, lack of knowledge, and expansion of human-sponsored habitats have all resulted in a sharp decline of ethno-medicinal plant flora, in addition to the issues of climate change. Recently, it has been predicted that about 25 to 50% of plant species on Earth will become extinct in the coming 30 years if current trends of plant species loss are not checked. Therefore, long-term sustainability in the management, protection, and preservation of these unique ethno-medicinal plants is crucial and will prove to be a sound strategy in the long run.

KEYWORDS

- **bioactive substances**
- **conservation**
- **ethno-medicinal plants**
- **ex situ conservation**
- **herbal medicines**
- **livelihood**
- **natural resources**
- **preservation**

REFERENCES

Abu-Rabia, A., (2005). Urinary diseases and ethnobotany among pastoral nomads in the Middle East. *J. Ethnobiol., 1*, 1–3.

Asase, A., Kokubun, T., Grayer, R. J., Kite, G., Simmonds, M. S., & Oteng-Yeboah, A. A., (2008). Chemical constituents and antimicrobial activity of medicinal plants from Ghana: *Cassia sieberiana, Haematostaphis barteri, Mitragyna inermis* and *Pseudocedrela kotschyi. Phytother. Res., 22*, 1013–1016.

Bentley, R., (2010). *Medicinal Plants* (pp. 23–46). London: Domville-Fife Press.

Biosphere Reserves in India, (2019). Ministry of Environment, Forest and Climate Change.

Calixto, J. B., (2005). Twenty-five years of research on medicinal plants in Latin America: A personal review. *J. Ethnopharmacol., 100*, 131–134.

Chacko, S. M., Thambi, P. T., Kuttan, R., & Nishigaki, I., (2010). Beneficial effects of green tea: A literature review. *Chin. Med., 5*, 13.

Chen, S. L., Yao, H., Han, J. P., Liu, C., Song, J. Y., Shi, L. C., Zhu, Y. J., et al., (2010). Validation of the ITS2 region as a novel DNA barcode for identifying medicinal plant species. *PLoS ONE, 5*, e8613.

Chiarucci, A., Maccherini, S., & De Dominicis, V., (2001). Evaluation and monitoring of the flora in a nature reserve by estimation methods. *Biol. Conserv., 101*, 305–314.

Conservation of Medicinal Plants, UNDP, India. Available from: https://www.in.undp.org/content/india/en/home/climateanddisasterresilience/successstories/conservation of medicinal plants.html (accessed on 22 April 2024).

De Albuquerque, U. P., & Hanazaki, N., (2009). Five problems in current ethnobotanical research—And some suggestions for strengthening them. *Hum. Ecol., 37*, 653–661.

Ding, F., Jin, S., Hong, N., Zhong, Y., Cao, Q., Yi, G., & Wang, G., (2008). Vitrification–cryopreservation, an efficient method for eliminating *Candidatus liberobacter* asiaticus, the citrus Huanglongbing pathogen, from in vitro adult shoot tips. *Plant Cell Rep., 27*, 241–250.

Engelmann, F., (2011). Use of biotechnologies for the conservation of plant biodiversity. *In Vitro Cell. Dev. Biol. Plant., 47*, 5–16.

Fabricant, D. S., & Farnsworth, N. R., (2001). The value of plants used in traditional medicine for drug discovery. *Environmental Health Perspectives (Supplement), 109*, 69–75.

Farzaei, M. H., Abbasabadi, Z., Ardekani, M. R. S., Rahimi, R., & Farzaei, F., (2013). Parsley: A review of ethnopharmacology, phytochemistry and biological activities. *J Trad. Chin. Medicine, 33*(6), 815–826.

Hamilton, A. C., (1997). Threats to plants: An analysis of centers of plant diversity. In: Touchell, D. H., & Dixon, K. W., (eds.), *Conservation into the 21st Century; Proceedings of the 4th International Botanic Gardens Conservation Congress (Perth, 1995)* (pp. 309–322). Kings Park and Botanic Garden, Perth, Australia.

Hamilton, A. C., (2004). Medicinal plants, conservation and livelihoods. *Biodivers. Conserv., 13*, 1477–1517.

Hamilton, A. C., (2008). Medicinal plants in conservation and development: Case studies and lessons learned. In: Kala, C. P., (ed.), *Medicinal Plants in Conservation and Development* (pp. 1–43). Salisbury: Plant life International Publisher.

Heinrich, M., (2000). Ethnobotany and its role in drug development. *Phytother. Res., 14*, 479–488.

Heywood, V. H., & Iriondo, J. M., (2003). Plant conservation: Old problems, new perspectives. *Biol. Conserv., 113*, 321–335.

https://www.cbd.int/ (accessed on 22 April 2024).

Jabbar, A., Zaman, M. A., Iqbal, Z., Yaseen, M., & Shamim, A., (2007). Anthelmintic activity of *Chenopodium album* (L) and *Caesalpinia crista* (L) against trichos-trongylid nematodes of sheep. *J. Ethnopharmacol., 114*, 86–91.

Joshi, B. C., & Joshi, R. K., (2014). The role of medicinal plants in livelihood improvement in Uttarakhand. *International Journal of Herbal Medicine., 1*(6), 55–58.

Kabir, M. H., Hasan, N., Rahman, M. M., Rahman, M. A., Khan, J. A., & Hoque, N. T., (2014). A survey of medicinal plants used by the Deb barma clan of the Tripura tribe of Moulvibazar district, Bangladesh. *J. Ethnobiol. Ethnomed., 10*, 19.

Kadir, M. F., Sayeed, M. S. B., & Mia, M. M., (2013). Ethnopharmacological survey of medicinal plants used by traditional healers in Bangladesh for gastrointestinal disorders. *J. Ethnopharmacol., 147*, 148–156.

Kala, C. P., (2003). Indigenous uses of plants as health tonic in Uttaranchal Himalaya, India. *Annals of Forestry, 2*, 249–254.

Kandari, L. S., Phondani, P. C., Payal, K. C., Rao, K. S., & Maikhuri, R. K., (2012). Ethnobotanical study towards conservation of medicinal and aromatic plants in upper catchments of Dhauli Ganga in the Central Himalaya. *J. Mt. Sci., 9*, 286–296.

Khan, M. S. I., Mannan, M. A., & Chowdhury, M. T. I., (2005). *Medicinal Plant Conservation Through Community Participation* (pp. 1–46). IUCN-The World Conservation Union, Bangladesh Country Office, Dhaka.

Laird, S. A., & Ten, K. K., (2002). Linking biodiversity prospecting and forest conservation. In: Pagiola, S., Bishop, J., & Landell-Mills, N., (eds.), *Selling Forest Environmental Services* (pp. 151–172). Earthscan, London.

Lulekal, E., Asfaw, Z., Kelbessa, E., & VanDamme, P., (2013). Ethnomedicinal study of plants used for human ailments in Ankober District, North Shewa Zone, Amhara Region. *Ethiopia J. Ethno. Biol. Ethnomed., 63*, 1–13.

Ma, J., Rong, K., & Cheng, K., (2012). Research and practice on biodiversity in situ conservation in China: Progress and prospect. *Sheng Wu Duo Yang Xing, 20*(5), 551–558.

Mesfin, K., Tekle, G., & Tesfay, T., (2013). Ethnobotanical study of traditional medicinal plants used by indigenous people of Gemad district, Northern Ethiopia. *J. Med. Plants, 1*, 32–37.

Mounce, R., Smith, P., & Brockington, S., (2017). Ex situ conservation of plant diversity in the world's botanic gardens. *Nat. Plants, 3*, 795–802.

Nadembega, P., Boussim, J. I., Nikiema, J. B., Poli, F., & Antognoni, F., (2011). Medicinal plants in baskoure, kourittenga province, Burkina Faso: An ethnobotanical study. *J. Ethnopharmacol., 133*, 378–395.

Nalawade, S. M., Sagare, A. P., Lee, C. Y., Kao, C. L., & Tsay, H. S., (2003). Studies on tissue culture of Chinese medicinal plant resources in Taiwan and their sustainable utilization. *Bot. Bull. Acad. Sin., 44*, 79–98.

Nasab, F. K., & Khosravi, A. R., (2014). Ethnobotanical study of medicinal plants of Sirjan in Kerman Province, Iran. *J. Ethnopharmacol., 154*, 190–197.

Nautiyal, S., Rajan, K. S., & Shibasaki, R., (2009). Interaction of biodiversity and economic welfare– a case study from the Himalayas of India. *Journal of Environmental Informatics, 6*, 111–119.

Negi, P. S., & Subramani, S. P., (2002). Ethnobotanical study in the village Chhitkul of Sangla valley, district Kinnaur, Himachal Pradesh. *J. Non-timber Forest Prod., 9*, 113–120.

Oljira, K., (2015). Ethnobotanical survey of plants traditionally used for malaria prevention and treatment in selected resettlement and indigenous Villages in Sasiga District, Western Ethiopia. *Journal of Biology, Agriculture and Healthcare, 5*(11), 2224–3208..

Qin, J., Li, R., & Raes, J., (2010). A human gut microbial gene catalogue established by metagenomic sequencing. *Nature, 464*(7285), 59–65.

Raj, A. J., Biswakarma, S., Pala, N. A., Shukla, G., Vineeta, Kumar, M., Chakravarty, S., & Bussmann, R. W., (2018). Indigenous uses of ethnomedicinal plants among forest-dependent communities of Northern Bengal, India. *Journal of Ethnobiology and Ethnomedicine, 14*, 8. Doi: 10.1186/s13002-018-0208-9.

Rashid, S., Ahmad, M., Zafar, M., Sultana, S., Ayub, M., & Khan, M. A., (2015). Ethnobotanical survey of medicinally important shrubs and trees of Himalayan region of Azad Jammu and Kashmir. *Pakistan. J. Ethnopharmacol., 166*, 340–351.

Rates, S. M. K., (2001). Plants as source of drugs. *Toxicon., 39*, 603–613.

Redzic, S., (2007). The ecological aspect of ethnobotany and ethnopharmacology of population in Bosnia and Herzegovina. *Collegium Antropologicum., 31*, 869–890.

Rodriguez, J. P., Brotons, L., Bustamante, J., & Seoane, J., (2007). The application of predictive modelling of species distribution to biodiversity conservation. *Divers. Distrib., 13*, 243–251.

Ross, I. A., (2005). *Medicinal Plants of the World: Chemical Constituents, Traditional and Modern Medicinal Uses* (Vol. 3, pp. 110–132). New Jersey: Humana Press Inc.

Sharma, J., Gairola, S., Gaur, R. D., & Painuli, R. M., (2012). The treatment of jaundice with medicinal plants in indigenous communities of the sub-Himalayan region of Uttarakhand, India. *J. Ethnopharmacol., 143*, 262–291.

Shukla, G., & Chakravarty, S., (2012). Ethnomedicinal plant use of Chilapatta reserved forest in West Bengal. *Indian Forester., 138*, 1116–1124.

Soule, M. E., Estes, J. A., Miller, B., & Honnold, D. L., (2005). Strongly interacting species: Conservation policy, management, and ethics. *Bioscience, 55*, 168–176.

Subrat, N., (2002). *Ayurvedic and Herbal Products Industry: An Overview.* Paper at a Workshop on Wise Practices and Experiential Learning in the Conservation and Management of Himalayan Medicinal Plants, Kathmandu, Nepal, supported by the Ministry of Forest and Soil Conservation, Nepal, the WWF-Nepal Program, MAPPA and PPI.

Sunderlin, W. D., Dewi, S., Puntodewo, A., Muller, D., Angelsen, A., & Epprecht, M., (2008). Why forests are important for global poverty alleviation: A spatial explanation. *Ecology and Society, 13*(2), 24.

Suriyavathana, M., Jeevitha, M., & Aranganathan, J., (2011). In vitro antioxidant profile of *Justicia tranquebariensis*. *Int. J. Pharm. Sci., 4*, 4259–4261.

Thakur, K., Bassi, K., & Sood, K., (2015). *Textbook of Ethnobotany* (pp. 1–70). S. Dinesh & Co: Jalandhar, India.

Thorpe, T., (2007). History of plant tissue culture. *J. Mol. Microbial Biotechnol., 37*, 169–180.

Van, W., B. E., Oudtshoorn, B. V., & Gericke, N., (1997). *Medicinal Plants of South Africa.* Johannesburg: Briza.

Venkataswamy, R., Mohamad, M. H., Doss, A., Ravi, T. K., & Sukumar, M., (2010). Ethnobotanical study of medicinal plants used by Malasar tribals in Coimbatore district of Tamil Nadu (South India). *Asian J. Exp. Biol. Sci., 1*, 387–392.

Volis, S., & Blecher, M., (2010). Quasi in situ: A bridge between ex situ and in situ conservation of plants. *Biodivers. Conserv., 19*, 2441–2454.

WHO, (2002). *Traditional Medicine Strategy 2002–2005.* WHO/EDM/TRM/2002.1, World Health Organization, Geneva, Switzerland.

World Bank India: Unlocking Opportunities for Forest-Dependent People, (2006). The World Bank, Agriculture and Rural Development Sector Unit, South Asia Region, Oxford University Press: New Delhi.

World Bank, (2001a). *A Revised Forest Strategy for the World Bank Group.* Washington, DC: World Bank.

Zerabruk, S., & Yirga, G., (2012). Traditional knowledge of medicinal plants in Gindeberet district, Western Ethiopia. *S. Afr. J. Bot., 78*, 165–169.

CHAPTER 3

Forests: Sustainable Uses and Their Role in Livelihood Security

MURAD MUHAMMAD,[1] ABDUL WAHAB,[2] SAQIB ULLAH,[3] ABDUL BASIT,[4] and KHUSHNOOD UR REHMAN[3]

[1]*Phyto-Ecology Lab, Department of Botany, University of Peshawar, Pakistan*

[2]*Shanghai Center for Plant Stress Biology, Center for Excellence in Molecular Plant Sciences, Chinese Academy of Sciences, Shanghai, China*

[3]*Department of Botany, Islamia Collage University Peshawar, Peshawar, Pakistan*

[4]*Department of Horticulture, Faculty of Crop Production Science, The University of Agriculture, Peshawar, Pakistan*

ABSTRACT

Plants, fungi, and animals that have been collected from the forest have, for a very long time, been a source of sustenance, health, financial worth, and cultural significance. The production of industrial wood has historically been the most important aspect of "forest management" across the world. The production of food based on forest resources is not even a secondary factor; rather, it is an everyday occurrence that is not considered in official statistics. For a great number of people, the presence of forests guarantees a reliable source of vital foods and nutrients. People who search for food in the woods have access to a variety of advantages, including those in the fields of culture and recreation, as well as the kitchen. Customers have a growing perception that these products are more "natural" and healthier than food that is produced using agricultural practices. The ever-increasing

Sustainable Forest Resources Management: Issues and Implications. Jahangeer A. Bhat, Gopal Shukla, Manmohan J. Dobriyal, Sumit Chakravarty, A. Arunachalam, and Rainer W. Bussmann (Eds.)

demand for "natural" food production has led to an increase in the number of multibillion-dollar firms that make extensive use of resources derived from forests and wild areas as essential components. According to consumer trends, the practice of harvesting food from forests, which involves biological processes and new sorts of culturally ingrained ties with the natural world, is becoming an increasingly popular pastime among consumers. There is also a growing demand for agricultural output to be "re-oriented," which presents an opportunity to strengthen the roles that forests and trees play in food production and to rethink how food systems are organized.

Through the utilization of a wide variety of plants, such as the baobab, forests, and trees may be able to provide food security and nourishment to people. The following diagram illustrates how woodlands and trees can be incorporated into everyday life. It is feasible to expand food systems and transition to a bio-economy through the utilization of the new opportunities presented by forests and trees, in addition to the challenges that these factors present. This transformation is important to meet the growing demand for food that is both safe and nutritious while also maintaining the biodiversity of the forest.

3.1 INTRODUCTION

An estimated 200–300 million people around the world are sustained by the food that is harvested from the world's forests each year (Residues et al., 1978). More than 2.6 billion people, or 62% of the world's population, live in rural areas and get the majority of their food from their immediate surroundings. These people are concentrated in developing countries. A small number of people in the region still live the hunter-gatherer lifestyle, while others engage in agricultural practices such as slash-and-burn, and still, others maintain permanent agricultural systems. Numerous applications can be found for both timber and non-timber products; many of these applications improve the economic well-being of individuals by making it possible for them to purchase food and other necessities. Deforestation occurs on approximately 15 million hectares of land across the globe on an annual basis (Surya et al., 2020). During this time, the population of the world is expanding at a startling rate, and the United Nations estimates that over 2 billion people are suffering from some form of malnutrition. According to the World Resources Institute, a tipping point has been reached regarding the depletion of forest resources in response to the growing demand for food among humans. As the global population continues to expand at an exponential rate, it is becoming increasingly challenging for the world's forests and food supplies to keep up. Over the past 15 years, there has been a dramatic reduction in the typical quantity of food that is produced for each individual. As more individuals are added to the population each year, the amount of food and forest resources that are accessible to each individual gradually diminishes. The fact that human population growth has a direct influence on forests and other natural resources, which are essential to the world's food supply, is also concerning (Antar et al., 2021; Muhammad et al., 2021). The production of food is only going to be able to continue into the foreseeable future if there is a sufficient supply of forests and other biological resources. Long-term food security is being put in jeopardy by several issues, some of which include an increasing global population, the depletion of resources in certain countries, and inefficient management of forests and other natural resources (Mushi et al., 2020). According to the World Bank, to guarantee the reliability of the supply of food in developing nations, an analysis of forests and goods made from non-timber raw materials was carried out. The production of forest-based meals, the conservation of natural resources and biodiversity, and the creation of jobs through forest-based food collection and hunting are some of the reasons why forests are important. Other reasons include the protection of natural resources and biodiversity. Forests are important for several other reasons as well, including the provision of food derived from forests, the protection of natural resources and biodiversity, and the creation of employment opportunities through the gathering of forest-derived foods and the pursuit of forest-based animals (Kopteva et al., 2019).

According to projections made by the Food and Agriculture Organization of the United Nations, the amounts of food consumed globally will more than triple by the year 2050. Over the next 40 years, it is anticipated that food production will increase by more than 70% points. It is unavoidable that agricultural infrastructure and land will be subject to an increased level of taxation as the global population continues to expand. The conversion of natural ecosystems and forests to agricultural production has been a long-term trend that has resulted in the loss of biological diversity (da Cunha Dias et al., 2021; Michal et al., 2019). Food production in the 21st century is a monumental challenge because the use of traditional methods to obtain food security has resulted in a loss of forest cover, which directly and indirectly supports the food and nutritional needs of 1 billion people, particularly in developing countries. Around 1 billion people depend on forests and trees for their daily survival, as well as on forest foods for their dietary and health requirements. This includes both the trees themselves and the products they produce. In rural communities close to forests, there is a broad range of food that can be purchased, which is especially advantageous to the residents of the area who are living in poverty (Pawera et al., 2020). Whether they are wealthy or impoverished, urban dwellers all over the world incorporate wild foods into their diets regularly (Kerr et al., 2021).

Forests contribute to food security and nutrition in a variety of different ways, according to the findings of an expert panel on food security and nutrition that was organized by the Committee on World Food Security (Gergel et al., 2020). This panel was organized by the Committee on World Food Security. Food that is produced directly from forests is extremely important to the health and well-being of families because it increases the amount of nutrition in the diet as well as the variety of foods that are consumed. Even though it accounts for less than 1% of the world's food supply, this type of food is extremely important. Micronutrients are typically in short supply for people living in poverty; however, foods like fruits and vegetables, which are plentiful in the forest, are good sources of micronutrients (Jekayinfa et al., 2020). The Food and Agriculture Organization (FAO) estimates that trees in developing nations produced approximately 390 million metric tons of food in the year 1997. Around 258 metric tons of forest foods are gathered from public lands across the United States every single year (Winchester & Reilly, 2020). A recent survey of households in Europe revealed that the average annual consumption of forest products is approximately 60 kilograms, with a median intake of 20 kilograms. This information was derived from the results of the survey. This data was extracted from a survey that was distributed to homes across Europe (Gergel et al., 2020).

The energy that is produced by the forest environment can be used as a source of fuel while also being used to produce goods for agriculture and the forest. However, in certain parts of Africa, wood fuel accounts for as much as 25% of the country's total energy consumption. Globally, wood fuel accounts for around 6% of the supply of primary energy. A by-product of urbanization, wood fuel is sold and processed into charcoal for use as a home energy supplement, creating jobs and revenue opportunities throughout the countryside. Charcoal can also be used as a residential energy supplement (Adamu et al., 2020). In many underdeveloped nations, the majority of urban families still rely on charcoal as their primary source of heating and cooking fuel. Families can cut their expenses by skipping meals, undercooking food, or boiling water for a shorter period, selling or trading food, using fuelwood of lower quality such as dung, or substituting lower-quality fuelwood like dung for the fuelwood they normally use. All of these things represent a danger to the well-being and ability of a family to provide for their food needs (Bessaad et al., 2021). The harvesting of fuelwood has a range of different effects on forests; nonetheless, there is evidence to show that in some circumstances, harvest levels may surpass what is considered to be the sustainable limit. The informal and unregulated wood fuel sector is responsible for a major amount of the environmental harm that is generated, and a large portion of this damage is attributed to value chains within the industry that lack proper enforcement and governance structures (Dhungana & Bhattarai, 2022).

According to the Food and Agriculture Organization, more than 20% of the income of rural households comes from forest crops and other forest products like fuelwood (Sheppard et al., 2020). More than 13 million people have jobs in the official forest industry, which gives them the ability to purchase food and other necessities, such as clothing and shelter. The formal forest sector is one of the largest employers in the world. The sale of non-wood forest products (NTFP), which is a subset of the category of non-timber forest products (NTFP), is anticipated to produce more than $88 billion in revenue in the year 2015 (Bahareldeen, 2019). Additional cash can be generated through the harvest of therapeutic and aromatic plants, as well as through the sale of animal goods such as bushmeat and animals. It is estimated that tens of thousands of individuals have jobs in the forest's informal economy, which includes activities like harvesting for direct consumption, engaging in bartering, and trading items without utilizing transaction records. It is estimated that hundreds of millions of people around the world make their living and get their nourishment from the world's forests. By searching for food in the wild, a family may be able to cut their monthly food expenses by as much as 60% (Chamberlain et al., 2020).

For the production of a vast number of different commodities and services all over the world, resources from forests are necessary. The hundreds of different kinds of forest plants, fungus, and animals provide people all around the world with a variety of benefits, including food, medicine, and other essentials and luxuries that they can use in their daily lives. Even while forests play a significant part in human nutrition and food security, we feel that the informal structure of markets and the use of forest products that are not included in official global statistics are particularly serious elements that should not be ignored. As we'll see in the next portions, forests can have a substantial impact on food systems on a global as well as a regional scale. This is something that we'll demonstrate in the following sections. Despite the significant advancements that have been made in this field, there is still room for development in this particular aspect (Clair et al., 2022). If the forest is to fulfilit's potential to sustain human life over the long term, these contributions must be acknowledged and appreciated. Only then can the forest fulfill its promise. According to the information presented, a review of the pertinent literature is currently taking place to combine the concepts into a more holistic approach to the production of food. The gathering of wild plants and fungi from the forest for food preparation receives a lot of attention, as does the role of forests as provisioning units for food systems and the bioeconomy (Clair et al., 2022). Gain an understanding of how important wood and trees have been to both ancient and modern civilizations, how consumers are increasingly seeking products that are friendly to the environment, and how some of these products may be harmful to human health. Wood and trees have been used in a wide variety of ways throughout history. Concerning the numerous production strategies available, a great deal of debate has been sparked by everything from gathering food in the wild to growing crops in a single-species monoculture. As a direct consequence of this conversation, we will next investigate how we might establish a bioeconomy that meets our food and dietary requirements through the utilization of forests and trees as a component of a more comprehensive framework for development (Lazos-Chavero et al., 2021).

3.2 FORESTS AND TREES MAKE IMPORTANT CONTRIBUTIONS TO PEOPLE IN TERMS OF FOOD SYSTEMS

3.2.1 *CULTURAL IDENTITIES AND NATURE-RELATED VALUES*

In both tropical and temperate regions, indigenous people have a deep cultural connection to forest foods. Indigenous communities have relied on the collection and use of forest plants and fungi to supplement their diets. For example, the hundreds of edible species discovered in Native American civilizations in temperate and boreal woodlands demonstrate that forest foods were an integral element of their traditional diet (Carson et al., 2018). For thousands of years, Native Americans have gathered mushrooms for both medicinal and culinary purposes. Wild foods from the forest, such as wild yams, could be gathered and stored for later use, allowing traditional hunter-gatherers of the Congo Basin to feed their families for long periods. In comparison to their Canadian and Mexican counterparts, Amazonian Indians ate fewer wild greens and favored wild fruits and tubers during their lives. Educating future generations about forest foods may be beneficial to their educational pursuits and to forest managers in their efforts to better manage forests for these resources (Manero et al., 2022).

Additionally, forest foods are crucial to urban cultures and lifestyles, as well as intangible values, in the same way as indigenous people rely on them. Over the past two decades, urban food foraging has become increasingly popular, and it is now regarded as a legal hobby. This sport is becoming increasingly popular on research sites. There is a wide range of reasons why urban Berliners use wild plants for everything from sustenance to personal pleasure to medicinal needs. In the last two decades, harvesting Appalachian Forest onions (Allium tricoccum Ait.), sometimes known as ramps or leeks and collected in huge quantities, has grown more popular among city inhabitants (Clair et al., 2022; Ahrland, 2019). A thriving society has grown up around foraging for wild mushrooms and other forest delicacies among Sweden's urban middle-class and hipster communities since the turn of the century, including educational and support services like evening classes, study circles, club meetings, and exhibitions (Garekae & Shackleton, 2020). Local economies and ecosystems that depend on forests are being revitalized as people in urban places around the world get back to gathering forest foods. Foragers of today's generation are more concerned about the source of their foraged foods than in past generations. Formal educational institutions play an important role in increasing the use of wild edible plants by humans, according to several studies (Łuczaj et al., 2021; Tauro et al., 2021).

Educated about nature and food production, children at a Berlin-area school that provided locally produced food altered their eating habits, grew their food in the school garden, and collected wild edible plants from a nearby empty lot, according to researchers. The Eastern Band of Cherokee Indians, for example, incorporates traditional feasts into its educational programs. To prevent the biocultural homogenization that will lead to a loss of household nutrition based on biodiversity, formal educational institutions in some Chilean traditional communities overlook local food plants and their traditional uses (Pulker et al., 2018).

3.2.2 AFFORDING HEALTHY, SUSTAINABLY SUPPLIED NATURAL FOODS TO CONSUMERS

Energy-dense forest foods have a high concentration of calories as well as macro- and micronutrients, all of which are needed for human nutrition. More broadly, they contribute to addressing the constantly shifting gastronomic requirements of a growing and increasingly diverse global population. The rise of the urban middle class due to economic expansion has resulted in substantial changes in diets, eating habits, and consumer preferences for food products in many countries. Consumer tastes and preferences are becoming increasingly diverse as a result of shifting socioeconomic conditions and modernizing lifestyles. Although this is true, one of the most significant food-related consumer trends is the growing desire for nutritious, sustainably produced, and convenient food products (Van Berkum et al., 2018; AlOudat et al., 2021). As the number of single-parent families grows and cooking abilities deteriorate, the market for ready-to-eat packaged, single-portioned, or pre-prepared food commodities expands to cover a broader range of fresh, ready-to-eat items, according to the International Food Information Council. Forest foods may be able to meet some of these nutritional requirements in some cases (Te Kanawa, 2021). According to Wiersum (Clair et al., 2022; Van Khuc et al., 2018), more and more Europeans are eating forest foods because they are interested in heritage-inspired forest methods of production and biocultural techniques.

Forest foods have been portrayed as out of date and only available to the poor (Kosgey, 2021). However, evidence shows that some wild edible plants have made their way into mainstream markets, even though this has been the case. As a result, wild foods are becoming more important in bioeconomics because they are healthy and sustainable sources of raw materials. As more people eat wild foods, they're becoming more common. There are a lot of people who think that forest foods like wild edible fruits and other forest foods, for example, are both nutritious and important. African traditional leafy vegetables, like tree leaves, are becoming more common in high-end restaurants and stores in Kenya and other African countries, says Aworh (Mujuka et al., 2021; Fischer et al., 2020). Following earlier research, forest fruits like the baobab (*Adansonia digitata* L.) are being turned into high-value food products and sold to people with a lot of money in Malawi (Irvin et al., 2019) through specialty and delicacy stores as well as in international markets. People have said that edible insects, shea (*Vitellaria paradoxa*), acai (*Euterpe oleracea* Mart.), and other plants have the same qualities as the edible insects (e.g., *Ruspolia differens* Scopoli). Xylitol, a low-calorie sweetener made from birch wood (*Betula pendula* Roth), has become popular in Finland because of its benefits for oral health and how easy it is to use in everyday cooking. In debates about "smart food," there is a lot of room for forests to produce nutritious, naturally grown food that people who care about their health and the environment would buy. This idea is being talked about in discussions about "smart food in the future." Bioeconomy may be able to help this change happen, but it's not certain. Public Health—Provisions and Risks (Wegner et al., 2022).

Even though forests and trees are good for people's health, there are certain risks to eating foods from the woods. Antioxidant, anti-hypercholesterolemic, antidiabetic, anti-inflammatory, anti-amyloidogenic, antimutagenic, and antibiotic properties of phytochemicals found in wild food plants have been proven in research (Mehrabi et al., 2022). To eliminate malnutrition, it is possible to include forest foods in food systems by adjusting consumer habits and preferences, agricultural practices, harvesting, and processing methods, as well as food value chain structure and governance mechanisms. More than 800 million people worldwide are suffering from undernutrition, food insecurity, and micronutrient deficiencies, which can be efficiently addressed using forest foods in agricultural systems. The frequency and severity of non-communicable diseases connected to a sedentary lifestyle as well as an unhealthy diet can be reduced by integrated food systems (Ahloowalia et al., 2004; Akem & Pemunta, 2020).

There is evidence that eating forest foods increases one's risk of contracting diseases. Accidental poisoning may become increasingly common if inedible or even dangerous species are mixed in with urban foragers' growing practice of gathering mushrooms and wild plants. People are putting themselves and their families at risk when they confuse wild garlic (*Allium ursinum* L.) with lily of the valley (*Convallaria majalis* L.), which contains a high concentration of digitalis glycosides (Ahloowalia et al., 2004). It has been shown that human use of bushmeat and the methods used to hunt, handle, and process it can all contribute to the spread of wildlife zoonotic diseases, which have been made more prevalent by the alterations to the climate, land use, and biological variety themselves. In many cases, the most vulnerable members of society are the ones who bear the brunt of these dangers. Women who prepared bat meat for consumption, as well as young, poorly educated, jobless hunters in Cameroon, were at risk of contracting Ebola virus infection (Friant et al., 2015). People living in poverty were more likely to engage in bushmeat hunting and consumption, which has the potential to harm wildlife conservation and increase the risk of infection with zoonotic illnesses, according to the study. For those who depend on wild meat as their primary source of nutrition, restrictions on the consumption of wildlife, especially considering the COVID-19 outbreak, could have unintended consequences. We need to stop demonizing foraged crops and instead educate people on the proper harvesting, handling, and ingesting ways. These kinds of dangers may be tolerated if they are necessary to meet the nutritional demands of the family (Heemstra & Randall, 1993).

3.2.3 FOOD FROM TREES AND FORESTS

A forest is the consequence of the cooperative efforts of animals and plants to create a win-win situation. One can find a wide variety of forest plants, including forbs, grasses, and herbaceous plants that make up the forest floor as well as shrubs and other woody species that thrive in an understory generated by canopy trees. Folklore has traditionally associated the term "woods" with large areas of land predominantly covered in trees. Things have changed since then. For the first time, the term "forest" encompasses trees in all their various habitats, including natural and managed ecosystems, agroforests, single-species plantings, and places where they aren't the dominating growth habit, including fields, motorway ditches, and urban landscapes. The biodiversity found in natural forests can be harvested for food on a local and global scale. Agroforests, which are a mix of trees and crops, are created to protect the environment and aid in the production of food, to provide food directly to people. For both internal use and export, a wide variety of food crops are grown in secondary forests and single-species plantations. Other than in emergencies, trees provide a valuable source of food for a wide range of people in both urban and rural areas. Rural and urban residents alike can benefit from the availability of food and other resource-producing trees (Chamberlain et al., 2020).

Forest foods can be produced in a variety of methods, ranging from unmanaged wild foraging to intensively managed single-species crop production. The appearance of some production systems is that they are haphazard collections of plants with minimal inputs, while others appear to be well-organized with a large amount of human involvement. The purpose and demand of the item have a considerable impact on the scale at which it is manufactured. This means that the number of forest foods necessary for direct home consumption will be less than the number of forest foods required for revenue production. Harvesting forest edibles has been a long-standing tradition in many countries, and because the products are considered public resources, there are no restrictions on who can harvest them (Foli et al., 2014).

Forest dwellers have been shaping their ecosystem for millennia. Since humans have been roaming the earth for thousands of years, most woodlands have undergone significant change over time due to "management" to meet the immediate food demands of families. Multi-story agroforests, home gardens, and rotating crops are just a few examples of neglected traditional forest management practices that include incorporating food production (Vinceti et al., 2013) (Figure 3.1).

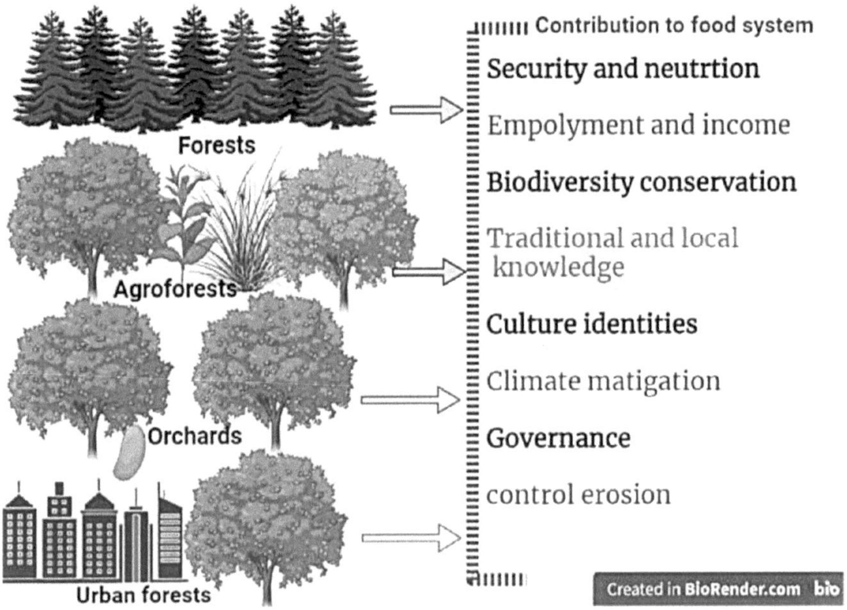

FIGURE 3.1 The significance of forests and trees in food production systems.

3.2.4 WILD HARVESTING/FORAGING

People have been scavenging for food in forests since the dawn of time, according to historical records. Before agricultural advancements made it feasible to plant food in fields, people and communities had relied on foraging for food in the forest for generations to feed themselves and their families. Indigenous people have been picking food from trees for thousands of years, even before agricultural methods were developed. Although technological improvements have made it possible to grow food in fields, forest foods continue to meet key nutritional requirements for families and communities. The identification of the people who rely on them and the contribution to the identities of communities all over the world are two characteristics of many forest food plants (Meneses-Romero, 2015), which are called cultural keystone species. If these species are exterminated or access to the plants is restricted, the resulting consequences for people and civilizations will be catastrophic (Provenza & Malechek, 1986).

To ensure the continued production of these forest staples, certain forager organizations have developed governance and management structures to ensure the continued production of these forest staples on open-access or community grounds. *Caryocar coriaceum* Wittn, a fruit harvested by pequi harvesters in northeastern Brazil, adheres to strict processes to ensure that the fruit is gathered without causing damage to future harvests or the tree's health. Several forest plants were shown to produce more biomass when they were cut or coppiced by Native Americans, according to the research (Nersesian et al., 2011). To maintain their traditional food production methods, certain Native American tribes, such as the Eastern Band of Cherokee Indians, have put in place a systematic forest management strategy. Many states in the United States have imposed official limits on the harvesting of American ginseng roots to boost conservation efforts. Traditional and local knowledge can be used to improve forest management for wild edibles, which can benefit everyone. The effectiveness of governance systems, on the other hand, may differ significantly or have unintended implications (Camacho et al., 2016).

Many people believe that foraging for wild foods is done without respect for scientific issues and that this is incorrect. To secure the long-term availability of their products, many modern foragers have developed management systems (Marriott et al., 2004). Ramp and leek harvesters in the United States who have been in the business for a long time may decide to rotate harvest sites and use larger leafed plants to help manage the resource. For example, traditional harvesting methods in northern European countries where berries are picked have been merged with scientific expertise to ensure that resources are conserved to meet current forest food demands (Riginos et al., 2012).

3.3 MANAGED FORESTS AND THEIR IMPORTANCE FOR CONSUMERS

In traditional forestry organizations, silvicultural activities (those that produce wood) may, on the other hand, be able to contribute to the provision of food for their members. By including food-producing plants in forest management, it is possible to promote biodiversity conservation. Ephemeral forest herbs such as Allium spp., which are harvested for food in the springtime, contribute to nitrogen cycling, which in turn helps to improve forest health while also supplying sustenance to humans. Wood rotations should be longer, invasive species should be reduced, and native understory plant development should be encouraged to effectively manage American ginseng forests. Some food plants, such as berries, may benefit from forest stand thinning, but others, such as mushrooms, may suffer as a result. Before making decisions on how to manage forests for forest foods, it is critical to evaluate the implications of silvicultural interventions on other resources (Gulbrandsen, 2006).

Kyrgyzstan's walnut (*Juglans regia* L.) fruit forests have traditionally been farmed primarily for food production. Since the Soviet era, these forests have been designated as protected forests because of the large number and genetic diversity of crop wild relatives they contain, as well as their environmental importance for stabilizing slopes and ensuring equal water flow for agricultural irrigation areas downstream. One of the primary objectives of forest management has been to increase access to foraged foods such as wild walnuts and other fruits and vegetables. As part of the management plans, hygienic felling and frequent thinning and felling were also used to encourage forest renewal. Walnut stands can benefit from silvicultural interventions and management that are focused on quality and stability. This forest's health has been compromised by a variety of human-caused activities such as overgrazing, illegal logging, poor reforestation methods, and the extraction and use of unsustainable quantities of fuelwood. Forest foods such as walnuts are negatively affected by this (Gibb et al., 2006).

The availability of resources must be taken into account while formulating forest food concepts. There can be both good and negative effects on a plant's productivity when it is exposed to changing light and moisture conditions. Berry and other plants that need a lot of sunlight may benefit from openings appearing in the forest canopy. For plants that prefer shade, this means that openings in the forest floor could result in less light reaching the forest floor, resulting in their demise (Schaetzl et al., 1988; Nowacki & Abrams, 2008). Hazelnuts (*Corylus* spp.) are one of the few forest species that can grow in the shade of a forest canopy, although they require lighter than most other plants to develop fruit. Thinning and spacing trees in a way that promotes maple syrup production helps to encourage the growth of large-crowned trees. If wood co-management is not an option, the employment of agroforestry techniques with food-producing plants like ramps could be an interesting alternative (Barton, 1993).

3.3.1 AGROFORESTS

On farmland with a modest bit of forest cover, you can find a wide variety of wild plants. When woody perennial plants and agriculture are intertwined, the result is a more productive system and a network of positive interactions. Food production methods known as agroforestry are included in this group. Over a billion hectares of agricultural land are thought to be used for agroforestry, a practice that involves retaining naturally occurring trees on farms to create multi-layered home gardens filled with annual and perennial herbs, bushes, trees, and vines. Central American agricultural land is classified as agroforestry in greater proportion than land in Southeast Asia or South America, where the percentage is almost 80% higher at nearly 96%. Agroforestry was used to grow more than 82 million hectares of land at the end of the decade ending in 2010, according to Hall et al. (2019).

Agroforestry systems can be used in a variety of ways to enhance food security and nutrition. When it comes to environmental and human well-being, they play a significant role. Trees native to the area can help farmers better withstand market and climate fluctuations. They can also look forward to receiving better-quality food (Quandt, 2021). Agroforestry practices can result in a variety of benefits, including enhanced soil qualities, increased crop yields, and an increase in the amount of food available to households. Indonesian agroforestry systems are linked to a higher frequency of healthy food intake because they cultivate a large number and manage wild plants. In contrast to the predicted changes in industrial agriculture, Ugandans cultivate their food in their backyards (Jamnadass et al., 2011).

As a result, agroforestry systems that include native fruit trees and traditional foods from the forests could help satisfy nutrition and food security goals (Mbow et al., 2014; Duffy et al., 2021). Traditional agroforestry systems are hindered by market insufficiencies and failures, such as inadequate demand, inadequate supply, marketing channels, or supply management tactics. Businesses and new technologies must be developed to meet the increased demand for traditional agroforestry systems' native fruit tree species to boost food security, livelihoods, and resilience in the face of future crises (Dawson et al., 2014).

3.3.2 ORCHARDS AND PLANTATIONS

Agricultural trees that produce food provide a substantial contribution to food security and nutrition. Hundreds of different tree species are grown by individuals for their personal use as well as for commercial purposes

(Reichard & White, 2001). According to estimates, more than 170 tropical and subtropical tree species are cultivated worldwide. Small landholders cultivate a subset of these crops for export, and a few are farmed on a large scale by major landholders for export to international commodity markets. There is a "dynamic process of domestication" in which several non-timber forest species are collected from the wild and then cultivated on farms at the same time, which is referred to as "synthetic domestication." From foraging for fruit in natural surroundings to producing it in private gardens, there is a logical progression from foraging to cultivation. Orchards and plantations would be established if the amount of fruit produced at home was no longer sufficient to meet the needs of the population (He et al., 2020). Baobab, for example, is now recognized to be more common in communities and farms than it is on natural plains, according to recent research. Even though several high-value NTFPs, such as rubber, oil palm, and walnuts, are cultivated in carefully managed plantations, many of these products are still primarily produced by small-scale and/or non-industrial farmers in their home countries. In some places in China, for example, smallholder farmers handle nearly 90% of the walnut harvest (Pantzar, 1997). With 3,90,000 hectares of walnut plantations and a production of 1.6 million tons of walnuts in the shell in 2018, China has surpassed the United States as the world's biggest producer of walnuts since the 1990s (Chamberlain Darr, 2020). Smallholders in Central Asia's highlands, as well as many countries along the ancient trading route known as the Silk Road, rely on the money from these orchards and plantations for their livelihood. Smallholders, via better management of this and other food-bearing trees, have the potential to make significant contributions to world food production. A total of 0.84 t/ha was produced on average by non-industrial walnut plantations in India, China, and Iran (Czwartkowski et al., 2022), while industrial walnut plantings in California, United States, produced an average of 4.2 t/ha. Many agronomic inputs, such as fertilizer, pesticides, and other modern propagation techniques, rootstocks, and cultivars, are underutilized, resulting in yield disparities between different crops (Aliano-González et al., 2022).

3.3.3 URBAN AND PERI-URBAN FORESTS

Greater than 15% of the food that is consumed on the planet is grown on less than 600,000 km^2 (105 square miles) of urban and peri-urban land. It is possible to grow food-producing trees on more than 6,400 square kilometers of the metropolitan area in the United States to assist in feeding the country.

It is possible to accomplish this. There are many other places than parks, campuses, street trees, urban woodlots, cemeteries, backyards, and allotment gardens where plants, fungi, and insects can be discovered. Examples of these additional places include urban woodlots. Foragers in metropolitan locations can use these food resources for a range of uses, ranging from a pastime activity to a reliable source of revenue, including both of these uses simultaneously. According to estimations, there is sufficient fruit on public trees to meet the requirements for the recommended daily consumption of fruit. It is not unusual for foraging on public lands to be disregarded, heavily restricted, or downright forbidden, particularly in more remote areas (Wartmann et al., 2021).

Urban food security must be balanced with food safety concerns because cities are both pollution hotspots and potential sources of free foods. It is possible for contaminated soil or airborne heavy metal emissions to be absorbed by plants through photosynthesis and then transferred to their edible sections. According to the findings of the study, 86% of the wild mushrooms collected in Berlin did not meet the limits set forth by the EU for lead and cadmium due to the high quantities found in them (Konijnendijk et al., 2004). Following a thorough rinsing with regular tap water, wild edible plants that were grown in soils that contained high concentrations of heavy metals are safe to ingest, according to the findings of a study conducted in the San Francisco East Bay region of the United States. There is a danger in urban and peri-urban regions, but there is a possibility that wild fish and meat as well as flora in the Kamchatka region of Russia's far east could be contaminated by heavy metals (Christopoulou et al., 2007).

3.4 MANIPULATING FOOD SYSTEMS FOR FORESTS AND TREES

Forests unquestionably contribute significantly to the provision of food, both in terms of direct consumption and in terms of generating cash that may be used to purchase additional food. When it comes to markets, Baobab, for example, provides a varied range of options ranging from small, local markets that serve individuals or small groups to enormous, multinational marketplaces that serve numerous countries. It is less developed and known that the value chains of forest foods are compared to the value chains of agricultural and timber-based industries. Certain operations offered by agriculture and forestry authorities may be foreign to individuals who write regulations, track production, and promote new ideas and market opportunities in the agricultural and forestry industries (e.g., pharmaceuticals, nutraceuticals; natural foods, health foods). Therefore, national and international governments as well as resource management and development organizations often neglect many of these challenges in their planning and operations (Powell et al., 2015). Acceptance and participation in relevant institutions may aid in raising awareness of the relevance of these individuals. African-grown products may be delivered to Europe following the approval of baobab fruit pulp as a safe new food. As part of the Nagoya Accord, "other" resources, such as the adoption and usage of "novel foods" or the corresponding access rights and benefit-sharing procedures with local populations, may be highlighted in international discussions. All of these things are required: new business models and goods, new markets driven by consumer demand for more sustainable food systems, and evidence-based governance at all levels by stakeholders in sustainable forest management and production (Figure 3.2). In the absence of these components, a bioeconomy that can meet the growing global need for nutritious and wholesome food while also addressing the increasing concern over the conservation of biodiversity is unlikely to be established (Nakamura et al., 2017).

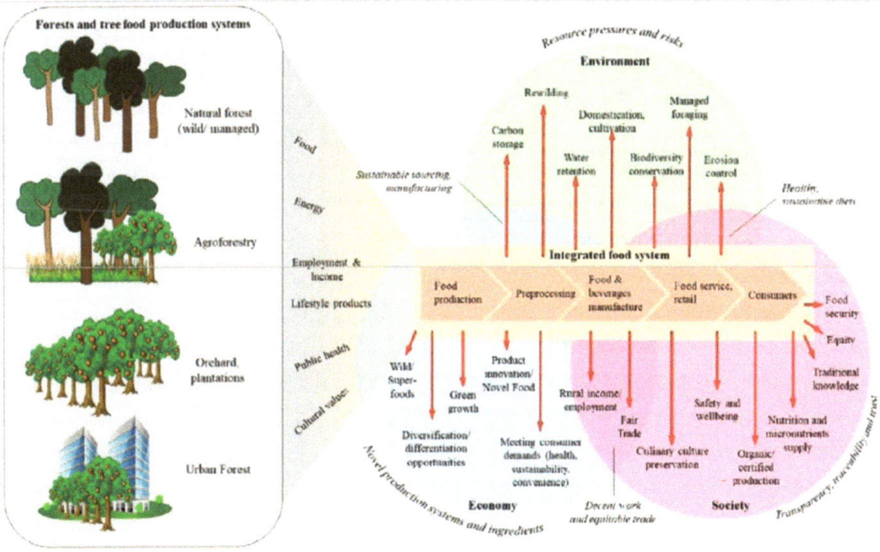

FIGURE 3.2 Forest and tree integration into food systems have many components that contribute to the creation of a framework for moving toward a bioeconomy. A version of the Institute of Food Science and Technology's food system framework was adapted for this project.

3.4.1 SHIFT TO A BIOECONOMY

Introducing edible plants and fungi into forest production and management systems will entail a significant shift in priorities in the forest sector. In bioeconomies, even though forest foods have a lower economic value than other resources, their contribution to food security and nutrition is taken into consideration when making economic decisions and managing resources. This means that their contribution to food security and nutrition is taken into consideration when making economic decisions and managing resources in bioeconomies (Kircher, 2021). When it comes to aiding the transition to a bioeconomy, the full understanding that edible forest resources are natural resources that demand equal management efforts as other forest products is a vital component of the equation. It is possible to map a course toward a more progressive bio-based economy armed with this information (Hermans, 2018).

In the 21st century, one of the most significant economic trends has emerged: the bioeconomy. The original motivating force behind this was the desire to use biotechnology to replace fossil fuels with alternative energy sources such as biofuels, particularly wood fuels, as a means of combating climate change (Kleinschmit et al., 2014). The major goal of a bioeconomy, according to some, is to ensure a supply of nutritious, contaminant-free food. It will have an impact on everyone in society, from consumers to producers, as the shift to a bioeconomy takes place. To decrease or eliminate negative environmental impact, it is necessary to source products in a manner that is supported by scientific evidence. As illustrated in Figure 3.2, planks of wood and trees contribute to integrated food systems, and the three foundational pillars of sustainable bio-based economies are depicted in Figure 3.3 (i.e., economics, society, and ecology) (Muscat et al., 2021).

Despite their importance to the world economy, edible forest products are still underappreciated in the forest-based economy, according to the Forest Stewardship Council. Throughout this economic paradigm, wood is used to manufacture the vast bulk of the products. Wood forest biodiversity, on the other hand, is harvested in massive quantities and makes major contributions to bio-based economies (Muscat et al., 2021). Forest foods, such as baobab, can contribute to the development of a bioeconomy that serves a wide range of markets, innovates to offer products produced from sustainable sources, and incorporates agreements that promote Fair Trade and other social-equity systems, among other things. To make the transition to a "green" economy simpler, it may be necessary to make investments in "green" market infrastructure (Robert et al., 2020).

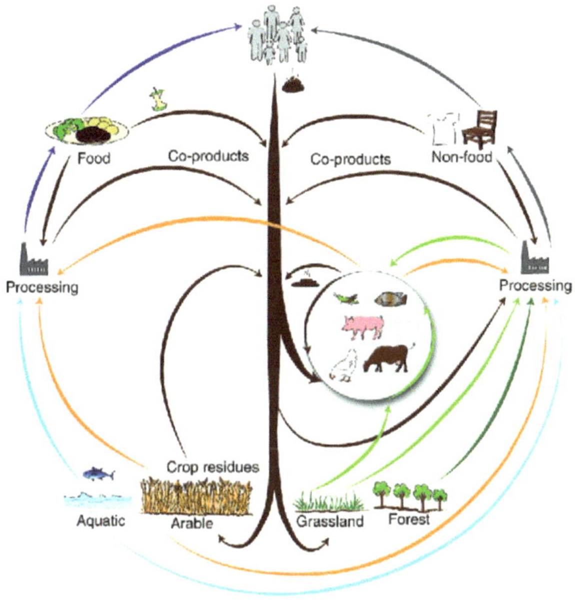

FIGURE 3.3 Biomass flows in a circular bioeconomy. Biomass from aquatic, arable, grassland, and forest production systems are processed or consumed. By-products and wastes are recycled back into the system.

Source: Reprinted with permission from Kumar et al., 2023. https://creativecommons.org/licenses/by/4.0/

3.4.2 PROVISIONING UNITS: FORESTS AND TREES

Forest production methods and management will expand beyond their "traditional" functions to include food production as the world transforms to a bioeconomy. Decisions about food production and distribution will be factored into management assessments, plans, and actions. Food and related services can be included in "multiple-use forest management" [multiple-use forest management]. Forest management would be a part of the ecosystem management paradigm (paradigms) (Knoke et al., 2021). Encouraging healthy food consumption and maximizing wood output while giving environmental services to society are only two examples of bioeconomy development efforts (Ferraz et al., 2014). To meet the bioeconomy's needs for forest foods, a wide range of forest food production systems are employed, including forested landscapes that are open to wild harvesting, forest landscapes that are managed for food production, and urban and peri-urban forested landscapes, and single-species plantings when required. With their guidance, the landscape may be transformed into a mosaic of habitats where a wide range of species can thrive (Francisco et al., 2022). Forests that are not typically thought of as forests, such as urban treescapes and rural farms, can provide habitat for a wide variety of species. Single-species plants should be strategically placed in agroforests, which provide food and other items for personal consumption as well as additional revenue. Tree maintenance in urban areas would fall on the shoulders of the people who live there (Francisco Cavane, 2022; Wagner et al., 2014). Forest ecosystems and biodiversity may suffer significantly if forest food species are harvested at a faster rate than in earlier generations. The change to a bioeconomy may necessitate the domestication and cultivation of crops outside of the natural forest. A bioeconomy, for example, would apply evidence-based production practices to produce products made from sustainably sourced raw materials, among other things (Vauhkonen & Ruotsalainen, 2017).

3.4.3 INNOVATIVE BUSINESS MODELS DRIVEN BY CONSUMER DEMAND

The bioeconomy will eventually be driven by consumer understanding, demand, and support, among other factors. A growing number of city dwellers are looking for environmentally friendly products, all-natural, sourced locally, and produced ethically. Because of this, Fair Trade products that are organic and environmentally friendly are more likely to be sought after by these clients. To meet customer demand, forest foods will need to be produced in a manner that is both environmentally, socially, economically, and ethically responsible (Bocken & Short, 2016). Social marketing could be used to develop the enthusiasm needed to break free from one's current situation and go forward in this demographic group. Because of the demand for these items, the industry must constantly develop to keep up with the demand (Bocken & Short, 2016).

The introduction of new rules at the global and national levels could speed up industry transitions. New and innovative business models and investments in product innovation and technical development are projected to be spurred by the rising demand for healthy, locally sourced foods (Trigkas et al., 2020). Promising plant species must be cultivated in appropriate settings to supply the growing demand for nutritious, sustainable foods derived from ingredients sourced properly. Regional and global economic revolution can be sparked by new processing, distribution, and marketing technologies as well as other components of the value chain (Toppinen et al., 2018).

3.4.4 INNOVATION OPPORTUNITIES

Functional food manufacturers, as well as those involved in the production of natural and homeopathic treatments, are constantly on the lookout for novel, as of yet uncharacterized, plant-based raw components that can be incorporated into their respective products. A natural ingredient that does not have to be mentioned on food labels because of its natural origin can be used to increase the storage shelf life of products, improve their technological properties, or provide other desired characteristics (e.g., taste). Companies can increase their product portfolios while still keeping their competitive advantage as a result of this. For biotech companies, the phytochemicals found in trees and forest plants are of particular interest, and additional research into these compounds has the potential to yield major benefits (Fan et al., 2012). Food colorants and flavors that are new to the market and made totally of natural components are becoming increasingly popular as a method to meet the growing demand for food products that are made entirely of natural ingredients. For the forward-thinking food sector, plant-based components, which consumers are increasingly tolerant of or perceive to be beneficial additions, are of particular importance. The biological effects of Phyto colorants originating from fruits, such as anthocyanins and betalains, go beyond their ability to color food (Blandon et al., 2007). They also have physiological effects on the body. However, it has been discovered that artificial chemicals might cause allergic reactions in children and even the development of attention deficit hyperactivity disorder in youngsters. Wild items are being used in a variety of businesses, including the food and pharmaceutical industries, as well as cosmetics, and their use is not sustainable nor ethically obtained, and in certain cases, it is contributing to the extinction of valuable plant and animal species (Blandon Cranfield, 2007).

According to scientists, because the phytochemical compositions of the vast majority of forest plant species have not yet been properly studied, the biodiversity of forest plants represents an essentially unlimited source of potentially beneficial compounds. It has been shown that extracts of the sweet chestnut (*Castanea sativa* Mill.) male flower can be utilized in baked goods instead of potassium sorbate to increase the number of free radical scavengers present while retaining the nutritional and chemical profiles of the baked products. Goji berry (*Lyciumbarbarum* L.), Acai (*Euterpe oleracea* Martius), and Acerola *(Malphigiaemarginata* D.C.), research into a number of these effects has been conducted both *in vitro* and *in vivo*, with some results being validated. As an example, noni berry juice from French

Polynesia was found to protect against cigarette smoke-induced DNA damage, blood lipid, homocysteine increases, and systemic inflammation, among other health benefits in human intervention studies. According to the findings of a study conducted by the University of California, bread enhanced with baobab slows both the glycemic response and the digestion of starch in people (Dike, 2010).

Foraging for forest foods is a relatively new phenomenon that Wiersum defines as having multiple dimensions. These dimensions include biological processes of "re-wilding" as well as newly culturally rooted relationships between humans and natural environments. It is a relatively new phenomenon that has been described as having multiple dimensions. These dimensions include biological processes of "re-wilding" as well as newly culturally rooted ties between humans and nature. Foraging for forest edibles is an example of this phenomenon. The ancient practice of going into the woods in search of food that grows naturally there is making a comeback as a trendy new hobby. Shea butter, argan oil (*Argania spinosa* L.), baobab (*Baobab demersifera*), moringa (*Moringa oleifera* L.), and pygeum (*Prunus africana*) are only a few examples of items that originate from forests (Chamberlain et al., 2020). Everything has been sourced to fulfill the rising demand for "natural" food production in several billion-dollar industries, which has resulted in a glut of available supplies since these industries have been able to get everything they need. The United Nations Food and Agriculture Organization believes that modifying customer expectations and behaviors, in addition to implementing an integrated food system that makes full use of the potential of products derived from orchards, will result in an improvement to the food production industry on a global scale. According to the findings of this study, the production of food in forests is on par with the production of food in agricultural settings in terms of its significance. The same is true for the structure and control of value chains for forest foods, as well as the quality and safety concerns associated with the distribution of forest foods (Sharma et al., 2018).

3.4.5 PRODUCTION, PROMOTION, AND DELIVERY

The processing, marketing, and distribution of forest foods may have to change as we move toward a bioeconomy. Local markets and complex global supply networks are two examples of how things are distributed to consumers around the world. Forest products such as maple syrup, acai powder, and baobab powder are used in a wide variety of culinary dishes that may be purchased on worldwide markets. To improve their nutritional content, most forest foods remain still unprocessed. Income and profitability could be improved by expanding the number of goods and services that are manufactured and processed in-house. Without the interference of intermediaries or elite traders, small-scale farmers must be able to receive the full advantages of their labor (Pynnönen et al., 2021). There are benefits and drawbacks to processing forest foods, both of which can help conserve them while also reducing their nutritional value in the long term. As a result, many forest foods are only available in local, regional markets with limited supply networks instead of bigger, national markets. Exporting forest foods to high-value markets may benefit some of them, but doing so may be difficult for smallholder farmers and extremely risky for less wealthy producers, depending on their conditions. A focus on local markets and short supply chains is thought to be connected with consumer demand for locally produced food. In any bioeconomy strategy, developing local markets for small-scale producers should be a top emphasis (Lu et al., 2015).

Sustainability in natural resource development and more equitable benefit distribution can both be achieved by utilizing environmentally friendly production methods such as Fair Trade, Fair Wild, and organic certification. Food items with features like local origin, organic or sustainable production, social equity, or geographic indication are highly sought by buyers, according to surveys. Providing information about a product's health benefits may encourage consumers to pay more for the product. For small and medium-sized industries and organizations, certification costs might be prohibitively expensive. Some organizations offer cost-sharing programs and other incentives to encourage involvement and the transition to a bioeconomy. Participation and the transition to a bioeconomy are the goals of these programs (van Noordwijk, 2019).

The quality and safety of food must be taken into consideration in some more sophisticated markets. To be part of the food chain, forest foods must meet quality and safety requirements similar to those used in farmed food systems. The Codex Alimentarius International Food Guidelines have specified standards for a few forest foods, such as shea and Arabic gum. The wide variety of wild forest foods and the inferior quality of their product attributes (such as size or sugar content) make it difficult to set standards for their consumption (Murdoch et al., 2017). There are several worries about the food safety of forest foods, such as the potential of microbiological contamination due to inadequate post-harvest treatment and the risk of zoonosis in bushmeat. These issues have limited the international market for forest foods. Forest foods' nutritional content may be compromised because of these worries. Several factors could contribute to a shift toward bio-based economies. For example, the growing demand for nutritious, natural, and sustainably sourced meals, as well as the high quality and safety of forest foods (Novoveská et al., 2019).

3.4.6 EMPLOYMENT: FORESTRY AND FOREST PRODUCTS

There are a wide variety of tree species available to Tribal people living in tropical forests. Around 30% to 50% of rural income is generated by forests, while forest and forest product-related employment provides for 20 to 45% of full-time employment in rural areas. Whiteman et al., (2015) estimates of the number of people participating and the types of work that can be performed. Amazonian households make $872.35 per year from the utilization of 37.2 hectares of forest, which works out to about $2.35 per hectare (Whiteman et al., 2015). Around 12.3% of the population is employed in forest-based sectors in Jamaica, Thailand, and Honduras, as well as countries such as Egypt, Sierra Leone, and Bangladesh (FAO, 1987). There are at least 50 million people who would be entirely employed if a similar situation existed in other developing countries and an additional 100–200 million people who would be working part-time in a variety of forest sectors. Though the forecast is considered conservative, it would have a significant impact on the company's bottom line (Pandey et al., 2016). There are an estimated 70 million forest dwellers in India alone (Pandey & Tripathi, 2016) and an estimated 18 million forest dwellers in the Philippines alone who live close to tropical forests (Sharma et al., 2015). We estimate that approximately 300 million people live in proximity to tropical forests and earn part or all of their income from the food provided by tropical forests, according to (Ninan & Kontoleon, 2016). There may be more women using forest resources than men, despite the lack of data on this. This is because women are often the ones responsible for gathering food and fuelwood.

3.4.7 KNOWLEDGE AND GOVERNANCE

Forest and tree incorporation into food systems acknowledge local knowledge and traditional practices. Science and policy work together to incorporate the knowledge of all relevant parties into a sustainable food system. Traditional land-use practices, which form the basis of some of our food systems, have been the subject of extensive scientific investigation. By combining our knowledge and skills, we can make more informed decisions that benefit the entire neighborhood (Marchetti et al., 2020). Stakeholder participation is essential for successful bioeconomy-based forest management and forest governance. A wide range of people, including the armed forces, would be able to weigh in on forest policy decisions. To ensure that the forest bioeconomy's value chains all share in the system's success, it's important to have vertically integrated management. Forests would be managed for the benefit of all people by a collaborative effort between governments, communities, and private organizations.

Consumers' interest in superfoods that are both nutritious and environmentally friendly may soon fade, providing another example of resources being overexploited to meet fleeting market demand. As a possible safeguard for customary knowledge and intellectual property rights, the Nagoya Protocol provides access and benefit-sharing agreements to protect the rights of local populations to their resources. It is possible to ensure that local producers and communities receive a reasonable portion of the benefits by increasing the volume of value-added activities in the countries of origin, even though there are now logistical and technological obstacles to doing so. Because of this, local communities are encouraged to consume forest foods that are grown in their area, which is a big step forward in the localization of food production (Issa et al., 2019).

3.5 CONCLUSION

Protecting the environment, which boosts the productivity of forestry and agriculture, providing a habitat for a wide variety of other organisms, and providing food for large numbers of people are all reasons why healthy forests are essential to ensuring that there will be enough food for everyone around the world. Since the global population is expanding at an alarming rate and, as a consequence, there is a pressing requirement for a substantially increased food supply, it is of the utmost importance to maintain healthy forest ecosystems to ensure our food security both now and in the future. This is a fact that applies not only to the present but also to the foreseeable future. A lot of the time, trees provide nutrition in the form of fruit, oil, seeds, nuts, gums, roots, and other components that are directly consumed by humans. Approximately 9% of the world's food supply comes from trees that are farmed in undeveloped countries, which provide 430 million tons of food per year. Additional food sources can be found in the big woodland setting as well as in the surrounding area. It should come as no surprise that many people in impoverished countries living near forests harvest wild plant products and capture animals for food. In the immediate neighborhood, we estimate that between 5 and 10% of their food originates from forested regions. In some cultures, and under favorable conditions, these foods can account for as much as 50 to 100% of a person's daily calorie intake, demonstrating the significance of forests in the provision of food. Trees and vegetation also serve as a source of food for several animals. They can provide up to 80% of the feed for cattle, allowing subsistence farmers to continue to raise livestock on their lands.

Forests and the myriad non-timber products they produce, in addition to the direct benefits they provide to a sustainable food system, also provide employment and cash. An estimated 300 million people in developing nations rely on forests for the majority or all of their income, according to estimates. The ability to buy a healthier diet and other necessities is made possible by this approach of generating cash.

Inappropriate amounts of healthy soil, water, and a diverse range of biological organisms provide the foundation for a stable and productive food system. Trees and shrubs used as shelterbelts or as part of a cropping system can assist in conserving all of these resources by providing shade and protection. Protecting valuable soil and water resources from rapid water runoff is one of the most significant benefits of shielding fertile soil in fields from the effects of wind and rain. This results in the addition of nitrogen to the soil by leguminous plants, while nutrients are transported to the surface by other trees' roots. The use of trees not only helps to protect the environment but also helps to boost crop yields when they are properly integrated into the system. By adding trees to the farming system, it is possible to increase food yields by 10 to 70% while also producing fuelwood. The development of systems that focus on forest integrity will ensure that the basic needs of humans and the biological resources of the environment are not jeopardized. In the long run, the health of our forest ecosystems will have a direct impact on our ability to feed ourselves in the future.

KEYWORDS

- **biodiversity**
- **cultural significance**
- **food insecurity**
- **forest food**
- **forest management**
- **sustainable use**
- **wild harvesting**

REFERENCES

Adamu, M., Adamu, H., Ade, S., & Akeh, G., (2020). Household energy consumption in Nigeria: A review on the applicability of the energy ladder model. Journal of Applied Sciences and Environmental Management, 24, 237–244.

Ahloowalia, B., Maluszynski, M., & Nichterlein, K., (2004). Global impact of mutation-derived varieties. Euphytica, 135, 187–204.

Ahrland, Å., (2019). Fields, meadows, and gardens–an integral part of the city: The example of södermalm in Stockholm, Sweden. Jahrbuch für Geschichte des Ländlichen Raumes, 16, 144–167.

Akem, E. S., & Pemunta, N. V., (2020). The bat meat chain and perceptions of the risk of contracting Ebola in the Mount Cameroon region. BMC Public Health, 20, 1–10.

Aliaño-González, M. J., Gabaston, J., Ortiz-Somovilla, V., & Cantos-Villar, E., (2022). Wood waste from fruit trees: Biomolecules and their applications in agri-food industry. Biomolecules, 12, 238.

AlOudat, M., Magyar, N., Simon-Sarkadi, L., & Lugasi, A., (2021). Nutritional content of ready-to-eat meals sold in groceries in Hungary. International Journal of Gastronomy and Food Science, 24, 100318.

Antar, M., Lyu, D., Nazari, M., Shah, A., Zhou, X., & Smith, D. L., (2021). Biomass for a sustainable bioeconomy: An overview of world biomass production and utilization. Renewable and Sustainable Energy Reviews, 139, 110691.

Backhouse, M., Lehmann, R., Lorenzen, K., Puder, J., Rodríguez, F., & Tittor, A., (2021). Contextualizing the bioeconomy in an unequal world: Biomass sourcing and global socio-ecological inequalities. Bioeconomy and Global Inequalities (pp. 3–22). Palgrave Macmillan, Cham.

Bahareldeen, S. S., (2019). Assessment of Contribution of Non-Wood Forest Products in Improving Livelihood of Local Communities-Wadi Saleh Locality, Central Darfur State, Sudan. Sudan University of Science and Technology.

Barton, A. M., (1993). Factors controlling plant distributions: Drought, competition, and fire in montane pines in Arizona. Ecological Monographs, 63, 367–397.

Bessaad, A., Bilger, I., & Korboulewsky, N., (2021). Assessing biomass removal and woody debris in whole-tree harvesting system: Are the recommended levels of residues ensured? Forests, 12, 807.

Blandon, J., Cranfield, J., & Henson, S., (2007). Functional food and natural health product issues: The Canadian and international context. International Food Economy Research Group Department of Food, Agricultural and Resource Economics, 1–55.

Bocken, N. M., & Short, S. W., (2016). Towards a sufficiency-driven business model: Experiences and opportunities. Environmental Innovation and Societal Transitions, 18, 41–61.

Camacho, L. D., Gevaña, D. T., Carandang, A. P., & Camacho, S. C., (2016). Indigenous knowledge and practices for the sustainable management of Ifugao forests in Cordillera, Philippines. International Journal of Biodiversity Science, Ecosystem Services & Management, 12, 5–13.

Carson, S. L., Kentatchime, F., Nana, E. D., Njabo, K. Y., Cole, B. L., & Godwin, H. A., (2018). Indigenous peoples' concerns about loss of forest knowledge: Implications for forest management. Conservation and Society, 16, 431–440.

Chamberlain, J. L., Darr, D., & Meinhold, K., (2020). Rediscovering the contributions of forests and trees to transition global food systems. Forests, 11, 1098.

Christopoulou, O., Polyzos, S., & Minetos, D., (2007). Peri-urban and urban forests in Greece: Obstacle or advantage to urban development? Management of Environmental Quality: An International Journal.

Clair, A. L. S., Zhang, G., Dolezal, A. G., O'Neal, M. E., & Toth, A. L., (2022). Agroecosystem landscape diversity shapes wild bee communities independent of managed honey bee presence. Agriculture, Ecosystems & Environment, 327, 107826.

Czwartkowski, K., Wierzbic, A., & Golimowski, W., (2022). Quality, key production factors, and consumption volume of niche edible oils marketed in the European Union. Sustainability, 14, 1846.

Da Cunha, D. T. A., Lora, E. E. S., Maya, D. M. Y., & Del, O. O. A., (2021). Global potential assessment of available land for bioenergy projects in 2050 within food security limits. Land Use Policy, 105, 105346.

Dawson, I. K., Leakey, R., Clement, C. R., Weber, J. C., Cornelius, J. P., Roshetko, J. M., Vinceti, B., et al., (2014). The management of tree genetic resources and the livelihoods of rural communities in the tropics: Non-timber forest products, smallholder agroforestry practices and tree commodity crops. Forest Ecology and Management, 333, 9–21.

Dhungana, A. R., & Bhattarai, B. P., (2022). Differential in utilization of forest products among the users of community forest in Kaski District. Prithvi Academic Journal, 38–45.

Dike, M., (2010). Proximate, phytochemical and nutrient compositions of some fruits, seeds and leaves of some plant species at Umudike, Nigeria. Journal of Agricultural and Biological Science, 5, 7–16.

Duffy, C., Toth, G. G., Hagan, R. P., McKeown, P. C., Rahman, S. A., Widyaningsih, Y., Sunderland, T. C., & Spillane, C., (2021). Agroforestry contributions to smallholder farmer food security in Indonesia. Agroforestry Systems, 95, 1109–1124.

Fan, T. P., Deal, G., Koo, H. L., Rees, D., Sun, H., Chen, S., Dou, J. H., et al., (2012). Future development of global regulations of Chinese herbal products. Journal of Ethnopharmacology, 140, 568–586.

Ferraz, S. F., Ferraz, K. M., Cassiano, C. C., Brancalion, P. H. S., Da Luz, D. T., Azevedo, T. N., Tambosi, L. R., & Metzger, J. P., (2014). How good are tropical forest patches for ecosystem services provisioning? Landscape Ecology, 29, 187–200.

Fischer, S., Jäckering, L., & Kehlenbeck, K., (2020). The baobab (Adansonia digitata L.) in southern Kenya–a study on status, distribution, use and importance in Taita–Taveta County. Environmental Management, 66, 305–318.

Foli, S., Reed, J., Clendenning, J., Petrokofsky, G., Padoch, C., & Sunderland, T., (2014). To what extent does the presence of forests and trees contribute to food production in humid and dry forest landscapes?: A systematic review protocol. Environmental Evidence, 3, 1–8.

Francisco, F., Cavane, E., & Sitoe, A., (2022). Modelling services provisioning through tree species in the Moribane Forest Reserve, Mozambique. Global Ecology and Conservation, 36, e02128.

Friant, S., Paige, S. B., & Goldberg, T. L., (2015). Drivers of bushmeat hunting and perceptions of zoonoses in Nigerian hunting communities. PLoS Neglected Tropical Diseases, 9, e0003792.

Garekae, H., & Shackleton, C. M., (2020). Urban foraging of wild plants in two medium-sized South African towns: People, perceptions and practices. Urban Forestry & Urban Greening, 49, 126581.

Gergel, S. E., Powell, B., Baudron, F., Wood, S. L., Rhemtulla, J. M., Kennedy, G., Rasmussen, L. V., et al., (2020). Conceptual links between landscape diversity and diet diversity: A roadmap for transdisciplinary research. BioScience, 70, 563–575.

Gibb, H., Pettersson, R., Hjältén, J., Hilszczański, J., Ball, J., Johansson, T., Atlegrim, O., & Danell, K., (2006). Conservation-oriented forestry and early successional saproxylic beetles: Responses of functional groups to manipulated dead wood substrates. Biological Conservation, 129, 437–450.

Gulbrandsen, L. H., (2006). Creating markets for eco-labelling: Are consumers insignificant? International Journal of Consumer Studies, 30, 477–489.

Hall, M. A., Nimmo, D. G., Cunningham, S. A., Walker, K., & Bennett, A. F., (2019). The response of wild bees to tree cover and rural land use is mediated by species' traits. Biological Conservation, 231, 1–12.

He, K., Lu, H., Zheng, H., Yang, Q., Sun, G., Zheng, Y., Cao, Y., & Huan, X., (2020). Role of dynamic environmental change in sustaining the protracted process of rice domestication in the lower Yangtze River. Quaternary Science Reviews, 242, 106456.

Heemstra, P. C., & Randall, J. E., (1993). Groupers of the world. FAO Fisheries Synopsis, 16, I.

Hermans, F., (2018). The potential contribution of transition theory to the analysis of bioclusters and their role in the transition to a bioeconomy. Biofuels, Bioproducts and Biorefining, 12, 265–276.

Irvin, L., Jackson, C., Hill, A. L., Bajaj, R., Mahmoudi, C., Vaidya, B. N., & Joshee, N., (2019). Skullcaps (Scutellaria spp.): Ethnobotany and current research. Medicinal Plants, 141–168.

Issa, I., Delbrück, S., & Hamm, U., (2019). Bioeconomy from experts' perspectives–results of a global expert survey. PloS One, 14, e0215917.

Jamnadass, R. H., Dawson, I. K., Franzel, S., Leakey, R., Mithöfer, D., Akinnifesi, F., & Tchoundjeu, Z., (2011). Improving livelihoods and nutrition in sub-Saharan Africa through the promotion of indigenous and exotic fruit production in smallholders' agroforestry systems: A review. International Forestry Review, 13, 338–354.

Jekayinfa, S. O., Orisaleye, J. I., & Pecenka, R., (2020). An assessment of potential resources for biomass energy in Nigeria. Resources, 9, 92.

Kerr, R. B., Madsen, S., Stüber, M., Liebert, J., Enloe, S., Borghino, N., Parros, P., et al., (2021). Can agroecology improve food security and nutrition? A review. Global Food Security, 29, 100540.

Kircher, M., (2021). Bioeconomy–present status and future needs of industrial value chains. New Biotechnology, 60, 96–104.

Kleinschmit, D., Lindstad, B. H., Thorsen, B. J., Toppinen, A., Roos, A., & Baardsen, S., (2014). Shades of green: A social scientific view on bioeconomy in the forest sector. Scandinavian Journal of Forest Research, 29, 402–410.

Knoke, T., Kindu, M., Schneider, T., & Gobakken, T., (2021). Inventory of forest attributes to support the integration of non-provisioning ecosystem services and biodiversity into forest planning—From collecting data to providing information. Current Forestry Reports, 7, 38–58.

Konijnendijk, C. C., Sadio, S., Randrup, T. B., & Schipperijn, J., (2004). Urban and peri-urban forestry in a development context-strategy and implementation. Journal of Arboriculture, 30, 269–276.

Kopteva, L., Shabalina, L., & Kashirina, E., (2019). Ecological Aspects of the World Food Security. E3S Web of Conferences; 2019: EDP Sciences.

Kosgey, G. C., (2021). Nutritional Quality and Anti-Oxidant Properties of Minimally Processed African Indigenous Leafy Vegetables. JKUAT-Agriculture.

Kumar, M., Nallapaneni & Haque, Md & Patwary, Sarif. (2023). Editorial: It Is Time to Synergize the Circularity of Circular Bioeconomy with Sustainability and Resiliency Principles. Sustainability. 15. 12239. 10.3390/su151612239.

Lazos-Chavero, E., Meli, P., & Bonfil, C., (2021). Vulnerabilities and threats to natural forest regrowth: Land tenure reform, land markets, pasturelands, plantations, and urbanization in indigenous communities in Mexico. Land, 10, 1340.

Lu, X., Withers, M. R., Seifkar, N., Field, R. P., Barrett, S. R., & Herzog, H. J., (2015). Biomass logistics analysis for large scale biofuel production: Case study of loblolly pine and switchgrass. Bioresource Technology, 183, 1–9.

Łuczaj, Ł., Wilde, M., & Townsend, L., (2021). The ethnobiology of contemporary British foragers: Foods they teach, their sources of inspiration and impact. Sustainability, 13, 3478.

Manero, A., Taylor, K., Nikolakis, W., Adamowicz, W., Marshall, V., Spencer-Cotton, A., Nguyen, M., & Grafton, R. Q., (2022). A systematic literature review of non-market valuation of Indigenous peoples' values: Current knowledge, best-practice and framing questions for future research. Ecosystem Services, 54, 101417.

Marchetti, L., Cattivelli, V., Cocozza, C., Salbitano, F., & Marchetti, M., (2020). Beyond sustainability in food systems: Perspectives from agroecology and social innovation. Sustainability, 12, 7524.

Marriott, C., Fothergill, M., Jeangros, B., Scotton, M., & Louault, F., (2004). Long-term impacts of extensification of grassland management on biodiversity and productivity in upland areas: A review. Agronomie, 24, 447–462.

Mbow, C., Van, N. M., Luedeling, E., Neufeldt, H., Minang, P. A., & Kowero, G., (2014). Agroforestry solutions to address food security and climate change challenges in Africa. Current Opinion in Environmental Sustainability, 6, 61–67.

Mehrabi, S., Perez-Mesa, J. C., & Giagnocavo, C., (2022). The role of consumer-citizens and connectedness to nature in the sustainable transition to agroecological food systems: The mediation of innovative business models and a multi-level perspective. Agriculture, 12, 203.

Meneses-Romero, M., (2015). The hunting game: Chasing the hare, killing the wild boar. Odysseys of Plates and Palates: Food, Society and Sociality (pp. 1–11). Brill.

Michalk, D. L., Kemp, D. R., Badgery, W. B., Wu, J., Zhang, Y., & Thomassin, P. J., (2019). Sustainability and future food security—A global perspective for livestock production. Land Degradation & Development, 30, 561–573.

Muhammad, M., Badshah, L., Shah, A. A., Shah, M. A., Abdullah, Bussmann, R. W., & Basit, A., (2021). Ethnobotanical profile of some useful plants and fungi of District Dir Upper, Tehsil Darora, Khyber Pakhtunkhwa, Pakistan. Ethnobotany Research and Applications, 21, 1–15.

Mujuka, E., Mburu, J., Ogutu, A., Ambuko, J., & Magambo, G., (2021). Consumer awareness and willingness to pay for naturally preserved solar-dried mangoes: Evidence from Nairobi, Kenya. Journal of Agriculture and Food Research, 5, 100188.

Murdoch, J., Marsden, T., & Banks, J., (2017). Quality, Nature, and Embeddedness: Some Theoretical Considerations in the Context of the Food Sector (pp. 107–126). The Rural: Routledge.

Muscat, A., De Olde, E. M., Ripoll-Bosch, R., Van, Z. H. H., Metze, T. A., Termeer, C. J., Van, I. M. K., & De Boer, I. J., (2021). Principles, drivers and opportunities of a circular bioeconomy. Nature Food, 2, 561–566.

Muscat, A., De Olde, E., Kovacic, Z., De Boer, I., & Ripoll-Bosch, R., (2021). Food, energy or biomaterials? Policy coherence across agro-food and bioeconomy policy domains in the EU. Environmental Science & Policy, 123, 21–30.

Mushi, H., Yanda, P. Z., & Kleyer, M., (2020). Socioeconomic factors determining extraction of non-timber forest products on the slopes of Mt. Kilimanjaro, Tanzania. Human Ecology, 48, 695–707.

Nakamura, A., Kitching, R. L., Cao, M., Creedy, T. J., Fayle, T. M., Freiberg, M., Hewitt, C., et al., (2017). Forests and their canopies: Achievements and horizons in canopy science. Trends in Ecology & Evolution, 32, 438–451.

Nersesian, C. L., Banks, P. B., & McArthur, C., (2011). Titrating the cost of plant toxins against predators: Determining the tipping point for foraging herbivores. Journal of Animal Ecology, 80, 753–760.

Ninan, K., & Kontoleon, A., (2016). Valuing forest ecosystem services and disservices–case study of a protected area in India. Ecosystem Services, 20, 1–14.

Novoveská, L., Ross, M. E., Stanley, M. S., Pradelles, R., Wasiolek, V., & Sassi, J. F., (2019). Microalgal carotenoids: A review of production, current markets, regulations, and future direction. Marine Drugs, 17, 640.

Nowacki, G. J., & Abrams, M. D., (2008). The demise of fire and "mesophication" of forests in the eastern United States. BioScience, 58, 123–138.

Pandey, A., Tripathi, Y., & Kumar, A., (2016). Non timber forest products (NTFPs) for sustained livelihood: Challenges and strategies. Research Journal of Forestry, 10, 1–7.

Pandey, V., Mittal, R., & Sharma, P., (2017). Documentation and application of indigenous traditional knowledge (ITK) for sustainable agricultural development. Asian Journal of Agricultural Extension, Economics & Sociology, 15, 1–9.

Pantzar, M., (1997). Domestication of everyday life technology: Dynamic views on the social histories of artifacts. Design Issues, 13, 52–65.

Pawera, L., Khomsan, A., Zuhud, E. A., Hunter, D., Ickowitz, A., & Polesny, Z., (2020). Wild food plants and trends in their use: From knowledge and perceptions to drivers of change in West Sumatra, Indonesia. Foods, 9, 1240.

Powell, B., Sandbrook, C., Sunderland, T., & Tu, T. N., (2015). The historical, environmental and socio-economic context of forests and tree-based systems for food security and nutrition. Forests and Food, 73.

Provenza, F., & Malechek, J., (1986). A comparison of food selection and foraging behavior in juvenile and adult goats. Applied Animal Behavior Science, 16, 49–61.

Pulker, C. E., Trapp, G. S., Scott, J. A., & Pollard, C. M., (2018). Global supermarkets' corporate social responsibility commitments to public health: A content analysis. Globalization and Health, 14, 1–20.

Pynnönen, S., Haltia, E., & Hujala, T., (2021). Digital forest information platform as service innovation: Finnish Metsaan.fi service use, users and utilization. Forest Policy and Economics, 125, 102404.

Quandt, A., (2021). Agroforestry trees for improved food security on farms impacted by wildlife crop raiding in Kenya. Trees, Forests and People, 4, 100069.

Ramcilovic-Suominen, S., & Pülzl, H., (2018). Sustainable development–a 'selling point' of the emerging EU bioeconomy policy framework? Journal of Cleaner Production, 172, 4170–4180.

Reichard, S. H., & White, P., (2001). Horticulture as a pathway of invasive plant introductions in the United States: Most invasive plants have been introduced for horticultural use by nurseries, botanical gardens, and individuals. BioScience, 51, 103–113.

Residues F. P. O. E. O. P., Environment t., Residues W. E. G. O. P., (1978). Pesticides Residues in Food, 1977 Evaluations: The Monographs: Data and Recommendations of the Joint Meeting of the FAO Panel of Experts on Pesticide Residues and the Environment and the WHO Expert Group of Pesticide Residues, Geneva, 6–15 December 1977: Food & Agriculture Org.

Riginos, C., Porensky, L. M., Veblen, K. E., Odadi, W. O., Sensenig, R. L., Kimuyu, D., Keesing, F., et al., (2012). Lessons on the relationship between livestock husbandry and biodiversity from the Kenya long-term exclosure experiment (KLEE). Pastoralism: Research, Policy and Practice, 2, 1–22.

Robert, N., Jonsson, R., Chudy, R., & Camia, A., (2020). The EU bioeconomy: Supporting an employment shift downstream in the wood-based value chains? Sustainability, 12, 758.

Schaetzl, R. J., Burns, S. F., Johnson, D. L., & Small, T. W., (1988). Tree uprooting: Review of impacts on forest ecology. Vegetatio, 79, 165–176.

Sharma, D., Tiwari, B., Chaturvedi, S., & Diengdoh, E., (2015). Status, utilization and economic valuation of non-timber forest products of Arunachal Pradesh, India. Journal of Forest and Environmental Science, 31, 24–37.

Sharma, Y. K., Yadav, A. K., Mangla, S. K., & Patil, P. P., (2018). Ranking the success factors to improve safety and security in sustainable food supply chain management using fuzzy AHP. Materials Today: Proceedings, 5, 12187–12196.

Sheppard, J. P., Chamberlain, J., Agúndez, D., Bhattacharya, P., Chirwa, P. W., Gontcharov, A., Sagona, W. C. J., et al., (2020). Sustainable forest management beyond the timber-oriented status quo: Transitioning to co-production of timber and non-wood forest products—A global perspective. Current Forestry Reports, 6, 26–40.

Surya, B., Syafri, S., Sahban, H., & Sakti, H. H., (2020). Natural resource conservation based on community economic empowerment: Perspectives on watershed management and slum settlements in Makassar City, South Sulawesi, Indonesia. Land, 9, 104.

Tauro, A., Ojeda, J., Caviness, T., Moses, K. P., Moreno-Terrazas, R., Wright, T., Zhu, D., et al., (2021). Field environmental philosophy: A biocultural ethic approach to education and ecotourism for sustainability. Sustainability, 13, 4526.

Te Kanawa, R., (2021). Mana Taonga Kākahu: Exploring Approaches to Reconnecting Taonga Kākahu to Tangatawhenua. Open Access Victoria University of Wellington| Te Herenga Waka.

Toppinen, A., Autio, M., Sauru, M., & Berghäll, S., (2018). Sustainability-driven new business models in wood construction towards 2030. Towards a Sustainable Bioeconomy: Principles, Challenges and Perspectives (pp. 499–516). Springer.

Trigkas, M., Anastopoulos, C., Papadopoulos, I., & Lazaridou, D., (2020). Business model for developing strategies of forest cooperatives. Evidence from an emerging business environment in Greece. Journal of Sustainable Forestry, 39, 259–282.

Van, B. S., Dengerink, J., & Ruben, R., (2018). The Food Systems Approach: Sustainable Solutions for a Sufficient Supply of Healthy Food. Wageningen Economic Research.

Van, K. Q., Tran, B. Q., Meyfroidt, P., & Paschke, M. W., (2018). Drivers of deforestation and forest degradation in Vietnam: An exploratory analysis at the national level. Forest Policy and Economics, 90, 128–141.

Van, N. M., (2019). Integrated natural resource management as pathway to poverty reduction: Innovating practices, institutions and policies. Agricultural Systems, 172, 60–71.

Vauhkonen, J., & Ruotsalainen, R., (2017). Assessing the provisioning potential of ecosystem services in a Scandinavian boreal forest: Suitability and tradeoff analyses on grid-based wall-to-wall forest inventory data. Forest Ecology and Management, 389, 272–284.

Vinceti, B., Termote, C., Ickowitz, A., Powell, B., Kehlenbeck, K., & Hunter, D., (2013). The contribution of forests and trees to sustainable diets. Sustainability, 5, 4797–4824.

Wagner, S., Nocentini, S., Huth, F., & Hoogstra-Klein, M., (2014). Forest management approaches for coping with the uncertainty of climate change: Trade-offs in service provisioning and adaptability. Ecology and Society, 19.

Wartmann, F. M., Baer, M. F., Hegetschweiler, K., Fischer, C., Hunziker, M., & Purves, R. S., (2021). Assessing the potential of social media for estimating recreational use of urban and peri-urban forests. Urban Forestry & Urban Greening, 64, 127261.

Wegner, G. I., Murray, K. A., Springmann, M., Muller, A., Sokolow, S. H., Saylors, K., & Morens, D. M., (2022). Averting wildlife-borne infectious disease epidemics requires a focus on socio-ecological drivers and a redesign of the global food system. eClinicalMedicine, 47, 101386.

Whiteman, A., Wickramasinghe, A., & Piña, L., (2015). Global trends in forest ownership, public income and expenditure on forestry and forestry employment. Forest Ecology and Management, 352, 99–108.

Winchester, N., & Reilly, J. M., (2020). The economic and emissions benefits of engineered wood products in a low-carbon future. Energy Economics, 85, 104596.

CHAPTER 4

Forest Resources Status, Utilization, and Conservation

ARIF YAQOOB MALIK

Department of Education, UT of Jammu and Kashmir, India

ABSTRACT

Forests are considered to be the lifelines of a nation because they contribute substantially to the economic development of the country, besides checking air pollution and soil erosion. Forests, also called green gold, are vast natural resources that have been providing a broad array of commodities, amenities, and other environmental services to mankind. Forests provide us with the air we breathe and the wood we use. Forests have been used for thousands of years by mankind, especially by the people who live in or near a forest, while others derive benefits indirectly that are necessary for the existence of life. Forests are a reservoir of non-wood forest products that include both animal and plant-based substances. Animal products include bush meat, skin, fur, hides, beeswax, honey, fish oils, and edible insects. Plant products include medicinal plants, raw materials for pharmaceutical industries, food, fodder, spices, leaves, herbs, fibers, dyes, aromatic compounds, seeds, oils, waxes, and resins. Besides providing non-wood and wood products, forests help provide solutions to modern challenges like energy crises, poverty, food security, environmental resilience, sustenance of agriculture, watershed management, clean water supplies, conservation of biodiversity, mitigation of climate change, desertification, land degradation, and disaster reduction. Millions of people worldwide depend on the existing forests for wood and non-wood products essential for their jobs and livelihood security. Poor households across countries are more dependent on forest resources, while rich households derive the maximum income from forest resources. Hence,

Sustainable Forest Resources Management: Issues and Implications. Jahangeer A. Bhat, Gopal Shukla, Manmohan J. Dobriyal, Sumit Chakravarty, A. Arunachalam, and Rainer W. Bussmann (Eds.)

it is impossible to think of life on this planet without forests, as forests are very important for local people in terms of livelihoods in most developing countries. This chapter depicts valuable content about the role of forests in livelihood security and their sustainable use, which could be important for designing effective conservation measures to improve ecosystem services and people's livelihoods.

4.1 INTRODUCTION

Forests are important storehouses for the livelihood of people in most developing nations of the world. The word forest has been derived from ancient French "Foris" which means outside. A forest is a biome chiefly consisting of trees with mean annual precipitation of about 76 cm where other vegetation like herbs and shrubs are also important besides other plant and animal life. It may also be defined as a community with a closed canopy predominantly consisting of trees, shrubs, and herbs or any other similar woody vegetation. Forests are usually dominated by trees and vary in species composition and density from region to region. Forests have been used for sustenance by mankind since times immemorial. Forests provide food security, livelihood, employment, livestock security, wood and non-wood products besides their role in ecosystem services. Mostly indigenous communities, about 200 million directly depend on forests for both their maintenance and survival and feeding of their livestock. Worldwide, about 1.6 billion people, which accounts for about a quarter of the population globally, depend on forests for medicines, fodder, fiber, fertilizer, food, fuel besides their livelihoods employment.

Over 80% of the global terrestrial biodiversity is possessed by forests and as a vital source of livelihood, energy, fresh water, and biodiversity repository, forests contribute indispensably by providing the ecosystem services for all mankind. Forests have proven to absorb and store $1/10^{th}$ of the global carbon emissions into soils, products, and their biomass. Most of the fresh water, about three-quarters, is provided through forested areas as these act as storehouses of glaciers. Forests help in slowing down soil erosion by locking the soil moisture and soil particles, help to prevent land degradation, reduce desertification, and decrease the incidences of landslides and floods. Healthy forests significantly contribute to soil production, enrichment, climate change mitigation, and shelter for wildlife. Besides, millions of people across the world depend on forests either directly or indirectly for their subsistence and income, especially those living in extreme poverty. Forests absorb greenhouse gases and also contribute to the natural environment like soil conservation, watershed management, and climate control on which the food production systems depend (Vizzari et al., 2015).

Over the years, there has been a rapid forest loss due to expanding agriculture and population explosion. To stop the deforestation trend, one requires a change in laws and policies, incentives, and institutions within and beyond the forestry sector. One such approach is the "landscape approach" that includes activities such as restoration of degraded forests, increasing agricultural productivity by modernizing agriculture, providing incentives for forests and farms to ensure no forest is being converted to farmland, including trees on farms and grasslands, and involving local communities in the conservation and protection of forest areas. Despite the consequences of biodiversity loss and forest degradation and the heavy dependence of locals on forest livelihoods, not much data is available on the role of forest resources in livelihood security. Hence, this chapter analyzes the forest dependence and the role of forests in local livelihood security.

4.2 DISTRIBUTION FOREST RESOURCE AND UTILIZATION

Forest resources are found in all countries and their conditions are changed due to anthropogenic pressure (Figure 4.1 and Table 4.1). In the developing world, 80% of the population uses non-wood forest products for nutritional

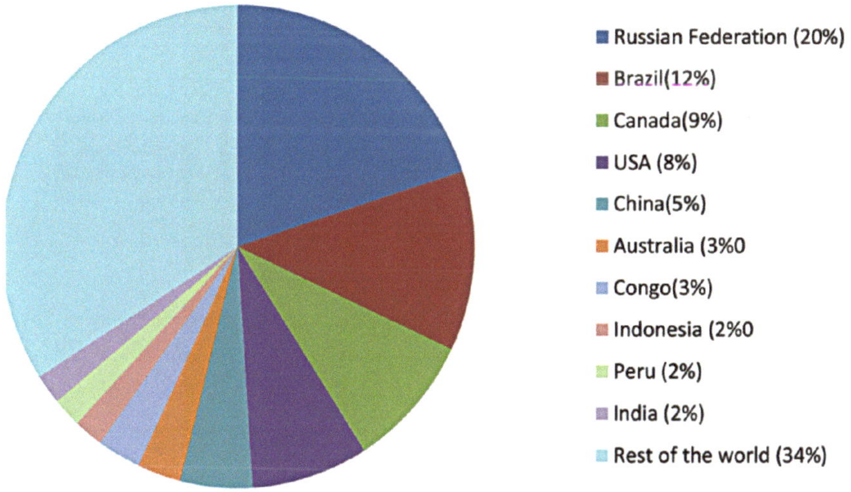

FIGURE 4.1 Global forest distribution.

Source: Adapted from FAO, 2020

needs and primary health care. Around 1.3 billion people worldwide have shelter provided directly by forest products, with an estimated income generated of about US$ 88 billion as of 2011 (Aggarwal et al., 2009). Forests make an immense contribution to food security, especially for ethnic communities and women, who tend to be vulnerable groups living in and around forests, making them a reliable means for sustaining rural livelihoods.

4.2.1 LIVELIHOOD EARNINGS

A series of forest products is collected by the people for their self-consumption and for sale in the local commercial market. Forests provide a unique source of livelihood for people in the form of employment. Local livelihoods are created through the sale of wood, fodder, grasses, medicinal plants, handicrafts, and wild edibles, which account for about 40–60% of their daily income (Figure 4.2). The indigenous communities collect these products for sale besides for their own consumption.

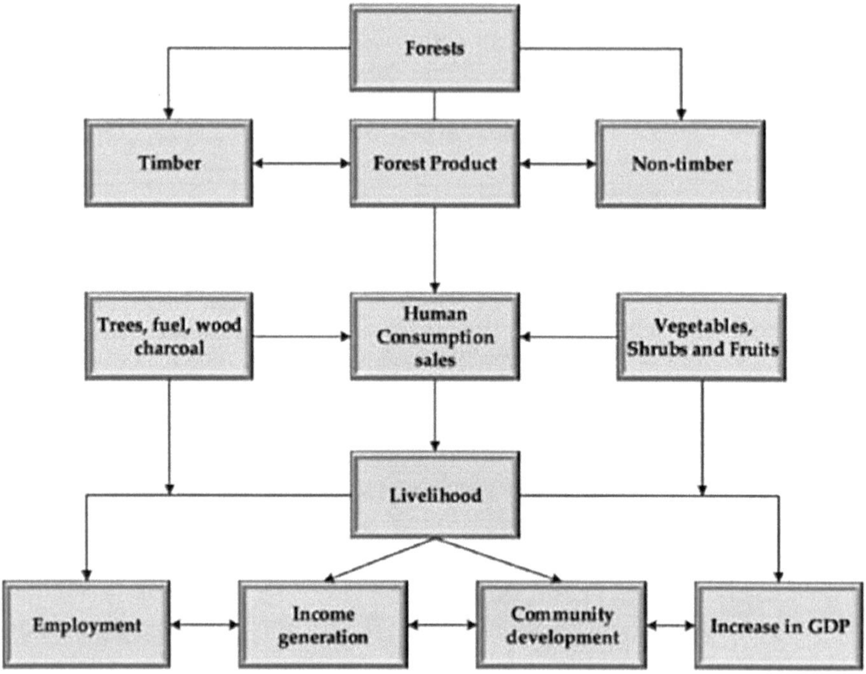

FIGURE 4.2 Livelihood security chart from forests.

A significant amount of income is generated from all forest types for the people, countries, and trade purposes, accounting for nearly 1% of global GDP, which amounts to about US $468 billion in the year 2009. Around 10 million people are formally employed in the forest sector around the world, in addition to a large number of seasonal workers in informal sectors (Bahuguna, 2000). The environmental and social contributions in terms of shelter, energy, and medicine are estimated to be significant components of GDP. Forests play a pivotal role in improving the livelihoods of the increasing population as the world faces challenges such as food and water shortages, climate change, desertification, and land degradation, along with a population explosion as the world will need to provide livelihoods to another 2 billion people if the predictions are correct (Loveleen et al., 2020). According to a World Bank report, global temperatures could rise by 4° at the end of this century, adversely impacting fresh water availability, agricultural produce, and other commercial facilities. By 2025, 2.4 billion people will not have sufficient water supplies for basic health, domestic use, and agriculture (Chao, 2012). However, for centuries, forests have acted as a natural safety net for most communities during times of natural hazards, famine, or other events that adversely affect agricultural and food production. Forests take care of people and their animals by providing food in times of crop failure (Bandi, 2013).

4.2.2 NON-WOOD FOREST PRODUCTS

These are forest byproducts of living origin other than forest-derived wood, which are used by mankind from times immemorial. These include food additives (herbs, spices, sweeteners, cinnamon, flavorings), food in terms of edibles (nuts, huckleberries, ferns, tree nuts, mushrooms, fruits, game, honey, gums); fodder (grasses and leaves), fiber (furniture, clothing, construction); plant and animal derivatives with medicinal value, ornamental plants and seeds; resins; gums, tree oils; rubber and Ginseng and other plant and animal products usually used for cultural, cosmetic, and pharmaceutical products. The tribal communities use NWFP for diverse applications in terms of edibles, medicinal plants, vegetables, spices, etc. About 50% of the forest income and 70% of export incomes come from NWFP in most of the states of India, which accounts for US 1.1 billion Dollars. These products are more beneficial to poorer sections of society and women. Proper and sustainable harvest and use of NWFP is responsible wise management of global forest resources, biodiversity conservation, and improvement in the income generation and livelihood security of people of forest fringe communities (Yaqoob et al., 2014).

4.2.3 MEDICINAL PLANTS AND WILD EDIBLES

Plant medicine has a wide range of applications and helps in the economic development of the communities living in and around the forests. People are shifting from chemical medicine to these therapies, and it is being used on a global scale now (Muthul et al., 2006). Plant medicine is an affordable means of treating diseases for a large number of people in rural areas of developing countries (Mesfin et al., 2013). Plant resources from forests are used for the treatment of various human diseases like jaundice, gastrointestinal disorders, asthma, tonic, cough, dermatitis, urinary disorders, irregular menstrual cycles, joint pain, snake bites, cancer, pneumonia, mumps, hair regrowth, and blood purification. Several livestock diseases like inflammation of hooves, indigestion, wounds, throat and eye infections, and delivery-related issues are addressed by the use of medicinal plants in a traditional way. About 75% of species are taken orally, 9% are applied externally for such purposes, and 16% have a dual mode of administration (Yaqoob et al., 2021).

There is a great tradition of using wild edibles as food by the people. So many species of herbs and shrubs in terms of their leaves, fruits, berries, nuts, etc., are used by these ethnic groups as wild edibles. Among different species, fruits of *Prunus spp., Berberis lyceum, Fragaria nubicola, Elaeagnus spp., Rosa webbiana, Prunus prostrata, Rubus ellipticus, Rubus fruticosus, Celtis spp., Ziziphus jujuba*, and *Viburnum grandiflorum* are eaten, providing a good supply of essential minerals and nutrients to ethnic groups (Yaqoob et al., 2017). Besides, many species like *Malva neglecta, Taraxacum officinale, Polygonum plebeium, Rumex acetosa, Plantago major, Oxalis corniculata*, and many species from *Rheum genera* are used as vegetables; their tender leaves, after collection from the wild, are washed and cooked. For some species like *Rheum webbiana, Dipsacus inermis*, dried leaves are cooked as vegetables. Forest dwellers use the plant species to derive nutritional benefits as they lack access to fruit and vegetable markets (Yaqoob et al., 2019).

4.2.4 FUEL WOOD, FODDER, AND TIMBER

Indigenous communities depend directly on forests for fuel, fodder, and house construction. The chief contribution to household income is fuelwood

(51%) and that of food products (27%). Due to a lack of access to the market and other income-generating opportunities, most of the forest communities construct their settlements using timber from the forests. Wood, bamboo, rocks, clay, and straw are used along with branches, twigs, leaves, and hay for home construction by forest communities (Langat et al., 2016). Forest departments also remove timber and wood annually, which also brings huge incomes to the local governing bodies. Fuelwood collected along with dry leaves is a chief source of energy used for cooking in developing countries. Due to a lack of modern cooking gas and kerosene, most of them use fuelwood and other dry forest products for cooking their food and for warming in the winters. Most of the tribal families have some kind of livestock like sheep, cow, goat, buffalo, etc., which represent a precious means of livelihood for them. Livestock rearing is done on the grasslands of the forests, and several 100 species of herbs are used as fodder in both tender and dried forms. This practice provides them with an additional source of income throughout the year. Forests provide quality, quantity, and sustainable stocks of fodder to these people for maximum periods of the year.

4.3 PATTERN AND REASONS OF FOREST DEGRADATION

Simultaneously, much of the world's remaining forest area is under increasing threat as a result of ruthless activities of humans, population pressures, mineral exploitation, agricultural expansion, growing energy demands, and climate change challenge. About 14.5 million hectares of forests are lost each year (Davidar et al., 2010). The forest area at the global level was reduced by 129 mha between the years 1990 and 2015 (Global Forest Resources Assessment, 2015), which accounts for about 3.1% net loss, and almost double this area – 239 Mha of forest was destroyed globally between 1990 and 2008.

However, the nature and extent of forest loss was uneven geographically. In boreal forests and temperate regions, the degree of forest degradation slowed and almost reversed until the beginning of the 20th century. An increasing trend was seen in the tropical regions, although at a slightly slower rate than earlier (Anonymous, 2019). Deforestation is a globally acknowledged fact, and despite the widespread forest ecosystem degradation, there is a lack of quantitative information on the status as well as the role of forest resources for sustainable use and livelihood security.

Sustainable practice of forest use involves managing forests to fulfill current needs and aspirations of mankind from existing forests in terms of ecological uses and services, products, and values, without depleting the resource availability for coming generations to meet their own (Abate et al., 2010). Forest trees and grasslands when managed sustainably make significant contributions both to people as well as to the planet in terms of sustaining livelihoods, supplying clean resources like water and air, conserving biological diversity, and reducing the extent of climate change (FAO, 2011). Forests provide an array of livelihoods to the people in terms of direct employment where forest departments, rural development, agriculture, and cooperatives employ a major section of society for protecting and managing forests, planning and research purposes, and other executive jobs (FAO, 2010). Indigenous communities earn through self-employment by selling forest products like wood, fodder, charcoal, medicinal plants, forest-based cottage industries, and handicrafts. These communities not only earn income but also ensure food security by having access to the food items using this forest-earned money besides consuming the forest edible products for their own (FSI, 2011). Although illegal, firewood collection for commercial market sale is also quite common in most parts of the world, especially in Indian forest areas that account for the livelihood of 11% of the population.

TABLE 4.1 Forest Area Region Wise (FAO, 2010a)

Region	Land Area (million Ha)	Total Forest (Natural + Plantations)			Natural forest (million Ha)	Plantation Forest (million Ha)
		Million Ha	Percent Land Area	Percent All Forests		
Africa	2,978	650	22	17	642	8
Asia	3,084	548	18	14	432	116
Europe	2,260	1,039	46	27	1,007	32
North and Central America	2,137	549	26	14	532	18
Oceania	849	198	23	5	194	3
South America	1,755	886	51	23	875	10
Total	13,064	3,869	186	100	3,682	187

Sometimes, local technology and the use of skills by these people in small-scale forest-based enterprises also provide secondary employment to a handful of people. The ethnic communities generate about 60% of their total income by selling the forest products in the commercial markets. These forests are the chief contributors to the rural economy besides providing subsistence income for millions of people (Islam et al., 2016). Lack of income opportunities, lack of access to markets, and large-scale poverty often make these communities solely dependent on their livelihoods, which sometimes causes over-exploitation of these resources. Nevertheless, a series of forest products have been harvested sustainably by indigenous communities for many years, which has been an indispensable source of regular income for many families (Kandwal, 2018).

There is an intense and devastating pressure on the existing forest resources of the globe as a result of which the biological capacity of these ecosystems to provide precious services to mankind is reduced (World Bank, 2013). This is known as forest degradation. About half of the global tropical forests was degraded since 1960 and more than one hectare of it is destroyed every second (FAO, 2016). The degradation does not remain restricted to tropics only because 3.7 mha of Europe's forests have also been damaged by livestock grazing, pests, diseases, forest fires, and other human-induced activities, especially cutting down of forests where forest lands are converted for non-forest purposes.

4.3.1 FOREST CHANGE AND FOREST DEGRADATION

Deforestation means removal or cutting down of forest vegetation by natural or anthropogenic means. Mass scale destruction of forests is caused by man for economic gains leading to long-term irreversible effects on mankind. Earlier in our country, deforestation was only done by the British, but after Independence the same trend was followed by the people of India (Loveleen, 2020). Since the main driving force behind forest destruction is population explosion, some important factors responsible for deforestation are briefly discussed:

1. **Forest Fires and Pest Attack:** The people residing in or around forests may be responsible for setting large areas of forests on fire for their short-term gains. Sometimes tourists may do so intentionally or unintentionally, resulting in the destruction of forests. Pest attacks are a natural process where a range of forest or a patch of forests gets diseased, ultimately resulting in the killing of plant species in that particular area.

2. **Shifting Cultivation:** This method is practiced by indigenous land-less people or tribes who live in or around forests. They cut down large areas of forests to grow crops without applying fertilizer or

manure. After a few years, the same areas are abandoned due to reduced yield, and farmers move to other areas to continue the same practice. This results in large areas of forests being left in a degraded condition.

3. **Expansion of Agriculture:** This is another factor responsible for forest destruction where large areas of forest land are converted to agricultural fields in order to meet the demands of the growing population, especially in tropical and sub-tropical regions. Around 70 million people are added to the globe annually, and most of these are added in developing countries (Kumar, 2001). Such an expanding population demands more food availability that can't be achieved from the existing farmland, and hence, more and more forests are cut to be converted to agricultural areas to meet the present increasing food demands.

Forest lands also get cleared for the construction of roads, dams, and other developmental projects which need to clear large areas of forests for their completion. Besides, the increased demand for wood for commercial and domestic purposes also brings the existing forests under tremendous pressure, hence causing irreversible damage to these vital resources (Kumar et al., 2011). Additionally, mining, quarrying, raw materials for industries, fuelwood gathering, overgrazing, and conversion of forests to pastures and other non-forest purposes also have a huge impact on these resources. Deforestation has both long-term as well as short-term adverse effects on the well-being of humans and ecosystem services. It gives rise to several environmental problems, like soil erosion, increased siltation of water bodies, increases in the frequency of floods, changes in precipitation patterns, etc. (Mahapatra et al., 2005). However, the main specific effects of deforestation include disturbance of ecological balance, scarcity of forest goods and raw material, destruction of wild habitats, lowering of the water table, melting of glaciers, increase in soil erosion and desert areas, and loss of biodiversity.

4.4 STRATEGIES OF FOREST CONSERVATION

The conservation, protection, and control of forest destruction is not only the need of the hour but also necessary for the economic development of the nation and maintenance of ecological balance (MOEF, 2006). There is a need to protect the existing forest cover from further destruction besides increasing the forest cover to cover the prescribed area (i.e., 33% of the geographical area).

Some important methods evolved in sustainable conservation as well as protection of forests are as follows:

1. **Afforestation and Reforestation:** In order to reduce the pressure of demand on forests, massive forestation programs need to be launched to increase the forest cover of the nation. The top priority should be given to afforestation in the open wastelands and reforestation of already degraded forest areas which are not fit for cultivation (MOEF, 2009). This shall ensure the increase in the existing forest cover that will surely act as security for the livelihood of the people.
2. **Social Forestry:** The National Commission on Agriculture gave the term in 1976. It was an initiative to decrease the pressure on existing forests by making use of all unused and fallow lands of the country. This type of forest program is done on private lands (public and common land), along railway lines, in and around farmlands, road and riversides, canal banks, village common lands, and government wastelands (Nayak et al., 2013). It aims at raising plantations by the common public to meet their own demands of 5F's. The local people as well as school children are involved to raise plants to produce in a decentralized way; the fuelwood, fodder, small timber for local poor people in order to reduce the pressure of such demands on existing forests (Pandey, 2002). This scheme has involved the public in the rehabilitation of degraded forests in terms of afforestation and reforestation that has encouraged public participation and recognition of the local community rights over the minor forest.
3. **Agroforestry:** When a land area is used for agriculture, forestry, and animal husbandry, i.e., using woody perennial plants on the same land management units as annual agriculture crops either sequentially or simultaneously for obtaining greater outputs in the growth of firewood, fodder, and small timber, it is referred to as agroforestry (Sadashivappa et al., 2006). Farmers plant trees on their farmlands to meet their domestic demands of fodder, fuel, etc., and to provide shade to their crops, use trees as wind shields, and to prevent their fields from soil erosion (Springate, 2010). Thus, this will also reduce the sole dependence of man on forests and hence will help in the conservation of existing forests.

4. **Community Forest Management:** The need to involve local communities' activity in protection, conservation action, and regulation of forests has become a growing concern. In this process, an arrangement between local communities and the forest department is made to share some economic benefits between the government and locals for the development of the community from the annual harvest of the forest (Vedeld et al., 2007). It is an initiative to channelize the participatory governance of forest resources by involving those communities which live in and around forests. It is a co-management practice between the forest department and the fringe communities on the basis of mutual trust and joint responsibilities. This program was initiated under the national forest policy of 1988 where necessary resolutions and guidelines are being issued to be acted upon for the successful operation of this program. Village forest committees (VFC) are formed to protect and manage the forest resources through mutual sharing of costs and benefits (Saha et al., 2012). The key achievement of JFM program is reforestation of the degraded forests with public participation in terms of Forest Protection Committees. It has proved to be a win-win program for both the government department and the local people where forests are protected, and people share access to minor forest products.

5. **Law Enforcement:** Forest law enforcement and governance first emerged in the 1990s in the world for its concern over the irregularities in the wood industry, forest degradation, and loss of revenue in poor countries, which is about US $5 billion a year (Saha et al., 2012). The national forest policy in India adopted in 1952 makes it obligatory that 33% of our country should be under forest cover. However, due to population explosion, there has been a continuous decrease in the forest cover of many nations. Keeping this thing in view, the existing forest resources could be protected by actively implementing the forest conservation Acts, where people have to respect and follow each and every provision of this act and anybody found guilty should have to pay penalties for any offense punishable under the act.

6. **National Programs:** National Afforestation and Eco-Development Board was developed by MOEF in the year 1992 to come up with specific schemes for afforestation, management strategies, and eco-development packages for forests to ensure biomass production

through a joint planning process. NAP was launched in 2002 as a flagship program by the National Afforestation Eco-Development Board, which involves plantations in degraded forest areas and helps Forest Development Agencies with capacity building support.

7. **Protected Area Networks and Conservation Efforts:** The adoption of the National Policy on Conservation of Wildlife in 1970 along with the Wildlife Protection Act of 1972 has led to a significant growth in strengthening the protected area network from a few national parks and sanctuaries to about 669 protected areas in the country, accounting for about 4.92% of the geographical area. In addition, biosphere reserves, sacred groves, conservation reserves, coastal protected areas, etc., have also played a pivotal role in this direction (IPCC, 2007). The launch of various projects like Project Tiger, Project Elephant, Project Hangul, Project Snow Leopard, etc., and cooperation in international conventions like CITES, CBD, etc., by the country have also proved to be indirect means for the conservation and protection of forest resources of the country.

Besides all these measures, the Forest Rights Act needs to be strengthened to ensure the rights of the community over the forest resources and to address the livelihood security of the people. Although significant progress has been made for sustainable forest conservation at the global level, the policy, law, and regulations implementation seems to be highly uneven with respect to geographical locations. Furthermore, the practice of agriculture that triggers land use change and deforestation seems to be more financially attractive than forest management in certain areas due to short-term gains.

4.5 CONCLUSION

This chapter has revealed the important role of forest resources and their linkage to household income in terms of livelihood. The forest income shares are usually higher for poor households, as is their dependence on forest resources, although most of the collection of forest products is illegal. About 33% of household annual income comes from the sale and consumption of these products. A little effort has been made to discuss the genuine reasons behind forest degradation and its impacts, highlighting the peculiar strategies and other government initiatives to protect and conserve forest resources. Due to rising human populations, demand for forest products is likely to increase manyfold in the surrounding areas of forests. It would not be wise to exclude local communities from accessing the forest area for their livelihood, as it would further worsen the condition of poverty. Pertinently, low-level sustainable collection of forest products and a licensing system of approach for the sustenance of the communities need to be encouraged rather than large-scale commercial extraction. Besides, increasing tree covers on the farms, improving agricultural efficiency, minimizing illegal exploitation, coupled with increasing income opportunities independent of forests, can reduce the pressure on forest resources and will be helpful in improving livelihoods, eradicating poverty, and conserving forest biodiversity.

KEYWORDS

- **animal products**
- **biodiversity**
- **conservation**
- **deforestation**
- **forest sustainability**
- **livelihood**
- **medicinal plants**

REFERENCES

Abate, R. S., & Wright, T. A., (2010). A green solution to climate change: The hybrid approach to crediting reductions in tropical deforestation. *Duke Environmental Law & Policy Forum, 20*(1).

Aggarwal, A., Paul, V., & Das, S., (2009). Forest resources: Degradation, livelihoods, and climate change. In: Datt, D., & Nischal, S., (eds.), *Looking Back to Change Track* (pp. 91–108, 219). New Delhi: TERI.

Anonymous, (2019). *Environment: Classroom Study Material.* Retrieved from: https://s3-us-west-2.amazonaws.com/visionresources/valueaddedmaterial/PT-365-ENVIRONMENT-2019.pdf (accessed on 6 May 2024).

Bahuguna, V. K., (2000). Forests in the economy of the rural poor: An estimation of the dependency level. *Ambio 29*(3), 126–129.

Bandi, M., (2013). Forest-dependant people, their livelihood challenges and forest rights act: A review. Mainstream, Vol, L. I., No 37.

Chao, S., (2012). *Forest People: Numbers Across the World, Forest Peoples Program*, Moreton-in Marsh, U. K.

Davidar, P., Sahoo, S., Mammen, P. C., Acharya, P., Puyravaud, J. P., Arjunan, M., Garrigues, J. P., & Roessingh, K., (2010). Assessing the extent and causes of forest degradation in India: Where do we stand? *Biological Conservation, 43*(12), 2937–2944.

Down to earth, (2019). *Rewind 2018: Development and Fire Eat up India's Forest Cover.* Retrieved from: https://www.downtoearth.org.in/news/forests/rewind-2018-development-and-fire-eat-upindia-s-forest-cover-62631 (accessed on 22 April 2024).

FAO (2010). *Policies, Strategies and Technologies for Forest Resource Protection.* Retrieved from http://www.fao.org/3/AC805E/ac805e0v.htm (accessed on 22 April 2024).

FAO, (2011). *Forests for Improved Food Security and Nutrition.* Rome.

FAO, (2014). *State of the World's Forests.* Rome.

FAO, (2016). *Integrated Policy for Forests, Food Security and Sustainable Livelihoods Lessons from the Republic of Korea.* Retrieved from: http://www.fao.org/3/a-i5444e.pdf (accessed on 22 April 2024).

Forest Survey of India, (2011). *India State of Forest Report* (p. 286). New Delhi: Ministry of Environment and Forests, Government of India.

IPCC (2007). In: Parry, M. L., Canziani, O. F., Palutikof, J. P., Van, D. L. P. J., & Hanson, C. E., (eds.), *Climate Change 2007: Impacts, Adaptation and Vulnerability* (p. 976). Contribution of Working Group II to the Fourth Assessment Report of the Intergovernmental Panel on Climate Change, Cambridge University Press, Cambridge UK.

Islam, M. A., Quli, S. M. S., & Baba, M. Y., (2016). Household drivers of forest dependence for employment support among tribes of Jharkhand, India. *Economic Affairs, 61*(2), 339–347.

Kandwal, P., (2018). *A Study on Forest-Based Resources for Livelihood in Lower Shivalik Hills.* Ph.D. dissertation, Punjab Agricultural University, Ludhiana.

Kumar, H. D., (2001). *Forest Resources Conservation and Management.* Affiliated East-West Press Pvt Ltd, New Delhi.

Kumar, L. B., Patil, B. L., Basavaraja, H., Mundinamani, S. M., Mahajanashetty, S. B., & Megeri, S. N., (2011). Participation behavior of indigenous people in non-timber forest products extraction in Western Ghats Forests. *Karnataka Journal of Agricultural Science, 24*(2), 170–172.

Langat, D. K., Maranga, E. K., Aboud, A. A., & Cheboiwo, J. K., (2016). Role of forest resources to local livelihoods: The case of east Mau Forest ecosystem, Kenya. *Int. J. of Forest Res. 2016.* Art.ID 4537354.

Lele, S., (2011). Rethinking forest governance. *The Hindu Survey of the Environment, 2011*, 95–103.

Loveleen, K., & Ritu, M., (2020). Forest conservation for livelihood security. *Int. Arch. App. Sci. Technol., 11*(4), 61–67.

Mahapatra, K., & Kant, S., (2005). Tropical deforestation: A multinomial logistic model and some country specific policy prescriptions. *Forest Policy and Economics, 7*, 1–24.

Malik, A. Y., Singh, D. P., Yunus, M., Bhat, G. A., Shukla, G., Bhat, J. A., Rather, M. M., & Chakarvarty, S., (2021). Diversity and structure of plant assemblages in open scrub vegetation patches of Dachigam National Park at Kashmir Himalayas. *Tree, Forests and People, 3.*

Malik, A., & Singh, D. P., (2019). *Ethnobotanical and Ethnoveterinary Importance of Plants of Scrub Areas of Dachigam National Park, Jammu and Kashmir, India* (Vol. 12, p. 582–586).

Mesfin, F., Demissew, S., & Teklehamanot, T., (2009). An ethnobotanical study of medicinal plants in Wonago Woreda, SNNPR. Ethopia. *J. Ethnobiol. Ethnomed., 5*(1), 28.
Ministry of Environment and Forests, (2006). *Report of the National Forest Commission* (p. 421). New Delhi: Ministry of Environment and Forests, Government of India.
Ministry of Environment and Forests, (2009). *Asia-Pacific Forestry Sector Outlook Study II: India Country Report* (p. 78). Working Paper No. APFSOS II/WP/2009/06. Bangkok: FAO.
Muthul, C., Muniappa, A., Nagapan, R. M., & Savarmuthu, I., (2006). Medicinal plants used by traditional healers in Kancheepuram District of Tamil Nadu. *J. Ethnobiol. Ethnomed., 2*, 43–51.
Nayak, B. P., Kohli, P., & Sharma, J. V., (2013). *Livelihood Of Local Communities and Forest Degradation in India: Issues for REDD+.* A Policy Brief of The Energy and Resources Institute (TERI). Retrieved from: https://www.teriin.org/projects/nfa/2008-2013/pdf/Policy_Brief_Livelihood_Local_Communities.pdf (accessed on 22 April 2024).
Pandey, S., Jain, S., & Gupta, R., (2002). Conservation model and modern techniques for sustainable wilderness. *International Journal of Research IJR, 2*(9), 2348–2355.
Sadashivappa, P., Suryaprakash, S., & Vijaya, K. V., (2006). *Participation Behavior of Indigenous People in Non-Timber Forest Products Extraction and Marketing in the Dry Deciduous Forests of South India.* Conference on International Agricultural Research for Development, Tropentag University of Bonn.
Saha, D., & Sundriyal, R. C., (2012). Utilization of non-timber forest products in humid tropics: Implications for management and livelihood. *Forest Policy and Economics, 14*, 28–40.
Shackleton, C. M., & Shackleton, S. E., (2006). Household wealth status and natural resource use in the Kat River valley, South Africa. *Ecological Economics, 57*(2), 306–317.
Springate-Baginski, O., & Wollenberg, E., (2010*). REDD. Forest Governance and Rural Livelihoods: The Emerging Agenda* (p. 279). Bogor: CIFOR.
Vedeld, P., Angelsen, A., Bojö, J., Sjaastad, E., & Berg, G. K., (2007). Forest environmental incomes and the rural poor. *Forest Policy and Economics, 9*(7), 869–879.
Vizzarri, M., Tognetti, R., & Marchetti, M., (2015). Forest ecosystem services: Issues and challenges for biodiversity, conservation, and management in Italy. *Forests, 6*(6), 1810–1838.
World Bank, (2013). *Sustaining Forests and Livelihoods in a Changing World.* Retrieved from: https://www.worldbank.org/en/news/feature/2013/01/29/sustaining-forests-livelihoods-changing-world. Forest Conservation for Livelihood Security (accessed on 22 April 2024).
Yaqoob, A., Singh, D. P., Yunus, M., & Bhat, G. A., (2017). Phytotherapeutic and ethnobotanical importance of plant biodiversity of Dachigam national park, Kashmir. *Am. J. Ethnomed., 4*(2), 12.
Yaqoob, A., Yunus, M., & Bhat, G. A., (2014). Comparative study of shrub diversity in lower Dachigam Kashmir Himalaya. *Int. J of Biod. and Conse., 6*, 12.

CHAPTER 5

Forest Health in a Changing Scenario of Climate Change

ABHAY KUMAR,[1] SWATI SHABNAM,[2] P. R. ORAON,[1] and M. S. MALIK[1]

[1]*Krishi Vigyan Kendra, Jehanabad, Bihar Agricultural University, Sabour, Bhagalpur, Bihar, India*

[2]*Department of Agronomy, Faculty of Agriculture, Birsa Agricultural University, Ranchi, Jharkhand, India*

ABSTRACT

Forest ecosystems are classified into boreal, temperate, and tropical forests. According to their relative location, climatic conditions, and sometimes geological substrate and their subsequent characteristics, these major forest groups can further be divided into subgroups. All forest ecosystems are significantly impacted by climate change, which is also becoming the main threat to tree species individually. In IUCN Red List assessments of tree species, changing climate and weather extremes are identified as a threat in 1,080 cases. But climate change affects each type of forest ecosystem differently. Trees of boreal, montane, and coastal ecosystems are getting more and more affected by climate change, with major impacts of higher-than-normal temperatures during summer and decreased precipitation during winter. The most common consequences of climate change on forest ecosystems are alterations and shifts in habitats of tree species, especially in tropical islands and intertidal zones, most likely because of melting ice resulting in a rise in sea levels. These coastal habitats are also getting threatened by storms and flooding more than any other habitats. In boreal ecosystems also, more than 20% of tree species are affected by habitat shifts driven by encroachment of temperate vegetation as boreal regions are becoming warmer. However,

Sustainable Forest Resources Management: Issues and Implications. Jahangeer A. Bhat, Gopal Shukla, Manmohan J. Dobriyal, Sumit Chakravarty, A. Arunachalam, and Rainer W. Bussmann (Eds.)

the true effect of climate change is likely to be more pervasive, since it also has impacts on the forest ecosystems by increasing the risk of forest fires, pests and diseases, emergence of pests, and spread of invasive species. All of these conditions contribute to losses of biodiversity in forest ecosystems and consequently the productivity.

5.1 INTRODUCTION

Forests are communities of plants, animals, and microorganisms interacting with each other as well as with abiotic components such as soil, water, and the climate, making the interactions complex. Trees are the most dominant species in a forest and to a large extent determine the plants and animals in the forest, protect soil from erosion, reduce runoff, improve water quality, and reduce pollution by cleaning the air and also making it cool. Similarly, a forest's biotic and abiotic components influence, or rather determine the composition of the forest and the tree species found. Humans are also an important and integral component of forest communities. Hence, a healthy forest must be able to accommodate and fulfill the present and future needs of people for a variety of products, values, and services.

Forest health, a qualitative term, indicates the general state and well-being of a forest. It is a condition of sustaining the structure, composition, processes, function, productivity, and resilience of forest ecosystems over time and space. A healthy forest is one that is largely free of insects, pests, diseases, exotic species, invasive weeds, and pollution. All the tree species composing the forest are capable of growing at rates proportionate with the local geographic position, climate, and soil resources to accomplish their life cycles. A forest is said to be healthy only if it is capable of resisting injury from disastrous events like sensitive insects, pests, fire, disease attacks, wind, and flooding, and fully recovers from the perturbations to continue its life history functions over centuries, decades, or millennia. Soil is one of the most significant elements that affect forest health in several ways, by strongly anchoring tree roots, by changing energy flow along with different ecosystem elements, and by regulating water, light, and nutrient availability for the advantage of the entire forest ecosystem. The forest habitat of soil microorganisms that play a pertinent role in the decomposition of litterfall and the process of nutrient cycling is also controlled by the type of soil. During dry periods, drought-prone soils may predispose forests to pest and disease attacks.

There is no single parameter or characteristic to assess the health of the forest. This is because forests are ever-changing systems. Over time, one tree species may flourish while another gets eliminated. Natural calamities like fires and insect infestations can destroy several thousands of hectares of natural forest. None of these things suggest an unhealthy forest. It is well known that forests evolve with a stable climate. Weather parameters like precipitation, snowfall, and temperatures differ every year; however, they are stable over long periods of time. Forests play essential roles in maintaining and sustaining the ecology of the planet, including climate regulation, and deliver a multitude of ecosystem services within their local landscape. But in the recent past, anthropogenic effects alter critical components of these natural ecosystems. Streams are being redirected or drained completely for agriculture and development. Air pollution from communities and industries is widespread. Climate change is resulting in increasing temperatures. Forests are suffering more as these factors that influence their health continue to alter.

Climate change will have mostly negative and some positive effects on forests, with varying regional and temporal patterns. The warming of the planet is impacting wide swaths of forests adversely. For example, higher temperatures promote fire, drought, and insect disturbances, and thus tree mortality will increase with climate change. Negative impacts are already being observed in many regions.

Around the globe, tree mortality has been rising over the past few years. Large bursts of tree mortality have been associated with warmer and drier than typical circumstances for forests in temperate and boreal biomes (Sommerfeld et al., 2018; Seidl et al., 2020). Similarly, long-term monitoring in tropical forests also shows that climate change has already begun to increase tree mortality and disrupt regeneration potentials (Hubau et al., 2020; Sullivan et al., 2020). Climate-related dieback has also been noted due to unique interactions between the life cycles of trees and pest species (Kurz et al., 2008; Lesk et al., 2017; Sambaraju et al., 2019). Forest dieback could significantly affect wood production among other impacts. A recent example of the impacts of climatic extremes is the 2018 European drought (Buras et al., 2020), which led to a significant browning of the vegetation and resulted in widespread tree mortality (Brun et al., 2020; Schuldt et al., 2020). In contrast, it has also been noted the increasing productivity in high-latitude forests such as those in Siberia.

5.2 GLOBAL STATUS OF FOREST

Forest is a Latin word 'Foris' meaning 'outside the village boundary.' A forest is defined as a plant community that consists primarily of trees and other woody vegetation and typically has a closed canopy (Ecology). It can also

TABLE 5.1 Top 10 Countries for Forest Areas

Country	Forest Area (000 ha)	Percentage of World Forest Area	Percentage of Country Area
Russia	8,15,312	20	49.8
Brazil	4,96,620	12	59.4
Canada	3,46,928	9	38.7
USA	3,09,795	8	33.9
China	2,19,978	5	23.3
Congo	1,34,005	3	17.4
Australia	1,26,155	3	55.6
Indonesia	92,133	2	49.1
Peru	72,330	2	56.5
India	72,160	2	24.3
Total	4,85,438	66	

Source: Adapted from: FAO (2020).

be described as a region preserved for the purpose of producing timber and other forest products or for some indirect benefits, such as climatic or protective. In addition to major products like timber, fuelwood, and minor products like NTFP, fruit, fodder, fiber, flosses, gum, resin, medicine, rubber, coffee, dye, spices, leaves, bamboo, cane, etc., forests also provide other ecosystem services like oxygen, rainfall, soil conservation, control of soil and water erosion, recreation, biodiversity conservation, improved water quality, health effects, windbreak, shelter break, flood control, pollution absorption, carbon sequestration, control of greenhouse gases, and mitigation of climate change.

Forests also play a major role in the Earth's biogeochemical processes, influencing the production of soil, hydrological, nutrient, and carbon cycles, and global climate. Forests contain about 50% of the world's terrestrial carbon stocks while over 75% of the world's accessible fresh water is obtained from catchment areas of forests (Newton, 2021). Forests also provide habitat for a number of other species supporting many of the Earth's known terrestrial plant and animal species (Newton, 2021). As per the reports, up to 80% of terrestrial plants and animals are reported to occur in forests; however, this cannot be verified with precision (FAO & UNEP, 2020). There are 58,497 species of trees globally (Beech et al., 2021). Global Tree Search, a database of tree species and their country distributions, provides a complete list of the world's tree species as a tool for assessing, monitoring, and managing tree species diversity and forests on a global, regional, and national level (Beech et al., 2017). Around 4.06 billion hectares or 30.8% of the planet's total surface area are covered in forests. Despite the fact that forests are not fairly distributed among the world's populations or geographical regions, this area is almost equivalent to 0.52 ha per person. The major top 10 countries having the largest forest areas in descending order are Russia, Brazil, Canada, USA, China, Congo, Australia, Indonesia, Peru, and India (Table 5.1).

About 93% (3.75 billion ha) of the total forest area in the world is under naturally regenerating forests, while only 7% (290 million ha) is planted forest, though the extent of forests that are naturally regenerating has been decreasing since 1990. Worldwide, primary forests have decreased by 81 million hectares, whereas planted forests have grown by 123 million hectares, though the rate of growth has slowed since 2010 (FAO, 2020). Between 1990 and 2020, the percentage of land that is covered by forests declined from 32.5 to 30.8% (Table 5.2). There is forest area loss in some countries while there is gain in others. In India, according to the India State of Forest Report (ISFR, 2021), the total forest and tree cover is 80.95 million hectares, which is 24.62% of the total geographical area.

TABLE 5.2 Change in World Forest Area (Deforestation)

Period	Net Changes (Million ha/yr)	Net Changes Rate (% year)
1990–2000	–7.84	–0.19
2000–2010	–5.17	–0.13
2010–2020	–4.74	–0.12

Source: Adapted from: FAO (2020).

Several biotic and abiotic factors such as temperature, moisture, atmospheric pressure, soil conditions, biodiversity, and a combination of these factors affect the forest. The rate of photosynthesis in plants is affected by the amount of light which in turn depends on topography, i.e., altitude, latitude, and many other geographical aspects. Similarly, the metabolic phenomena of the trees are also impacted by the prevalent temperature and moisture conditions.

5.3 ROLE OF MAJOR CLIMATIC FACTORS IN FOREST

According to the World Meteorological Organization (WMO), climate is the average pattern of the weather in an area. Climate is characterized by temperature, moisture, wind, humidity, pressure conditions, and solar radiation of an area, and that directly and greatly influence the growth and development of plants and forests. It also includes the concentration of atmospheric gases, i.e., O_2, N_2, CO_2, NH_3, HNO_3, water vapor, and solid particles, having direct and indirect effects on several biological, physical, and chemical processes occurring in the forest. The variation in climatic conditions causes differences in plant species. For example, tree species observed in cold temperatures of high altitudes of Himalayan regions are distinct from those identified in lower altitudes or tropical areas.

5.3.1 ATMOSPHERIC GASES

Among the important atmospheric gases are O_2, N_2, CO_2, NH_3, HNO_3, and water vapor. Out of all these gases, N_2 composes 78% of the Earth's atmosphere while 21% of it is oxygen, which is essential for energy generation in all living organisms, including plants. Nitrogen is a crucial element required to make amino acids, nucleic acids, and various other enzymes. Plants take up nitrogen from the soil in the form of nitrate, and through the food chain, plants pass on other nitrogen compounds to other species. Plant growth and development are influenced by the amount of nitrogen present in the soil. Carbon dioxide is present in very low amounts compared to these two gases, about 0.04%, yet is very important as it is taken up by the plant community for photosynthesis and serves as a greenhouse gas that maintains the temperature of the earth's atmosphere. Ammonia (NH_3) and nitric acid (HNO_3) are brought to the soil by rainfall. Plant species belonging to the Leguminosae family fix atmospheric nitrogen to the soil.

5.3.2 LIGHT OR SOLAR RADIATIONS

Light is very important to sustain life on Earth as the process of photosynthesis in green plants occurs in the presence of light. Light is thus one of the most crucial elements that affect the growth, development, and reproduction of plants.

Most green plants show phototropism, which is controlled by both the intensity and quality of light. Trees and their branches, like most green plants, tend to bend towards light. For example, in a forest, stems, and branches elongate in response to light. Low light intensity causes the elongation of trees. For this reason, plants persisting in shade are taller than other plants growing in light under similar conditions. Blue and UV light cause dwarfness, while red and infrared light aid in the increment of height. In the presence of sunlight, the tree crown also takes on a larger shape, resulting in fast growth and development and a clear bole in trees. Thus, light plays a significant role in developing the proper form and structure of trees, as well as improving resistance to cold, drought, disease, insect attacks, and other unfavorable factors. It also influences certain physiological functions.

5.3.3 AIR TEMPERATURE

Temperature has a diverse effect on various metabolic processes of vegetation and forests. It increases the rate of respiration, photosynthesis, and transpiration, which result in better growth and development. Also, it increases soil microbial activity, which accelerates the recycling of organic wastes and releases nutrients into the soil. The higher the temperature, the greater the diversity of species, for example, in tropical or warmer regions there is a very high diversity of species compared to temperate, colder, or alpine regions. However, very high temperatures can adversely affect plant growth, causing seedling mortality, cracking of stems, etc., while excessively low temperatures below 5°C result in reduced growth and frost injury. Both of these phenomena are harmful to plants.

5.3.4 SOIL TEMPERATURE

It affects microbial activities in the soil as well as absorption of moisture and nutrients. Soil temperature is one of the significant climatic factors that influence germination, emergence, and subsequent establishment of seedlings, under appropriate soil and irrigation situations.

5.3.5 SNOWFALL

Heavy accumulation of snow on the canopy causes bending of bole and branches of the tree due to weight, rupture of tree crown and branches, erosion of soil and uprooting of trees results due to sliding snow.

5.3.6 MOISTURE

Moisture is present in the atmosphere as well as in soil. The moisture source in the atmosphere is evaporation from water bodies and wet soil as well as transpiration from plants. Plants obtain water from the soil via the roots.

Water is required by all living organisms since it is the universal solvent and is found in all living cells, protoplasm, cell walls and vacuoles. It is an important constituent of carbon/food assimilation process during photosynthesis, respiration, and ascent of sap in stems resulting in the absorption of nutrients and their transfer to various organs.

5.3.7 WIND

Strong winds have the capacity to break the leading shoots or side branches and sometimes bring whole woods to the ground. In warm areas, dry winds regularly reduce the moisture of a region to a harmful level which poses a threat to the forest; cold and moist winds may also reduce the temperature, and thus, obstruct the healthy growth and development of the trees.

5.4 FORESTS BY CLIMATIC DOMAIN AND ECOLOGICAL ZONE

Forests can be found from arid regions to wetlands and tidal areas. The largest portion of the forest (45%) is found in the tropics, followed by the boreal (27%), temperate (16%), and subtropical (11%) domains. There are five major climatic domains in the world (Figure 5.1). These categories are further divided into 20 worldwide ecological terrestrial zones. Around 10 global ecological zones were found to have had a net decrease in tree cover between 1992 and 2015, while the other 10 saw a net increase. The tropical rainforest, which covers a large portion of Central Africa (Congo), the Amazon Basin (Brazil), Indonesia, and Papua New Guinea, saw the greatest reduction in tree cover, but the boreal tundra woodland, which is found in Canada and Russia, saw the greatest rise.

Brazil has the highest number of tree species (8,847), followed by Colombia (5,868 spp.) and Indonesia (5,716 spp.). The 10 nations with the largest reported number of indigenous trees contain a total of 15,748 endemic tree species. The countries having the maximum endemic trees are Brazil, Madagascar, Australia, China, Indonesia, Malaysia, New Caledonia, Mexico, Colombia, Papua New Guinea, and the Philippines. The six categories—extinct, threatened, near threatened, not threatened, data inadequate, and not evaluated—were applied to every tree species. Combining all the results, 17,510 tree species which comes at 29.9% of tree species are considered threatened. The larger mega diverse countries (Brazil, China, Indonesia, etc.) having a higher number of tree species also have a large number of threatened species. Madagascar also stands out as one of the countries with the highest number of threatened tree species (Beech et al., 2021).

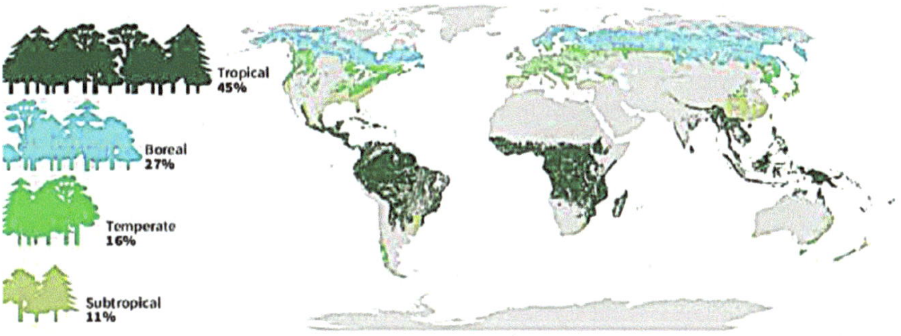

FIGURE 5.1 World forest map, 2020.

Source: Adapted from FAO 2020, from data by United Nations World Map 2020

5.5 FOREST HEALTH

Forest health is a very important driving factor of an ecosystem, and the phenomenon of climate change is directly and indirectly dependent on it. Forest health refers to the conditions of forest productivity which directly or indirectly satisfy human needs and, by recurrence, resilience, persistence, and biophysical processes, lead to sustainable ecological conditions. It can also be defined as a condition of forest ecosystems that maintains their complexity along with meeting the requirements of human needs (Sampson et al., 1994). Forest health is mainly maintained by sustainable biodiversity on the earth's surface along with socio-economic improvement and better livelihood, predominantly in rural areas. Forest health is mainly affected by fragmentation of forest areas in our terrestrial ecosystem, which is mainly concerned with forest degradation and deforestation, basically by a change of forestland into farmland and for other industrial or commercial demands. Therefore, assessment and regular monitoring of forest health are necessary.

Forest health monitoring and implementation programs have multiple aspects. In the current scenario, forest health monitoring has been incorporated into programs related to issues such as carbon sequestration, biodiversity conservation, climate change, SDG, long-term ecological research, international processes dealing with sustainable forest management programs, and long-term resource management. Forest health monitoring, diagnosis, implementation, and evaluation are areas of concern for resource managers, politicians, decision-makers, technicians, extension workers, scientists, and humankind.

5.6 GLOBAL SCENARIO OF CLIMATE CHANGE

A statistically significant variation in the climate's mean state or in its variability that lasts for a long time—typically decades or longer—is referred to as climate change. In the recent few decades, changes in climate are faster, widespread, intensifying, and unprecedented. It is evident that human activities are the major cause of climate change, making extreme climate events more frequent and severe, including heavy rainfall, heat waves, cyclones, floods, and droughts. Presently, climate change is a major issue affecting all regions on Earth's surface, in several ways.

The atmosphere, oceans, cryosphere, and biosphere have changed, which is proof that the world has warmed. Between 1850 and 1900, the global mean surface temperature (GMST) increased by 0.85°C, and between 1900 and 2020, it increased by 1.09°C. Over the past few decades, indicators of climate change have increased at levels which have not been seen in centuries to millennia, and the changes in climate are at unprecedented rates since the last 2000 years. Greenhouse gases (GHGs) have been rising at rates since 1850 that haven't been seen in at least the last 800,000 years in centennial time frames. Carbon dioxide (CO_2), methane (CH_4), and nitrous oxide (N_2O) concentrations all increased from 1750 to 2019 by corresponding amounts of 131.6 ± 2.9 ppm (47.3%), 1,137 ± 10 ppb (156%), and 62 ± 6 ppb (23.0%), respectively. CO_2, CH_4, and N_2O concentrations in the atmosphere have reached values of 409.9 ± 0.4 ppm, 1866.3 ± 3.3 ppb, and 332.1 ± 0.4 ppb, respectively, in 2019. As a result of compound pressures and events, climate change is having a growing impact on marine, freshwater, and terrestrial ecosystems and ecosystem services, as well as on water and food security, settlements and infrastructure, health and well-being, economies, and culture. The frequency and intensity of forest fires have increased due to climate change, aridity, and drought. Fire, altered land use patterns, notably deforestation, and climate change combine to have a direct influence on people's health, ecosystem function, forest structure, food security, and the livelihoods of communities who depend on natural resources.

Climate change impacts are negatively affecting agriculture, forestry, fisheries, and aquaculture. The post-industrial revolution climate change scenario causing global warming, aberrant weather conditions, and weather extremes is resulting in losses equivalent to at least 5% of global GDP each year. IPCC indicated a probability of a 10–40% loss in crop production in India due to droughts and floods. Rising temperature is negatively affecting crop and grassland productivity as well as quality. Warming of colder climates has increased tree mortality and forest area disturbances in many boreal and temperate forest biomes, also negatively affecting provisioning services. Climate change is also causing the loss of native species, an increase in disease and pest infestation, inversion of weed species, and mass mortality events of vegetation and animals, resulting in the first climate-driven ecosystem restructuring, extinctions, increases in areas burnt by wildfire, and declines in biodiversity and other ecosystem services. Climate change also has effects on the physical aspects of water security through increasing water scarcity and exposing more people to water-related extremes like floods and droughts; consequently, existing water-related vulnerabilities brought on by other socio-economic factors would be exacerbated.

5.7 MAJOR CAUSE OF CLIMATE CHANGE

5.7.1 ANTHROPOGENIC GREENHOUSE GASES (GHG)

As with greenhouse glass, certain gases in the Earth's atmosphere trap heat and prevent it from escaping back into space, hence called the greenhouse gases. Greenhouse gases in the atmosphere include water vapor, carbon dioxide (CO_2), methane (CH_4), nitrous oxide (N_2O), fluorinated gases, and sulfur hexafluoride (SF_6). These gases in the atmosphere emit radiation in all directions, resulting in warming the atmosphere of the planet. This phenomenon is known as the greenhouse effect. Most of these gases occur naturally and are necessary for maintaining the temperature on Earth, but the exponential increase in population and subsequent human activities have increased the concentrations of several of them in the atmosphere. CO_2 is the greenhouse gas most produced by anthropogenic activities and has a 64% share in global warming.

5.7.2 POPULATION INCREASE

The current world population is 7.97 billion as of August 2022. The exponential growth of the human population increases anthropogenic GHGs emissions through the burning of fossil fuels, clearing of forests, forest degradation, increased material extraction, industrial agriculture, transportation, and other manufacturing processes.

5.7.3 INDUSTRIALIZATION

By burning fossil fuels to create energy to produce goods like cement, iron, steel, electronics, plastics, clothing, and other items, industries produce emissions. Mining, extracting petroleum, metals, and other industrial processes also release harmful gases, as does the construction sector. Heavy machinery used in the manufacturing process runs on coal, gas oil; and some materials, like medical equipment, and plastics, are made from hazardous chemicals sourced from fossil fuels. The manufacturing industry is one of the biggest contributors to greenhouse gas emissions.

5.7.4 TRANSPORTATION

Main sources of transportation like trains, buses, trucks, cars, ships, and planes run on fossil fuels, making transportation a main contributor of greenhouse gases, especially carbon dioxide, carbon monoxide, and nitrogenous gas emissions. Due to the internal combustion engines' use of petroleum-based fuels including gasoline, diesel, kerosene, and other gasoline, road transportation makes up the majority. Nonetheless, at this time, emissions from ships and aircraft are still rising. About 25% of the world's energy-related carbon dioxide emissions come from transportation.

5.7.5 DEFORESTATION

The United States Environmental Protection Agency defines deforestation as the permanent removal of standing forests. Deforestation results in soil erosion, desertification, flooding, increased concentration of greenhouse gases in the atmosphere, which become a host of problems for indigenous people. Around 25% of the world's greenhouse gas emissions are caused by deforestation, along with agriculture and other changes in land use.

5.7.6 EXPLOITATION OF NATURAL RESOURCES

The indiscriminate use of fossil fuels, including coal, gas, and oil, is the biggest contributor to climate change. As fossil fuels are used, carbon dioxide is released into the atmosphere, warming the earth.

5.7.7 DESERTIFICATION

Land degradation in arid, semi-arid, and dry sub-humid environments caused by a variety of sources such as deforestation, faulty agricultural practices, and other human activities, known as desertification, pose a serious threat to the environment and human welfare (Veron et al., 2006).

5.8 IMPACTS OF CLIMATE CHANGE

- **Increased Temperature:** The period from 2011 through 2020 was the warmest on record, with the average global temperature in 2019 rising by 1.1°C from pre-industrial values. At the moment, human-caused global warming is accelerating at a rate of 0.2°C per decade.
- **Rise in Sea Level:** Since 1901, the sea level has risen by 0.20 m, and it has recently begun to increase more quickly. From 1901 to 1971, the average annual rate of sea level rise was 1.3 mm, and between 2006 and 2018, it increased to 3.7 mm.
- **Ocean Acidification:** The pH of ocean water has decreased by 0.1.
- **Warmer Oceans:** From 1850 to 2020, there is a 0.88°C increase in the temperature of the oceans.
- **Shrinking Arctic Sea Ice:** Both in the summer and the winter, there is less ice in the Arctic. Arctic sea ice extent dropped by a decadal mean from 6.23 million km^2 in 1979–1988 to 3.76 million km^2 in 2010–2019.
- **Melting of Glaciers:** The loss of ice was 147 Gt yr^{-1} between 2002 and 2011.
- Bleaching of coral reefs.
- Loss of planktons.
- Increased population resulting in habitat loss of several species.
- Spread of diseases.
- Degradation of ecosystem and loss of biodiversity.
- Problems related to agriculture and food security.
- Change in rainfall patterns.
- Extreme weather and increased flooding, drought, hurricanes, cyclones, etc.

5.8 IMPACTS OF CLIMATE CHANGE ON FOREST HEALTH

The response of trees, plants, and forests as a whole to global warming is multifaceted. The continuing changes in Earth's atmosphere and environment will have deep effects on future forest area distribution, composition of forests, growth, and development of trees and other plants as well as on the microorganisms and animals that live and depend on forests. Total loss of forest area due to climate change can be accounted for 4% worldwide. It is predicted that changes in temperature and rainfall patterns will cause alterations in the life cycles of insects and pathogens, increasing their host ranges and virulence, and trees will become more susceptible to biotic factors like insects, pests, diseases, and abiotic factors like forest fires, posing increased stress on forests. In addition to these effects, forest health is also threatened by invasive species. Model-based projections suggest that climate change presents both an opportunity and a threat to the forests in the state. The chance arises from the projections of increased forest productivity, increased biomass or carbon stock, and increased soil organic carbon in some parts of the earth's surfaces. On the other hand, the negative risk comes from the projections of shifting cultivation. Shifting cultivation, in combination with the lack of biodiversity and disturbed and fragmented habitats, poses serious threats to forest ecosystems. The IPCC (2014) suggests that phenology, growth rates, productivity, shifting distributions of species, and biomes are the most prominent indicators of changes in forest ecosystems in response to changing climate.

Bala et al. (2013) estimated that Net Primary Productivity (NPP) over the Indian region during the period of 1982 to 2006 has increased by 3.9% per decade, driven mainly by increased atmospheric CO_2 concentration in different ecosystems. However, it is also predictable that due to changes in precipitation patterns, temperature, wind, and land degradation, increasing concentrations of atmospheric carbon dioxide are less likely to contribute positive modifications in natural and planted forest areas. Forest fires, insect pest outbreaks, diseases, injurious plants, damage by wind, and other extreme incidents are resulting in significant economic damage to forests. In a continuously changing climate, maximum direct and indirect risks are caused by more frequent extreme events, affecting forest productivity, timber supplies, quality of timber, market prices, and insurance costs (De Walle et al., 2003). On some sites, more adapted species will replace the earlier species that are no longer suited to the changing climate. Also, the plantation of forests can be relocated to many new regions according to more suitable climates. In general, it is observed that the plantation of forests and related forest operations tend more toward higher latitudes of subtropical climate regions, particularly from some tropical site regions, if they continue to warm substantially due to climate change.

5.8.1 IMPACTS DUE TO ENVIRONMENTAL STRESSES

Plants constantly face a wide range of environmental stresses which create restrictions to growth and development. The environmental stresses faced by forests can be classified as abiotic and biotic stress. Abiotic stresses include drought, flood, salinity, extremes in temperature, radiation, and heavy metals, which are leading factors causing the loss of forests globally. Although there are still many unknown factors, it is anticipated that extreme occurrences like heat waves and severe droughts would occur more frequently in a warmer environment. Biotic stress, on the other hand, includes attacks by living organisms, pathogens such as bacteria and fungi, insects, pests, nematodes, oomycetes, other microorganisms, and herbivores. Infections produced due to these organisms are responsible for major loss of forests worldwide.

5.8.2 IMPACTS DUE TO CHANGES IN TEMPERATURE

The rise in the global temperature will result in the spread of unpredictable and dangerous weather outcomes like floods, heat waves, droughts, and hurricanes. Changes in temperature and precipitation patterns are quite likely to have a significant direct influence on both natural and artificial forests. Many biogeographical models show that the potential vegetation for the CO_2 climate will move from boreal zones to polar regions by at least 500 km. According to some dynamic vegetation models and plant equilibrium models, because of temperature changes, vegetation will shift towards newly accessible regions with favorable climate conditions, which will eventually lead to the replacement of forests and an expansion of up to 50% of the current tundra area. Yet, there is fear that this forest migration could result in a significant loss of natural forests, increased deforestation at the southern edge of the boreal forests, and an associated high carbon pulse. There are indications that some forest insects, like the mountain pine beetle, have already changed their normal habitats as a result of warming temperatures.

5.8.3 IMPACTS DUE TO CHANGES IN PRECIPITATION

Since the 1950s, many regions have experienced an increase in precipitation intensity. Substantially more people (709 million) reside in areas where annual maximum precipitation has increased than in regions where it has decreased (86 million). At the same time, more people (700 million) have been experiencing longer dry spells since the 1950s, leading to compound hazards related to both warming and precipitation extremes in a major proportion of the world. Many major forest insects, like the gipsy moth, are more sensitive to changes in precipitation.

5.8.4 IMPACTS DUE TO CHANGES IN CO_2 CONCENTRATION

Global CO_2 concentration in the atmosphere has increased from 256 ppm in the pre-industrial era to 410 ppm today. Multiple studies have confirmed increasing net primary productivity and biomass productivity due to rising CO_2 concentrations in the atmosphere, including in South Asia (Bala et al., 2013).

5.8.5 IMPACTS DUE TO CHANGES IN SALINITY

Due to raised sea level, there is and will be an increase in soil salinity in the coastal areas. Salinity and osmotic stress affect germination and seedling development. Thus, an increase in salinity is leading to a reduction of forest population, growth, and development. Therefore, an increase in salinity has become the most important abiotic factor limiting plant growth and productivity.

5.8.6 IMPACTS DUE TO CHANGES IN WINDS

Even in the absence of fires or insect damage, changes in the frequency of extreme events like strong winds, winter storms, droughts, etc., can cause significant loss to commercial forestry. Such effects of climate change extremes are area specific and include reduced access to forestland areas, increased costs for road construction, facility maintenance, direct damage to trees by wind, frosts, snow, hail, ice, effects of moister winters and early melts on logging, etc. Very high wind speed can damage trees through trunk breakage, breaking of branches, crown loss, complete stand destruction, especially caused in fast-growing stocks of trees in a warmer climate region. For instance, in January 2005, Hurricane Gudrun with maximum gusts of 43 m/s damaged 60 million m^3 of timber in Sweden.

5.8.7 IMPACTS DUE TO CHANGES IN FOREST FIRES

In some forest ecosystems, natural fires are crucial for preserving ecosystem dynamics, biodiversity, and production. It is also an important and broadly used tool to meet land management goals. Most forest fires are caused by humans, but sometimes they get out of control. Every year, intentionally set forest fires burn millions of hectares of forests, agricultural stubble, and other types of vegetation. A global inventory of the total forest area affected by forest fires identified approximately 67 million hectares burnt per year between 2003 and 2012 (van Lierop et al., 2015). Forest fires affected 98 million hectares of land in the United States in 2015 (FAO, 2020). These fires mostly occur in warmer or tropical regions, affecting about 4% of the forest area. The most common fire phenomena, leaving more than two-thirds (66%) of the total forest area burnt, were located in South America and Africa. About 90% of forest fires are readily contained and account for only 10% or less of the total area burnt. However, the remaining 10% of forest fire events account for the other 90% of the burnt area. These dramatic and high-profile wildfire events, such as those occurring in Australia, Brazil, Greece, Russia, and the United States of America (California) in 2018 and 2019, caused huge loss of human, plant, and animal lives, wealth of nations, infrastructure, as well as bringing massive environmental and economic damage, both in terms of destroyed natural or artificial resources and the costs of suppression. Longer fire seasons and more intense fires are predicted to spread throughout a large portion of the earth due to climate change, even in some locations where fire has not historically been a major issue. Although it is impossible to completely prevent forest fires, it is possible to significantly reduce their frequency and effects by implementing integrated fire management and fire-smart forest management practices, as well as by taking sociocultural realities and ecological imperatives into account in the landscapes where fire occurs (FAO, 2006).

5.8.8 IMPACTS DUE TO LOSS OF HABITAT

Habitat loss is presently the greatest threat to tree species. Habitat loss can be defined as the total removal of vegetation or forest area, as well as fragmentation and degradation of persisting habitats. Global forest cover has shrunk by around 40% over the previous three centuries, and 29 countries have lost more than 90% of their forest cover (Newton, 2021).

5.8.9 IMPACTS DUE TO MIGRATION OF SPECIES

The climate change period from the pre-industrial era displayed that tree species migrate along with climatic zones, but their moving velocity is projected to be 4–200 km per 100 years and turn out to be at high risk of extinction because of climate change (Davis, 1989).

5.8.10 *IMPACTS DUE TO INVASIVE SPECIES*

The worldwide health, sustainability, and productivity of natural and planted forests are under threat from invasive species (exotic insect pests, pathogens, vertebrates, and plants), as well as from outbreaks of native insect pests and illnesses. Around 1,356 tree species listed on the IUCN Red List are considered threatened by invasive species. One invasive pest species in North America, Russia, and Eastern Europe is the emerald ash borer (*Agrilus planipennis*), which is a wood-boring beetle in East Asia.

5.8.11 *IMPACTS DUE TO INCREASED PESTS AND DISEASES*

A vast variety of pests and diseases that are transmitted by both natural and man-made means affect different tree species. Over 35 million hectares of forests are harmed each year by outbreaks of forest insect pests (FAO, 2010b). For instance, between 1998 and 2002, insects negatively impacted 14 million hectares of forest in Canada and 5 million ha of forest in the United States, totaling 1.7% and 2.9% of the continent's total forest area, respectively (GFRA, 2005). Climate change is affecting survival opportunities for many pests and diseases in new environments. Fraxinus (ash) trees have become critically endangered as a result of predation by a beetle which was accidentally introduced through infested shipping pallets (Barstow et al., 2018). A further threat to ash species across Europe is the disease Ash dieback, which is caused by the invasive fungal pathogen (*Hymenoscyphus fraxineus*) which blocks the water transport systems in trees, causing crown dieback, lesions, and the eventual death of the tree. Defoliation and growth loss, wood degradation, and severe forest diebacks are only a few of the repercussions.

5.9 IMPACTS DUE TO OTHER FACTORS LIKE

- Climate change due to decrease of forest density and productivity;
- Climate change due to massive extinction of wildlife;
- Damage to forests due to heavy rainfall;
- Increase soil erosion due to high frequency of rains cause loss of forest;
- Land degradation due to increase areas affected by drought;

- Increase livestock deaths/wildlife due to droughts;
- Damage of tree crops due to cyclonic storms mainly in coastal area.

5.10 CONCLUSION

Due to climate change, there is an increase in temperature, global warming, humidity, vapor pressure deficit, evaporation, changes in rainfall patterns and frequency, delays in monsoon, etc. These factors directly or indirectly have a deep impact on forest health. Forests are susceptible to climate change, affecting their structure, function, composition of deciduous/evergreen trees, processes, productivity, forest biomass, water and nutrient cycling, and resilience of forest ecosystems. Climate change also increases the risk of forest fires, pests and diseases, the emergence of pests, and the spread of invasive species, which have harmful effects on forest health.

KEYWORDS

- **climate change**
- **deforestation**
- **ecosystem**
- **forest ecosystem**
- **forest health**
- **global warming**
- **tree species**

REFERENCES

Bala, G., Joshi, J., Chaturvedi, R. K., Gangamani, H. V., Hashimoto, H., & Nemani, R., (2013). Trends and variability of AVHRR-derived NPP in India. *Remote Sens., 5*, 810–829.

Beech, E., Rivers, M. C., Oldfield, S. F., & Smith, P. P., (2021). *Global Tree Search Download 1.5.* https://doi.org/10.13140/RG.2.2.33593.90725.

Beech, E., Rivers, M., Oldfield, S., & Smith, P. P., (2017). Global tree search: The first complete global database of tree species and country distributions. *J. Sustain. For., 36*, 454–489. https://doi.org/10.1080/10549811.2017.1310049.

BGCI, (2021). *State of the World's Trees* (p. 18). BGCI, Richmond, UK.

Brun, P., Psomas, A., Ginzler, C., Thuiller, W., Zappa, M., & Zimmermann, N. E., (2020). Large scale early-wilting response of Central European forests to the 2018 extreme drought. *Global Change Biology*.

Buras, A., Rammig, A., & Zang, C. S., (2020). Quantifying impacts of the 2018 drought on European ecosystems in comparison to 2003. *Biogeosciences, 17*, 1655–1672.

Davis, M., (1989). Lags in vegetation response to greenhouse warming. *Climatic Change, 15*(1), 75–82. https://doi.org/10.1007/BF00138846.

De Walle, D. R., Buda, A. R., & Fisher, A., (2003). Extreme weather and forest management in the mid-Atlantic region of the United States. *Northern Journal of Applied Forestry, 20*(2), 61–70.

FAO and UNEP, (2020). *State of the World's Forests 2020: Forestry, Biodiversity and People*. FAO and UNEP, Rome, Italy.

FAO, (2006). *Fire Management Global Assessment 2005* (p. 151). Food and Agriculture Organization Forestry Paper.

FAO, (2010b). *Global Forest Resources Assessment 2010– Main Report* (p. 163). FAO Forestry Paper No. Rome. http://www.fao.org/3/i1757e/i1757e00.htm.

FAO, (2020). *Global Forest Resources Assessment 2020 – Main Report*. FAO (p. 15). https://doi.org/10.4060/ca9825en (accessed on 6 May 2024).

GFRA, (2005). *Global Forest Resources Assessment-2005*. Food and Agriculture Organization, Rome Food and Agriculture Organization Forestry Paper 147.

Hubau, W., Lewis, S. L., Phillips, O. L., Affum-Baffoe, K., Beeckman, H., Cuní-Sanchez, A., et al., (2020). Asynchronous carbon sink saturation in African and Amazonian tropical forests. *Nature, 579*(7797), 80–87. https://doi.org/10.1038/s41586-020-2035-0.

IPCC, (2014). *Climate Change 2014: Impacts, Adaptation, and Vulnerability; Part A: Global and Sectoral Aspects* (1st edn.). 47 Contribution of Working Group II to the Fifth Assessment Report of the Intergovernmental Panel on Climate 48 Change, Cambridge, UK and New York, NY, USA: Cambridge University Press.

ISFR, (2021). *India State of Forest Report, Forest Survey of India*. Ministry of Environment & Forests, Dehradun, India.

Kurz, W. A., Stinson, G., & Rampley, G., (2008). Could increased boreal forest ecosystem productivity offset carbon losses from increased disturbances. *Phil. Trans. Roy. Soc. B., 363*, 2285–2299.

Lesk, C., Coffel, E., D'Amato, A., et al., (2017). Threats to North American forests from southern pine beetle with warming winters. *Nature Clim. Change, 7*, 713–771. https://doi.org/10.1038/nclimate3375.

Newton, A. C., (2021). *Ecosystem Collapse and Recovery* (1st edn.). Cambridge University Press. https://doi.org/10.1017/9781108561105.

Sambarajua, K. R., Carrollb, A. L., & Aukemac, B. H., (2019). Multiyear weather anomalies associated with range shifts by the mountain pine beetle preceding large epidemics. *Forest Ecology and Management, 438*, 86–95.

Sampson, R. N., & Adams, D. L., (1994). *Assessing Forest Health in the Inland West* (p. 461). Food Products Press, Binghamton N. Y.

Schuldt, B., et al., (2020). A first assessment of the impact of the extreme 2018 summer drought on Central European forests. *Basic Appl. Ecol.* doi: 10.1016/j.baae.2020.04.003.

Seidl, R., et al., (2020). Globally consistent climate sensitivity of natural disturbances across boreal and temperate forest ecosystems. *Ecography, 43*, 967–978.

Sommerfeld, A., et al., (2018). Patterns and drivers of recent disturbances across the temperate forest biome. *Nat. Commun.*, *9*, 4355.

Sullivan, M., Lewis, S., Affum-Baffoe, K., Castilho, C., Costa, F., Sanchez, A., et al., (2020). Long-term thermal sensitivity of Earth's tropical forests. *Science, 368*(6493), 869–874. https://doi.org/10.1126/science.aaw7578.

Van, L. P., Lindquist, E., Sathyapala, S., & Franceschini, G., (2015). Global forest area disturbance from fire, insect pests, disease and severe weather events. *Forest Ecology and Management*, *352*, 78–88.

Veron, S. R., Paruelo, J. M., & Oesterheld, M., (2006). Assessing desertification. *Journal of Arid Environments, 66*(4), 751–763. https://doi.org/10.1016/j.jaridenv.2006.01.021.

CHAPTER 6

Fate of Plant Species in Response to Changing Climatic Scenarios with Special Reference to Habitat Suitability in the Himalayan Region

QAMER RIDWAN, NAHILA ANJUM, SHAISTA RASHID, and MOHD. HANIEF

Department of Botany, Baba Ghulam Shah Badshah University, Rajouri, Jammu and Kashmir, India

ABSTRACT

Plant species, particularly those growing in mountains, are extremely susceptible to climatic alterations, as their distribution is predominantly determined by climatic factors. Changes in the climatic conditions influence the habitat suitability of plant species, and the species may adapt to it by changing their habitats in order to better match pre-existing climatic conditions. The Himalayan region, which symbolizes the greatest bioclimatic altitudinal gradient and provides the highest altitudinal boundaries for vascular plant species in the world, is extremely susceptible to climate change because of predominant topographic inclinations. In the region, suitable habitat areas of many plant species are dwindling due to warming and changes in precipitation regimes. It is anticipated that the currently suitable habitats of such species will become unsuitable and vice versa in the near future. Furthermore, many high-elevation plant species that are susceptible to high temperatures are relocating to higher altitudes. To comprehend the behavior of the species and make plans for the protection of biodiversity, it is imperative to monitor the distribution and habitat suitability of plant species with respect to climate

Sustainable Forest Resources Management: Issues and Implications. Jahangeer A. Bhat, Gopal Shukla, Manmohan J. Dobriyal, Sumit Chakravarty, A. Arunachalam, and Rainer W. Bussmann (Eds.)

change. For that purpose, the species distribution modeling (SDM) technique is quite helpful. This chapter explores the impact of climate change on the habitat suitability of plant species growing in the Himalayan region and the role of SDM in forecasting species' response to climate change. Additionally, implications for the conservation and management of plant species in the region are discussed.

6.1 INTRODUCTION

The climate of the Earth system has changed throughout the geological past, and the changes can be seen in the form of fluctuations in temperature and precipitation, and rapidity in the prevalence of other climatic extremes. Climate change has been proven to be one of the main factors that have a detrimental impact on biodiversity and related products and services (MEA, 2005). Studies have demonstrated that climate change has caused range shifts and sporadic extinctions of various plant species in different ecological settings, including mountains (Grabherr et al., 1994; Parmesan & Yohe, 2003; Thuiller et al., 2005). Rapid warming, increased precipitation, and the lengthening of the growing season are all affecting mountain ecosystems (Shrestha et al., 2012). Plant species distribution in mountainous regions is largely governed by climatic factors that predominate at high altitudes, thus making floras of such areas susceptible to climate change (Korner, 1999; Wani et al., 2022a). Along altitudinal and latitudinal courses (Wilson et al., 2005; Hickling et al., 2006), warming-induced upward shifts of species take place. As a result, mountains are thought to be especially excellent observation locations for tracking the impacts of climate change (Korner, 1999; Diaz et al., 2003).

The Himalayan Mountain range, which is the world's longest bioclimatic elevational gradient and home to the greatest elevations (6,400 m) of any vascular plant species (Grytnes & Vetaas, 2002), was created as a consequence of the Eurasian and Indian plates colliding gently in the early Eocene and more forcefully in the Miocene (Gansser, 1964; Rowley, 1996). The step-wise development of the Himalayas during a 45–60-million-year timeframe offered floral and faunal components fresh possibilities to colonize the rapidly developing landscapes and eventually diversify into distinct ecological units (Das, 1966; Singh & Singh, 1987). Many fragile endemic species are found in this youngest and one of the key mountain ranges of the world (Price, 1981; Liu & Zhang, 1998). Due to its prevalent geomorphic propensities, the Himalayan region is particularly prone to natural disasters, and there is ongoing fear regarding the detrimental consequences of climate change, including unexpected floods, dry spells, and landslides, which ultimately lead to the loss of biodiversity and pose a threat to food security (Cruz et al., 2007; Xu et al., 2009). Based on observations from the ground and space, it is clear that the vegetation in the Himalayan bio-zones has changed in recent years. There is evidence that in the Himalayan region, many plant species that are sensitive to elevated temperatures are moving to higher altitudes; however, the degree of shift varies greatly depending on the species and the location. By creating protected zones across the region, significant measures have lately been taken for the *in-situ* protection of species. However, to assess the relative susceptibility of species to climatic factors in the region, long-term and species-specific ecological investigations are still insufficient. Therefore, such empirical investigations with a wider perspective are necessary to establish suitable biodiversity management policies in the region (Yadav et al., 2021). To understand the behavior of species and prepare for the preservation of floristic diversity in the Himalayas, surveillance of floristic diversity with respect to climate change requires special consideration (Samant, 2021). A promising source of information on the distribution of species for emphasizing conservation and management measures is provided by projections using spatial models such as species distribution models (SDMs), linking species occurrences or richness to environmental parameters (Ferrier et al., 2002; Jetz et al., 2012). This chapter explores the impacts of climate change on the habitat suitability of plant species growing in the Himalayan region and the role of SDM in predicting species response to climate change. Further implications for the conservation and management of plant species in the region are discussed.

6.2 CLIMATE CHANGE IMPACTS HABITAT SUITABILITY OF PLANT SPECIES GROWING IN THE HIMALAYAN REGION

The vegetation of any area and the climate that prevails there are in constant balance. However, any change in climatic factors influences the colonization and regeneration dynamics of plant species, which induces a competitive shift in the natural range of species that has an impact on local and provincial species diversity (Bell, 2001; Camarero & Gutierrez, 2004; Hofmann, 2012; Burrows et al., 2014). Climate change may lead to alterations in species assemblages with the introduction of new species, including invasive ones and even disease-causing agents (Diez et al., 2012). Many plant species in mountainous regions are anticipated to relocate to considerably higher altitudes because of changing climatic conditions (Braunisch et al., 2014). Despite evidence supporting this projection (Alexander et al., 2015), efforts to estimate how plant species of mountainous ecosystems would respond to climatic alterations have produced conflicting results. For example, some models (Chen et al., 2011) anticipate range expansions, while others predict range reductions (Zhu et al., 2015). Given that climatic alterations have already had an impact on the habitats of species, including those of various forest ecosystems globally (Koralewski et al., 2015), it is anticipated that it will also have an influence on how plants are distributed in the Himalayan region (Mainali et al., 2015). Altitudinal modifications in the flora of the Himalayan Mountain chain have been observed recently (Schickhof, 2005; Negi, 2012; Rawat, 2012; Telwala et al., 2013; Schickhof et al., 2015). It is becoming more and more likely that the continued warming that is changing the dynamics of the upper ecotonal vegetation would significantly alter the biodiversity of the vulnerable Himalayan environment. The studies carried out thus far in the Himalayan region, although fragmentary, demonstrate that the plant species' responses to the climatic changes are inconstant throughout various ecological configurations (Gaire et al., 2014; Schickhof et al., 2015; Yadava et al., 2017); necessitating the investigation of species distribution patterns throughout various ecological configurations.

6.2.1 IMPACT ON TREELINE SPECIES

One of the most obvious vegetation limits is the high-elevation boundaries of forests often termed as treeline (Korner, 1998). The high-elevation treeline is seen to be the perfect early-warning indicator that reacts spatially, physically, and compositionally to climatic changes (Kullman, 2007; Kirdyanov et al., 2012). In the past, vegetation in the Himalayan treeline sector had changed with respect to long-term climatic alterations (Wani & Pant, 2021). According to several dendroecological studies, trees near the treeline frequently respond to climate warming by increasing the recruitment of tree density and also extending the position of the treeline upwards (Camarero & Gutierrez, 2004; Batllori & Gutierrez, 2008; Kullman & Oberg, 2009; Leonelli et al., 2011). Although not constant and ubiquitous, treeline displacement in response to climate change is commonly seen across the world (Gaire et al., 2017). It is claimed that plant species would move to higher altitudes in response to shifting climatic circumstances, and treelines have been favored for supervising the influence of climatic alterations (Pauli et al., 2003). In the Himalayas, uphill treeline movement due to changing climatic conditions is not a one-way reaction; instead, it depends greatly on the magnitude and methodology of the study, the sites and species included, and the area's climatic history (Gaire et al., 2017). In the western Himalayas, a significant part of the upper treeline is occupied by the Himalayan pine (*Pinus wallichiana*). This species is found sporadically all along the Himalayas (except Sikkim) in the temperate regions within an altitudinal range of 1,800–3,000 m asl, reaching up to Bhutan towards the east (Raizada & Sahni, 1960). However, in the west, it sometimes grows up to 3,900 m asl or even higher, combining with *Juniperus* and *Betula utilis* in the temperate altitude zone (Troup, 1921). Investigations on Himalayan pine range change in the monsoon and monsoon shadow areas of Uttarakhand and Himachal Pradesh over 12 high-altitude sites indicated site-specific variance in the frequencies of altitudinal shifts in the 20th century (Yadava et al., 2017). The results clearly demonstrated that the Himalayan pine treeline has moved upward across a variety of locations, albeit at variable rates (11–54 m/10 years), mostly as a result of site-specific biotic and microclimatic variables. Similarly, a rise in plant density and an upward transition in the *Abies spectabilis* and *B. utilis*-dominated treeline ecotones were seen in the central Nepal Himalaya (Gaire et al., 2014). Another study on *B. utilis* predicted that some areas that are presently less suitable for *B. utilis* will become suitable in the future (2050 and 2070) due to changing climatic circumstances. Meanwhile, few regions which are predicted to be greatly suitable in the existing climatic circumstances will turn out to be less suitable. According to predictions, in the future, a highly appropriate habitat for *B. utilis* will shift towards the eastern Himalayas, with the western Himalayas becoming less favorable. In projected scenarios, the climatic domain of this treeline species is shifting at a moderate pace (Hamid et al., 2019).

6.2.2 IMPACT ON SUB-ALPINE AND BOREAL PLANT SPECIES

Climate change has resulted in compositional modifications of sub-alpine forest seedlings and saplings of species like *Quercus semecarpifolia*, *Abies pindrow*, etc., and elevational shifts of *Juniperus polycarpos* in (Lahaul Valley, Himachal Pradesh, Pinus gerardiana in Pooh Block of district Kinnaur and Pangi Valley, Himachal Pradesh), *Rhododendron campanulatum, P.*

wallichiana, Populus ciliata, A. pindrow, etc., have been reported (Samant, 2021). Similar to this, the presence of *Taxus wallichiana* in the Lahaul Valley, Himachal Pradesh as seedlings and saplings indicated that this species had only recently arrived there. This demonstrates unequivocally how altering environmental factors, in particular climate change, are causing the species to move up or down in elevation (Lal et al., 2020). The existence of seedlings and saplings of *J. polycarpos* in avalanche-prone locations and above the altitudinal range of the species in upper areas in Lahaul valley is conclusive proof of the influence of climate change (Singh & Samant, 2020). Additionally, the presence of *P. wallichiana* in the elevational zone of *B. utilis* and *Rhododendron campanulatum* in Sangla Valley and Nanda Devi Biosphere Reserve, Himachal Pradesh (Samant & Joshi, 2003) gives the confirmation of effects of the climatic alterations (Samant, 2021). In the central Himalayas, Chakraborty et al. (2016) investigated the possible impact of climatic alterations on the future trajectories of four prominent tree species, namely *P. roxburghii, Q. leucotrichophora, Q. semecarpifolia,* and *Q. floribunda.* All future climatic projections revealed both the expansion and contraction in the extent of suitable habitats for these species. The overall area that the tree species were anticipated to cover seemed to stay roughly stable, but there were shifts in the geographical distribution of low, medium, and high-appropriate regions. For *P. roxburghii* (pine), the highest growth was seen from 2375.04 km^2 to 2671.32 km^2 (by 2070) for high appropriate regions, while a small increase was observed for medium suitable sites. Conversely, less suitable *P. roxburghii* regions decreased with expected climatic changes. This suggests that as we get closer to the worst effects of climatic alterations, there are more regions that are extremely favorable for pine species. On the other hand, Quercus species exhibited a decline in locations that were appropriate with predicted climatic changes. For example, *Q. leucotrichophora*, also known as the ban oak, exhibits a drop in the area from 788.56 km^2 to around 775 km^2 (near-term) and 765 km^2 (long-term scenario), respectively for high appropriate area, and up to 990.37 km^2 (long-term scenario) for medium suitable regions. Higher elevations of *Q. semecarpifolia* (Kharsu oak) exhibited similar tendencies, with a reduction in the suitable area up to 30 km^2 and an expansion in low suitable area up to 25 km^2. Similar declining trends were seen in suitable regions of *Q. floribunda* (Moru oak) up to around 30 km^2, but increases were observed in low zones up to 70 km^2. These findings demonstrate that, depending on the greenhouse gas emission scenario, alterations in future climatic conditions may cause minor to significant decline in the climatic ranges of tree species

(Chakraborty et al., 2016). According to a study conducted in the Sikkim Himalaya, 16% and 18%, respectively, of indigenous angiosperm species are predicted to lose their prospective habitat by the year 2050 and 2070. By 2050 and 2070, respectively, meadows may lose around 1% and 3% of their existing geographic distribution to shrublands. The future range of all endemic species was significantly predicted to migrate uphill or northward based on MaxEnt modeling forecasts. Upland species will likely be gradually displaced by lowland species as they move their habitats and altitudinal ranges upward. A considerable expansion in the habitable zones of shrubs, and their expansion northward, progressively displacing herbaceous ecosystems was also demonstrated (Manish et al., 2016).

6.2.3 IMPACT ON ALPINE VEGETATION

Alpine species are thought to be particularly susceptible to climate change (Lenoir et al., 2008), and climate change is thought to be a major factor influencing range changes in alpine species because of their comparatively low biotic complexity (Pauli et al., 2007). Additionally, species found at upper ranges frequently encounter a "nowhere-to-go" situation since cold habitats are only found at mountain peaks. They are typically restricted to particular places and have very narrow climatic tolerances (Nogs-Bravo et al., 2007). According to recent studies, climatic warming has induced species range changes in alpine settings, including plant colonization at higher altitudes. Plant characteristics and regional ecological circumstances, which serve as filters by identifying species with characteristics suited for the conditions of the area, are the key factors influencing successful colonization at higher elevations (Hamid et al., 2020). For instance, the plant populations in the alpine Sikkim Himalaya have changed as a result of continuous warming. Due to a rise in average temperature, about 90% of the indigenous plant species in the alpine Sikkim Himalaya, which have diverse taxonomic connections, have moved the boundaries of their elevational range. As a result of species expanding their upper ranges, the upper alpine zone now has more species than it did in the 19th century. In particular, at mountain peaks, the continuous trend of warming is anticipated to cause persistent altitudinal range reductions and finally, species extinction (Telwala et al., 2013). Seven species of the genus *Meconopsis*—commonly known as Himalayan blue poppies—that are native to subnival regions and alpine meadows in the Himalayan-Hengduan Mountain were examined for their climatic change responses. The results revealed a 302.3 m upward shift in average elevation between the pre-1970s (1922–1969) and post-1970s (1970–2016). Additionally, it was expected that by 2070, the area with a favorable climate will keep moving northward and westward in latitude as well as upward in elevation (He et al., 2016). A study on the distributional responses to climate change of the two endemic alpine species (*Cyananthus* and *Primula)* of the Himalaya-Hengduan Mountain revealed that under a variety of climate change projections, the distribution of the species will shift northward in latitude and upward in elevation; in addition, under such conditions, the species will increase the size of their habitats (He et al., 2019). In Ladakh, the western Himalaya, plants growing between 5,500 and 6,150 m asl have seen a vertical elevation change of around 150 m between 2003 and 2013 (Dolezal et al., 2016).

6.3 ROLE OF SPECIES DISTRIBUTION MODELLING (SDM) IN FORECASTING SPECIES RESPONSE TO CLIMATE CHANGE

For many years, forest managers have employed climate data in multiple ways for planning and decision-making processes, and this data has been comprehensively used to describe and map different kinds of vegetation and ecological locations (Daubenmire, 1978; Pojar et al., 1987; Thackway & Cresswell, 1992); to identify species and lineage (Booth, 1988, 1990); to determine the seed domain (Johnson et al., 2004); to assess the risk of forest fires and model fire behavioral patterns (Carvalho et al., 2008); to model forest productivity (Battaglia et al., 2004) and to evaluate the dynamics of a wide series of ecological progressions (Anderson, 1991). However, forecasting species reactions to climatic changes poses a distinct series of challenges as it involves factoring climatic projections that are frequently beyond the historic scope of so many species (Keenan, 2015). Projections from spatial models that associate species distributions or richness to environmental variables provide useful data on species distributions with great implications for organizing management and conservation efforts (Jetz et al., 2012). In order to enable adaptive conservation efforts, SDM approach also known as ecological niche modeling (ENM), is a very promising technique. The extensive usage of SDM has accelerated the establishment of applications in several fields of ecological and biogeographical research, together with biodiversity protection (Anderson & Meyer, 2004; Guisan et al., 2013). Recently, the data availability on species distribution has increased, and

GIS tools have been created to process and analyze spatial data, which has increased interest in this subject (Jetz et al., 2012). There are many software and algorithms available for SDM which work with various operating systems (OS) like Linux, Mac OS, and Windows (Table 6.1). SDMs are characterized as statistical algorithms that may link geographic data concerning the presence or absence of a particular species with a number of environmental factors defining the habitation of species. The ultimate goal of this method is to characterize the ecological traits of various species (Pecchi et al., 2019).

TABLE 6.1 Available Software's for SDM

Software	Operating System	Algorithms Implemented
Biomod 2 (R package)	Windows, Mac OS, Linux	RF, CART, GARP, ANN, MaxEnt, MARS, GAM, GLM, SRE
SDM (R package)	Windows, Mac OS, Linux	RF, ENFA, domain, GLM, mahalanobis distance, GAM, ANN, CART
Dismo (R package)	Windows, Mac OS, Linux	Mahalanobis distance, bioclim
SDM toolbox (ArcGIS)	Windows	MaxEnt
Species	Windows	ANN
MOPA (R package)	Windows, Mac OS, Linux	RF, MaxEnt, CART, GLM, MARS,
ENiRG (GRASS+R)	Windows, Mac OS, Linux	ENFA

Abbreviations: RF: Random forest; CART: classification trees; GARP: genetic algorithm; ANN: artificial neural network; MaxEnt: maximum entropy; MARS: multivariate adaptive regression spline; GLM: generalized linear model; GAM: generalized additive model; SRE: surface range envelope; and ENFA: environmental niche factor analysis.

Source: Adapted from: Pecchi (2019).

Since policymakers now seek information on the spread of forest species and the influence of various biological forces on them in order to identify current and future habitats of the species (Falk & Hempelmann, 2013), and when it comes to forecasting the past, present, and future distributions of species in mountain ranges across the world, these concomitant modeling strategies are becoming quite popular nowadays (Dullinger et al., 2004; Parolo et al., 2008). SDMs are currently used to refer to a broad spectrum of ideas, including models for species distribution, ecological niches, habitat appropriateness, preferred habitats, and climatic envelopes (Zurell et al., 2020). Even though this union appears to be plausible at first, there are variations in the goals, perception, and difficulties involved with the modeling (Peterson & Soberon, 2012; Peterson et al., 2015). The assessments of basic environmental conditions in connection to the ecology of the species, which in turn form a fundamental prerequisite for comprehending present, historical, and future species distribution ranges, are a common feature of all SDM procedures. To estimate niches of species and distributions through time and space, climatic variables are usually utilized, along with additional abiotic and occasionally biotic characteristics (Elith & Leathwick, 2009). In order to formulate a model and produce predictions with predictive power, knowing the facts of the species occurrences is a prerequisite, regardless of inter and extrapolation goal of the model (Holtmeier & Broll, 2019; Bobrowski, 2021a). SDMs have been widely acknowledged as useful methods in several restoration evaluations to confront significant conservation issues (Franklin, 2010). Improved understanding of the actual distribution of species provides important ecological information and excellent forecasting abilities and may be used to define species distribution and evaluate sampling tactics for rare and threatened species (Guisan et al., 2006; Marcer et al., 2013), to create networks of natural reserves and identify conservation target regions (Fajardo et al., 2014; Hermoso et al., 2015), or to direct efforts towards ecological restoration and species conservation (Angelieri et al., 2016; Villero et al., 2017). Additionally, SDMs can be extended to different geographical or temporal instances as part of risk evaluations of invasive species (Roura-Pascual et al., 2009; Jimenez-Valverde et al., 2011), as well as to hindcast or forecast possible impacts of climate change (Martin et al., 2013; Runge et al., 2016; Regos et al., 2016). Practical methods for creating strategies that effectively leverage SDMs through comprehensive decision-making systems have received less scientific attention than would be ideal in order to increase the likelihood of execution and eventually maximize conservation programs (Villero et al., 2017).

SDM technique has been extensively employed in the Himalayan region over the past decade to forecast how different species would respond to climate change (Table 6.2). However, it is a little bit difficult to simulate species distribution and ecological niches over large areas in isolated, high mountain zones such as the Himalayas. The two key obstacles are – (i) accumulating substantial high-quality data, fine-scale environmental indicators; and (ii) massive surveying of species distributions. The challenging terrain's accessibility and the vast geographic area of the Himalayan Mountain system are the main causes of sampling deficiencies. The importance of the model, however, is determined by the caliber and amount of occurrence data. The fundamental principle of the model is based on the random or uniform sampling of the investigation region. The choice of an acceptable predictor variable set depends on the geographical extent, species autecology, and the complexity of the ecological niche in addition to the accessibility of high-grade environmental data (Bobrowski et al., 2021b). The key to choosing pertinent and significant predictor variables is having a grasp of the connection between the ecological restrictions of species and their geographic spread. If ignored, the disconnect between model performance and ecology might grow in support of high model performance measures at the expense of declining real model accuracy and inadequate ecological justification of the habitat demands of the species (Peptitpierre et al., 2017; Santini et al., 2021).

6.4 IMPLICATIONS FOR CONSERVATION AND MANAGEMENT

Irregular trends in the shifting of spatial distributions of climatic niches suggest the immediate need for dynamic forest management practices. The recovery of the dwindling populations of the species and the reintroduction of those species must be done with careful attention in order to prevent unforeseen effects on the ecosystem's general functioning. One of the fundamental methods for protecting species at risk of collapsing may be to reduce the severity and scope of all potential threats in their natural environment. Focus must be placed on reducing biotic stress, restoration, the creation of protected area networks, proper management techniques, and connecting pathways to rebuild fragmented habitats. To understand the status and development trends, it is necessary to survey and monitor the biodiversity of both unprotected and protected regions throughout the Himalayan region (Tiwari et al., 2017). In addition to performing a delicate job in assessing the influence of climate change on the habitat appropriateness and shift in the range of species, the technique of SDM may be useful in identifying the existing distribution of various species. This might also be used to assess the interconnectivity of the landscape and offer extremely useful details about the ecological niches over the whole extent of its distribution. The few appropriate sites must be preserved even if climate change predictions predict a significant decline in habitat suitability. These places will serve as roosts for improved survival of the plant species in the event of adverse climatic circumstances (Wani et al., 2021a). The encouragement of *ex-situ* reproduction in carefully regulated conditions like natural habitats, botanical gardens, and other preservation institutions might significantly help to increase the recoverability of the species. Community, non-governmental organizations (NGOs), scientific institutions, as well as various government entities, must participate in order

TABLE 6.2 Some of the Niche Modeling Studies on Plant Species under Climatic Change Scenarios in the Himalayan Region

Plant Species	Model Used	Most Significant Bioclimatic/ Environmental Variable(s)	Range Change (%) Under Future Climatic Scenarios. +(Gain) and –(Loss)				References
			2050		2070		
			Current-RCP4.5	Current-RCP8.5	Current-RCP4.5	Current-RCP8.5	
Betula utilis	Biomod2	Precipitation of driest month, precipitation seasonality and precipitation of warmest quarter.	–9.76	–14.74	–12.01	–13.27	Hamid et al. (2019)
Cyananthus hookeri	MaxEnt	Precipitation of driest month, mean diurnal temperature and precipitation seasonality.	–	–	+66.2	+84.1	He et al. (2019)
Cyananthus incanus	MaxEnt	Mean diurnal temperature, precipitation seasonality, precipitation of driest month.	–	–	+130.6	+150.1	
Cyananthus inflatus	MaxEnt	Precipitation of driest month, precipitation seasonality and mean diurnal temperature.	–	–	+14.4	–11.2	
Primula poissonii	MaxEnt	Mean diurnal temperature, precipitation seasonality, precipitation of driest month.	–	–	+77.3	+88.6	
Primula skkimmensis	MaxEnt	Mean diurnal temperature, precipitation seasonality, precipitation of driest month.	–	–	+192.9	+224.6	
Primula tangutica	MaxEnt	Mean diurnal temperature, precipitation seasonality, precipitation of driest month.	–	–	+11.8	–11.8	
Lilium polyphyllum	MaxEnt	Precipitation of driest month, temperature seasonality and elevation.	–	–40.9	–	–37.9	Dhyani et al. (2021)

TABLE 6.2 *(Continued)*

Plant Species	Model Used	Most Significant Bioclimatic/ Environmental Variable(s)	Range Change (%) Under Future Climatic Scenarios. +(Gain) and –(Loss)				References
			2050		2070		
			Current-RCP4.5	Current-RCP8.5	Current-RCP4.5	Current-RCP8.5	
Valeriana wallichii	Biomod2	Precipitation of driest quarter, annual mean precipitation, and annual mean temperature.	–40.539	–80.225	–79.687	–94.070	Kumari et al. (2022)
Aconitum heterophyllum	Biomod2	Annual precipitation, mean temperature of wettest quarter and temperature seasonality.	–38.33	–52.58	–52.70	–56.59	Wani et al. (2022b)
Dactylorhiza hatagirea	Biomod2	Annual mean temperature, isothermality, mean temperature of wettest quarter and precipitation of driest month.	–27.98	–31.42	–29.93	–51.34	Wani et al. (2022)
Rheum webbianum	Biomod2	Annual mean temperature, isothermality, mean temperature of wettest quarter, precipitation of driest month and precipitation seasonality.	–44.39	–48.49	–53.47	–70.57	

Abbreviations: MaxEnt: Maximum entropy; and RCP: representative concentration pathways.

to protect plant species in their natural environment and restore ecosystems. For the gathering and long-term storage of specimens and genetic material, the construction of biobanks and cryo-conservation facilities can also be taken into consideration as a viable *ex-situ* option. Ecological niche modeling employing occurrence data and several environmental factors will be very beneficial for reintroducing *ex-situ* cultivated plant material in natural settings (Wani et al., 2021b). Due to rising concern about their unfavorable consequences, hydroelectric projects that have been developed recently in the area need to have their environmental repercussions reviewed. For the purpose of recommending conservation plans for Himalayan micro-watersheds under changing climatic circumstances, studies on the current watersheds are also necessary. Within international cooperation networks and a productive capacity-building program, these efforts may succeed more readily. Compared to other ecosystems, the susceptible mountain ecosystems are anticipated to be more prone to the impacts of climate change. Therefore, coordinated efforts are needed to create efficient mitigation and adaptation solutions (Tiwari et al., 2017).

6.5 CONCLUSION

Changing climatic conditions pose a significant impact on the species' ecological niches and distribution. The Himalayan region, which symbolizes the greatest bioclimatic altitudinal gradient and constitutes the highest altitudinal boundary of vascular plant species on Earth, is highly susceptible to climate change due to topographic inclinations. The region contains many endemic species with less resilience to changing climatic factors. The Himalayan ecosystems are undergoing rapid warming and experiencing changes in precipitation patterns, which influence plant diversity and especially affect the habitat suitability of the species. Plant species have been shown to shift their existing locations due to climatic alterations, and many species have been predicted to lose their suitable habitats in the near future. Therefore, urgent management strategies are required to prevent the extinction of species in their natural habitats. Reducing the emission of greenhouse gases and other hazardous substances that are major contributors to climate change is the need of the hour. Moreover, monitoring the distribution of species in response to climate change, establishing protected areas, and implementing *ex-situ* conservation strategies are significant in preventing species extinction.

KEYWORDS

- **climate change**
- **conservation**
- **habitat suitability**
- **Himalayan region**
- **pre-existing climatic conditions**
- **species distribution modeling**

REFERENCES

Alexander, J. M., Diez, J. M., & Levine, J. M., (2015). Novel competitors shape species' responses to climate change. *Nature*, *525*(7570), 515–518.

Anderson, J. M., (1991). The effects of climate change on decomposition processes in grassland and coniferous forests. *Ecological Applications*, *1*(3), 326–347.

Anderson, R. P., & Martınez-Meyer, E., (2004). Modeling species' geographic distributions for preliminary conservation assessments: An implementation with the spiny pocket mice (*Heteromys*) of Ecuador. *Biological Conservation*, *116*(2), 167–179.

Angelieri, C. C. S., Adams-Hosking, C., Ferraz, K. M. P. M. D. B., De Souza, M. P., & McAlpine, C. A., (2016). Using species distribution models to predict potential landscape restoration effects on puma conservation. *PloS One*, *11*(1), e0145232.

Batllori, E., & Gutiérrez, E., (2008). Regional tree line dynamics in response to global change in the Pyrenees. *Journal of Ecology*, *96*(6), 1275–1288.

Battaglia, M., Sands, P., White, D., & Mummery, D., (2004). CABALA: A linked carbon, water and nitrogen model of forest growth for silvicultural decision support. *Forest Ecology and Management.*, *193*(1, 2), 251–282.

Bell, G., (2001). Neutral macroecology. *Science*, *293*(5539), 2413–2418.

Bobrowski, M., Weidinger, J., & Schickhoff, U., (2021a). Is new always better? Frontiers in global climate datasets for modeling treeline species in the Himalayas. *Atmosphere*, *12*(5), 543.

Bobrowski, M., Weidinger, J., Schwab, N., & Schickhoff, U., (2021b). Searching for ecology in species distribution models in the Himalayas. *Ecological Modeling*, *458*, 109693.

Booth, T. H., (1990). Mapping regions climatically suitable for particular tree species at the global scale. *Forest Ecology and Management*, *36*(1), 47–60.

Booth, T. H., Nix, H. A., Hutchinson, M. F., & Jovanic, T., (1988). Niche analysis and tree species introduction. *Forest Ecology and Management*, *23*(1), 47–59.

Braunisch, V., Coppes, J., Arlettaz, R., Suchant, R., Zellweger, F., & Bollmann, K., (2014). Temperate mountain forest biodiversity under climate change: Compensating negative effects by increasing structural complexity. *PloS One*, *9*(5), e97718.

Burrows, M. T., Schoeman, D. S., Richardson, A. J., et al., (2014). Geographical limits to species-range shifts are suggested by climate velocity. *Nature*, *507*(7493), 492–495.

Camarero, J. J., & Gutiérrez, E., (2004). Pace and pattern of recent treeline dynamics: Response of ecotones to climatic variability in the Spanish Pyrenees. *Climatic Change, 63*(1), 181–200.
Carvalho, A., Flannigan, M. D., Logan, K., Miranda, A. I., & Borrego, C., (2008). Fire activity in Portugal and its relationship to weather and the Canadian fire weather index system. *International Journal of Wildland Fire, 17*(3), 328–338.
Chakraborty, A., Joshi, P. K., & Sachdeva, K., (2016). Predicting distribution of major forest tree species to potential impacts of climate change in the central Himalayan region. *Ecological Engineering, 97*, 593–609.
Chen, I. C., Hill, J. K., Ohlemüller, R., Roy, D. B., & Thomas, C. D., (2011). Rapid range shifts of species associated with high levels of climate warming. *Science, 333*(6045), 1024–1026.
Cruz, R. V., Harasawa, H., Lal, M., Wu, S., Anokhin, Y., Punsalmaa, B., Honda, Y., et al., (2007). Asia, climate change 2007: Impacts, adaptation and vulnerability. In: Parry, M. L., Canziani, O. F., Palutikof, J. P., Van Der Linden, P. J., & Hanson, C. E., (eds.), *Contribution of Working Group II to the Fourth Assessment Report of the Intergovernmental Panel on Climate Change* (pp. 469–506). Cambridge University Press, Cambridge.
Das, S. M., (1966). Palaearctic elements in the fauna of Kashmir. *Nature, 212*(5068), 1327–1330.
Daubenmire, R., (1978). *Plant Geography: With Special Reference to North America* (p. 338). Academic Press, New York.
Dhyani, A., Kadaverugu, R., Nautiyal, B. P., & Nautiyal, M. C., (2021). Predicting the potential distribution of a critically endangered medicinal plant *Lilium polyphyllum* in Indian Western Himalayan Region. *Regional Environmental Change, 21*(2), 1–11.
Diaz, H. F., Grosjean, M., & Graumlich, L., (2003). Climate variability and change in high elevation regions: Past, present and future. *Climatic Change, 59*(1), 1–4.
Diez, J. M., D'Antonio, C. M., Dukes, J. S., et al., (2012). Will extreme climatic events facilitate biological invasions? *Frontiers in Ecology and the Environment, 10*(5), 249–257.
Dolezal, J., Dvorsky, M., Kopecky, M., et al., (2016). Vegetation dynamics at the upper elevational limit of vascular plants in Himalaya. *Scientific Reports, 6*(1), 1–13.
Dullinger, S., Dirnböck, T., & Grabherr, G., (2004). Modelling climate change-driven treeline shifts: Relative effects of temperature increase, dispersal, and invasibility. *Journal of Ecology., 92*(2), 241–252.
Elith, J., & Leathwick, J. R., (2009). Species distribution models: Ecological explanation and prediction across space and time. *Annual Review of Ecology, Evolution, and Systematics., 40*(1), 677–697.
Fajardo, J., Lessmann, J., Bonaccorso, E., Devenish, C., & Munoz, J., (2014). Combined use of systematic conservation planning, species distribution modelling, and connectivity analysis reveals severe conservation gaps in a megadiverse country (Peru). *PloS One, 9*(12), e114367.
Falk, W., & Hempelmann, N., (2013). Species favorability shift in Europe due to climate change: A case study for *Fagus sylvatica* L and *Picea abies* (L.) Karst. based on an ensemble of climate models. *Journal of Climatology*, 1–18.
Ferrier, S., Watson, G., Pearce, J., & Drielsma, M., (2002). Extended statistical approaches to modelling spatial pattern in biodiversity in northeast New South Wales. I. Species-level modelling. *Biodiversity & Conservation, 11*(12), 2275–2307.
Franklin, J., (2010). *Mapping Species Distributions: Spatial Inference and Prediction.* Cambridge University Press.

Gaire, N. P., Koirala, M., Bhuju, D. R., & Borgaonkar, H. P., (2014). Treeline dynamics with climate change at the central Nepal Himalaya. *Climate of the Past, 10*(4), 1277–1290.

Gaire, N. P., Koirala, M., Bhuju, D. R., & Carrer, M., (2017). Site-and species-specific treeline responses to climatic variability in eastern Nepal Himalaya. *Dendrochronologia, 41*, 44–56.

Gansser, A., (1964). *Geology of the Himalayas*. New York: John Wiley.

Grabherr, G., Gottfried, M., & Pauli, H., (1994). Climate effects on mountain plants. *Nature, 369*, 448.

Grytnes, J. A., & Vetaas, O. R., (2002). Species richness and altitude: A comparison between null models and interpolated plant species richness along the Himalayan altitudinal gradient, Nepal. *The American Naturalist, 159*(3), 294–304.

Guisan, A., Broennimann, O., Engler, R., Vust, M., Yoccoz, N. G., Lehmann, A., & Zimmermann, N. E., (2006). Using niche-based models to improve the sampling of rare species. *Conservation Biology, 20*(2), 501–511.

Guisan, A., Tingley, R., Baumgartner, J. B., et al., (2013). Predicting species distributions for conservation decisions. *Ecology Letters, 16*(12), 1424–1435.

Hamid, M., Khuroo, A. A., Charles, B., Ahmad, R., Singh, C. P., & Aravind, N. A., (2019). Impact of climate change on the distribution range and niche dynamics of Himalayan birch, a typical treeline species in Himalayas. *Biodiversity and Conservation, 28*(8), 2345–2370.

Hamid, M., Khuroo, A. A., Malik, A. H., Ahmad, R., & Singh, C. P., (2020). Assessment of alpine summit flora in Kashmir Himalaya and its implications for long-term monitoring of climate change impacts. *Journal of Mountain Science, 17*(8), 1974–1988.

He, S. L., Wang, Y. S., Li, D. Z., & Yi, T. S., (2016). Environmental and historical determinants of patterns of genetic differentiation in wild soybean (*Glycine soja* Sieb. et Zucc). *Scientific Reports, 6*(1), 1–11.

He, X., Burgess, K. S., Gao, L. M., & Li, D. Z., (2019). Distributional responses to climate change for alpine species of *Cyananthus* and *Primula* endemic to the Himalaya-Hengduan Mountains. *Plant Diversity, 41*(1), 26–32.

Hermoso, V., Kennard, M. J., & Linke, S., (2015). Assessing the risks and opportunities of presence-only data for conservation planning. *Journal of Biogeography, 42*(2), 218–228.

Hickling, R., Roy, D. B., Hill, J. K., Fox, R., & Thomas, C. D., (2006). The distributions of a wide range of taxonomic groups are expanding pole wards. *Global Change Biology, 12*(3), 450–455.

Hoffmann, M. H., (2012). Not across the North Pole: Plant migration in the arctic. *New Phytologist., 193*(2), 474–480.

Holtmeier, F. K., & Broll, G., (2019). Treeline research—From the roots of the past to present time. A review. *Forests, 11*(1), 38.

Jetz, W., McPherson, J. M., & Guralnick, R. P., (2012). Integrating biodiversity distribution knowledge: Toward a global map of life. *Trends in Ecology &Evolution, 27*(3), 151–159.

Jimenez-Valverde, A., Peterson, A. T., Soberon, J., Overton, J. M., Aragón, P., & Lobo, J. M., (2011). Use of niche models in invasive species risk assessments. *Biological Invasions, 13*(12), 2785–2797.

Johnson, G. R., Sorensen, F. C., St Clair, J. B., & Cronn, R. C., (2004). Pacific northwest forest tree seed zones a template for native plants? *Native Plants Journal, 5*(2), 131–140.

Keenan, R. J., (2015). Climate change impacts and adaptation in forest management: A review. *Annals of Forest Science, 72*(2), 145–167.

Kirdyanov, A. V., Hagedorn, F., Knorre, A. A., et al., (2012). 20th century tree-line advance and vegetation changes along an altitudinal transect in the Putorana Mountains, northern Siberia. *Boreas, 41*(1), 56–67.

Koralewski, T. E., Wang, H. H., Grant, W. E., & Byram, T. D., (2015). Plants on the move: Assisted migration of forest trees in the face of climate change. *Forest Ecology and Management, 344*, 30–37.

Korner, C., (1998). A re-assessment of high elevation treeline positions and their explanation. *Oecologia, 115*(4), 445–459.

Korner, C., (1999). *Alpine Plant Life: Functional Plant Ecology of High Mountain Ecosystems.* Berlin: Springer.

Kullman, L., & Oberg, L., (2009). Post-little ice age tree line rise and climate warming in the Swedish Scandes: A landscape ecological perspective. *Journal of Ecology, 97*(3), 415–429.

Kullman, L., (2007). Tree line population monitoring of *Pinus sylvestris* in the Swedish Scandes, 1973–2005: Implications for tree line theory and climate change ecology. *Journal of Ecology, 95*(1), 41–52.

Kumari, P., Wani, I. A., Khan, S., Verma, S., Mushtaq, S., Gulnaz, A., & Paray, B. A., (2022). Modeling of *Valeriana wallichii* habitat suitability and niche dynamics in the Himalayan region under anticipated climate change. *Biology, 11*(4), 498.

Lal, M., Samant, S. S., & Singh, A., (2020). New distributional records of the Himalayan yew in north Western Himalaya, India-a consequence of climate change. *Indian Forester, 146*(12), 1131–1136.

Lenoir, J., Gegout, J. C., Marquet, P. A., De Ruffray, P., & Brisse, H., (2008). A significant upward shift in plant species optimum elevation during the 20th century. *Science, 320*(5884), 1768–1771.

Leonelli, G., Pelfini, M., Morra Di, C. U., & Garavaglia, V., (2011). Climate warming and the recent treeline shift in the European Alps: The role of geomorphological factors in high-altitude sites. *Ambio, 40*(3), 264–273.

Liu, X., & Zhang, M., (1998). Contemporary climatic change over the Qinghai-Xizang plateau and its response to the green-house effect. *Chinese Geographical Science, 8*(4), 289–298.

Mainali, J., All, J., Jha, P. K., & Bhuju, D. R., (2015). Responses of montane forest to climate variability in the central Himalayas of Nepal. *Mountain Research and Development, 35*(1), 66–77.

Manish, K., Telwala, Y., Nautiyal, D. C., & Pandit, M. K., (2016). Modelling the impacts of future climate change on plant communities in the Himalaya: A case study from Eastern Himalaya, India. *Modelling Earth Systems and Environment, 2*(2), 1–12.

Marcer, A., Saez, L., Molowny-Horas, R., Pons, X., & Pino, J., (2013). Using species distribution modelling to disentangle realized versus potential distributions for rare species conservation. *Biological Conservation, 166*, 221–230.

Martin, Y., Van, D. H., Dendoncker, N., & Titeux, N., (2013). Testing instead of assuming the importance of land use change scenarios to model species distributions under climate change. *Global Ecology and Biogeography, 22*(11), 1204–1216.

MEA: Millennium Ecosystem Assessment, (2005). *Synthesis*. Island Press, Washington, D.C, Ecosystems and Human Well-Being.

Negi, P. S., (2012). Climate change, alpine treeline dynamics and associated terminology: Focus on northwestern Indian Himalaya. *Tropical Ecology, 53*(3), 371.

Nogues-Bravo, D., Araújo, M. B., Errea, M. P., & Martinez-Rica, J. P., (2007). Exposure of global mountain systems to climate warming during the 21st century. *Global Environmental Change*, *17*(3, 4), 420–428.

Parmesan, C., & Yohe, G., (2003). A globally coherent fingerprint of climate change impacts across natural systems. *Nature*., *421*(6918), 37–42.

Parolo, G., Rossi, G., & Ferrarini, A., (2008). Toward improved species niche modelling: *Arnica montana* in the alps as a case study. *Journal of Applied Ecology*, *45*(5), 1410–1418.

Pauli, H., Gottfried, M., & Grabherr, G., (2003). Effects of climate change on the alpine and nival vegetation of the alps. *Journal of Mountain Ecology*, *7*(Suppl), 9–12.

Pauli, H., Gottfried, M., Reiter, K., Klettner, C., & Grabherr, G., (2007). Signals of range expansions and contractions of vascular plants in the high alps: Observations (1994–2004) at the GLORIA* master site Schrankogel, Tyrol, Austria. *Global Change Biology*, *13*(1), 147–156.

Pecchi, M., (2019). *A Model-Based Assessment of the Potential Impact of Climate Change on Italian Forest Systems* (pp. 1–189). Ph.D. Thesis, University of Florence, Italy.

Pecchi, M., Marchi, M., Giannetti, F., Bernetti, I., Bindi, M., Moriondo, M., Maselli, F., et al., (2019). Reviewing climatic traits for the main forest tree species in Italy. *iForest-Biogeosciences and Forestry*, *12*(2), 173.

Peterson, A. T., & Soberon, J., (2012). Species distribution modeling and ecological niche modelling: Getting the concepts right. *Natureza and Conservação*, *10*(2), 102–107.

Peterson, A. T., Papeş, M., & Soberon, J., (2015). Mechanistic and correlative models of ecological niches. *European Journal of Ecology*, *1*(2), 28–38.

Petitpierre, B., Broennimann, O., Kueffer, C., Daehler, C., & Guisan, A., (2017). Selecting predictors to maximize the transferability of species distribution models: Lessons from cross-continental plant invasions. *Global Ecology and Biogeography*, *26*(3), 275–287.

Pojar, J., Klinka, K., & Meidinger, D. V., (1987). Biogeoclimatic ecosystem classification in British Columbia. *Forest Ecology and Management*, *22*(1, 2), 119–154.

Price, L. W., (1981). *Mountains and Man: A Study of Process and Environment*. University of California Press.

Raizada, M. B., & Sahni, K. C., (1960). Living Indian gymnosperms. Part 1 (Cycadales, Ginkgoales and Coniferales). *Indian Forest Records (Botany)*, *5*(2), 73–150.

Rawat, D. S., (2012). Monitoring ecosystem boundaries in the Himalaya through an 'eye in the sky.' *Current Science*, *102*(10), 1352–1354.

Regos, A., D'Amen, M., Titeux, N., Herrando, S., Guisan, A., & Brotons, L., (2016). Predicting the future effectiveness of protected areas for bird conservation in Mediterranean ecosystems under climate change and novel fire regime scenarios. *Diversity and Distributions*, *22*(1), 83–96.

Roura-Pascual, N., Brotons, L., Peterson, A. T., & Thuiller, W., (2009). Consensual predictions of potential distributional areas for invasive species: A case study of Argentine ants in the Iberian Peninsula. *Biological Invasions*, *11*(4), 1017–1031.

Rowley, D. B., (1996). Age of initiation of collision between India and Asia: A review of stratigraphic data. *Earth and Planetary Science Letters*, *145*(1–4), 1–13.

Runge, C. A., Tulloch, A. I., Possingham, H. P., Tulloch, V. J., & Fuller, R. A., (2016). Incorporating dynamic distributions into spatial prioritization. *Diversity and Distributions*, *22*(3), 332–343.

Samant, S. S., & Joshi, H. C., (2003). Floristic diversity, community patterns and changes of vegetation in nanda devi national park. *Biodiversity Monitoring Expedition Nanda Devi.*, 39–54.

Samant, S. S., (2021). Floristic diversity of the Himalaya in relation to climate change: Status, values and conservation. *Journal of Graphic Era University*, 31–54.

Santini, L., Benítez-López, A., Maiorano, L., Čengić, M., & Huijbregts, M. A., (2021). Assessing the reliability of species distribution projections in climate change research. *Diversity and Distributions*, *27*(6), 1035–1050.

Schickhoff, U., (2005). The upper timberline in the Himalayas, Hindu Kush and Karakorum: A review of geographical and ecological aspects. *Mountain Ecosystems*, 275–354.

Schickhoff, U., Bobrowski, M., Böhner, J., Bürzle, B., Chaudhary, R. P., Gerlitz, L., Heyken, H., et al., (2015). Do Himalayan treelines respond to recent climate change? An evaluation of sensitivity indicators. *Earth System Dynamics*, *6*(1), 245–265.

Shrestha, U. B., Gautam, S., & Bawa, K. S., (2012). Widespread climate change in the Himalayas and associated changes in local ecosystems. *PloS One*, *7*(5), e36741.

Singh, A., & Samant, S. S., (2020). Population and community structure pattern of *Juniperus polycarpos* K. Koch with climate change effect in the cold desert trans Himalayan Region, India. *Arid Ecosystems*, *10*(1), 17–26.

Singh, J. S., & Singh, S. P., (1987). Forest vegetation of the Himalaya. *The Botanical Review*, *53*(1), 80–192.

Telwala, Y., Brook, B. W., Manish, K., & Pandit, M. K., (2013). Climate-induced elevational range shifts and increase in plant species richness in a Himalayan biodiversity epicenter. *PloS One*, *8*(2), e57103.

Tewari, V. P., Verma, R. K., & Von, G. K., (2017). Climate change effects in the Western Himalayan ecosystems of India: Evidence and strategies. *Forest Ecosystems*, *4*(1), 1–9.

Thackway, R., & Cresswell, I. D., (1992). *Environmental Regionalization of Australia: A User-Oriented Approach*. Environmental Resources Information Network.

Thuiller, W., Lavorel, S., & Araújo, M. B., (2005). Niche properties and geographical extent as predictors of species sensitivity to climate change. *Global Ecology and Biogeography*, *14*(4), 347–357.

Troup, R. S., (1921). *The Silviculture of Indian Trees* (Vol. III). Clarendon.

Villero, D., Pla, M., Camps, D., Ruiz-Olmo, J., & Brotons, L., (2017). Integrating species distribution modelling into decision-making to inform conservation actions. *Biodiversity and Conservation*, *26*(2), 251–271.

Wani, I. A., Khan, S., Verma, S., Al-Misned, F. A., Shafik, H. M., & EL-Serehy, H. M., (2022). Predicting habitat suitability and niche dynamics of *Dactylorhiza hatagirea* and *Rheum webbianum* in the Himalaya under projected climate change. *Scientific Reports*, *12*, 13205.

Wani, I. A., Verma, S., Kumari, P., Charles, B., Hashim, M. J., & El-Serehy, H. A., (2021a). Ecological assessment and environmental niche modeling of Himalayan rhubarb (*Rheum webbianum* Royle) in northwest Himalaya. *PLoS One*, *16*(11), e0259345.

Wani, I. A., Verma, S., Mushtaq, S., Alsahli, A. A., Alyemeni, M. N., Tariq, M., & Pant, S., (2021b). Ecological analysis and environmental niche modelling of *Dactylorhiza hatagirea* (D. Don) Soo: A conservation approach for critically endangered medicinal orchid. *Saudi Journal of Biological Sciences*, *28*(4), 2109–2122.

Wani, Z. A., & Pant, S., (2021). Betula utilis D. Don: An ecologically and economically important timberline species of Himalayan region in jeopardy. *The Botanical Review*, *87*(3), 377–391.

Wani, Z. A., Ridwan, Q., Khan, S., Pant, S., Siddiqui, S., Moustafa, M., Ahmad, A. E., & Yassin, H. M., (2022). Changing climatic scenarios anticipate dwindling of suitable habitats for endemic species of Himalaya—Predictions of ensemble modeling using *Aconitum heterophyllum* as a model plant. *Sustainability*, *14*(14), 8491.

Wani, Z. A., Satish, K. V., Islam, T., Dhyani, S., & Pant, S., (2022a). Habitat suitability modelling of *Buxus wallichiana* Bail.: An endemic tree species of Himalaya. *Vegetos*., 1–8.

Wilson, R. J., Gutierrez, D., Gutierrez, J., Martinez, D., Agudo, R., & Monserrat, V. J., (2005). Changes to the elevational limits and extent of species ranges associated with climate change. *Ecology Letters*, *8*(11), 1138–1146.

Xu, J., Grumbine, R. E., Shrestha, A., Eriksson, M., Yang, X., Wang, Y. U. N., & Wilkes, A., (2009). The melting Himalayas: Cascading effects of climate change on water, biodiversity, and livelihoods. *Conservation Biology*, *23*(3), 520–530.

Yadav, R. R., Negi, P. S., & Singh, J., (2021). Climate change and plant biodiversity in Himalaya, India. *Proceedings of the Indian National Science Academy*, *87*(2), 234–259.

Yadava, A. K., Sharma, Y. K., Dubey, B., Singh, J., Singh, V., Bhutiyani, M. R., Yadav, R. R., & Misra, K. G., (2017). Altitudinal treeline dynamics of Himalayan pine in western Himalaya, India. *Quaternary International*, *444*, 44–52.

Zhu, K., Woodall, C. W., & Clark, J. S., (2012). Failure to migrate: Lack of tree range expansion in response to climate change. *Global Change Biology*, *18*(3), 1042–1052.

Zurell, D., Franklin, J., Konig, C., Bouchet, P. J., Dormann, C. F., Elith, J., Fandos, G., et al., (2020). A standard protocol for reporting species distribution models. *Ecography*, *43*(9), 1261–1277.

CHAPTER 7

Forest Fire Susceptibility and Sustainability in Relation to Climate Change: A Geospatial Evaluation of Global Forests

MOHAMMAD AFSAR ALAM and SAKUL KUNDRA

Department of Social Sciences, Fiji National University, Natabua Campus, Lautoka, Fiji

ABSTRACT

Human activity, the demand for forest products, and global warming are all elements that have led to a buildup of pressure on forests throughout the past few decades. Climate change and increasing temperatures have had a considerable influence on the world's forests and wildlife. The evaporation of soil water is exacerbated by high temperatures, resulting in drier soils and water loss in forest vegetation. Since 1984, the number of forest fires has increased, and these fires are directly connected to climate change. Due to a lack of water, drought causes changes in forest plants that contribute to fires. Temperature and drought stress during the plant's developmental stage cause species shifts, reducing the forest's resilience. Wildfire vulnerability and resilience at national, regional, and international levels are the focus of this research. A primary goal of this study is to discern the characteristics that are geographically connected to the occurrence and incidence of forest fires. Secondary data is used in this investigation. Data from historical forest fire reports were used to gather this information. Forest fires may be avoided using sound control techniques such as self-sustaining, community-based, and urban landscape maintenance. Stress precursor monitoring, sustainable

Sustainable Forest Resources Management: Issues and Implications. Jahangeer A. Bhat, Gopal Shukla, Manmohan J. Dobriyal, Sumit Chakravarty, A. Arunachalam, and Rainer W. Bussmann (Eds.)

forest use, environmental studies, and the introduction of relatively close and fresh native plant species are all tools for good forest management.

7.1 INTRODUCTION

Several biotic and non-biotic variables influence the incidence of forest fires, but the impacts of each element differ among ecosystems, as well as across geographical and temporal scales, which complicates our knowledge of forest fire dynamics (Yang et al., 2007). Avila et al. (2010) found that the incidence of forest fires in Durango State, Mexico, is influenced by factors that are not consistent across geography. Human activities have a substantial impact on forest fire frequency and geographical patterns, even though combustible bed availability is one of the most often addressed variables in preventive initiatives. This means that knowing both the human and natural elements that influence forest fire patterns might assist in identifying the areas most likely to be affected by these occurrences (Syphard et al., 2008).

The recent flames in Greece, North Macedonia, and Italy are still raging, but Turkey has managed to extinguish about 300 of the blazes. According to Greenpeace Russia, Russia's wildfires in 2021 are already the country's biggest in satellite measurements, covering 17.08 million hectares of land. With only a few weeks remaining in a terrible wildfire season, Russia has established a new record for the largest number of hectares destroyed by wildfires. Russian Greenpeace wildfire chief Grigory Kuksin blamed the extraordinary blazes this year on climate change, which is making Russia's vast swaths of forest drier and hotter. According to the group's statement, "the unprecedented magnitude of fire-damaged regions is partly the outcome of climate change, and the fires themselves are partly driving climate change" (Eckel, 2021). As a result of this new reality, it must consider what steps must be taken immediately in such nations to help avoid such catastrophes in the future or at the very least minimize their magnitude.

Aside from a lack of funds for extinguishing operations, environmentalists criticize Russia's firefighting policy, which permits areas to overlook blazes when combating them would do more harm than good. There are presently 170 fires raging throughout Russia, from the western Kostroma area to the eastern Magadan region, according to government statistics (The Moscow Times, 2021). The biggest and coldest part of the nation, the Republic of Sakha (Yakutia), is home to more than 100 of these blazes, which are on course to become the greatest single inferno in human history if they continue. The current size of the wildfires in Siberia dwarfs that of the rest of the world's combined. It is a rising concern among climate scientists that the severity of Siberia's annual wildfires, along with increased emissions from those flames, might push global warming even higher. For the first time ever, the 2020 Russian forest fires released more carbon into the atmosphere than ever before (The Moscow Times, 2021).

It was the largest wildfire in California's history when the Carlton Compound Wildfire erupted in 2014. It was estimated that the Western United States lost 3.6 million acres to wildfires in 2015, a year marked by an exceptional lack of precipitation and poor snowpack (Marlier et al., 2017). Drier soils and fuels, large-scale conflagrations, and expanded burn areas are all consequences of the extended fire and growing seasons (Westerling, 2016; Stavros et al., 2014; Littell et al., 2010; McKenzie et al., 2004). The amount of land expected to be burnt is expected to rise because of less summertime precipitation (Holden et al., 2018).

Ecological changes may be triggered by interplay involving fires as well as other disturbance agents as the temperature warms (such as drought or insect outbreaks). Drought's synergistic effects on wildfire intensity (plant and soil destruction) might make the situation much worse (Reilly et al., 2017; Littell et al., 2016; Stavros et al., 2014; McKenzie et al., 2008). Alterations to vegetation conditions (such as those caused by previous management practices) might interact with stressors associated with climate change, affecting both current and future forest conditions (e.g., fire frequency, area, and intensity) (Keeley & Syphard, 2016). This study has manifold objectives described below:

- To determine which factors are related to the occurrence and prevalence of wildfires.
- To explain the geographical distribution of wildfires at national, regional, and global levels; and
- To identify the mechanisms to implement regulations of key strategies for forest fire mitigation and prevention.

7.2 MATERIALS AND METHODS

This study is descriptive in nature and is based on forest fire susceptibility and sustainability data from secondary sources. These data are collected from government offices, published books, articles, and unpublished soft copies sourced from the internet and software. Various website publications at different times have also been consulted. But the primary source of forest fire susceptibility and sustainability data is the books and research articles published recently. Information gathered from various sources has been presented in the form of tables, maps, and diagrams.

7.3 THE HISTORICAL AND CURRENT CLIMATE-FIRE INTERACTION

Charcoal deposited in lake sediments may be used to find specific wildfire incidents and calculate the occurrence of fire over a period of hundreds to thousands of years (Itter et al., 2017). Sediment pollen and charcoal records may help us better understand how climate change has influenced vegetation and fire frequency and intensity in the past (Gavin et al., 2007). When reconstructing the Northwest Paleoecology, pollen and charcoal evidence found in lakes west of the Cascade Range is primarily relied upon, but research into the region's arid core is scarce (Kerns et al., 2017).

In terms of temperature, the early Holocene in the Northwest (nearly 10,500 to 5,000 years BP) was the hottest post-glacial period (Whitlock, 1992). Droughts were more common in summers of the early Holocene, with more severe droughts occurring on a regular basis (Briles et al., 2005; Whitlock, 1992). Hotter and drier weather contributed to an increase in summer wildfires in various parts of the Northwest United States this year (Walsh et al., 2008, 2015; Whitlock, 1992). As early as the Holocene, sediment charcoal analysis revealed that fires occurred frequently in eight different places, such as the North Cascade Range (Prichard et al., 2009), Olympic Peninsula, Puget Lowlands, Southwest Washington (Walsh et al., 2008), Oregon Coast Range (Long et al., 1998), Willamette Valley (Walsh et al., 2010), and the Siskiyou Mountains (Briles et al., 2009; Brunelle & Whitlock, 2003). There was a correlation between greater fire frequency in these locations and the prevalence of fire in Douglas-fir, lodgepole pine, and Oregon white oaks are common because there are fire-adapted tree species. Other pollen investigations (without a simultaneous charcoal analysis) confirm the early Holocene development of these plants (e.g., Whitlock & Worona, 1995; Whitlock, 1995).

Paleoecological studies of the early Holocene and earlier historical eras have evidence of climate having a substantial impact on wildfire for millennia in the Northwest. During times of severe climate fluctuation, fire conditions (such as those of the early Holocene) caused wide-ranging changes in the shape, size, and structure of the forest (Crausbay et al., 2017; Prichard et al., 2009). People who have lived through times of rapid change know that certain species such as the Oregon white oak, red alder, the ponderosa pine, the lodgepole pine, and the Douglas fir have helped them endure in settings that are constantly disturbed (Brubaker, 1988; Whitlock, 1992). In a warmer temperature, some animals may be able to thrive (Prichard et al., 2009; Whitlock, 1992).

7.4 AN OVERVIEW OF THE CURRENT SITUATION

The fire epidemic since July has continued to spread to new countries and to engulf greater territories, although firefighting operations are deploying more resources to try to control the blazes. For the first time in decades, the Mediterranean area is sweltering in record temperatures. The EU Civil Protection Mechanism continues to support Greece and the Mediterranean area in the battle against extraordinary forest fires, according to a statement from the European Commission (European Commission, 2021). After requests from Greece, North Macedonia, Albania, Italy, and Turkey, more than 1,300 rescuers and 250 vehicles have been recruited. There are already nine aircraft, about 1,000 firefighters, and 200 vehicles battling the flames alone in Greece. So far this year, about 23 million acres in Siberia have burned down, making it one of Russia's major climate-related issues. The flames in Greece, North Macedonia, and Italy are still raging, but Turkey has managed to extinguish about 300 of the blazes.

Although numerous fires in Albania, Bulgaria, Italy, Greece, and Turkey have been brought under control by firefighting planes and volunteers from Europe, there are still several smoldering blazes. All the deaths and serious injuries are having a long-term effect on the families of those who were affected, although the majority of those who were hurt are just suffering minor ones. Fears of additional temperature rise and unprotected residents from flooding are growing in Greece after the hilly terrain surrounding Athens was demolished. Wildfires have long been a part of the Mediterranean ecology, and to some degree they are essential to its functioning. However, this year's wildfires have decimated vast swaths of forest and other ecosystems, and the damage will take years to repair. A significant effect may be seen on the European economy if forest fires continue to grow in severity and frequency. Another long-term effect of wildfires is desertification, particularly in southern Europe. As a result, hundreds of thousands of families would see their food supply shrink, a problem already exacerbated by shortages. European Forest Fire Information System estimates that the number of wildfires in 2021 will be much higher than the yearly norm (Wildfires Information Bulletin 2, 2021).

7.5 GLOBAL FOREST FIRE SCENARIO AT A GLANCE

North and South America have been affected by wildfires, as have Madagascar, the eastern Arabian Peninsula, and the Mediterranean coast of Europe, according to the National Aeronautics and Space Administration (NASA's) Fire Information for Resource Management System reports (FIRMS). India, Sri Lanka, and Russia's Siberia area are among the Asian countries where wildfires have been noticed. Eurasia and Oceania are also impacted by forest fires, including Ukraine, Georgia, and Australia. Forest fires are also a problem in Mediterranean countries such as Italy, Spain, Greece, Tunisia, Egypt, and Algeria. An emergency map published by the European Commission shows that large and small forest fires have been raging throughout Europe, almost reaching Russia's far north. Satellite images show that several fires of varied sizes have been burning over Europe for at least a week.

Large and minor forest fires are raging over Italy's northern and southern regions. More than a 1,000 square kilometers of land have been destroyed by flames on Sicily Island, as well as in the Calabria, Pugliese (Southern Italy), and Campania regions in recent weeks, according to local media. Italy urged EU members to provide firefighting planes in addition to its own. Thousands of people were evacuated from their homes in the north of Athens because of forest fires. In 2021, five persons were wounded and almost 60 villages and communities were evacuated because of flames in Greece's southern region. Due to Greece's ongoing heatwave, Prime Minister Kyriakos Mitsotakis said that the nation was experiencing "unprecedented" circumstances. In addition, the Greek army was ordered to assist in putting out forest fires. According to Nikos Hardalias, the Greek civil defense and crisis management deputy minister, there have been 118 forest fires in the preceding 24 hours and 60,000 acres of forest had burned in the last four days. Albania has also had to deal with a rash of forest fires in the last few days. According to the Albanian Defense Ministry, there are presently six active fires throughout the nation. More than 2,000 hectares of land were burnt by a forest fire that broke out near the city of Monchique in southern Portugal on July 17, 2021 (Oz, Emel, 2021). The Catalonia area of northeastern Spain was hit by the largest wildfire in the previous two years on July 26, 2021. The Aude area in southern France was devastated by a wildfire on July 25, and a major highway between Toulouse and Narbonne was shut down as a result. Lines carrying energy to France, Spain, and Portugal were damaged because of the fire.

7.5.1 BURNED AREA AND AVERAGE LOSSES IN EUROPE

The average burned area for the years 2008–2020 has already been surpassed in Italy, France, and Romania, according to data from the European Forest Fire Information System. While Italy's fires consumed 53,797 hectares this year, the 12-year average was 40,781 hectares. A total of 22,807 hectares of forest were destroyed in France in 2017, compared to an annual average of 9,283 hectares. As of 2021, 18,620 hectares of land were burnt in Romania, compared to the previous 12 years when the average was 16,082. In comparison to the average loss of non-EU nations between 2008 and 2020, Bosnia & Herzegovina lost 25,583 hectares in the fire. This year's fires in Tunisia have scorched an area larger than the average for the previous 12 years: 17,599 hectares.

Firefighters in northern Russia's Yakutia region are still battling forest fires that began in July. More than 1.3 million hectares of forests have been devastated, and a state of emergency was declared by the authorities in this area.

The National Interagency Fire Center reports that nearly 96% of forest fires are burning across the United States, resulting in the destruction of over 1.8 million acres of land. One historic town in California, which has been in flames for three weeks, has been destroyed. Known as "Dixie," on July 13–14, a large fire broke out in the north of the state. Several villages have been threatened by the fast spread of the fire in the previous two days. According to the Canadian Wildland Fire Information System, there are now 292 active wildfires in Canada. There were 4,891 fires in the United States this year, up from an average of 3,753 in the prior decade. There was a yearly average loss of 1,736 hectares of land from the fires totaling 2,758 hectares (Oz, Emel, 2021). Brazil had the highest number of wildfire outbreaks in South America in 2020, with approximately 223,000. With over 74,000 wildfires, Argentina was the second most affected country in the region that year. Over 105,000 hectares of forest were burned in the Amazon area in 267 major fires in 2021. Fires in the Brazilian Amazon accounted for more than 70% of these incidents.

Rainfall in India and China during the summer months reduces the risk of forest fires since the rainy season occurs throughout the summer months. Nevertheless, deforestation in these areas is exacerbating the devastation caused by downpour floods. Floods killed at least a dozen people in both nations last week. Approximately 10 million hectares of land and more than a billion wild animals were destroyed by forest fires in Australia during the summers of 2019 and 2020. Since the turn of the century, wildfires have wreaked havoc in 20 nations, as seen in Table 7.1 and Figure 7.1. Following the United States in terms of economic losses are Indonesia and Canada.

TABLE 7.1 Twenty Countries Have Been Hit the Worst by Wildfire Disasters in Terms of Financial Loss (1900 to 2013)

Countries	Number of Incidents	Loss in Millions of Dollars
Spain	14	2,754
Australia	30	2,645
Russia	22	2,183
South Africa	9	440
United States	64	17,787
Indonesia	9	9,329
Chile	9	880
Canada	20	6,463
Mexico	3	91
Malaysia	4	302
Greece	13	2,425
Portugal	8	3,475
Argentina	5	100
Italy	7	1,700
Nacaragua	3	80
Croatia	5	38
Israel	3	315
Mongolia	3	1,823
Yugoslavia	2	200
China	6	110

Source: Adapted from Emergency Events Database (EM-DAT, 2013).

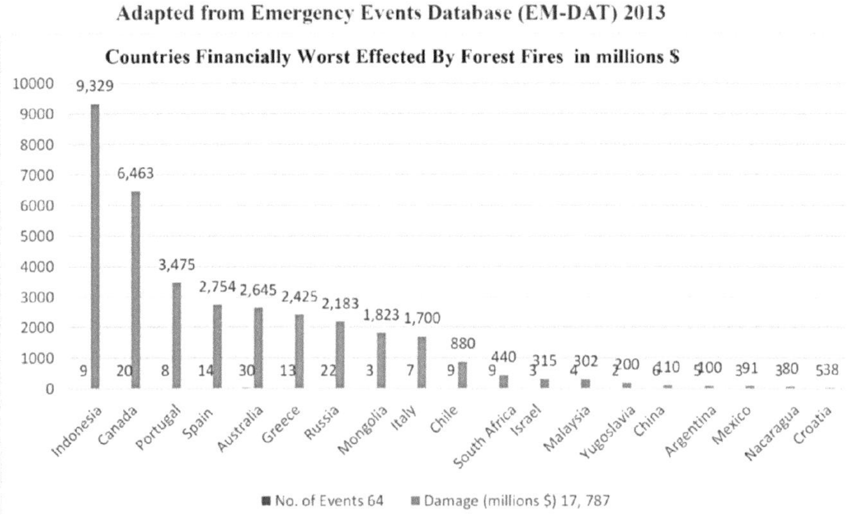

FIGURE 7.1 Twenty most affected countries of the world by wildfire.

7.5.2 *MOIST CONIFEROUS FORESTS AND FIRE RISK*

Risks of climate-induced fire in humid coniferous forests are typically less (Figure 7.2). Many of the forests in Oregon and Washington west of the Cascade Mountains are dominated by Douglas-fir and western hemlock. It is rare for new trees to develop in damp coniferous woods because of the great severity of the flames (Agee, 1993). Even if fire frequency and severity increase due to changing climate, the fire frequency in wet ecosystems is expected to remain at a low level. Fire-climate interactions are shown in Figure 7.3 in a low- to mid-elevation dry coniferous forest and woodlands (east bank of the Columbia River, southern Idaho, and the drier regions of the Rocky Mountains in the United States). Low, medium, and high are all possible degrees of probability and certainty. More than 99% likelihood (approximately 66 to 100% probability). In the middle between the two extremes is moderate confidence, whereas high confidence is characterized by a broad scientific agreement and a large quantity of evidence.

7.6 FORECASTS CLIMATE CHANGE IMPACTS ON WILDFIRES

Increasing temperatures, less summer precipitation, and deteriorating drought conditions are expected to fuel more wildfires in the Pacific

Northwest. Forecasting severity and occurrences of fire using a variety of models is essential for better understanding the potential impact of changing climate on both (McKenzie et al., 2004). The emphasis was on the use of empirical statistical modeling and data visualization for assessing in the Northwest.

As shown in Figure 7.2, risk assessment for the consequences of climate-fire interactions in wet coniferous forests at low to mid-elevations (Olympics, Westside Cascades in northern Idaho, and the West-side Rocky Mountains in the United States) in the middle to late of the 21st century can be found in the following chart. There are three grades for likelihood and confidence: low, medium, and high. Consequences with a low likelihood (about 0 to 33% chance), a moderate likelihood (roughly 33 to 66% probability), and a high likelihood (approximately 66 to 100% probability) are all defined by their likelihood (approximately 66 to 100% probability). Minimal scientific agreement and weak evidence constitute low confidence, while a great scientific deal and substantial evidence define high confidence, with intermediate confidence lying somewhere in the middle.

7.6.1 DRY CONIFEROUS FOREST FIRE DANGER

There is a considerable danger of climate-induced fires in dry coniferous forests and woods (Figure 7.3). Low-altitude dry coniferous forests and woodlands may be found across Oregon and Washington as well as the Rockies in Idaho and Montana. The intensity and frequency of fires in dry coniferous forests and woodlands range from mild to frequent. The most frequent tree species in the region are the white-fir, Douglas-fir, Ponderosa pine, and grand-fir. Regeneration failures after forest fires are also likely to occur in certain areas, as well as interacting disturbances and hydrologic change in some forests on global climate model temperatures and precipitation. These models cannot effectively estimate future burn hazards because they disregard current burn scars and long-term changes in plant flammability as a consequence of changing climate (Littell et al., 2018; McKenzie & Littell, 2017; Parks et al., 2015). Furthermore, they disregard the significance of human involvement in the first stages of a fire's ignition (Syphard et al., 2017).

Empirical models have been developed by a number of studies to forecast likely burned regions or future burned areas at both regional and global scales (Moritz et al., 2012; Krawchuk et al., 2009) (e.g., western US; Kitzberger et al., 2017; Yue et al., 2013; Littell et al., 2010; McKenzie et al., 2004). Western US wildfires and burned areas are predicted to rise because of warming climate models over time. Below are a few occasions in which the Pacific Northwest is expressly mentioned. As seen in these cases, variations in the charred area may be observed across landscapes in future fire projections.

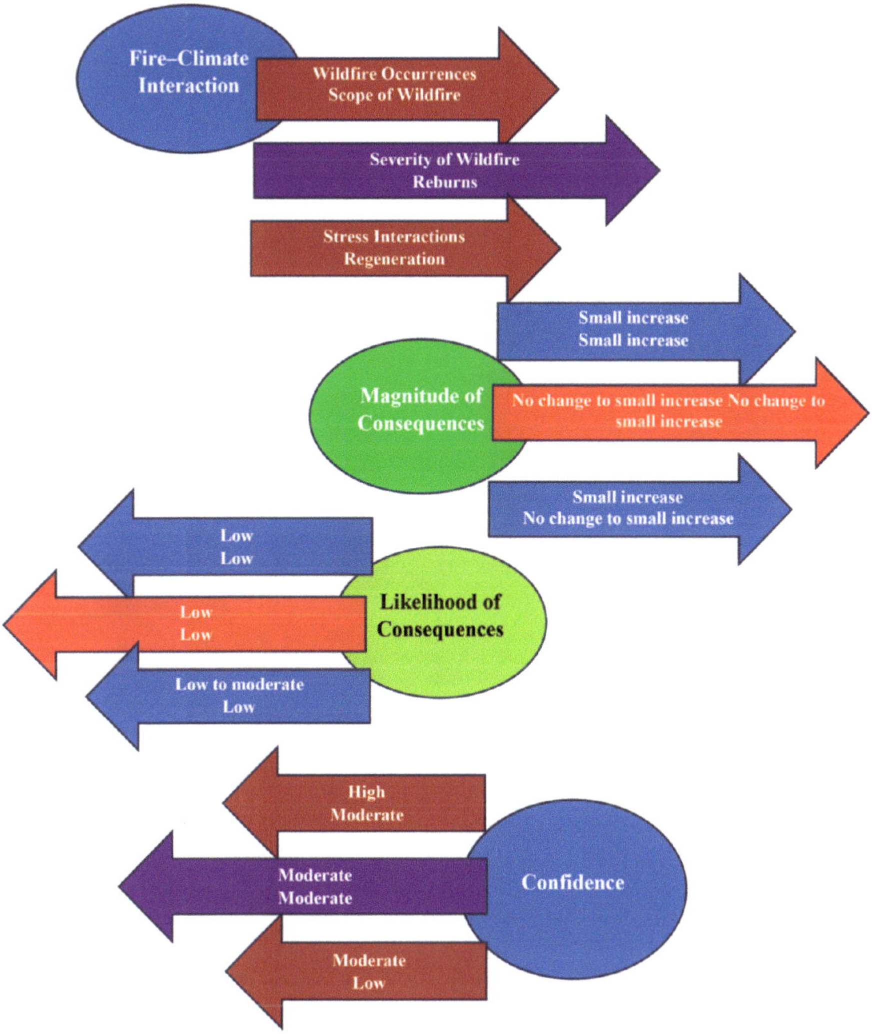

FIGURE 7.2 Climate-fire interactions in wet coniferous forests.

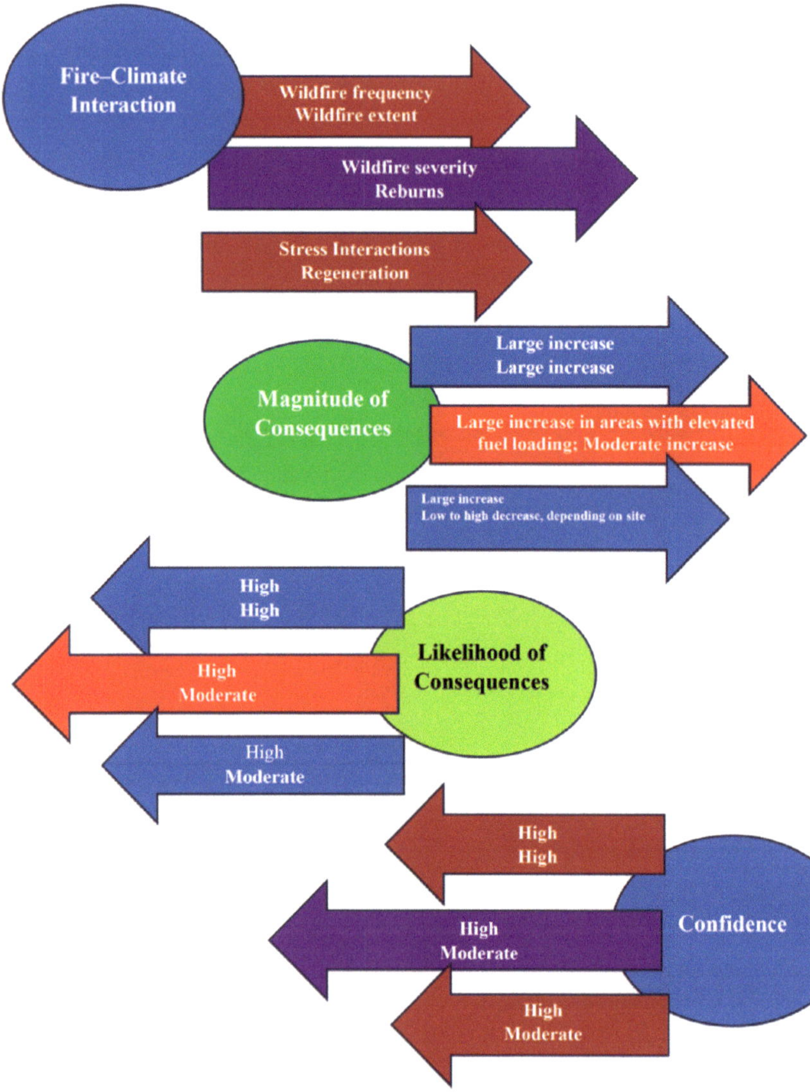

FIGURE 7.3 Climate-fire interactions in a parched coniferous forests.

A study by McKenzie et al. (2004) predicts that the amount of land burnt by wildfires in the western states will increase by a factor of 1.4 to 5 if the average temperature rises by 2°C. By 2039, the annual burned area in 11 Western states would be five times more than it was between 1961 and 2004, according to research by Kitzberger et al. (2017). Two global climate models, used by Littell et al. (2010), estimated that burned area will triple or quadruple by the mid-2080s in Idaho, Montana, Oregon, and Washington. The average burned area is predicted to grow from 0.2 million to 0.3 million ha in 2020, and subsequently to 0.5 million in 2040 and 0.8 million in 2080. Even within the projections' bounds, the exact quantity of land that will be destroyed is likely to vary substantially given that these are just apex estimates. Littell et al. (2010) developed empirical models for the state of Washington on a finer (eco-section) scale. A lack of frequent fires in coastal woods makes empirical models harder to create. A lack of water balance and potential evapotranspiration was the most significant determinant in determining how much area burnt in drier woods. By 2040, forest ecosystems (including the Okanogan Highlands and Blue Mountains Eco sections) were expected to see an increase in average burnt area of 3.8 times their 1980–2006 levels. Forested Eco portions of the Northwest's forested Eco regions, updated versions of these models suggest that burnt area would rise in the future, although the increased burned area may be moderated or even decrease in places that are more fuel constrained (Littell et al., 2018).

Empirical models are also used to forecast the future incidence of extremely large fires, which are commonly characterized as the largest 5 to 10% of all fires or fires with an area more than 5,000 hectares. Between 2041 and 2070, Barbero et al. (2015) anticipated that the yearly likelihood of big wildfires will increase by a factor of four compared to the period from 1971 to 2000. More than 20% of forests in Oregon and Washington will be highly susceptible to fires of more than 40 hectares during the next century, according to Davis et al. (2017), although this increase will be less prominent in the Coast Range and Puget Lowlands in Washington. The East Cascades, the Klamath Mountains, and the Blue Mountains are expected to see the most growth. A rise is also likely in the number of fires that do not succumb to an initial attack (Fried et al., 2008).

Future fire severity can only be predicted using a limited number of empirical model forecasts. Parks et al. (2016) used empirical models to imply that the severity of fires in the Northwest may not change considerably in a warmer environment because fuels restrict the severity of fires. Climate change is anticipated to have both a direct and an indirect influence on changes in plant composition and structure, Figure 7.4 (which affect fuels), which in turn will influence fire intensity. Empirical data-based models do not account for the probability of changes in vegetation and fuel (among other limitations; *see* McKenzie & Littell, 2017). Within the next several months, dry, historically frequent-fire forests may see an increase in fire intensity due to high stem density caused by fire exclusion and previous management (Haugo et al., 2019).

7.7 TOOL FOR TACKLING LARGE WILDFIRES

Although this study is intended to provide an international perspective, some background on the introduction and use of aerial fire-fighting systems within Australia is instructive as Australia is one of the worst forest fire victimized countries of the world. In recent years, there has been a greater recognition within Australia that Large Air Tankers (LATs) provide a useful tool for tackling large wildfires in a variety of terrains (Heffernan, 2018). Indeed, Victoria and New South Wales (NSW) have operated a variety of LATs including the C-130 for several years.

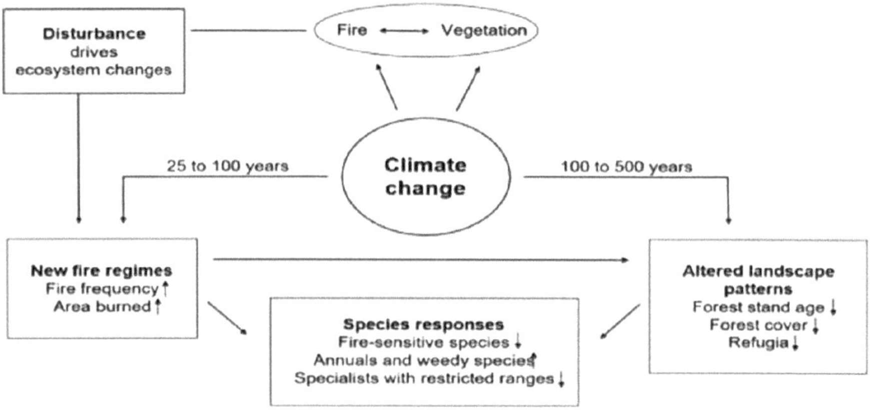

FIGURE 7.4 Simulation of climate change's impacts on vegetation.

Source: Reprinted with permission from: Halofsky et al. (2020).

However, instead of using the MAFFS, the currently shared fixed-wing LATs are all fully dedicated aircraft with permanent airframe modifications. Aside from lower operational costs, another possible reason that these permanently converted aircraft were presumably preferred over the MAFFS is that they permit higher drop rates. In particular, the Coulson "RADS XXL" system can disperse water at 6,000 liters per second using a simple gravity-drop arrangement; however, it should be noted that this necessitates the structural modification of the floor of the C-130. Consequently, old C-130 airframes were chosen for these permanent-dedicated conversions.

7.7.1 USE OF EXPANDED MODULAR AERIAL FIRE FIGHTING SYSTEMS (MAFFS)

There are ~20 MAFFS (including both MAFFS I and II) configurations currently flying in the USA, Brazil, Thailand, Tunisia, Morocco, and Turkey. MAFFS II is still in production in the USA for the C-130 (Figure 7.5) and in prototype development for the Brazilian Embraer KC390. Within Australia, 12 C-130J aircraft are operated by the RAAF (37th Air Squadron), but they are solely used for military airlift in accordance with the current mandate of the RAAF. MAFFS II (Benz, 2013) is compatible with the C-130J and capable of delivering ~11,300 liters without the need for any aircraft modification. Therefore, if MAFFS II-C130J aircraft could be operated in Australia under a modified RAAF mandate embracing homeland civil protection, then it would add 12 LATs to the existing NAFC fleet, more than doubling the current Australian LAT delivery capability, without the need to purchase any new aircraft. Similar expanded capabilities could also be realized in Europe, Africa, and Southeast Asia.

FIGURE 7.5 MAFFS installed in C-130 currently operated by US Air National Guard.

Source: Public domain

7.7.2 OTHER MULTI-ROLE MILITARY TRANSPORT AIRCRAFT

Aside from the C-130 MAFFS, many different military transport aircraft have been used for aerial firefighting (Drew, 2015). The advantages of using contemporary production-line military cargo aircraft include state-of-the-art avionics, especially night vision capability and ground terrain collision avoidance systems. Night vision capability permits a significant extension in operating time (EMV, 2018), and therefore improves overall effectiveness. Rear loading ramps also permit quick roll-on and roll-off of various payloads. In particular, the multirole Airbus C295W can be quickly converted to a fire bomber by installing a removable gravity feed system jointly developed with Coulson (USA) as well as being used for medevac or cargo/troop transport (Gabbert, 2015).

No firefighting agency has yet proposed using the Airbus A400M (Figure 7.6) or the Boeing C-17 Globemaster as fire bombers, although this appears to be a logical progression – following recent past developments. Since the C-17 has a payload capacity of ~80 tons, in principle, it would be possible to fit a larger version of MAFFS II (or a similar alternative system) into a C-17 to deliver ~65,000 liters of water (allowing ~15 tons for the dry system*). Another possible delivery method is to consider the adoption of the so-called containerized aerial firefighting system (CAFFS) and similar container systems (Cooke, 2014; Drew, 2015).

FIGURE 7.6 Impression of Airbus A-400M interchangeable firebomber with MAFFS-type system.

Source: Reprinted with permission from Dorrington (2020).

According to this study, fire frequency and severity are strongly related not only to changes in land use, but also to precipitation. The intensity variables of land use change, such as agriculture, forestry, and tourism, indicate human-induced effects on wildfire intensities.

7.8 CONCLUSION

According to the findings of this research, land use modification is one of the chief factors in explaining the geographical pattern of forest fires. Other important determinants are vegetation type and precipitation. Forest fires are caused by a specific mix of natural conditions, and when these elements are combined with human activity, they are more likely to occur. Forest fires are caused mostly by human activity, with vegetation type and precipitation ranking second and third, respectively. These factors should be considered while developing forest fire prevention strategies. For wildfire inhibition, control and supervision of activities are essential, particularly in locations where fire use is an essential element of the productive cycle. In addition, there is good evidence that Large and Very-Large Air Tankers (LATs and VLATs) are a useful tool in mitigating the unacceptable destruction caused by wildfires (e.g., Heffernan, 2018; Kline, 2017; Keating et al., 2012; Loane & Gould, 1986). However, there are currently too few of these aircraft available world-wide. Moreover, many of the existing airframes currently in use are relatively old and lack state-of-the-art systems that permit night operation capability.

Based on past decisions over recent decades, the costs of establishing large fleets of LATs/VLATs (especially new airframes fitted with state-of-the-art avionics) appear to be prohibitively high to fund on a state/national basis, despite the history and future threat of sporadic multi-billion-dollar insurance claims (Doerr & Santin, 2013). Consequently, it appears sensible to consider the formation of an international body that provides an internationally shared fleet of LAT/VLATs, operated under a set of uniform standards and recommended procedures. When properly implemented and coordinated, the deployment of such an emergency-use fleet should not only result in enhanced wildfire protection capabilities for under-equipped agencies in developing nations, but also for well-established firefighting agencies such as EMV in Australia.

7.9 RECOMMENDATIONS

Preventing forest fires is made easier by enforcing rules on what people may and may not do. Following a brief discussion of past, current, and possible future aerial firefighting systems, it is proposed that an international fleet of air tanker aircraft should be deployed to mitigate the anticipated destruction resulting from future wildfires. It is suggested that these aircraft should be large, i.e., capable of delivering ~15,000–75,000 liters of water or retardant per sortie, as well as being interchangeable as cargo transports. It is also suggested that this fleet be managed by an appropriate United Nations office and funded by international contributions. In the event of any wildfire emergency occurring, any subscribing member nation would have optional recourse to this fleet operated in coordination with their own national/state fire agencies. Furthermore, to provide enhanced early warning, it is proposed that a constellation of mini/micro/nano-satellite platforms be deployed in low Earth orbit to remotely detect any incipient wildfire. It is also suggested that the data from the satellite constellation should result in the free worldwide dissemination of appropriate automatic warnings and that this service be managed by the UN, similarly funded by international contributions. There is a pressing need to incorporate fire control into global natural resource management policies as our world continues to endure the cumulative effects of global warming and urban development.

KEYWORDS

- **climate change**
- **forest fire**
- **forest vegetation**
- **global warming**
- **stress precursor monitoring**
- **susceptibility**
- **sustainability**

REFERENCES

Agee, J. K., (1996). *Fire Ecology of Pacific Northwest Forests*. Island Press.

Avila, F. D. Y., Pompa, G. M., & Vargas, P. E., (2010). Spatial analysis of fire occurrence in the Durango State. *Revista Chapingo. Serie Ciencias Forestales, 16*(2), 253–260.

Barbero, R., John, T. A., Narasimhan, K. L., Crystal, A. K., & Brian, S., (2015). Climate change presents increased potential for very large fires in the contiguous United States. *International Journal of Wildland Fire, 24*(7), 892–899. https://doi.org/10.1071/WF15083.

Benz, S., (2013). *The MAFFS II Aerial Firefighting System.* C-130 Hercules Operators Council. https://wildfiretoday.com/documents/P3s_Blue_Aerospace.pdf (accessed on 22 April 2024).

Briles, C. E., Cathy, W., & Patrick, J. B., (2005). Postglacial vegetation, fire, and climate history of the Siskiyou Mountains, Oregon, USA. *Quaternary Research, 64*(1), 44–56. https://doi.org/10.1016/j.yqres.2005.03.001.

Brubaker, L. B., (1988). Vegetation history and anticipating future vegetation change. *Ecosystem Management for Parks and Wilderness*, 41–61. https://doi.org/10.1016/j.yqres.2003.07.009.

Cooke, P. L., (2014). *Special Report: Colorado Firefighting Air Corps Report to the Governor and General Assembly on Strategies to Enhance the State's Aerial Firefighting Capabilities, Colorado Division of Fire Prevention and Control,* Lakewood, USA. See also: Containerized Aerial Fighting System (CAFFS), Guardian System. https://caylym.com/guardian (accessed on 22 April 2024).

Crausbay, S. D., Philip, E. H., Douglas, G. S., & Linda, B. B., (2017). Fire catalyzed rapid ecological change in lowland coniferous forests of the Pacific Northwest over the past 14,000 years. *Ecology, 98*(9), 2356–2369. https://doi.org/10.1002/ecy.1897.

Davis, R., Zhiqiang, Y., Andrew, Y., Cole, B., & Warren, C., (2017). The normal fire environment—Modeling environmental suitability for large forest wildfires using past, present, and future climate normals. *Forest Ecology and Management, 390*, 173–186. https://doi.org/10.1016/j.foreco.2017.01.027.

Doerr, S. H., & Cristina, S., (2013). *Wildfire: A Burning Issue for Insurers.* Lloyds: London, UK.

Dorrington, G. E. (2020) Global Wildfire Protection: a Proposal for an International Shared Aerial Wildfire-Fighting Aircraft Fleet and Satellite Early Warning System; Conference Paper · October 2020.

Drew, B., (2015). *Military Aerial Fire Fighting – a Drop in the Ocean.* Aerial Fire Fighting Conference, Croatia.

Eckel, M., (2021). *Russia Is Burning. Here's What You Need to Know.* RadioFreeEurope Radio Liberty. Retrieved from: https://www.rferl.org/a/russian-wildfires-climate-change/31429880.html (accessed on 22 April 2024).

Emel, O., (2021). *Wildfires Ravaging Forestlands in Many Parts of World: Form North America to Australia, World Grappling with Massive Forest Fires.* Retrieved from: https://www.aa.com.tr/en/world/wildfires-ravaging-forestlands-in-many-parts-of-world/2327146 (accessed on 22 April 2024).

EMV, (2018). *Night-Time Firebombing Approved.* Emergency Management Victoria, Port Melbourne, State of Victoria. https://www.emv.vic.gov.au/news/night-time-firebombing-approved (accessed on 22 April 2024).

European Commission, (2021). *Forest Fires: EU Continues to Mobilize Assistance for Greece and Other Countries in Region.* Retrieved from: https://ec.europa.eu/commission/presscorner/detail/en/ip_21_4121 (accessed on 22 April 2024).

Fried, J. S., Gilless, J. K., Riley, W. J., Moody, T. J., De Blas, C. S., Hayhoe, K., Moritz, M., et al., (2008). Predicting the effect of climate change on wildfire behavior and initial attack success. *Climatic Change, 87*, Torn, 251–264 https://doi.org/10.1007/s10584-007-9360-2.

Gabbert, B., (2015). *Coulson to Supply Retardant Systems for Airbus C295W.* Fire Aviation, 10 https://fireaviation.com/tag/c295/ (accessed on 22 April 2024).

Gavin, D. G., Douglas, J. H., Feng, S. H., Kenneth, P. L., Susan, J. P., Kendrick, J. B., Jason, A. L., et al., (2007). Forest fire and climate change in western North America: Insights from sediment charcoal records. *Frontiers in Ecology and the Environment, 5*(9), 499–506. https://doi.org/10.1890/1540-9295(2007)5[499:FFACCI]2.0.CO;2.

Gavin, D. G., Linda, B. B., & Noah, G. D., (2013). Postglacial climate and fire-mediated vegetation change on the western Olympic Peninsula, Washington (USA). *Ecological Monographs, 83*(4), 471–489. https://doi.org/10.1890/0012-9623-94.4.386.

Halofsky, J. E., Peterson, D. L., & Harvey, B. J. (2020). Changing wildfire, changing forests: the effects of climate change on fire regimes and vegetation in the Pacific Northwest, USA. Fire Ecology, 16(1), 1–26.

Haugo, R. D., Bryce, S. K., Alina, C. C., Crystal, A. K., Kerry, B. K., James, C. R., Kerry, L. M., et al., (2019). The missing fire: Quantifying human exclusion of wildfire in Pacific Northwest forests, USA. *Ecosphere, 10*(4), e02702. https://doi.org/10.1002/ecs2.2702.

Heffernan, J., (2018). *Large Air Tanker Operations in New South Wales, Aerial Firefighting Asia Pacific Conference, Wollongong, Australia.*

Holden, Z. A., Alan, S., Charles, H. L., Matt, J. W., Marco, M., Jared, W. O., Dyer, A. W., et al., (2018). Decreasing fire season precipitation increased recent western US forest wildfire activity. *Proceedings of the National Academy of Sciences, 115*(36), E8349–E8357.https://doi.org/10.1073/pnas.1802316115.

https://reliefweb.int/report/turkey/europe-wildfires-information-bulletin-2-13-august-2021 (accessed on 22 April 2024).

https://www.pnas.org/doi/pdf/10.1073/pnas.1802316115 (access on 6 May 2024).

https://www.themoscowtimes.com/2021/08/17/russias-2021-wildfires-now-largest-in-its-recorded-history-a74814 (accessed on 22 April 2024).

Itter, M. S., Andrew, O. F., Mevin, B. H., Philip, E. H., Jennifer, R. M., Ryan, K., & McLachlan, J. S., (2017). A model-based approach to wildland fire reconstruction using sediment charcoal records. *Environmetrics, 28*(7), e2450.

Keating, E. G., Andrew, R. M., Carter, C. P., Dulani, W., Daniel, M. N., Christina, P., Evan, S., & Ricardo, S., (2012). *Air Attack Against Wildfires: Understanding US Forest Service Requirements for Large Aircraft.* ISBN: 978-0-8330-7677-9.

Keeley, J. E., & Alexandra, D. S., (2016). Climate change and future fire regimes: Examples from California. *Geosciences, 6*(3), 37. https://doi.org/10.3390/geosciences6030037.

Kerns, B. K., Powell, D. C., Mellmann-Brown, S., Carnwath, G., & Kim, J., (2017). Effects of climatic variability and change on upland vegetation in the blue mountains [Chapter 6]. In: Halofsky, J. E., & Peterson, D. L., (eds.), *Climate Change Vulnerability and Adaptation in the Blue Mountains* (Vol. 939, pp. 149–250). Gen. Tech. Rep. PNW-GTR-939. Portland, OR: US Department of Agriculture, Forest Service, Pacific Northwest Research Station.

Kitzberger, T., Donald, A. F., Anthony, L. W., & Thomas, W. S., (2017). Direct and indirect climate controls predict heterogeneous early-mid 21st century wildfire burned area across western and boreal North America. *PloS One, 12*(12), e0188486. https://doi.org/10.1371/journal.pone.0188486.

Kline, J. S., (2017). *Hidden Decision Traps Affecting Aerila Fire Fighting Use and Effectiveness (VLAT Example).* ResearchGate, Working paper. doi: 10.13140/RG.2.2.20691.07209; https://www.researchgate.net/publication/321167497_Hidden_Decision_Traps_Affecting_Aerial_Firefighting_Use_and_Effectiveness_VLAT_Example; https://www.researchgate.net/publication/321167497_Hidden_Decision_Traps_Affecting_Aerial_Firefighting_Use_and_Effectiveness (accessed on 22 April 2024).

Krawchuk, M. A., Max, A. M., Marc-André, P., Jeff, V. D., & Katharine, H., (2009). Global pyrogeography: The current and future distribution of wildfire. *PloS One, 4*(4), e5102. https://doi.org/10.1371/journal.pone.0005102.

Littell, J. S., David, L. P., Karin, L. R., Yongquiang, L., & Charles, H. L., (2016). A review of the relationships between drought and forest fire in the United States. *Global Change Biology, 22*(7), 2353–2369. https://doi.org/10.1111/gcb.13275.

Littell, J. S., Elaine, E. O., McKenzie, D., Jeffrey, A. H., James, A. L., Robert, A. N., & Marketa, M. E., (2010). Forest ecosystems, disturbance, and climatic change in Washington State, USA. *Climatic Change, 102*(1), 129–158. https://doi.org/10.1007/s10584-010-9858-x.

Littell, J. S., McKenzie, D., Ho, Yi, W., & Samuel, A. C., (2018). Climate change and future wildfire in the western United States: An ecological approach to nonstationarity. *Earth's Future, 6*(8), 1097–1111. https://doi.org/10.1029/2018EF000878.

Loane, I. T., & James, S. G., (1986). *Aerial Suppression of Bushfires: Cost-Benefit Study for Victoria*. National Bushfire Research Unit, CSIRO Division of Forest Research.

Long, C. J., Cathy, W., Patrick, J. B., & Sarah, H. M., (1998). A 9000-year fire history from the Oregon coast range, based on a high-resolution charcoal study. *Canadian Journal of Forest Research, 28*(5), 774–787. https://doi.org/10.1139/cjfr-28-5-774.

Marlier, M. E., Mu, X., Ruth, E., Ben, L., John, T. A., & Dennis, P. L., (2017). The 2015 drought in Washington State: A harbinger of things to come? *Environmental Research Letters, 12*(11), 114008. https://doi.org/10.1088/1748-9326/aa8fde.

McKenzie, D., & Jeremy, S. L., (2017). Climate change and the eco-hydrology of fire: Will area burned increase in a warming western USA? *Ecological Applications, 27*(1), 26–36. https://doi.org/10.1002/eap.1420.

McKenzie, D., David, L. P., & Jeremy, J. L., (2008). Global warming, and stress complexes in forests of Western North America. *Developments in Environmental Science, 8*, 319–337. https://doi.org/10.1016/S1474-8177(08)00015-6.

McKenzie, D., Ze'ev, G., David, L. P., & Philip, M., (2004). Climatic change, wildfire, and conservation. *Conservation Biology, 18*(4), 890–902. https://doi.org/10.1111/j.1523-1739.2004.00492.x.

Moritz, M. A., Marc-André, P., Enric, B., Meg, A. K., Jeff, V. D., David, J. G., & Katharine, H., (2012). Climate change and disruptions to global fire activity. *Ecosphere, 3*(6), 1–22. https://doi.org/10.1890/ES11-00345.1.

Parks, S. A., Carol, M., John, T. A., Lisa, M. H., Marc-André, P., & Solomon, Z. D., (2016). How will climate change affect wildland fire severity in the western US? *Environmental Research Letters, 11*(3), 035002. https://doi.org/10.1088/1748-9326/11/3/035002.

Parks, S. A., Lisa, M. H., Carol, M., & Cara, R. N., (2015). Wildland fire as a self-regulating mechanism: The role of previous burns and weather in limiting fire progression. *Ecological Applications, 25*(6), 1478–1492. https://doi.org/10.1890/14-1430.1.

Prichard, S. J., Ze'ev, G., Wyatt, O. W., & David, L. P., (2009). Holocene fire and vegetation dynamics in a montane forest, North Cascade Range, Washington, USA. *Quaternary Research, 72*(1), 57–67. https://doi.org/10.1016/j.yqres.2009.03.008. https://doi.org/10.1002/env.2450.

Reilly, M. J., Christopher, J. D., Garrett, W. M., Thomas, A. S., Robert, E. K., John, D. B., & Kevin, B., (2017). Contemporary patterns of fire extent and severity in forests of the Pacific Northwest, USA (1985–2010). *Ecosphere, 8*(3), e01695.https://doi.org/10.1002/ecs2.1695.

Sea, D. S., & Cathy, W., (1995). Postglacial vegetation and climate of the Cascade Range, Central Oregon. *Quaternary Research, 43*(3), 370–381.https://doi.org/10.1006/qres.1995.1043.

Stavros, E. N., John, A., Narasimhan, K. L., McKenzie, D., & Ashley, S. E., (2014). Climate and very large wildland fires in the contiguous western USA. *International Journal of Wildland Fire, 23*(7), 899–914. https://doi.org/10.1071/WF13169.

Syphard, A. D., Jon, E. K., Anne, H. P., & Ken, F., (2017). Human presence diminishes the importance of climate in driving fire activity across the United States. *Proceedings of the National Academy of Sciences, 114*(52), 13750–13755. https://doi.org/10.1073/pnas.1713885114.

Syphard, A. D., Volker, C. R., Nicholas, S. K., Robert, S. T., Todd, J. H., Susan, I. S., & Murray, K. C., (2008). Predicting spatial patterns of fire on a southern California landscape. *International Journal of Wildland Fire, 17*(5), 602–613. doi: 10.1071/WF07087.

The Moscow Times, (2021). *Russia's 2021 Wildfires Now Largest in Its Recorded History.* https://www.themoscowtimes.com/2021/08/17/russias-2021-wildfires-now-largest-in-its-recorded-history-a74814 (accessed on 6 May 2024).

Walsh, M. K., Cathy, W., & Patrick, J. B., (2008). A 14,300-year-long record of fire–vegetation–climate linkages at Battle Ground Lake, Southwestern Washington. *Quaternary Research, 70*(2), 251–264. https://doi.org/10.1016/j.yqres.2008.05.002.

Walsh, M. K., Jennifer, R. M., Simon, J. G., Kendrick, J. B., & Daniel, G. G., (2015). A regional perspective on Holocene fire–climate–human interactions in the Pacific Northwest of North America. *Annals of the Association of American Geographers, 105*(6), 1135–1157. https://doi.org/10.1080/00045608.2015.1064457.

Walsh, M. K., Whitlock, C., & Bartlein, P. J., (2010). 1200 years of fire and vegetation history in the Willamette Valley, Oregon and Washington, reconstructed using high-resolution macroscopic charcoal and pollen analysis. *Palaeogeography, Palaeoclimatology, Palaeoecology, 297*(2), 273–289. https://doi.org/10.1016/j.quascirev.2010.02.011.

Westerling, A. L., (2016). Increasing western US forest wildfire activity: Sensitivity to changes in the timing of spring. *Philosophical Transactions of the Royal Society B: Biological Sciences, 371*(1696), 20150178. https://doi.org/10.1098/rstb.2015.0178.

Whitlock, C., (1992). Vegetational and climatic history of the Pacific Northwest during the last 20,000 years: Implications for understanding present-day biodiversity. *Northwest Environmental Journal, 8*, 1–24.

Wildfires Information Bulletin 2, (2021). **The Situation – Overview.**

Worona, M. A., & Cathy, W., (1995). Late quaternary vegetation and climate history near Little Lake, central Coast Range, Oregon. *Geological Society of America Bulletin, 107*(7), 867–876. https://doi.org/10.1130/0016-7606(1995)107%3C0867:LQVACH%3E2.3.CO;2.

Yang, J., Hong, S. H., Stephen, R. S., & Eric, J. G., (2007). Spatial patterns of modern period human-caused fire occurrence in the Missouri Ozark Highlands. *Forest Science, 53*(1), 1–15.

Yue, X., Loretta, J. M., Jennifer, A. L., & Jed, O. K., (2013). Ensemble projections of wildfire activity and carbonaceous aerosol concentrations over the western United States in the mid-21st century. *Atmospheric Environment, 77*, 767–780. https://doi.org/10.1016/j.atmosenv.2013.06.003.

CHAPTER 8

Traditional Healthcare System in Rural Communities: An Overview from the Himalayan Protected Area

SHOWKAT AZIEM,[1] JAHANGEER A. BHAT,[2] ZUBAIR A. MALIK,[3] AJEET KUMAR NEGI,[1] and N. P. TODARIA[1]

[1]*Department of Forestry and Natural Resources, H.N.B. Garhwal University, Srinagar, Garhwal, Uttarakhand, India*

[2]*Piramal Foundation for Education Leadership, Piramal University, Mumbai, India*

[3]*Department of Biology, Government HSS Hardutooru, Anantnag, Jammu and Kashmir, India*

ABSTRACT

Medicinal plants are used to treat diseases primarily in remote areas of developing countries. When compared to modern medicine, they are readily available and have few to no side effects. The current study focused on the documentation of traditional medicinal plant uses in health care in remote villages of Govind Pashuvihar Wildlife Sanctuary (GPVWS). Information was gathered through interviews with 290 people from the Wildlife Sanctuary's 20 villages. We evaluated commonly used ethnobotanical indices such as informant consensus factor (Fic) and use value based on the data obtained (UV). A total of 87 taxa were reported by informants to have medicinal importance and 69% of the reported genera belong to herb strata. The Asteraceae was found to be the dominant family in terms of the number of medicinal plants. About 63% of the plants or products were used alone to treat a particular ailment. The most commonly used plant part for herbal preparation

Sustainable Forest Resources Management: Issues and Implications. Jahangeer A. Bhat, Gopal Shukla, Manmohan J. Dobriyal, Sumit Chakravarty, A. Arunachalam, and Rainer W. Bussmann (Eds.)

was the leaves. The informant consensus factor values ranged from 0.00 to 0.98. Jaundice had the highest value reported. The current research reveals that medicinal plant use serves as primary health care for people living in GPVWS's remote area. The people of this region are extremely knowledgeable about ethnomedicinal plants. The traditional knowledge among the rural folk should be given patronage at the government scale that will help in the conservation of important high-valued medicinal plants that are endemic to the region and are facing a serious threat from a varied degree of extinction.

8.1 INTRODUCTION

Medicinal and aromatic plants embody a consistent part of the natural biodiversity bequest of many countries (Okigbo et al., 2008). The assenting contribution of medicinal plants to rural economies as well as their role as an alternative to Western medicine has been well-recognized throughout the world (Cunningham, 1989; Mzimela, 1995; van Wyk et al., 1997; Williams et al., 2000; Malik et al., 2015). The Indian Himalayan region is tantamount to the treasure trove that offers expedient ways for harvesting medicinal plants (Rao et al., 2004). The contribution of medicinal plants to the health of rural people in the Himalayan region is extremely important because most of the population still relies on traditional healthcare systems (Nautiyal et al., 2005; Negi et al., 2010). Govind Wildlife Sanctuary (GWLS) forms a part of the Indian Himalayan region situated in Uttarkashi district of Uttarakhand, India. The region supports luxuriant diversity including a huge number of high medicinal plants (Balodi & Kumar, 2002). The region, owing to its remoteness and lack of infrastructure, makes people rely on natural products to fulfill their daily-based life supporting necessities. Due to the lack of proper medical infrastructure, people use traditional medicine for treating most ailments. In the wake of the significance of medicinal plants to the communities living in the remote pockets of this immense ecosystem, the present study was carried out to ascertain the significance of medicinal plants in the remote areas of Govind Pashuvihar Wildlife Sanctuary.

8.2 MATERIALS AND METHODS

8.2.1 STUDY AREA

The current study was conducted in the Govind Pashuvihar Wildlife Sanctuary, which is one of Uttarakhand's most important protected areas. The sanctuary, which covers an area of approximately 958 km^2 and has elevations ranging from 1,290 to 6,387 m asl (Figure 8.1), is one of the most biologically diverse areas of the Central Himalaya. The vegetation ranges from tropical deciduous forest in the foothills to temperate forest in the middle elevations, coniferous, subalpine, and alpine forest at higher elevations, alpine grasslands and high-altitude meadows, and finally scrublands leading up to the permanent snowline (Negi et al., 2010). The sanctuary forms specific microhabitats for many rare, endangered, and threatened taxa (Singh et al., 2014). The sanctuary has 42 adjacent villages wherein the inhabitants are dependent on its forest resources for their basic needs, including ethnomedicine (Negi et al., 2010).

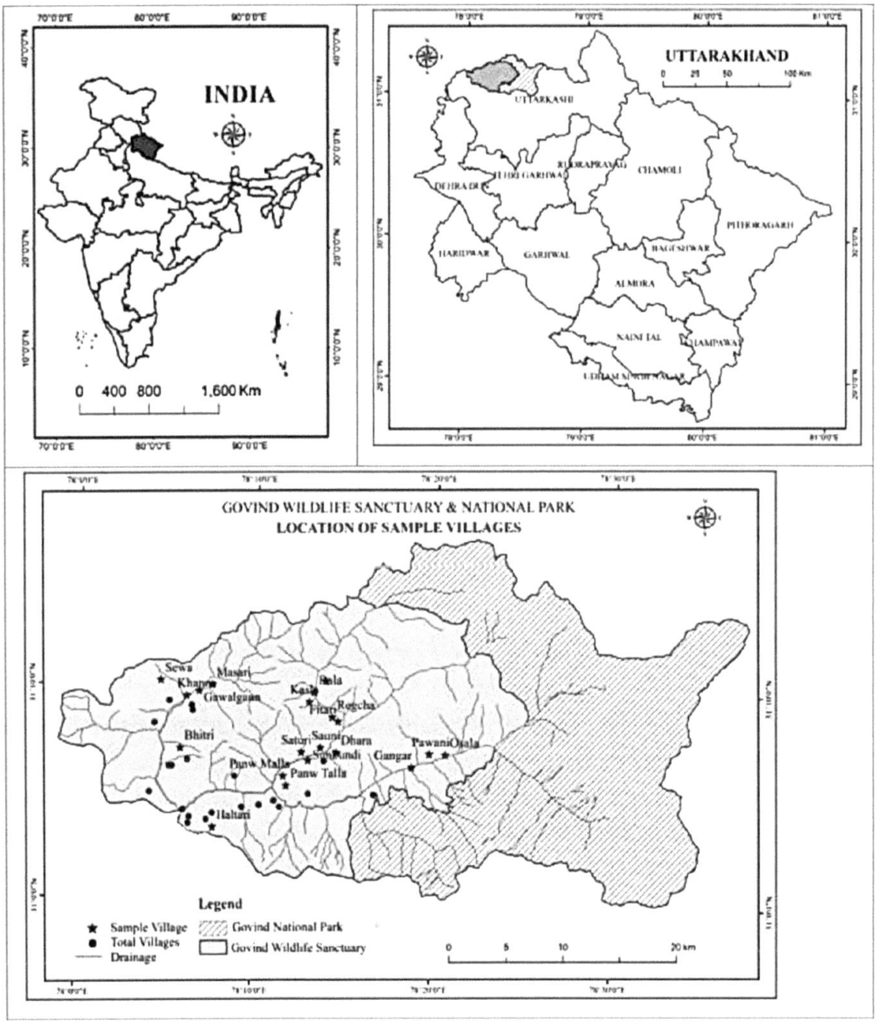

FIGURE 8.1 Map of study area showing the location of sampled villages.

8.2.2 METHODS

Ethno-medicinal survey was carried out in the sanctuary to quantify the ethno-medicinal use of plants to cure different diseases. The questionnaire was devised after going through the literature, and all the questions were framed to assess the ethno-medicinal knowledge among the rural folk living within the boundaries of the sanctuary. The stratified random sampling was followed for sampling villages. Out of a total of 42 villages, 20 villages were selected randomly across all the three ranges viz, Rupin, Supin, and Sankri in a ratio of 2:3:1 to make the sample representative. From these villages, a total of 290 respondents were selected randomly from whom the required information was gathered as per the questionnaire. Information about the plant local names, uses, parts used, and collection methods of each species was recorded. The respondents were asked to list the medicinal plants they knew and their uses to cure the diseases. During this exercise, the plants were identified by the local names and were further cross-checked with the available literature by referring to the local flora. The plants that were not found in the literature were taken as specimens for identification. Voucher specimens were identified using the related literature including Flora of Tons Valley (Rana et al., 2003), Flowers of the Himalaya (Polunin & Stainton, 1997). To confirm identification, the specimens were compared to authentic specimens housed at the Herbaria Botanical Survey of India in Dehradun (BSD). The specimens were submitted to the BSI Herbarium, and accession numbers were obtained.

8.2.3 STATISTICAL ANALYSIS

8.2.3.1 INFORMANT CONSENSUS FACTOR (F_{IC})

To investigate the total usage of plant species based on cultural applicability, the factor informant consensus (F_{IC}) was calculated. The acronym F_{IC} or ICF is used to represent the consistency of information among informants. Health disorders were classified into groups, such as plants with a high F_{IC} value being more pharmacologically active than plants with a low F_{IC} value (Trotter & Logan, 1986). The F_{IC} value can be calculated by the formula:

$$Fic = \frac{nur\text{-}nt}{nur\text{-}1}$$

where; F_{ic} is the informants consensus factor; nur is the number of use citation; and nt is the number used species.

8.2.3.2 USE VALUE (UV)

The use value (UV), as adapted by Ferreira et al. (2009) from the proposal of Phillips et al. (2002), was calculated for each species. Based on the relative use of each medicinal species among informants, this quantitative method assesses its relative importance. The following formula was used to calculate the use value:

$$UV = \frac{\Sigma U}{n}$$

where; 'U' represents the number of times a species is mentioned and n represents the number of informants. The use value of each species is thus objectively based on the importance assigned by the informants and does not depend on the researcher's opinion.

8.3 RESULTS

8.3.1 MEDICINAL PLANTS DIVERSITY, LIFE FORM, AND HABITAT

A total of 87 plant species belonging to 46 families and 78 genera were reported by the informants to have medicinal properties and used as ethnomedicines to cure different ailments (Table 8.1). The six dominant families in terms of the number of species were Asteraceae, Rosaceae, Polygonaceae, Ranunculaceae, Lamiaceae, and Liliaceae (Figure 8.2). The most commonly used plants were mostly herbs (69%) followed by shrubs (16%), trees (10%), and climbers (5%) (Figure 8.3).

8.3.2 FREQUENCY OF PLANT PARTS USED, WAYS OF PREPARATION AND ADMINISTRATION

Among the parts, leaves, roots, fruits, and flowers were mainly utilized by the informants (Figure 8.4). A detailed analysis concluded that leaves (31%), followed by roots (29%), seeds (4%), stems (4%), flowers (4%), fruits (7%), bark (4%), and in some cases the whole plant (11%) were used for ethnomedicinal purposes. Rhizome, stalk, and latex were the least used plant parts (Figure 8.4).

TABLE 8.1 Ethno-Medicinal Plant Species Used in Traditional Healthcare in Govind Wildlife Sanctuary and National Park

Scientific Name	Family	Common Name	Habitat	Part Used	Diseases Cured	Extent Use (%)
Achyranthes aspera	Amaranthaceae	Chamra	Herb	Root	Eye infection, pregnancy, tooth cleaning.	4.69
Aconitum falconeri Holmes ex Stapf	Ranunculaceae	Meetha	Herb	Leave, root, whole plant	Asthma, gastrointestinal disorders, immunity, poison, tooth cleaning, swealing.	9.38
Aconitum heterophyllum Wall. ex Royle	Ranunculaceae	Atees	Herb	Leave, root, whole plant	Abdominal cramps, dehydration, digestive disorders, fever, fever, and stomach ache, gastrointestinal disorders, immunity, multiple, rheumatism, vomiting, cuts, and wounds, cough, and cold.	20.31
Acorus calmus L.	Acoraceae	Boj	Herb	Root	Cough and cold, gastrointestinal disorders, worms, snake bite, menstrual disorder, bronchitis.	9.38
Actaea acuminata Wall. Ex Royle	Ranunculaceae	–	Herb	Whole plant, leaves, flower	Antirheumatic, purgative, nerve seditive, body inflimation.	7.81
Betula utalis	Acoraceae	BhojPatra	Tree	Bark	Psychological disorder, cut-wounds, jaundice, cough-cold.	3.13
Aesculus indica (Wall. ex Cambess.) Hook. f.	Hippocastanaceae	Pangar	Tree	Fruit	Burns	1.56
Ageratum conyzoides L.	Asteraceae	Pudinia	Herb	Whole plant	Digestive disorders	1.56
Ajuga parviflora Wall. ex Benth	Lamiaceae	Neelkanti	Herb	Leave, whole plant	Burns, fever, gastrointestinal disorders, intestinal infection, multiple, urinary infection.	9.38
Anaphalis busua (Buch-Ham. ex D. Don) DC.	Asteraceae	Rangdu grass	Herb	Leave	Cuts and wounds	1.56
Angelica glauca Edgew.	Apiaceae	Chora	Herb	Root	Abdominal cramps, cough, and cold, digestive disorders, gastrointestinal disorders, insect replent, kidney stone, worms.	10.94

TABLE 8.1 *(Continued)*

Scientific Name	Family	Common Name	Habitat	Part Used	Diseases Cured	Extent Use (%)
Argemone neplensis (Don) Nakai	Papaveraceae	Kandara	Herb	Root	Aprodiacic	1.56
Argimonia pilosa Ledeb.	Rosaceae	Kur/Kuri	Herb	Leave	Leprocy	1.56
Artemisia nilagirica C.B. Clarke	Asteraceae	Chamra/Panti	Herb	Leave, whole plant, root.	Gastrointestinal disorders, intestinal infection, respiratory, intestinal worms, worms, psychological disorders.	4.69
Aster penduncularis Wall. ex Nees	Asteraceae	–	Herb	Flower	Hiccups	1.56
Berberis lyceum Royle	Berberidaceae	Kashmol	Shrub	Bark, root	Alergy, diabetes, eye infection, likoria, migrane, fever, stomach ache, diaphoretic.	12.50
Bergenia ciliata (Haw.) Sternb.	Saxifragaceae	Dokodu	Herb	Root, whole plant	Cuts and wounds, fracture, kidney stone, swealing, tooth cleaning, diabetes, tuberculosis, hemorrhoides, asthama, stones.	10.94
Bistorta macrophylla (D. Don) Sojak	Polygonaceae	Kukri	Herb	Root	Mouth ulcer	1.56
Cannabis sativa L.	Cannabaceae	Bhang	Herb	Leave	Insect bite	1.56
Ceiltus australis L.	Cannabaceae	Kharak	Tree	Bark	Swealing, cuts, and wounds	3.13
Chaerophyllum acuminatum L.	Apiaceae	Shan	Herb	Root	Fever, intestinal infection	3.13
Cotoneaster bacillaris Wall. Raush.	Rosaceae	Narik	Shrub	Seed oil	Worms	1.56
Dactylorhiza hatagirea (D. Don)	Orchidaceae	Panja	Herb	Leave, root, tuber	Burns, cut, and wound, digestive disorders, fever, gastrointestinal disorders, imunity, throat irritation, stomach disorder, cough.	14.06

TABLE 8.1 *(Continued)*

Scientific Name	Family	Common Name	Habitat	Part Used	Diseases Cured	Extent Use (%)
Delphinium denudatum Wallich ex Hk.F	Ranunculaceae	Sampjadi	Herb	Leave, root	Antivenome, tonic, toothache.	1.56
Deutzia staminea R.Br. ex Wall	Asteraceae	Dalchi	Shrub	Root, leave	Throat irritation, stomache ache.	1.56
Datura stramonium L.	–	–	–	Leaf, seed, fruit	Skin disease, joint pain, locoderma.	12.50
Dioscorea bulbifera L.	Dioscoreaceae	Ganj	Climber	Root, tuber	Asthama, digestive disorders, fever.	3.13
Drimia indica (Roxb.) Jessop	Liliaceae	Jamosh	Herb	Leave	Gastrointestinal disorders	3.13
Elscholtizia fruticosa (D.Don) Rehder.	Lamiaceae	–	Shrub	Seed oil	Likoria	1.56
Elsholtzia strobilifera Benth	Lamiaceae	Potyi	Herb	Leave	Fever, tooth cleaning	3.13
Galium aparine L.	Rubiaceae	Kushkusha	Herb	Leave, root, whole plant	Cuts and wounds, eye infection, gastrointestinal disorders, pimples, pus.	7.81
Geranium wallichianum D. Don. Ex. Sweet	Geraniaceae	–	Herb	Leave	Urinary infection	1.56
Gerardinia diversifolia (Link) Friis	Urticaceae	Dodi kala	Herb	Leave, root	Asthma, cuts, and wounds, pimples, swealing.	6.25
Hedra neplensis	Araliaceae	Salam mishri	Climber	Root	Aprodiacic	1.56
Hedychium spicatum Buch-Ham. Ex Sm.	Zingiberaceae	Kapoor kachri	Herb	Leave, tuber	Cuts and wounds, fitague, stimulate appetite.	6.25
Juglans regia L.	Juglandaceae	JungliAkhroot	Tree	Leave, root	Cuts and wounds, fever, throat irritation, tooth cleaning.	1.56

TABLE 8.1 *(Continued)*

Scientific Name	Family	Common Name	Habitat	Part Used	Diseases Cured	Extent Use (%)
Allium stracheyi Baker	Alliaceae	Jungli Lassan/ Seemor phool	Herb	Root, tuber	–	23.44
Jurinea dolomiaea Boiss	Asteraceae	Masi	Shrub	Root	Phlegm	1.56
Jurinella macrocephala (Royle) Aswal and Goel	Asteraceae	Dooplakad	Herb	Root	Incense	1.56
Litsea glutinosa (Lour.) Robin	Lauraceae	Singora	Tree	Leave	Plaster	1.56
Mentha arvensis L.	Lamiaceae.	Pudeena	Herb	Leave, whole plant	Rheumatism, vomiting.	3.13
Morina longifolia Wall. ex.DC.	Dipsacaceae	Bishkandari	Herb	Leave	Cuts and wounds, ear infection, eye infection.	1.56
Origanum vulgare L.	LAMIACEAE	Van tulsi	Herb	Leave	Rheumatism	1.56
Oxalis corniculata	Oxalidaceae	Oxalis corniculata	Herb	Leave	Ear infection	3.13
Paris polyphylla Sm.	Liliaceae	Satwa	Herb	Leave, root, rhizome	Cough and cold, cuts, and wounds, digestive disorders, fever, multiple, pus, anthelmintic.	4.69
Parthenocissus semicordata (Wall) DC.	Fabaceae	Laioda	Climber	Latex, root, stalk	Cataracts, eye infection, pimples.	4.69
Phytolocca acinosa (Roxb)	Phytolaccaceae	Jalga	Herb	Fruit, leave, root, seed oil.	Anteseptic, aprodiacic, constipatiensy, gastrointestinal disorders, tooth cleaning.	7.81
Picrorhiza kurrooa Royle ex Benth.	Plantaginaceae	Kadwai	Herb	Leave, rhizome, whole plant, flower.	Cough, cough, and cold, cuts, and wounds, digestive disorders, fever, gastrointestinal disorders, headache, jaundice, migrane, nausea, typhoid, intestinal infection.	18.75
Plantago major L.	Plantaginaceae	–	Herb	Rhizome	Gastrointestinal disorders, headache, Jaundice	17.19

TABLE 8.1 *(Continued)*

Scientific Name	Family	Common Name	Habitat	Part Used	Diseases Cured	Extent Use (%)
Polygonatum cirrhifolium (Wall.) Royle	Liliaceae	Made	Herb	Root, whole plant	Tonic, blood purification	3.13
Polygonatum verticillatum (L.) All.	Liliaceae	–	Herb	Rhizomes	Tonic, aprodiacic	9.38
Polygonum amplexicaule D. Don	Polygonaceae	Nenai	Herb	Root	Eye infection, mouth ulcer, tooth cleaning, ear infection.	6.25
Polygonum polystachyum Wall. Ex Meissn	Polygonaceae	Bout	Herb	Leave	Cuts and wounds, fever	3.13
Prinsepia utilis Royle	Rosaceae	Bekal	Shrub	Seed oil	Worms	1.56
Prunus armeniaca L.	Rosaceae	Chulau	Tree	Seed oil	Rheumatism	1.56
Quercus semicarpifolia Smith	Fagaceae	Kharsu	Tree	Leave	Cuts and wounds	1.56
Rheum austral D.Don	Polygonaceae.	Archa	Herb	Fruit, leaf, root, whole plant	Blood clot, cuts, and wounds, hemorrhoids, indigestion, pus, relaxing agent/fatigue, scars, swealing.	12.50
Rosa macrophylla Lindl.	Rosaceae	Tatolu	Shrub	Root	Blood purifier	1.56
Rubus niveus Thunb.	Rosaceae	Kala	Shrub	Leave, root	Cuts and wounds, fever, pimples	4.69
Rubus spp.	Rosaceae	Katrada	Shrub	Root, fruit	Sleep apnea, stomachache, dysentery.	1.56
Rumax dentatus L.	Polygonaceae	Chadi patri	Herb	Leave	Burns	1.56
Rumex hastatus D.Don	Polygonaceae	Amloodi	Herb	Leave	Cough and cold, cuts, and wounds, headache	4.69
Rumex nepalensis Spr.	Polygonaceae	Chadi patri	Herb	Leave	Colic	1.56
Sassurea obvallata (DC.) Edgew	Asteraceae	Bramakamal	Herb	Whole plant	Fever, relaxing agent/fatigue	3.13

TABLE 8.1 *(Continued)*

Scientific Name	Family	Common Name	Habitat	Part Used	Diseases Cured	Extent Use (%)
Saussurea costus (Fale.) Lipsch	Asteraceae	Garyakut/kut	Herb	Root	Cold, Cough, and cold, cuts, and wounds, dehydration, fever, imunity, jaundice, kidney stone, snakebite, and skin diseases, toothache.	15.63
Saussurea gossypiphora D.Don	Asteraceae	Baidgada	Herb	–	Sexual disorder	1.56
Saussurea obvallata (DC.) Edgew.	Asteraceae	Bramakamal	Herb	Leave	Blood purifier	1.56
Skimmia anquetilia N.P. Taylor and Airy Shaw	Rutaceae	–	Shrub	Leaf	Scabies	7.81
Smilax aspera L.	Smilacaceae	–	Climber	Roots	Diuretic	18.75
Solanum nigrum L.	Solanaceae	Bumbla	Herb	Leave	Cuts and wounds, fever	3.13
Sonchus oleraceus L.	Asteraceae	Dudimori	Herb	Root	Cough and cold, head ache, imunity.	4.69
Spiraea canescens D.Don.	Rosaceae	Chaklate	Shrub	Root	Gastrointestinal disorders	1.56
Stegnogramma mollisima (Kunzes)	Thelypteridaceae	Nirbishi	Herb	Root	Epilepsy, sorosis, and skin disease	3.13
Symplocos racemose Roxb.	Symplocaceae	–	Tree	Bark	Blood purification	1.56
Swertia chirayita (Roxb. ex Fleming) Karsten	Gentianaceae	Chirayata	Herb	Whole plant	Fever	1.56
Taxus baccata L.	Taxaceae	Thunayar	Tree	Bark, root	Cancer, bone fracture, cuts, and wounds, peptic ulcer.	4.69
Thalictrum foliolosum DC.	Ranunculaceae	Mamri	Herb	Root	Eye infection	1.56
Trichosanthestri cuspidate Lour.	Cucurbitaceae	Indrayan	Herb	Fruit	Chicken pox, constipatiensy, mouth ulcer.	4.69
Trifolium repens L.	Fabaceae	Amlodi	Herb	Leave	Ear infection	1.56

TABLE 8.1 *(Continued)*

Scientific Name	Family	Common Name	Habitat	Part Used	Diseases Cured	Extent Use (%)
Trillidium govanianum Kunth	Melanthiaceae	Jatamasi/Nagchetri	Herb	Leave, root, tuber, whole plant.	Cough and cold, cuts, and wounds, fever, gastrointestinal disorders, imunity, incense, poision.	4.69
Urtica dioica L.	Urticaceae	–	Herb	Leaves	Galactogog	1.56
Urtica parviflora Roxb.	Urticaceae	Kali	Shrub	Leave, latex	Abscess, bleeding, menstruation	1.56
Verbascum thapsus L.	Scrophulariaceae	Verbescumthapsus	Herb	Root	Fever	1.56
Viburnum cotinifolium D.Don	Caprifoliaceae	–	Shrub	Bark	Digestive disorder	1.56
Viola canescens Wall. Ex Roxb.	Violaceae	Benaksha	Herb	Flower, leave, flower, whole plant	Cuts and wounds, fever, gastrointestinal disorders, multiple, pneumonia, poison, vomiting.	10.94
Zanthoxylum armatum DC.	Rubiaceae	Timru	Shrub	Latex	Tooth cleaning	1.56

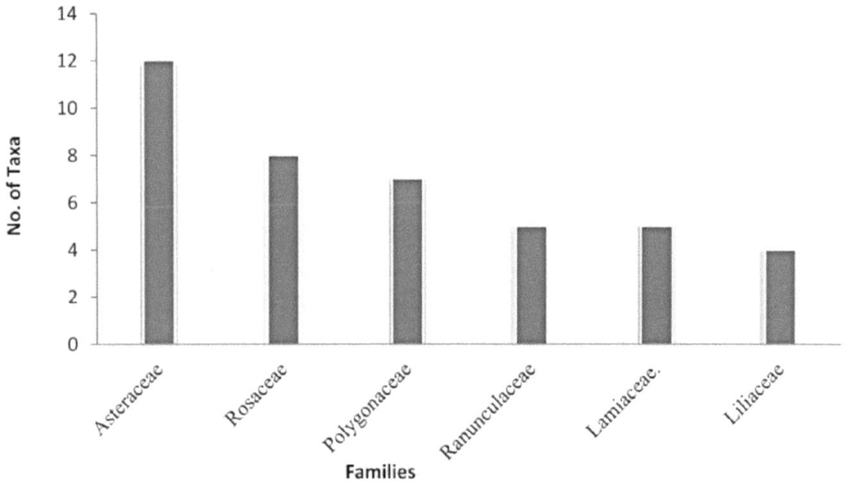

FIGURE 8.2 Dominant families with maximum number of medicinal plants.

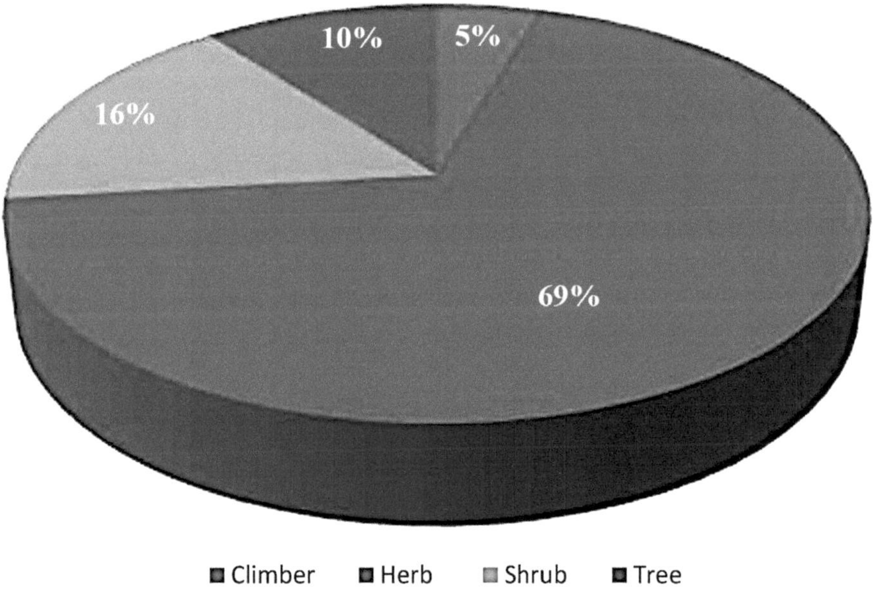

FIGURE 8.3 Percentage of medicinal plants under different life forms in the study area.

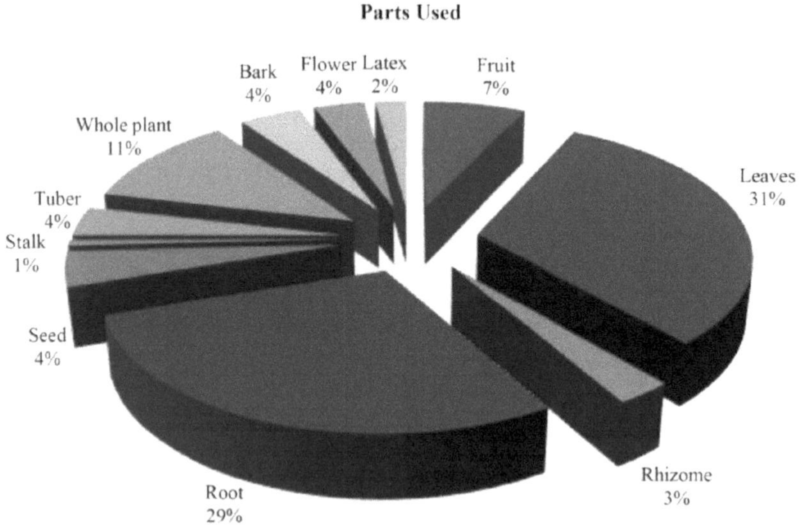

FIGURE 8.4 Percentage of plant parts used in preparing medicines in the study area.

The mode of formulation preparation included either administration of plant part as solo or mixture of two or more plants/parts. The present study reported that 62.79% of the plant products were used alone to treat a particular ailment, while 37.21% were used in a mixture with other plants or mediums. The modes of preparation in use are crushing (22.09%), decoction (12.79%), paste (13.95%), powder (11.63%), fresh part (10.47%), latex (8.14%), and boiling (5.81%). The dominant modes of administration of medicinal recipes included oral (50%) and topical (33.72%), with the rest being infusion (4.65%), smoke inhalation (4.65%), chewing, eating, and rubbing of oil for massage (2.33%).

8.3.3 INFORMANT'S CONSENSUS FACTOR AND USE EXTENT FOR THERAPEUTIC PURPOSES

The ethno-medicinal plants were reported to be used for treating 74 different ailments that were grouped into 16 categories. The highest informant's consensus factor (ICF) value was 0.98 for jaundice, followed by general health (0.86), urological problem (0.84), injuries (0.78), respiratory and bronchial (0.75), gastrointestinal (0.64), fever and headache (0.63), genital and sexual (0.57), ophthalmological (0.44), dental (0.29) categories indicated moderate consistency of informant knowledge in the use of these medicinal plant species while the low ICF value of dermatological disorder (0.08) and rest indicates the less consistency of the informant's knowledge in the plants used for treatment (Table 8.2).

TABLE 8.2 Informant Consensus Regarding Use of Taxa in the Study Area

Category of Illness	Number of Taxa Used (nt)	Number of Use Reports (nur)	Informant's Consensus Index Factor (Fic)*
Gastro-intestinal disorder	37	102	0.64
Jaundice	2	47	0.98
Fever and headache	26	69	0.63
Respiratory and bronchial	15	57	0.75
Dermatological disorder	12	13	0.08
Urological problem	6	33	0.84
Ophthalmological	10	17	0.44
Dental	11	15	0.29
Bites and venomous	5	8	0.43
Genital and sexual diseases	4	8	0.57
Nerve disorders	8	14	0.46
Injuries	36	159	0.78
Infectious/topical diseases	4	4	0.00
Diabetes	2	2	0.00
Pregnancy	2	2	0.00
General health	16	109	0.86

*Fic = nur – nt/(nur – 1), providing a value between 0 and 1, where high value indicates a high rate of informant consensus.

8.3.4 RARE AND THREATENED MEDICINAL PLANTS

The conservation status of these medicinal plants was assessed through secondary literature (Manikandan & Shirasvatav, 2015; IUCN 4.0, 2015). Among the 87 medicinal plant species documented in the present study, about 29% of the species are listed in different threatened categories viz., rare (R), vulnerable (VU; *Aconitum falconeri* and *Allium stracheyi*), endangered (E; *Sassurea bvalata* and *S. costus*) and critically endangered (CR).

8.4 DISCUSSION

Plant and plant products are a valuable source of medicines across the globe, whether it is for traditional use or as a source for the synthetic drug industries. The Indian Himalayan Region is well known for its wealth of medicinal plants, as it supports 1,748 medicinal plants. These plants are used as a source of traditional medicine by the inhabitants living in the remote areas of the Himalayan region (Samant et al., 1998, 2007). The contribution of medicinal plants to the health of rural people in the Himalayan region is indispensable, as most of the population still relies on traditional health-care systems (Nautiyal et al., 2005). The current study provides detailed information on the diversity, life form, habitat, parts used, and indigenous uses for the treatment of several diseases. The present study documented the ethnomedicinal importance of 87 plant species belonging to 46 families and 78 genera. In the same area, Negi et al. (2010) had reported 33 plant species belonging to 32 genera and 28 families that were traditionally used to cure 28 diseases. The dominant family as a source of traditional medicine was Asteraceae, which is in agreement with Rathore et al. (2015).

Intake of oral administration and external topical formulation were the main modes of administration used in traditional herbal medicines, which have also been previously reported (Gaur et al., 2010; Yirga et al., 2012; Laldingliani et al., 2022). Most of the traditional remedies are given through oral routes followed by dermal routes, which permit rapid physiological reactions with pathogens (Zeraburk & Yirga, 2012). The leaves were used more frequently (31%) as compared to roots (29%), seeds (4%), stems (4%), flowers (4%), fruits (7%), and bark (4%). Similar results were reported by Keter & Mutiso in 2012 from Kenya; Bhat et al. in 2013; Malik et al. in 2015; and Rathore et al. in 2015; Singh et al. in 2019 from Western Himalaya, and Akhtar et al. in 2013 from Pakistan. The majority of traditional studies reported that leaves were the most frequently used plant parts (Semwal et al., 2010). Mostly, the medicine is prepared by forming a paste, decoction, or powder. For decoction, the plant part was washed thoroughly and boiled with water, administering the juice orally, whereas for paste, the materials were crushed or rubbed within palms and applied topically. To make fine powder, plant parts were shade dried and ground.

Informants Consensus Factor (ICF) is used to represent the consistency of the information among the informants, indicating whether there was shared knowledge and concurrence in the use of plants for treating the ailment category among the plant's users in the study area (Laldingliani et al., 2022). The ICF reveals the agreement on the use of species by the informants of the region regarding the cure of a particular disease. The index can be an indicator for the chemical analysis of particular species for confirmation and mass-scale use in the drug industry. In the present study, the informant consensus ranged between 0 and 0.98, with the highest value corresponding to Jaundice (0.98), followed by general health (0.86) and Urological (0.84). This indicates that the use of these species is widely known in the study area, and most of the respondents concurred regarding the particular usage. The present findings revealed that three major categories show the ICF value of 0 and thus show disagreement among the general population regarding the use of species to treat these diseases in the communities living in the area. However, it is evident from the present study that 81.25% of the agreements of ICF fall in the range of 0.29 to 0.98. These figures are in agreement with Kumar et al. (2011) who reported a range of consensus factor index (*Fic*) from 0.33 to 1.0 in the lower belts of the Himalayas. Similar findings were reported for the Himalayas, with the ICF value ranging between 0.36 and 0.76 (Hussain et al., 2019).

8.5 CONCLUSION

The current study concluded that the residents of the study area have a unique knowledge of using medicinal plants to treat various ailments. The ethnomedicinal importance of 87 plant species from 46 families and 78 genera was documented in this study. These medicinal plants provide significant primary healthcare for the study area's rural population. These medicinal plants treat approximately 74 different ailments. Identification of active ingredients in plants used by locals may provide some useful leads for the development of new drugs. The calculated informant consensus factor was extremely high, indicating that the collected data can be used as a reference and reliable source for future ethnopharmacological research. Many of the reported species are rare and endangered, necessitating immediate action to conserve such vital resources in order to maximize their use in the primary health care system.

KEYWORDS

- **aromatic plants**
- **ethno-botany**
- **Govind Pashuvihar wildlife sanctuary**
- **herbal preparation**
- **informant consensus factor**
- **medicinal plants**
- **traditional uses**

REFERENCES

Balodi, B., & Kumar, S., (2002). Phytodiversity of Govind wildlife sanctuary, Uttaranchal with special reference to ethnomedicinal information. Phytotaxonomy, 2, 100–105.

Bhat, J. A., Kumar, M., & Bussmann, R. W., (2013). Ecological status and traditional knowledge of medicinal plants in Kedarnath wildlife sanctuary of Garhwal Himalaya, India. Journal of Ethnobiology and Ethnomedicine, 9(1), 1–18. https://doi.org/10.1186/1746-4269-9-1/TABLES/5.

Cunningham, T., (1998). Herbal medicine trade: A hidden economy. Indicator SA, 6(3), 51–54. https://doi.org/10.10520/AJA0259188X_370.

Ferreira, F. S., Brito, S. V., Ribeiro, S. C., Saraiva, A. A., Almeida, W. O., & Alves, R. R., (2009). Animal- based folk remedies sold in public markets in Crato and Juazeiro Do Norte, Ceará, Brazil. BMC Complement Alternative Medicine, 9(1), 17. https://doi.org/10.1186/1472-6882-9-17.

Gaur, R. D., Sharma, J., & Painuli, R. M., (2010). Plants used in traditional healthcare of livestock by Gujjar community of sub-Himalayan tracts, Uttarakhand, India. Indian Journal of Natural Products and Resources, 1(2), 243–248.

Keter, L. K., & Mutiso, P. C., (2012). Ethnobotanical studies of medicinal plants used by traditional health practitioners in the management of diabetes in lower eastern province, Kenya. Journal of Ethnopharmacology, 139(1), 74–80. https://doi.org/10.1016/j.jep.2011.10.014.

Kumar, M., Sheikh, M. A., & Bussmann, R. W., (2011). Ethnomedicinal and ecological status of plants in Garhwal Himalaya, India. Journal of Ethnobiology and Ethnomedcine, 7(1), 32–45. https://doi.org/10.1186/1746-4269-7-32.

Laldingliani, T. B. C., Thangjam, N. M., Zomuanawma, R., Bawitlung, L., Pal, A., & Kumar, A., (2022). Ethnomedicinal study of medicinal plants used by Mizo tribes in Champhai District of Mizoram. India. Journal of Ethnobiology and Ethnomedicine., 18(1), 1–29.

Malik, Z. A., Bhat, J. A., Ballabha, R., Bussmann, R. W., & Bhatt, A. B., (2015). Ethnomedicinal plants traditionally used in health care practices by inhabitants of Western Himalaya. Journal of Ethnopharmacology, 172, 133–144. https://doi.org/10.1016/j.jep.2015.06.002.

Mzimela, T., (1995). The muthi business. Indicator SA, 12(2), 81–85.

Namsa, N. D., Mandal, M., Tangjang, S., & Mandal, S. C., (2011). Ethnobotany of the monpa ethnic group at Arunachal Pradesh, India. Journal of Ethnobiology and Ethnomedicine, 7, 31.

Nautiyal, S., Rajan, K. S., & Shibasaki, R., (2005). Interaction of biodiversity and economic welfare – a case study from the Himalayas of India. Journal of Environmental Informatics, 6(2), 111–119.

Negi, V. S., Maikhuria, R. K., Phondania, P. C., & Rawat, L. S., (2010). An inventory of indigenous knowledge and cultivation practices of medicinal plants in Govind Pashu Vihar wildlife sanctuary, Central Himalaya, India. International Journal of Biodiversity Science, Ecosystem Services and Management, 6, 96–105.

Phillips, O., Gentry, A. H., Reynel, C., Wilkin, P., & Galvez-Durand, C. B., (2002). Quantitative ethnobotany and amazonian conservation. Conservation Biology, 8, 225–248.

Polunin, O., & Stainton, A., (1997). Flowers of the Himalaya. Oxford University Press.

Rana, T. S., Datt, B., & Rao, R. R., (2003). Flora of tons valley, Garhwal Himalaya (Uttaranchal). Bishen Singh Mahendra Pal Singh, Dehra Dun, 410.

Rao, M. R., Palada, M. C., & Becker, B. N., (2004). Medicinal and aromatic plants in agroforestry systems. Agroforestry Systems, 61(1), 107–122.

Rathore, S., Tiwari, J. K., & Malik, Z. A., (2015). Ethno medicinal survey of herbaceous flora traditionally used in health care practices by inhabitants of Dhundsir gad watershed of Garhwal Himalaya, India. Global Journal of Research in Medicinal Plants & Indigenous Medicine, 4(4), 65–78.

Samant, S. S., Dhar, U., & Palni, L. M. S., (1998). Medicinal Plants of Indian Himalaya: Diversity Distribution and Potential Values. Gyanodaya Prakashan: Nainital.

Samant, S. S., Pant, S., Singh, M., Lal, M., Singh, A., Sharma, A., & Bhandari, S., (2007). Medicinal plants in Himachal Pradesh. North Western Himalaya, India, International Journal of Biodiversity Science and Management, 3, 234–251.

Semwal, D. P., Kala, C. P., & Bhatt, A. B., (2010). Medicinal plants and traditional health care system of vaidyas, palsi and others, A case study from Kedarnath Valley of Uttarakhand, India. Medicinal Plants, 2(1), 51–57.

Singh, H., Husain, T., Agnihotri, P., Pande, P. C., & Khatoon, S., (2014). An ethnobotanical study of medicinal plants used in sacred groves of Kumaon Himalaya. Uttarakhand, India Journal of Ethnopharmacology, 154, 98–108.

Singh, S., Bhat, J. A., Malik, Z. A., Youssouf, M., Bussmann, R. W., & Kunwar, R. M., (2019). Sacred groves in Western Himalaya. India: Community-Managed Nature Refuges for Conservation of Biodiversity and Culture, 18, 1–21.

Trotter, R., & Logan, M., (1986). Informant consensus: A new approach for identifying potentially effective medicinal plants. In: Etkin, N. L., (ed.), Plants in Indigenous Medicine and Diet: Bio Behavioral Approaches (pp. 91–112). Bedfort Hills: Redgrave Publishers: New York.

Van, W. B., Van, O. B., & Gericke, N., (1997). Medicinal Plants of South Africa. Briza Publications: Pretoria.

Williams, V. L., Balkwill, K., & Witkowski, E., (2000). Unraveling the commercial market for medicinal plants and plant parts on the Witwatersrand, South Africa. Economic Botany, 54(3), 310–327.

Yirga, G., Teferi, M., Gidey, G., & Zerabruk, S., (2012). An ethnoveterinary survey of medicinal plants used to treat livestock Diseases in Seharti-Samre District. Northern Ethiopia, 10(5897), 113–119.

Zeraburk, S., & Yirga, G., (2012). Traditional knowledge of medicinal plants in Gindeberet District, Western Ethiopia. South African Journal of Botany, 78, 165–169.

CHAPTER 9

Review on Traditional Knowledge and Conservation Status of Medicinal Plants in Uttarakhand: Opportunities, Threats, and Challenges

AJAY MALETHA,[1] VEDANGINI BISHT,[1] SUNIL NAUTIYAL,[2] SATYA KUMAR,[3] JANMEJAY SETHY,[1] MURALI KRISHNA CHATTAKONDA,[1] RAKESH KUMAR MAIKHURI,[4] SURENDRA SINGH BARGALI,[5] and LAKHPAT SINGH RAWAT[1]

[1]*Amity Institute of Forestry and Wildlife, Amity University, Noida, Uttar Pradesh, India*

[2]*G.B. Pant National Institute of Himalayan Environment (NIHE), Kosi-Katarmal, Almora, Uttarakhand, India*

[3]*Center for Ecological Economics and Natural Resources (CEENR), Institute for Social and Economic Change, Nagarbhavi, Bangalore, Karnataka, India*

[4]*Department of Environmental Science, H.N.B. Garhwal University, Srinagar, Garhwal, Uttarakhand, India*

[5]*Department of Botany, DSB Campus, Kumaun University, Nainital, Uttarakhand, India*

ABSTRACT

The medicinal plants have a long historical record of serving humanity in almost all civilizations across the world. Traditional knowledge is the practice

Sustainable Forest Resources Management: Issues and Implications. Jahangeer A. Bhat, Gopal Shukla, Manmohan J. Dobriyal, Sumit Chakravarty, A. Arunachalam, and Rainer W. Bussmann (Eds.)

of sharing knowledge that is developed, sustained, and passed on from generation to generation within communities, including their cultural and spiritual identities. However, with the course of time, this traditional knowledge of indigenous people/communities has started disappearing due to the lack of written evidence, high poverty, and low-income resources. Although, in the past few decades, the traditional medicinal system has regained recognition due to the surging faith in herbal medicines that possess low side effects and affordable prices when compared with allopathic medicines. In Uttarakhand, several medicinal and aromatic plants (MAPs) have been listed under the category of endangered and threatened, and their collection has been banned after India became a signatory to CITES. Unsustainable collection and over-exploitation of some high-value MAPs by local people and migrant laborers pose serious challenges to the ecology and socio-cultural setup of the region, resulting in 14% of total plant species and MAPs (3.4%) being threatened in the Indian Himalaya. This review chapter therefore explores the potential in medicinal plant resources with special emphasis on the state of Uttarakhand and aims to understand the challenges and opportunities in the medicinal plants sector. It also suggests some recommendations based on the current state of knowledge to establish the smooth functioning of the medicinal plants sector and improve the socio-economic conditions of marginalized communities.

9.1 INTRODUCTION

Himalaya is considered one of the global biodiversity hotspots with unique geographical locations, ecological diversity, and evolutionary processes for the floral diversity and species richness which support approximately 18,440 species, of which 25.3% are endemic to the region (Singh & Hajra, 1996; Samant et al., 1998). Himalayan forests provide ecosystem services to the mountainous communities for their well-being, including fresh air, water, timber and wood, food, genetic resources, biomass, sustainable livelihood, etc. The usage of medicinal and aromatic plants for curing ailments and diseases is as ancient knowledge as human civilization, and with time, the usage of medicinal plants for the treatment of diseases was being discovered (De Pasquale, 1984; Rates, 2001; Suneetha, 2010; Choudhary et al., 2013; Kuniyal et al., 2015; Chandra & Sharma, 2018). Currently, as per the World Health Organization (WHO, 2003), as many as 80% of the world's population depends on traditional medicine and considers it as their principal health care system (Yadav & Misra, 2012). With the increasing demand for traditional health care and medicines in the contemporary world, immense pressure is exerted on the usage of MAPs resources from the wild as well as in cultivation for several mountainous communities living in the regions (Lubbe & Verpoorte, 2011; Chandra & Sharma, 2018; Chandra et al., 2019; UNDP, n.d.). The overall estimated herbal market in India was around US$ 1.75 billion in 2008, with exports accounting for about US$ 900 million. These numbers were expected to double by 2012 (Sharma, 2008). In 2012, as per the global export values, many medicinal plant species are suspected of having medicinal values that can be equivalent to 2.2 billion USD. The global demand for plant extracts and other medicines was estimated to be worth several 100 billion dollars in the year 2017 (Ahn, 2017). There is an increase in the number of nations, including China, Nigeria, Thailand, and Mexico, that have decided to integrate the traditional medicine system into their primary healthcare. The trade and extraction of many traditional medicines remain uncontrolled in many nations, but the WHO (World Health Organization) maintains a systematized network in order to facilitate healthy and fair use of MAPs. Some of the most common threats to medicinal plants are climatic and environmental changes, as well as the loss of habitat. A particular danger is the over-collection to meet consumer needs (Ahn, 2017). Uttarakhand is a major contributor of medicinal and aromatic plants (40.10%) resources with a huge traditional knowledge of the local people in the Indian Himalayan region (Kuniyal et al., 2015; Samant & Pal, 2003; UNDP, n.d.; Semwal, Saradhi, Kala, & Sajwan, 2010). It was estimated that several (n=132) species are active in trade; however, approximately 300 species are partially traded, of which 28 plant species are listed as Globally Significant Medicinal Plants (GSMP) by the Uttarakhand Government (Kuniyal et al., 2015; Kuniyal & Sundriyal, 2013; Ved & Goraya, 2008), and this demand is growing in the state at a rate of 15–20% annually (Chandra & Sharma, 2018). Around 90% of MAPs are extracted from the wild in destructive manners (Bisht et al., 2016; Olsen, 1998; UNDP, 2016; WHO, 2003). The indigenous communities inhabiting the mountainous region of Uttarakhand largely depend on diverse medicines provided in the form of provisioning services obtained from the forest ecosystem and their vicinity. However, unsustainable collection and overexploitation of medicinal and aromatic plants (*Aconitum balfourii, A. heterophyllum, Angelica glauca, Arnebia benthamii, Bergenia stracheyi, Dactylorhiza hatagirea, Dendrobium normale, Hedychium spicatum, Heracleum candicans, Juniperus communis, Malaxis acuminata, M. mucifera, Meconopsis aculeata, Nardostachys grandiflora, Picrorhiza kurrooa, Podophyllum hexandrum, Polygonatum cirrhifolium, P. verticilatum, Rheum austral, Saussurea costus,* and *Swertia chirata*) by local people and migrant laborers pose serious challenges to the ecology and sociocultural setup of the region. As a result, 30% of the total plant species of Uttarakhand Himalaya are recorded in different categories (RET; rare, endangered, and threatened) of threats (Samant et al., 1998; Kuniyal et al., 2015; Bisht et al., 2016).

Keeping in mind the significant role they play not only in the traditional sector but also in their increasing demand in the pharmaceutical and other drug-based industries, medicinal plants can be easily looked upon as a potential wellspring of renewable natural resources. Therefore, the proper conservation and utilization of these medicinal plants in a sustainable manner must involve an integrated, long-term, and scientifically oriented action plan that focuses on the interaction between ecological, social, and economic systems. This plan should necessitate the pertinent aspects of preservation, protection, balanced exploitation, conservation, and sustainable utilization. Therefore, this study highlighted the importance, threats, challenges, and conservation efforts that are being made, looks for new opportunities, and addresses ways to bridge the gap.

9.2 MATERIALS AND METHODS

9.2.1 STUDY AREA

Since time immemorial, the state of Uttarakhand has been well known for its rich diversity of medicinally essential plants and corresponding traditional knowledge. Uttarakhand is a part of the North-Western Himalayas and is located between 77° 34′–81° 02′ E longitudes and 28° 43′–31° 27′ N latitudes. The total forest cover area of this state is 3.47 million hectares (71% of the total geographical area of the state), ranking it 6th among all the other states in India in terms of the percentage of recorded forest area. According to the Forest Survey of India (FSI), forests in the state belong to nine different forest types (tropical moist deciduous forests, tropical dry deciduous forests, sub-tropical pine forests, Himalayan moist temperate forests, Himalayan dry temperate forests, sub-alpine forests, moist alpine scrub, dry alpine scrub, and tree outside forests), which are further classified into 43 forest types (Champion & Seth, 1968). The whole forested area harbors rich floral and faunal diversity, aromatic and herbal plants, rare and endangered species, and pristine fragile ecosystems. The majority of the people living in this state rely on their natural environment, with over 3/4 of the total population dependent on agriculture for their livelihood. As per the census of the year 2011, the state of Uttarakhand is home to approximately 291,903 tribal people, which constitutes about 2.9% of the total population of the state. Uttarakhand only contributes 0.83% to the total population and 0.3% to the total tribal population of India (Anon, 2011). The habitation of any specific tribe is confined to a particular location in the state. The tribal denizens of Uttarakhand are concentrated in the remote forested areas of Bhabar and Terai and the higher reaches of Chamoli, Pithoragarh, Uttarkashi districts, and the Jaunsar-Bawar area of Dehradun district. Among all the 13 districts of the state, Udham Singh Nagar with a tribal population of about 43% is at the top, closely followed by Dehradun (38.3%), Pithoragarh (6.7%), and Chamoli (4.2%).

9.2.2 METHODOLOGY

The present study is the synthesis of a systematic review of literature available online (public domain) related to the status, availability, threats, and opportunities of medicinal plants in the mountainous region of Uttarakhand. Various research papers, articles, reports, books, book chapters, and proceedings were explored using Google Scholar, ResearchGate, PubMed, Science Direct, and Springer Link. Various keywords such as medicinal plants, traditional beliefs, conservation of MAPs, sustainable wild harvesting or collection, livelihoods, economic importance of MAPs, threats and opportunities in the medicinal plant sector, RET medicinal plant species, health care system, traditional knowledge system, mountain policies, etc., were used for searching the literature. The literature pertaining to Uttarakhand state was further used for synthesis and analysis. Initially, a total of 90 studies were collected, out of which 15 duplicate articles were removed after close scrutiny. In the second stage, we screened out the articles/papers that carried irrelevant information not supporting the objectives. After the final screening, we selected 50 research papers/articles/book chapters/books (Bhat et al., 2013; Bisht et al., 2016; Chauhan et al., 2010; Choudhary et al., 2013, 2014; Dhar et al., 2000; Dhyani et al., 2011; Dubey et al., 2004; Gupta, 2016; Joshi & Joshi, 2016; Kala, 2006, 2009, 2015; Kala et al., 2006; Kuniyal et al., 2014; Maikhuri et al., 2003, 2016; Negi et al., 2010; Phondani et al., 2016; Samant & Pal, 2003; Sati, 2013; Silori & Badola, 2000) for the final review for the synthesis of this document (Figure 9.1).

9.3 RESULTS

9.3.1 CONTRIBUTION OF MEDICINAL PLANTS TO THE HEALTHCARE AND LIVELIHOOD OF UTTARAKHAND

A total of 74 species of vascular plants (angiosperms, gymnosperms, and pteridophytes) were identified within the study area. These species belonged to 56 families and 71 genera. Among the flowering plants (angiosperms), the proportional percentage of life forms was recorded as 16.21% trees, 22.98%

shrubs, and the remaining 60.81% as herbs and forbs. The native communities of the state thus rely on traditional knowledge for their primary healthcare system. Most of the plant species that are used by these communities are commonly found around their surroundings, nearby forest areas, and some wastelands. Many women use these drugs to cure common ailments, and apart from this, these medicinal plants play a very important role in improving the subsistence economy of rural households, especially human folks and children who collect these herbs and sell them in the market (Table 9.1).

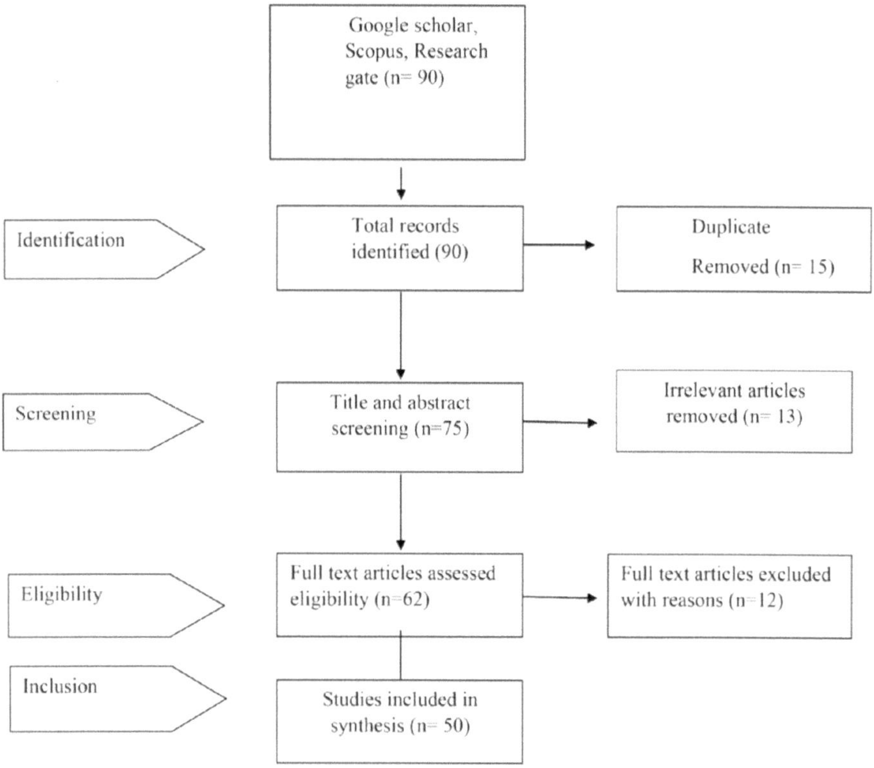

FIGURE 9.1 Flow diagram for database search of publications for systematic reviews.

9.3.2 *IDENTIFICATION OF THREATS ON MEDICINAL PLANTS AND THEIR CONSERVATION STATUS*

For threat assessment, a checklist of identified plant species was prepared to find out the possible reasons that pose a threat to the medicinal plants (Table 9.2). Out of the 37 species, the Government of Uttarakhand had

TABLE 9.1 Traditional Healthcare Practices of Plants Used in Household Remedies

Scientific Name	Vernacular Name	Family	Ailment	Mode of Uses
Abies pindrow Royle	Raga	Pinaceae	Wounds	Leaf paste is applied externally on wounds.
Abrus precatorius L.	Rattidana	Fabaceae	Rheumatic pain, ulcer	Roots are used.
Acacia catechu Will.	Kher	Fabaceae	Ulcer	Root paste is used to cure ulcer.
Aconitum balfourii Stapf.	Bauva	Ranunculaceae	Arthritis	Tuber powder boiled with butter is applied externally in arthritis.
Aconitum heterophyllum Wall. ex Royle	Ateesh	Ranunculaceae	Diarrhea, body pain	Paste of tuber with sugar taken to treat diarrhea, body ache.
Acorus calamus L.	Bach	Acoraceae	Dysentery, fever, asthma, throat infection.	Rhizome is used for the treatment.
Adhatoda zeylanica Medik.	Basinga	Acanthaceae	Cough and cold	Leaves and flowers paste mixed with honey taken in cough and cold fever.
Aegle marmelos (L.) Corr.	Bel	Rutaceae	Fever and cold	Dried fruit with sugar taken to treat fever and cold.
Ajuga bracteosa Wall	Neelkanth	Lamiaceae	Acidity, indigestion	Leaf extract is useful in acidity and indigestion.
Allium cepa L.	Piaz	Liliaceae	Ear pain, headache, high blood pressure.	Extract of bulb and Mentha leaves is useful in ear pain, headache, high blood and pressure epilepsy.
Allium sativum L.	Lahsun	Liliaceae	Joint pain	Extract of bulb taken with mustard oil applied on joint pain.
Amaranthus spinosus L.	Kadya sagoti	Amaranthaceae	Burns	Paste of leaves applied on burns.
Angelica glauca Edgew	Choru	–	Cough and cold, stomachache	Root powder useful in cough, cold, and stomachache.
Arisaema intermedium Bl.	Meen	Araceae	Fever, skin infections, body pain.	Root powder with water orally taken in fever, vomiting, body pain while stem extract applied on skin infections.

TABLE 9.1 *(Continued)*

Scientific Name	Vernacular Name	Family	Ailment	Mode of Uses
Arnebia benthamii	Balchad	Apiaceae	Piles	Root mixed with mustard oil used as hair tonic, antiseptic, and in throat problems. Root powder with butter given in piles.
Artemisia vulgaris (L.)	Kunju	Astraceae	Diabetes	Leaf powder with *Paeonia emodi* leaf powder given in diabetes.
Asparagus adscendens Roxb.	Ghirunn	Asparagaceae	Fever	Plant paste is used in fever.
Asparagus curillus Buch.-Ham	Jhirna	Asparagaceae	Urinary infection	Root extract with milk given as tonic and in urinary infection.
Barleria cristata L.	Kularkattya	Acanthaceae	Skin disease, pyorrhea.	Leaves paste are used in skin disease. Root paste used in pyorrhea.
Bauhinia variegata L.	Kachnar	Fabaceae	Blood pressure	Leaf paste used in skin diseases and bark powder is used in blood pressure.
Berberis lycium Royle	Kingod	Berberidaceae	Blood purification, skin diseases.	Root is used as blood purifier and in skin diseases.
Bergenia ciliata (Haw.) Sternb.	Silphodi	Saxifragaceae	Kidney stone, diabetes.	Root extract used in the treatment of kidney stone and diabetes.
Betula utilis D.Don	Bhojpatra	Betulaceae	Fever, body pain	Gum or bark extract applied externally to relieve from fever and body ache.
Boerhavia diffusa L.	Pundari	Nyctaginaceae	Cut and wound	Roots are used on wound and swelling.
Brassica campestris L.	Sarshaon	Brassicaceae	Fever, joint pain, jaundice.	Oil is used as ear drop to prevent cold, applied externally to relieve from fever, joint pain and jaundice.
Callicarpa macrophylla Vahl.	Daiya	Verbenaceae	Rheumatic pain	Fruit extract used in rheumatic pain and mouth ulcer.
Cannabis sativa L.	Bhang	Cannabaceae	Curing fever, bronchitis.	Seed and leaves extract with pepper, cumin seed and cardamom is useful in curing fever and bronchitis.

TABLE 9.1 *(Continued)*

Scientific Name	Vernacular Name	Family	Ailment	Mode of Uses
Carum carvi L.	Kalajeera	Apiaceae	Fever, headache	Seed boiled with salt taken orally is useful in fever, head ache.
Cassia tora L.	Chakunda	Fabaceae	Stomachache, cough, and cold.	Seed powder is given in tea for the treatment of stomachache, cough, and cold.
Cinnamomum tamala (Buch.-Ham.) T.Nees and Eberm.	Kikhadu, Tejpat	Lauraceae	Blood pressure, digestion.	Leaf is used as spice, to reduce blood pressure and digestion.
Cicerbita macrorhiza (Royle) Beauverd	Karatu	Asteraceae	Headache	Leaf juice taken orally during headache.
Colebrookea oppositifolia Sm.	Bindu	Lamiaceae	Eye problems	Leaf juice is useful in eye problems.
Cuscuta europaea L.	Alasjbail	Convolvulaceae	Skin diseases	Plant extract used in skin diseases.
Cynodon dactylon (L.) Pers	Doob	Fabaceae	Nose bleeding and anemia.	Plant extract is useful dysentery, nose bleeding and anemia.
Datura stramonium L.	Dhatura	Solanaceae	Bronchitis, asthma, cough	Flowers and seeds are used for the treatment of bronchitis, asthma, and cough.
Delphinium denudatum Wall.	Nirbisi	Ranunculaceae	Intestinal problem	Leaf powder with milk taken orally in intestinal problem.
Dioscorea bulbifera L.	Genthi	Dioscoreaceae	Diabetes, skin diseases, burns	Tubers used in diabetes, skin diseases and burns.
Eupatorium adenophorum Sprengel	Kala binda	Asteraceae	Wounds	Leaf paste applied externally on wounds.
Euphorbia royleana Boiss	Sullu	Euphorbiaceae	Ear problem	Bark extract useful as ear drop.
Gloriosa superba L.	Langali, Kalihari	Colchicaceae	Chronic ulcer, skin diseases	Rhizome and seeds are used for the treatment of chronic ulcer, skin diseases.
Glycine max (L.) Merr	Kala bhatt	Fabaceae	Eye tonic	Seed pastes useful in eyesores.
Habenaria pectinata Smith	Ridhi	Orchidaceae	Body weakness	Root powder useful as tonic.

TABLE 9.1 *(Continued)*

Scientific Name	Vernacular Name	Family	Ailment	Mode of Uses
Hordeum vulgare L.	Jau	Poaceae	Cataract, eye diseases	Leaf juice useful in cataract and eye diseases.
Macrotyloma uniflorum (Lam.) Verdc.	Gahath	Fabaceae	Kidney stone	Seed extract useful in kidney stone.
Megacarpaea polyandra Benth.	Barmauo	Brassicaceae	Asthma, fever, stomach pain	Root extract with butter useful in asthma, fever, and stomach pain.
Myrica esculenta Buch.-Ham.	Kaphal	Myricaceae	Fever, headache, body pain	Fruit juice with salt useful in fever, headache, and body pain.
Nardostachys jatamansi DC.	Jatamansi	Caprifoliaceae	Heart disease, high blood pressure.	Root powder with oil taken orally in heart disease and high blood pressure.
Nicotiana rustica L.	Hamaku	Solanaceae	Skin sores	Root powder applied externally in skin sores.
Orchis chusua D. Don	Hatha	Orchidaceae	Fever and cough	Tuber powder with sugar taken orally is useful in fever and cough.
Origanum vulgare L.	Van tulsi	Lamiaceae	Skin diseases, insect bites.	Leaf paste is useful in skin diseases and insect bites.
Oxalis corniculata L.	Khatti buti	Oxalidaceae	Pimples, skin disease.	Leaf extract taken orally in pimples and skin disease.
Peonia emodi Wall.	Chandra	Peoniaceae	Intestinal pain, dysentery, piles	Root decoction taken orally is useful during intestinal pain, dysentery, and piles.
Phaseolus vulgaris L.	Sem	Fabaceae	Skin irritation	Leaf paste applied externally to relieve in skin irritation.
Phyllanthus emblica L.	Aonla	Phyllanthaceae	Diabetes, eye problems, body weakness.	Fruits with the fruits of *Terminalia chebula* and *T. bellirica* as *Triphla* used as tonic to recover body weakness, diabetes, etc.
Prunus cerasoides Don.	Painya	Rosaceae	Sweling and Joint pains.	Boiled bark in water is useful for swelling.
Potentilla fulgens Wall.	Bajradanti	Rosaceae	Tooth ache, pyorrhea.	Plant decoction useful as tooth cleaning and pyorrhea.
Pyrus pashia Buch.-Ham	Molu	Rosaceae	Eye injury	Fruit juice used in eye injury.

TABLE 9.1 *(Continued)*

Scientific Name	Vernacular Name	Family	Ailment	Mode of Uses
Rheum emodi Wall.	Archu, Dolu	Polygonaceae	Rheumatism	Roots paste with turmeric powder mixed with mustard oil is applied externally to relieve muscular pain.
Rhododendron arboreum Sm.	Burans	Ericaceae	Blood pressure	Flower juice is useful in reducing blood pressure.
Rubus ellipticus Sm.	Hisool	Rosaceae	Skin diseases	Root paste applied on skin diseases.
Ricinus communis L.	Arand	Euphorbiaceae	Bronchitis, skin diseases, jaundice.	Roots and bark are used to cure asthma, bronchitis, and skin diseases. Fruits are useful in jaundice.
Saussurea obvallata (DC.) Edgew.	Brahmkamal	Asteraceae	Headache and body pain	Whole plant paste used for the treatment of headache and other body pain.
Solanum nigrum L.	Geahwai, Mako	Solanaceae	Jaundice and fever	Leaf and branch paste are useful in jaundice and fever.
Swertia chirata Grisebach	Chirayita	Gentianaceae	Fever, blood purifier.	Root powder or extract is given in fever and as blood purifier.
Terminalia bellirica (Gaertn.) Roxb.	Bahera	Combretaceae	Stomach disorder, fever	Fruits used in cough fever and stomach disorder.
Taraxacum offici-nale (L.) Weber ex F.H.Wigg.	Kadvae	Asteraceae	Fever	Roots are used in fever
Tinospora cordifolia (Thunb.) Miers	Giloya	Menispermaceae	High blood pres-sure, weakness.	Root powder is useful in high blood pressure and weakness.
Valeriana jatamansi Jones ex Roxb.	Sumaya	Caprifoliaceae	Stomach pain, nervous disorder.	Root extract given in nervous disorder and fits. Leaf juice used in stomach pain.
Verbascum thapsus L.	Akalveer	Scrophulariaceae	Intestinal pain	Root powder with milk is useful in intestinal pain.
Viola pilosa Blume	Banfsha	Violaceae	Kidney diseases, cough, liver disorder.	Flowers are used to treat cough, liver disorder and in kidney diseases.
Vitex negundo L.	Siwali	Lamiaceae	Wounds	Leaves extract applied on wounds.

TABLE 9.1 *(Continued)*

Scientific Name	Vernacular Name	Family	Ailment	Mode of Uses
Woodfordia fruticosa (L.) Kurz	Dhaula	Lythraceae	Dysentery	Dried flowers used in dysentery.
Zanthoxylum armatum DC	Timru	Rutaceae	Toothache, pyorrhea	Twigs used for teeth cleaning. Seeds useful in pyorrhea.
Zingiber officinale Rosc.	Adrak	Zingiberaceae	Cough and cold	Rhizome extract with honey is useful in cough and cold.

banned the collection of about 34 species of medicinal plants from the wild due to destructive harvesting and other abiotic and biotic reasons, which make these species vulnerable to possible extinction in the region. It is evident that medicinal plants are most threatened by habitat loss and habitat destruction (39.1%), closely followed by unscientific and destructive harvesting (27.5%), while overexploitation/harvesting accounts for (20.3%) and grazing/fodder collection (13.4%), respectively (Figure 9.2).

TABLE 9.2 Status of RET (Rare, Endangered, and Threatened) Species in Uttarakhand

Botanical Name	Common Name	Status	Uses	Threats
Aconitum balfourii Stapf.	Ativisha	CR	Roots are used to treat anorexia and act as analgesic.	Population declines due to overharvesting and loss of habitat.
Aconitum hetero-phyllum Wall.ex: Royle	Atis	CR	Roots are used in the treatment of fever, rheumatism, and stomachache.	Destructive harvesting, Habitat destruction.
Angelica glauca Edgew.	Gandhrayan	CR	The root is considered as cardio active and stimulant, also used in constipation.	Over harvesting.
Arnebia benthamii (Wall. ex G. Don) L M. Johnston	Balchar	CR	Roots act as antibacterial, antifungal, anti-inflammatory.	Habitat destruction, grazing, and un scientific harvesting.
Arnebia euchroma (Royle) Johnston	Laljari	EN	The root is antipyretic, cancer, contraceptive, emollient, and vulnerary.	Destructive harvesting, Habitat destruction.
Berberis aristata DC.	Daruhaldi	NA	Hypoglycemic, anti-diabetic, anti-bacterial.	Over harvesting due to host of black rust, harvest for medicine, and trade.

TABLE 9.2 *(Continued)*

Botanical Name	Common Name	Status	Uses	Threats
Berberis lycium Royle	Kasaundi	EN	They are used in the treatment of eye complaints, menorrhagia, chronic diarrhea and piles.	Over and unscientific harvesting.
Bergenia ligulata Wall.	Paashaanbhed	VU	Anti-diabetic, hepatoprotective, and diuretic.	Over harvesting and grazing.
Bergenia stracheyi Hook.f. and Thorns.	Pasanbhada	VU	To cure urinary disorders and gall/kidney stones.	Destructive harvesting, habitat destruction.
Betula utilis D. Don	Bhojpatra	LC	Antimicrobial, anti-inflammatory, anticancer.	Deforestation due to overuse for fuel.
Corydalis govaniana Wall.	Bhootkesi	NT	Root decoction is used to cure acute stomach-ache, fever, and toothache.	Destructive harvesting, habitat destruction.
Dactylorhiza hatagirea (D.Don) Soo	Salam panja	CR	Aphrodisiac, demulcent, nutritive, astringent.	Over harvesting and grazing.
Delphinium denudatum Wall	Jadwar	CR	Sedative, analgesic, brain, and nerve tonic.	Loss of habitat, over harvest for medicine and trade.
Dendrobium normale Falc.	Arkapushpi	VU	Vulnerable	Over harvesting, loss of habitat.
Dioscorea deltoidea Wall. ex Griseb.	Vidarikand	VU	Digestive disorders, sore throat for struma.	Indiscriminate collection (over harvesting).
Fritillaria roylei Hook.	Ksheer-kakoli	CR	Antipyratic, expectorant, and lactogog.	Destructive harvesting, habitat destruction.
Gentiana kurrooa Royle	Gentian	CR	To treat digestive problems, fever, hypertension.	Destructive harvesting, habitat destruction.
Gloriosa superba L.	Kalihari	NT	To treat acute gout, intestinal worms, infertility.	Destructive harvesting, habitat destruction.
Habenaria edgeworthii	Vridhi	VU	Tonic to cure general debility.	Destructive harvesting, habitat destruction.
Habenaria intermedia	Ridhi	VU	Tonic to cure general debility.	Habitat loss, fodder collection.
Hedychium spicatum	Kapoor kachri	VU	Treating fevers, vomiting, diarrhea, pains.	Destructive harvesting, habitat destruction.
Heracleum candicans	Kakriya	EN	Nerve tonic. Antimicrobial.	Destructive harvesting, habitat destruction.

TABLE 9.2 *(Continued)*

Botanical Name	Common Name	Status	Uses	Threats
Juniperus communis L.	Hauber	LC	Digestive disorders plus kidney and bladder problems.	Forest fire, afforestation, and climate change.
Malaxis acuminata D.Don	Risbhak	VU	An important constituent of Astaverga.	Habitat loss, fodder collection.
Malaxis muscifera (Lindl.) Kuntze	Jeevak	EN	An important constituent of Astaverga.	Illegal collection pressures, Habitat loss.
Meconopsis aculeata Royle	Kanda	CR	Narcotic, treat inflammation	Destructive harvesting, habitat destruction.
Nardostachys grandiflora (D.Don) DC.	Jatamasi	CR	Antiseptic, antispas-modic, appetizer.	Over harvesting and grazing.
Peonia emodi Wall. ex Royle	Chandayra	VU	Hypertension, palpita-tions, asthma.	Destructive harvesting, habitat destruction.
Picrorhiza kurroa Royle ex Benth.	Kutki	EN	Dyspepsia, asthma, biliousness, fever.	Destructive harvesting, habitat destruction.
Podophyllum hexandrum Royle	Aindri	CR	Form the source of medicinal resin.	Over harvesting and grazing.
Polygonatum cirrhifolium (Wall.) Royle	Mahameda	VU	Treat pain, fever, inflammation,	Loss of habitat, over exploitation.
Polygonatum verticillatum (L.) All.	Vulnerable (IUCN)	VU	Synthesizing steroidal hormones.	Destructive harvesting, habitat destruction.
Rheum australe D.Don	Dolu	VU	Anticholesterolemic, antiseptic, tonic.	Habitat destruction and grazing.
Rhododendron anthopogon D.Don	Ratpa	VU	Coughing and various skin disorders.	Destructive harvesting, habitat destruction.
Saussurea costus (Falc.) Lipsch.	Kuth	CR	Astringent, stimulant, digestive.	Over harvesting
Swertia chirata Buck.-Ham. ex Wall.	Chirata	EN	Toxic fever, malarial fever, urinary disorders.	Destructive and over harvesting.
Taxus baccata L.	Thuner	LC	Cardiotonic, diaphoretic, emmenagogue.	Destructive harvesting, habitat destruction.

Abbreviations: CR: Critical endangered; EN: endangered; VU: vulnerable; LC: least concern; NT: NEAR threatened; and NA: not available.

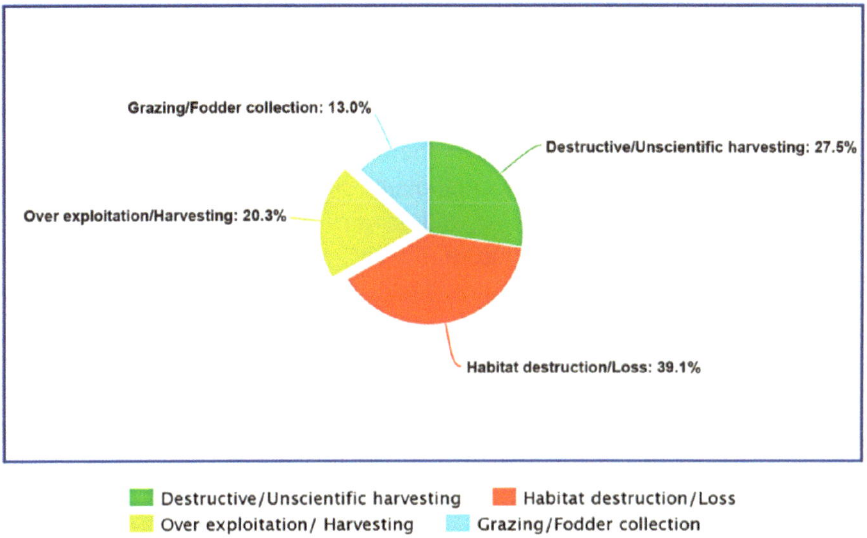

FIGURE 9.2 Possible threats on medicinal and aromatic plant species in Uttarakhand.

9.4 DISCUSSION

Medicinal plants are indispensable and a prime necessity for the sustainability of traditional health care systems (Kandari, 2005). The medicinal plant sector can provide an important source of income to the rural inhabitants of the region. At present, resource-poor people in the region used to collect plants from the wild in the traditional healthcare system and complement their meager incomes. Besides illegal collection, there are many threats like land use disturbances, heavy livestock grazing, changing micro-climatic conditions, etc. The conversion of socio-cultural traditions and indigenous knowledge into livelihood means and economic opportunities also has the advantage of preserving the rapidly eroding cultural knowledge and practices, which are increasingly threatened due to globalization and the homogenization of people (Nautiyal et al., 2005). Medicinal plants constitute the natural health care system for the people. This study indicated various medicinal plants with their medicinal value used by local communities to cure different ailments and diseases but are also sold at local and regional markets. The natural vegetation across the globe is disappearing and being altered at a fast-alarming rate. On top of that, the increasing demand for herbal drugs has further induced more pressure on medicinal plants, and with that, the people belonging to various cultures who have lived in close proximity to nature and who are dependent on it are facing major livelihood, economic, and cultural issues. There are other challenges in this sector too that can be attributed to the increase in the population, which is leading to the continuous erosion of forest resources. The weakening of the laws concerned with the harvesting of medicinal plants has increased the threat faced by medicinal plants. Despite the important role medicinal plants play, they are seldom handled in a regulated and organized sector, and most of them are still exploited with very little or no regard for the future. The damage is very severe when all the parts like seeds, flowers, bark, etc., are completely removed, which ensures the survival of the entire plant. It's a high priority for us to start protecting these rare medicinal plants. The samples that are being collected today can combat dreadful ailments in the future, but there is no guarantee of their existence till then. Conservation strategy aims to support development in a sustainable manner by utilizing the biological resources in a manner that doesn't deplete the variety of species globally or destroy their ecosystems. This involves different measures such as collection, propagation, evaluation, disease identification and elimination, storage, and distribution. The conservation of the medicinal plants and their genetic resources can be tackled by "*in-situ*" and "*ex-situ*" conservation. In order to satiate the increasing demands of the pharmaceutical industries, cultivation of these medicinal plants has become crucial and the only way to reduce the pressure on the naturally growing medicinal plants. The conservation strategy involves the imposition of certain regulations on the collection of medicinal plants from the wild because if such unmonitored collection continues, many of the plant species which are extensively available will become endangered and some of them may even go extinct. Therefore, appropriate scientific methods of cultivation must be encouraged for conservation of these medicinal plants in the wild and to fulfill the global health requirements on a sustainable basis.

The policies such as the Indian Forest Act (19270, the Convention on International Trade in Endangered Species of Wild Fauna and Flora, National Biodiversity Strategy and Action Plan (2002), Negative list of MAPs Collection from Wild (2004), Export Import Policy (2015), etc., enacted for protection, conservation, and legal/illegal extraction strictly need to be followed. However, India has signed a few international agreements such as the "Convention on International Trade in Endangered Species of Wild Fauna and Flora" (CITES); export-import policy (EXIM) focused on ensuring the conservation, protection, and trade of plants and wild animals. The SWOT (Strengths, Weaknesses, Opportunities, and Threats) analysis of the study highlighted that the state has a huge potential of knowledge, opportunity, technologies, and infrastructure to cultivate medicinal plants unless people should involve in the cultivation practices (Figure 9.3). The value chain marketing can be developed for farmers to directly link to the market, generate income, and improve their livelihood. As mountain regions have their own mountain specificities, it is necessary to have a different mountain value chain policy rather than implementing a one strategy one goal concept.

STRENGTH	WEAKNESS
☐ Products are inherently safe. ☐ Generalised information available in various traditional manuscripts. ☐ Friendly stance of government (i.e: NMPB, SMPB). ☐ Low cost manufacturing of high quality of generic drugs in the global market.	☐ Yield is nature dependent. ☐ High investment cost and low profit margin. ☐ Information gap. ☐ Poor marketing infrastructure. ☐ Inefficient processing techniques.
OPPORTUNITY	**THREAT**
☐ Employment generation. ☐ Alternate source of income. ☐ India has rich diversity and global market.	☐ Threat of biodiversity. ☐ Global competition. ☐ Lack of interest in cultivation. ☐ Lack of scientific methods according to regions.

FIGURE 9.3 SWOT analysis of medicinal plant sector in Uttarakhand.

9.5 CONCLUSION

The demand for medicinal plants is significantly rising across the globe at the rate of 7% per annum. This, in turn, is providing opportunities to developing nations that are rich in biodiversity to employ their natural resources innovatively and prudently in order to earn foreign exchange through the export of crude drugs and value-added products. This will also create jobs for both skilled and unskilled youth. India's potential in the growing market for medicinal plants is clearly evident since the medicinal plants required to prepare almost 50% of the drugs mentioned in the British Pharmacopoeia are documented in the western region of the Himalayas alone. Unfortunately, the understanding of the trade in medicinal plants in India is nowhere near satisfactory. The medicinal plant sector in India is extremely secretive, complex, confusing, traditional, highly underestimated, unregulated, and badly organized. Another setback is that there is no systematic data regarding the number of species being traded at either the national, regional, or local level. The state of Uttarakhand has great potential to be developed as a model AYUSH state, but to make this happen, the state is required to identify and prioritize the cultivation of plant species that have good commercial value, supply good quality and certified planting material, equip farmers with field-tested agro-techniques, establish well-built infrastructure and processing facilities at a primary scale, promote cluster-based cultivation, provide financial help during cultivation, assure a minimum support price, and issue high-yielding varieties, etc. Some of these activities are already taking place in the state, but they need more focus and a disciplined management approach. Although the practice of collecting medicinal plants from the wild has been done for centuries, it is tough to assess the exact value or the accurate volume of the trade because it is secretive in nature and only partially regulated. Apart from the official trade, the illegal trading of raw materials is also suspected. Thus, it is really necessary to upgrade and customize the existing herbal markets. An online trade portal should be opened to disseminate the market-related information, to help in the establishment of community-owned enterprises, the value chain assessment of selected plants, the branding of herbal products, and controlling the illegal activities, etc. With the growing sector come other problems like the various anthropogenic pressures and developmental activities in the forests, unscientific and unscrupulous exploitation of the medicinal plants due to the exponential increase in global and national trade, the inadequacy of inputs in regards to management, the gradual erosion of the role of local communities in the management of the resource, and also the lack of enabling legislation have triggered severe depletion of these medicinal plant resource bases in the state, thus affecting the livelihood and healthcare options of the rural poor. Since these medicinal plants are a source of income for many marginalized people like herders, shepherds, landless people, and women, it is important to regulate the over-harvesting of naturally occurring populations and to protect the local communities from overexploitation by traders and exporters. The local communities have to be trained with scientific methods in the collection and marketing of plants. The conservation of biodiversity and safeguarding of indigenous knowledge are crucial aspects that have to be addressed to avert the chances of biopiracy. The practice of cultivating MAPs is arising as an economic opportunity that needs to be judiciously utilized. Governments have to devise viable working policies for the protection and sustainable utilization of biodiversity and the wealth creation through this bio-resource. The proper harmonization of laws, policies, and regulations in biodiversity, conservation, forestry, trade, marketing, and IPR sectors will definitely result in significant benefits to the community and would also encourage the growth of fair trade and development of a robust industry. Although modern drug-based healthcare systems may be accessible to the more affluent sections of society, traditional medical practices will indeed continue to remain the basis for healthcare of the economically weaker sections of society in the developing world (Bodeker et al., 2005). Thus, it can be concluded safely that medicinal plants have good future prospects and they will eventually one day emerge as appropriate substitutes or even better alternatives for the synthetic chemicals-based allopathic drugs or possibly can even replace them. Till then all we can do is focus on the sustainable cultivation and management of this sector.

KEYWORDS

- **conservation status**
- **medicinal and aromatic plants**
- **socio-economic conditions**
- **threats**
- **traditional knowledge**
- **Western Himalaya**

REFERENCES

Ahn, K., (2017). The worldwide trend of using botanical drugs and strategies for developing global drugs. *BMB Reports, 50*(3), 111–116.

Bhat, J. A., Kumar, M., & Bussmann, R. W., (2013). Ecological status and traditional knowledge of medicinal plants in Kedarnath Wildlife Sanctuary of Garhwal Himalaya, India. *Journal of Ethnobiology and Ethnomedicine, 9*(1), 1–18.

Bisht, V. K., Negi, J. S., & Bhandari, A. K., (2016). Check on extinction of medicinal herbs in Uttarakhand: No need to uproot. *National Academy Science Letters, 39*, 233–235. https://doi.org/10.1007/s40009-016-0437-1.

Bodeker, C., Bodeker, G., Ong, C. K., Grundy, C. K., Burford, G., & Shein, K., (2005). *WHO Global Atlas of Traditional, Complementary and Alternative Medicine.* World Health Organization, Geneva.

Champion, H. G., & Seth, S. K., (1968). *A Revised Survey of the Forest Types of India.* Manager of Publications, Govt. of India Press, Delhi, India.

Chandra, P., & Sharma, V., (2018). Strategic marketing prospects for developing sustainable medicinal and aromatic plants businesses in the Indian Himalayan region. *Small-Scale Forestry, 17*(4), 423–441.

Chandra, P., Sharma, V., & Kant, S., (2019). From commodity to brand: The country of origin branding perspective for Indian medicinal and aromatic plants. *Business Strategy & Development, 2*(1), 4–12.

Chauhan, R. S., (2010). Socioeconomic improvement through medicinal and aromatic plants (MAPs) cultivation in Uttarakhand, India. *Journal of Sustainable Agriculture, 34*(6), 647–658. https://doi.org/10.1080/10440046.2010.493390.

Choudhary, D., Kala, S. P., Todaria, N. P., Dasgupta, S., & Kollmair, M., (2014). Drivers of exploitation and inequity in nontimber forest products (NTFP) value chains: The case of Indian Bay Leaf in Nepal and India. *Development Policy Review, 32*(1), 71–87.

Choudhary, D., Kala, S. P., Todaria, N. P., Rawat, R. B. S., Kunwar, M. S., & Kollmair, M., (2013). Upgrading mountain people in medicinal and aromatic plants value chains: Lessons for sustainable management and income generation from Uttarakhand, India. *International Journal of Sustainable Development & World Ecology, 20*(1), 45–53. https://doi.org/10.1080/13504509.2012.738434.

De Pasquale, A., (1984). Pharmacognosy: The oldest modern science. *Journal of Ethnopharmacology, 11*(1), 1–16.

Dhar, U., Rawal, R. S., & Upreti, J., (2000). Setting priorities for conservation of medicinal plants—A case study in the Indian Himalaya. *Biological Conservation, 95*(1), 57–65. https://doi.org/10.1016/S0006-3207(00)00010-0.

Dhyani, D., Maikhuri, R. K., & Dhyani, S., (2011). Seabuckthorn: An underutilized resource for the nutritional security and livelihood improvement of rural communities in Uttarakhand Himalaya. *Ecology of Food and Nutrition, 50*(2), 168–180. https://doi.org/10.1080/03670244.2011.552375.

Dubey, N. K., Kumar, & Tripathi, P., (2004). Global promotion of herbal medicine: India's opportunity. *Current Science, 86*(1), 37–41.

Gupta, S., (2016). Forest sustainability and development in hills of Uttarakhand, India: Can they move together? *Environment, Development and Sustainability, 18*(1), 279–294. https://doi.org/10.1007/s10668-015-9648-8.

Joshi, B. C., & Joshi, R. K., (2014). The role of medicinal plants in livelihood improvement in Uttarakhand. *International Journal of Herbal Medicine, 1*(6), 55–58.

Kala, C. P., (2006). Medicinal plants: Potential for economic development in the state of Uttaranchal, India. *International Journal of Sustainable Development & World Ecology, 13*(6), 492–498. https://doi.org/10.1080/13504500609469698.

Kala, C. P., (2009). Medicinal plants conservation and enterprise development. *Medicinal Plants-International Journal of Phytomedicines and Related Industries, 1*(2), 79–95.

Kala, C. P., (2015). Medicinal and aromatic plants: Boon for enterprise development. *Journal of Applied Research on Medicinal and Aromatic Plants, 2*(4), 134–139.

Kala, C. P., Dhyani, P. P., & Sajwan, B. S., (2006). Developing the medicinal plants sector in northern India: Challenges and opportunities. *Journal of Ethnobiology and Ethnomedicine, 2*(1), 32.

Kandari, L. S., (2005). *Eco-physiological and Socio-economic Studies of Some Rhizomatous Medicinal and Aromatic Plant Species*. D. Phil. Thesis. [Srinagar (India)]: HNB Garhwal University.

Kuniyal, C. P., & Sundriyal, R. C., (2013). Conservation salvage of *Cordyceps sinensis* collection in the Himalayan Mountains is neglected. *Ecosystem Services, 3*, e40–e43. https://doi.org/10.1016/j.ecoser.2012.12.004.

Kuniyal, C. P., Bhatt, V. P., Bhatt, V. P., Butola, J. S., & Sundriyal, R. C., (2014). Promoting nursery enterprise in high altitude villages: A participatory approach for conservation and commercialization of Himalayan threatened medicinal plants. *Journal of Medicinal Plants Research, 8*(48), 1399–1407. https://doi.org/10.5897/JMPR2014.5472.

Kuniyal, C. P., Bisht, V. K., Negi, J. S., Bhatt, V. P., Bisht, D. S., Butola, J. S., & Singh, S. K., (2015). Progress and prospect in the integrated development of medicinal and aromatic plants (MAPs) sector in Uttarakhand, Western Himalaya. *Environment, Development and Sustainability, 17*(5), 1141–1162. https://doi.org/10.1007/s10668-014-9595-9.

Lubbe, A., & Verpoorte, R., (2011). Cultivation of medicinal and aromatic plants for specialty industrial materials. *Industrial Crops and Products, 34*(1), 785–801. https://doi.org/10.1016/j.indcrop.2011.01.019.

Maikhuri, R. K., Phondani, P. C., Rawat, L. S., Jha, N. K., Maletha, A., Bahuguna, Y. M., & Kandari, L. S., (2016). Conservation and management strategies of medicinal plant resources through action research approaches in Indian Himalaya. *Iranian Journal of Science and Technology, Transactions A: Science, 41*(3), 771–777. https://doi.org/10.1007/s40995-016- 0057-0.

Maikhuri, R. K., Rao, K. S., Chauhan, K., Kandari, L. S., Prasad, P., & Rajasekaran, C., (2003). Development of marketing of medicinal plants and other forest products-can it be a path way for effective management and conservation? *Indian Forester, 129*(2), 169–178.

Nautiyal, S., Rao, K. S., Maikhuri, R. K., Semwal, R. L., & Saxena, K. G., (2005). Traditional knowledge related to medicinal and aromatic plants in tribal societies in a part of Himalaya. *Journal of Medicinal and Aromatic Plant Sciences, 23*, 28–41.

Negi, V. S., Maikhuri, R. K., Phondani, P. C., & Rawat, L. S., (2010). An inventory of indigenous knowledge and cultivation practices of medicinal plants in Govind Pashu Vihar Wildlife Sanctuary, Central Himalaya, India. *International Journal of Biodiversity Science, Ecosystem Services & Management, 6*(3, 4), 96–105. https://doi.org/10.1080/21513732.2011. 575385.

Olsen, C. S., (1998). The trade in medicinal and aromatic plants from central Nepal to northern India. *Economic Botany, 52*(3), 279–292.

Phondani, P. C., Bhatt, I. D., Negi, V. S., Kothyari, B. P., Bhatt, A., & Maikhuri, R. K., (2016). Promoting medicinal plants cultivation as a tool for biodiversity conservation and livelihood enhancement in Indian Himalaya. *Journal of Asia-Pacific Biodiversity, 9*(1), 39–46. https://doi.org/10.1016/j.japb.2015.12.001.

Rates, S. M. K., (2001). Plants as source of drugs. *Toxicon, 39*(5), 603–613.

Samant, S. S., & Pal, M., (2003). Diversity and conservation status of medicinal plants in Uttaranchal State. *Indian Forester, 129*(9), 1090–1108.

Samant, S. S., Dhar, U., & Palni, L. M. S., (1998). *Medicinal Plants of Indian Himalaya: Diversity Distribution Potential* Value. Almora: G.B Pant Institute of Himalayan Environment and Development.

Sati, V. P., (2013). Cultivation of medicinal plants and its contribution to livelihood enhancement in the Indian Central Himalayan Region. *Advancement in Medicinal Plant Research, 1*(2), 17–23.

Semwal, D. P., Saradhi, P. P., Kala, C. P., & Sajwan, B. S., (2010). Medicinal plants used by local Vaidyas in Ukhimath block, Uttarakhand. *Indian Journal of Traditional Knowledge, 9*(3), 480–485.

Sharma, A. B., (2008). *Indian Herbal Market to Grow by 20%.* Financial Express. https://www.financialexpress.com/archive/indian-herbal-market-to-grow-by-20/292575/ (accessed on 22 April 2024).

Silori, C. S., & Badola, R., (2000). Medicinal plant cultivation and sustainable development: A case study in the Buffer Zone of the Nanda Devi Biosphere Reserve, Western Himalaya, India. *Mountain Research and Development, 20*(3), 272–279. https://doi.org/10.1659/0276-4741(2000)020[0272:MPCASD]2.0.CO;2.

Singh, D. K., & Hajra, P. K., (1996). Floristic diversity. In: Gujral, G. S., & Sharma, V., (eds.), *Changing Perspective of Biodiversity Status in the Himalaya.* British Council Division, British High Commission Publ. Wildlife Youth Services, New Delhi.

Suneetha, M. S., (2010). Sustainability issues for biodiversity business. *Sustainability Science, 5*(1), 79–87.

United Nations Development Program (UNDP), (n.d.). *Conserving Medicinal Plants, Sustaining Livelihoods.* Retrieved from: http://www.in.undp.org/content/india/en/home/climate-and-disaster-reslience/successstories/conserving-medicinalplants-sustaining-livelihoods.html (accessed on 22 April 2024).

Ved, D. K., & Goraya, G. S., (2008). *Demand and Supply of Medicinal Plants in India.* Dehradun, Uttarakhand: Bishen Singh Mahendra Pal Singh.

WHO (World Health Organization), (2003). *WHO Guidelines on Good Agricultural and Collection Practices (GACP) for Medicinal Plants.* Geneva: World Health Organization.

Yadav, M., & Misra, S., (2012). Sustainable development: A role for market information systems for non-timber forest products. *Sustainable Development, 20*(2), 128–140.

CHAPTER 10

Traditional Ecological Knowledge for the Management of Medicinal Plants with Particular Emphasis on Northeast India

SUBHASHREE PATRA,[1] SHUBHAM ANURAG,[1] PURABI SAIKIA,[1] AMIT KUMAR,[2] and MOHAMMED LATIF KHAN[3]

[1]*Department of Environmental Sciences, Central University of Jharkhand, Ranchi, Jharkhand, India*

[2]*Department of Geoinformatics, Central University of Jharkhand, Ranchi, Jharkhand, India*

[3]*Department of Botany, Dr. Harisingh Gour Vishwavidyalaya, Sagar, Madhya Pradesh, India*

ABSTRACT

Traditional ecological knowledge (TEK), when supplemented with modern understanding, helps to maintain biodiversity and ecosystem benefits. Indigenous tribal groups use their TEK to design various resource management and utilization strategies, norms, and practices to ensure a continuous supply of ecosystem goods and services, including food, medications, and sustenance needs. Northeast India is known for its abundant biological diversity due to its favorable climatic, edaphic, and topographical features. Besides, traditional homegardens of Northeast India serve as replicas of natural forests, supporting a wide range of indigenous varieties of plant species that serve a variety of ecological, socio-economic, and essential sustenance needs. A range of medicinal plants is grown in these traditional agroecosystems that are protected and used for different ailments of humans

Sustainable Forest Resources Management: Issues and Implications. Jahangeer A. Bhat, Gopal Shukla, Manmohan J. Dobriyal, Sumit Chakravarty, A. Arunachalam, and Rainer W. Bussmann (Eds.)

and livestock generation after generation. Ethnomedicinal plant species hold significant ecological and socio-economic significance globally as traditional medicines (TMs) have been a part of the world's indigenous health care systems, playing a vital role in protecting humankind from several life-threatening diseases. The proliferation of medicinal plants combining tissue culture with conventional cultivation methods may help in their conservation from the ever-increasing human exploitation. Further, to effectively plan the conservation and management of rare, endangered medicinal plant species to meet the growing demand for medicinal plants, population assessments using standard ecological methods are required.

10.1 INTRODUCTION

Traditional medicines (TMs) have been a part of the world's indigenous health care systems and practices, playing a vital role in protecting humankind from several life-threatening diseases, including mental and physical disorders, since prehistoric times (Bent, 2008). Humans have been utilizing plants, animals, microorganisms, and other natural resources to develop medicines to prevent and treat various ailments (Fabricant & Farnsworth, 2001). TMs, also known as complementary and alternative medicines or ethnic medicines, play an important role in many modern societies of developing countries (Abdullahi, 2011; Fabricant & Farnsworth, 2001). TMs are the process and practice of treating diseases with locally available and accessible natural resources based on orally transmitted experience and knowledge generated by communities and handed down from generation to generation (Payyappallimana, 2010). TMs are the solutions that societies have developed to deal with material and moral disorders, dating back to the mythical period and continuing today (Sever, 2015), and include medical procedures that have been in use for hundreds of years before the advent of modern clinical medicine and are still in use with no known side effects (WHO, 2002). These medical systems have a remarkable comprehension of physiology, pathogenesis, pharmacology, and pharmaceuticals, which are not the same as western biomedicines (Telles et al., 2014). There are certain allied modes of well-being-related strategies that are not purely medical systems but are identified with prosperity such as Yoga, Tai Chi, Qi-gong, and breathing techniques (Payyappallimana, 2010). Acupuncture is another type of traditional practice to manage poor health (Kayne, 2010). Traditional cure practices, also called folk medicines or TMs, are formed by a group of individuals in a specific area agreeing on something, or by the sum of a person's knowledge and behaviors about health, which are of two types, *viz.*, natural and religious folk medicine (Eric, 1986). Historically, different cultures have developed various useful remedial measures to combat multiple diseases, which threatened health and life (Pan et al., 2014). TMs used to be the dominant medical practice and were practiced by millions of people throughout the developing world until modern medicine was adopted (Yuan et al., 2016). Sophisticated treatment of evidence-based medicine or modern allopathic medicines has resulted from scientific research, testing, and verification of plants, animals, and microbes worldwide (Patwardhan, 2014). TMs are defined as "*the sum of indigenous to different cultures' knowledge, abilities, and practices based on their theories, beliefs, and experiences*" (WHO, 2022). It is employed in the management, protection, detection, improvement, or diagnosis of both physical and mental health by relying on different transmission modes from generation to generation (Sofowora et al., 2013). Most of the TMs are backed by large volumes of reviewed literature, records, and other information based on theoretical principles and different practical skills, while some others are passed by verbal instruction through successive generations (Khatib et al., 2021). Therefore, the traditional medicinal health care system is classified into three main categories, *viz.*, (i) Well-codified written medical systems; (ii) Allied forms of health knowledge; and (iii) Folk medicine (Payyappallimana, 2010). Well-codified written medical systems are common throughout the world and are not unique to a place or language. Well-codified written treatises that include great traditions that have evolved over 3–4 centuries, like Unani, Ayurveda, Siddha in the subcontinents of India, Kampo in Japan, and Chinese traditional medicines and acupuncture in China (Telles et al., 2014). These processes can be included in a country's healthcare system and are backed up by codified written treatises and through formal consultation with a practitioner or healer (Haque et al., 2018). Practitioners of these systems used to be mentored in earlier days, but now we have schools of medicine where they can be trained (Sheri et al., 2019). It is the use of home remedies and procedures by the public when they are unable to reach a doctor due to various economic and societal reasons (Parisius et al., 2014). Moreover, folk medicines are region-specific and are commonly practiced by limited ancestors in different communities, orally transmitted more informally from a knowledgeable family member or friend, and there is no written proof of it (Sever, 2015). Traditional medicines are not really formalized but based on the changing contexts, they are diverse and adaptable (Payyappallimana,

2010). There are several similarities between widely differing, geographically separate cultures in folk/indigenous medicine (Pan et al., 2014). Traditional medicines are characterized by folkloristic, traditional societies, local health, natural remedies, and practitioners in treating poisonous bites, birth attendants, and bonesetters (Eric, 1986).

Natural folk medicine involves the use of different plant species, minerals, and numerous animal substances to treat specific diseases, while religious folk medicine involves magic and witchcraft practiced in societies that have initially arisen from the beliefs of locals, morals of the families, thoughts of the ancestors, and speeches of the communities, in which diseases are cured by healers and magicians (Ozioma & Chinwe, 2019) and also known as magic folk medicine (Hand, 1985). In India, tribal communities mainly practice folk medicine (Shankar et al., 2012). The variability in nature, duration, and intensity of treatment causes a potential dilemma in signifying the efficacy of traditional medicines, and the absence of standardization makes traditional medicinal interventions difficult (Ekor, 2014). Reports on the uses of medicinal plants for different diseases are given in most cases without mentioning the parts of the plant that are used, the mode of their preparation, dosages, and duration gap (Sofowora et al., 2013). This makes it challenging to comprehend the efficacy of the reported medicinal plants. Alternative medicine practitioners face a significant and difficult scenario in which no sponsorship is available to build the research proof necessary to register a product or technique over which they would never take ownership or property rights (WHO, 2019). This is due to a complex interaction of several factors, including the high cost of gathering scientific proof for an unpatented product (Yakoot, 2013). Furthermore, there are no precise, practical, or comprehensible herbal medicine rules or laws (Ekor, 2014). Government committees and research ethics committees that assess files filed for new ayurvedic medicine registration typically find it difficult to grant permission for any investigator to begin a human clinical study on the full herbal extract (Yakoot, 2013). It requires very experienced pre-clinical and pharmacokinetic studies on every individual component of a herb, an essential requirement for approval, even for those edible plants with a long history of applications based on traditional uses (Moreira et al., 2014). This condition will persist for years until systematic actions are properly taken and can simultaneously put forward several difficulties to study and gather data based on the scientific basis of Complementary Alternative Medicine (CAM) (Yakoot, 2013). Problems with the safeguarding of indigenous traditional knowledge and issues with guaranteeing its right usage must be addressed quickly (Hossain & Ballardini, 2021). Local populations in every habitat, from the trans-Himalayas to the coastal plains, have identified the medicinal applications of hundreds of plants found locally in their ecosystem, according to oral tradition (Akshay et al., 2014). Indians have a strong affinity for medicinal plants and have utilized them for everything from treating a common cold to improving memory and treating deadly snake bites to curing muscular dystrophy and increasing immunity (DeFilipps & Krupnick, 2018). Roughly 25,000 effective plant-based formulations utilized in folk medicine and known to rural communities across India are estimated, with around 10,000 developed formulations available in indigenous medical textbooks (Akshay et al., 2014).

10.2 PROSPECTS OF TRADITIONAL MEDICINES

Throughout the globe, scientists are emphasizing folk medicine as an alternative medicine from natural resources and their commercial potential in health care (Mukherjee & Wahile, 2006). The Traditional Medicine Program of the WHO, in their analysis of the connection between the current use of various chemicals and their use in TMs, noted that TMs had played a significant role in producing effective new pharmaceuticals in 2013 (WHO, 2013). Around 80% of compounds out of the total 122 compounds extracted from 94 plant species exhibited pharmaceutical effects in folk medicine (Fabricant & Farnsworth, 2001). People in developing nations such as India (70%), Tanzania (60%), Benin (80%), Ethiopia (90%), Rwanda (70%), and Uganda (60%) continue to rely on their own traditional medicine to address their primary health care needs (WHO, 2019). Plants and plant-based products are used in many traditional and alternative treatments around the world, and a significant percentage of the population in developed countries such as Belgium (31%), the United States (42%), Australia (48%), France (49%), and Canada (70%) has used traditional and alternative remedies at least once for health care (Fabricant & Farnsworth, 2001). Natural biological products or chemical entities generated from natural resources account for over 42% of the top-selling pharmaceutical drugs, valued at $17.5 billion in the United States (Raskin et al., 2002). According to the World Health Organization (WHO), phototherapeutic medications are used by approximately 85% of the world's population for basic health care (Oliveira et al., 2006). Medicinal herbs are used in self-help methods by millions of rural households. The global market for medicinal plant commerce is projected to be worth over $60 billion per year and is expanding at a 7% annual rate, with different shares between developed and developing countries (Raskin et al., 2002). Predictably, natural chemicals and derivatives account for over 60% of all pharmaceuticals in clinical use, with medicinal plants accounting for less than 25% (Gurib-Fakim, 2011). Traditional medicine and the herbal industry rely heavily on medicinal plants, which give livelihood and health security to a huge portion of the Indian people (Ngo et al., 2013).

10.3 DISTRIBUTION OF MEDICINAL PLANTS IN INDIA

Medicinal plants can be found in a variety of environments and landscape components (Lakshman, 2016). The Western and Eastern Ghats, the Vindhyas, the Chotta Nagpur plateau, the Aravallis, and the Himalayas are home to over 70% of India's medicinal plants (Khumbongmayum et al., 2005). Although temperate and alpine zones at higher altitudes contain less than 30% of medicinal plants, they do contain species with considerable medicinal value (Maiti, 2004). The bulk of medicinal plants are taller blooming plants, with trees accounting for one-third, shrubs for another, and herbs, grasses, and climbers for the remaining one-third (Sharma & Kumar, 2013). Lower plants such as lichens, ferns, algae, and other lower plants make up a small percentage of therapeutic plants (GOI, Planning Commission, 2000). Medicinal plants are an important part of India's healthcare system and a valuable national resource (Ravishankar & Shukla, 2007). There are over 8,000 kinds of medicinal plants in India, accounting for roughly half of all higher flowering plant species (Jain et al., 2006). According to WHO, the amount and quality of data on traditional medicines' safety and efficacy are insufficient to meet the requirements required to support their use globally (WHO, 2019). It is quite evident that most of the modern medicines used today have some extracts of traditional and folk medicine, which are formulated and further stabilized with other phytochemical constituents to get rid of variable side effects and thus increase their worldwide acceptability (Yuan et al., 2016; Ekor, 2014). One of the most potential emphases of traditional medicine extracts can be witnessed in several dentifrices where the potential phytochemical constituents like xanthan gum and strontium acetate are widely derived and stabilized from traditional medicine with other constituents to upgrade their relieving potential (Mensah et al., 2011).

Though India has diverse biodiversity, rising demand is putting pressure on existing resources, particularly medicinal plants, which are becoming increasingly endangered in their natural habitats (Sharma et al., 2010). The growth of medicinal plants should be encouraged in order to satisfy future needs. According to an ethnobiological survey conducted by the Ministry of Environment and Forests of the Government of India, India's people use over, 8,000 varieties of plants (Ramawat et al., 2009). Various tribal and local people use the devoted medicinal plants to treat a variety of problems, including simple injuries, wounds, cuts, fever, diarrhea, ulcers, swelling, bone fractures, potency, antidote, skincare, night blindness, toothache, asthma, cough, and cold (Rajkumar et al., 2011). As a result, there is a critical need to conserve the diversity of medicinal plant richness for current and future generations by implementing an acceptable strategy and using the most effective conservation method.

10.4 TRADITIONAL MEDICINES AS A PART OF INDIAN CULTURE

Many plant-derived medications utilized in clinical medicine in India are developed from traditional medicines (Fabricant & Farnsworth, 2001), which is one of the world's oldest medical disciplines (Shi et al., 2020). India has the privilege of having six recognized systems of medicines in traditional medicines (Ravishankar & Shukla, 2007) that are Ayurveda, Yoga, Unani, Siddha, and Homeopathy; in short, it is known as AYUSH (Shi et al., 2020). Ayurveda, Unani, and Siddha medicine are examples of drug applications in AYUSH medicine (Sen & Chakraborty, 2017) explained in Figure 10.1. In traditional and other systems of medicine, such as Unani, Homeopathy, Ayurveda, and Siddha, it is estimated that 6,000–7,000 species of flowering plants have therapeutic uses (Mukherjee et al., 2014). AYUSH encourages natural remedies, which consider body, mind, and spirit in total (Shi et al., 2020). Medicinal plants are used by about 1.5 million traditional practitioners and 7,000 pharmaceutical manufacturing units for the prevention and treatment of various illnesses (Sen et al., 2011). Various rural groups in India use about 25,000 effective plant-based compositions as folk medicines (Nema et al., 2011). Based on the Vedic culture, there are mainly three humoralisms that are vata, pitta, and kapha and the five elements' theories are Jala, Agni, Akasha, Vayu, and Prithvi in Indian medicine which are used for the description of human physiology (Shi et al., 2020) as shown in Figure 10.2.

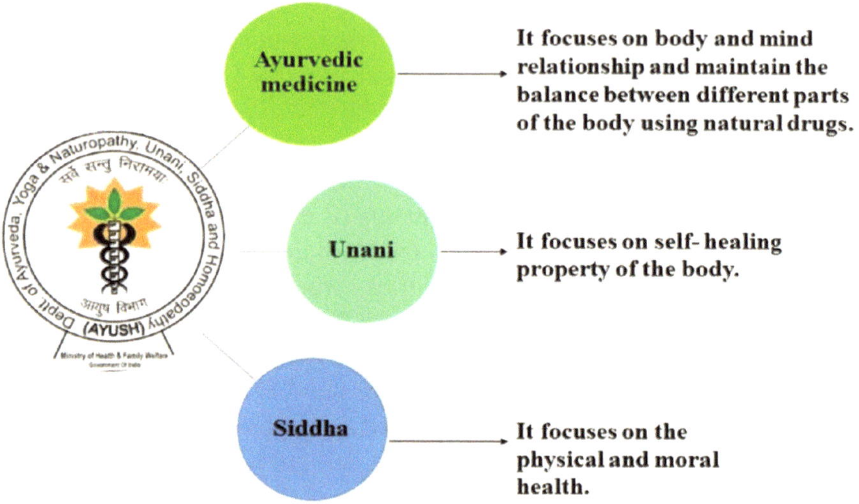

FIGURE 10.1 Drug application in AYUSH medicine.

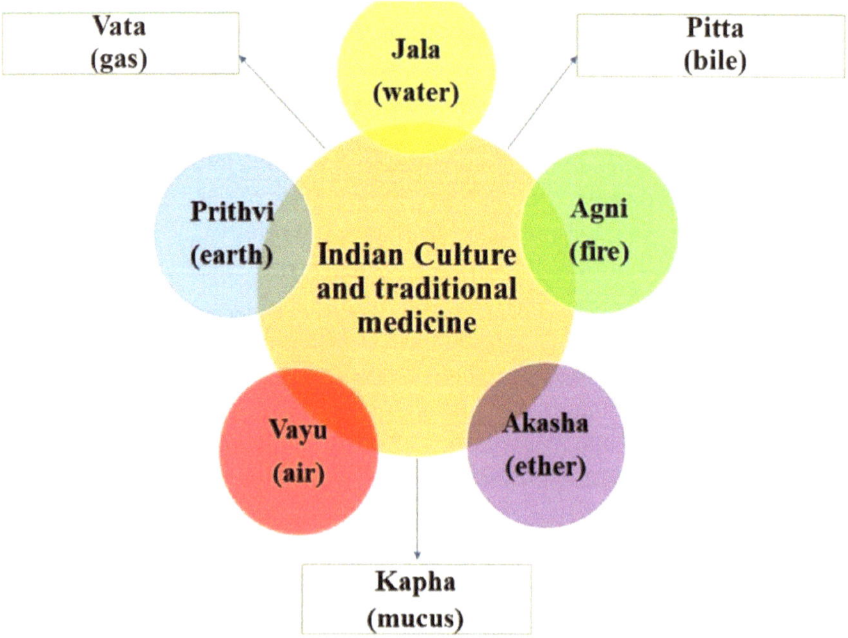

FIGURE 10.2 Interaction between traditional medicine and Indian culture.

10.4.1 TRADITIONAL MEDICINES OF NORTHEAST INDIA

Northeast India comprises eight states, namely Assam, Arunachal Pradesh, Manipur, Meghalaya, Mizoram, Nagaland, Tripura, and Sikkim, holding nearly 50% of the reservoirs of plant diversity in India (Mao & Hynniewta, 2000). It is part of the Indo-Burma and Himalaya biodiversity hotspots containing world-recognized important biomes with unique floral and faunal diversity (Chen et al., 2016). The region is home to more than 250 tribes of various ethnic groups speaking more than 200 dialects with distinct cultural entities. The majority of medicinal plants utilized by Northeast people and tribes are indigenous and unknown to the broad realm of phyto-chemical science and research (Sharma & Das, 2018). Agriculture is their primary occupation, and the different tribes possess rich traditional indigenous knowledge (Sharma et al., 2010). The northeast states have a diverse topology with rich varieties of medicinal plants in which the traditional system of medicines plays a vital role in the healthcare of rural people of Assam, and 286 species of plants are used in traditional medicines as it holds 50% of India's entire biodiversity (Sharma & Das, 2018). During the migration of Mishing communities, they expanded their skills by learning from other adjacent cultures and treating various ailments using herbs found in and around their settlements (Shankar et al., 2012a). Adi, Apatani, and Nyishi tribal people in Arunachal Pradesh, which are similar to Mishings, employ locally accessible medicines to heal illnesses (Srivastava et al., 2010). Herbs in the form of fresh medicines, crushed juice, boiling of drug parts, powdered medicine for oral consumption, and paste for local application on skin illnesses and wounds are among the traditional Mishing tribal medicinal procedures (Shankar et al., 2012a). An extensive literature survey related to traditional knowledge of folk medicine associated with different communities and tribes in northeast India is carried out to identify the various folk medicinal practices, research gap, and scope for further ethnomedicinal research. The active components and powerful phytochemicals found in such plants with intriguing pharmacological activities have yet to be discovered (Mao & Hynniewta, 2000). Malaria and jaundice are the most common ailments in North East India, and traditional healers have been known to use 68 plants to cure malaria and 88 herbs to treat jaundice (Shankar et al., 2012b). The various ethnomedicinal plants used in folk medicines of the world including the North-eastern states of India and their parts of uses, and modes of uses, as per the available literature are shown in detail in Table 10.1.

TABLE 10.1 Plant Species with the Greatest Medicinal Values with Multipurpose Uses

Plant Species	Family	Medicinal Values with Multipurpose Uses	Parts Used	References
Abelmoschus moschatus Medik.	Malvaceae	Treatment of leucorrhea, diarrhea, asthma, vomiting, neural disorders, leukoderma, nausea, constipation.	Root, seed	Dwivedi et al. (2019); Joshi & Pant (2012)
Acorus calamus L.	Acoraceae	Digestive problems, anti-bacteria, fever, asthma, bronchitis, cough.	Root	Balakumbahan et al. (2010)
Adenostemma lavenia (L.) Kuntze	Asteraceae	Mouth dryness treatment, wound healing.	Flower, leaf	Prasad et al. (2013)
Aegle marmelos (L.) Correa	Rutaceae	Treatment of diarrhea, dysentery, constipation.	Fruit, root	Kala (2006)
Aesculus indica (Wall. ex Cambess.) Hook.	Sapindaceae	Rheumatism	Fruit	Rajasekaran & Singh (2009)
Ageratum conyzoides L.	Asteraceae	Wound healing, leprosy, intestinal colic.	Leaf	Kamboj & Saluja (2008); Joshi & Pant (2012)
Allium cepa L.	Amaryllidaceae	Abdominal pain, skin diseases, and antioxidant properties.	Bulb	Okusa et al. (2007)
Amelanchier canadensis (L.) Medik.	Rosaceae	Diarrhea	Fruit	Sharma et al. (2017)
Anaphalis triplinervis (Sims) C.B.Clarke	Asteracese	Eye infections, skin diseases, wounds, flu, fever, and nausea.	Leaf	Khan et al. (2011)
Anisomeles indica (L.) Kuntze	Lamiaceae	Cough, cold, chronic rheumatism, psoriasis, and snake bites.	Leaf	Batish et al. (2007); Joshi & Pant (2012)
Asparagus curillus Buch.-Ham. ex Roxb.	Asparagaceae	Dysuria, diabetes, and dysentery.	Root	Negi et al. (2010)
Asparagus racemosus Willd.	Asparagaceae	Stomach problems, urinary infections, neural disorders, swelling and redness, hepatic diseases, peptic ulcers, and cancer.	Rhizome, Root	Goyal et al. (2003)
Bergenia ciliata (Haw.) Sternb.	Saxifragaceae	Kidney stones and ulcers	Root	Uniyal et al. (2006)
Boenninghausenia albiflora (Hook.) Rchb. ex Meisn.	Rutaceae	Hepatoprotective, antioxidative, anti-inflammatory, and immunomodulating.	Whole plant	Wang et al. (2007)

TABLE 10.1 *(Continued)*

Plant Species	Family	Medicinal Values with Multipurpose Uses	Parts Used	References
Boerhavia diffusa Sw.	Nyctaginaceae	Liver problem, jaundice, and asthma with antimicrobial activity.	Root	Kumar et al. (2014)
Bombax ceiba L.	Malvaceae	Improve gynecological disorders.	Leaf, root	Patil & Bhaskar (2006)
Boswellia serrata Roxb.	Burseraceae	Chronic inflammatory diseases.	Bark	Siddiqui (2011)
Buchanania cochinchinensis (Lour.) M.R. Almeida	Anacardiacea	Asthma and cancer	Leaf	Ekka & Dixit (2007)
Butea monosperma (Lam.) Kuntze	Fabaceae	Gastritis problems and ulcer treatment.	Flower, gum	Joshi & Pant (2012)
Cajanus scarabaeoides (L.) Thouars	Fabaceae	Hepatoprotective and immense medicinal potential.	Leaf	Rai & Rahman (2018)
Callicarpa macrophylla Vahl	Lamiaceae	Skin diseases, urinary problems, and hair nourishment.	Fruit	Hossan et al. (2009)
Calotropis gigantea (L.) W.T. Aiton	Apocynaceae	Epilepsy, wound healing, and toothache.	Leaf	Kumar et al. (2010)
Carissa spinarum L.	Apocynaceae	Rheumatism	Root	Parmar & Kaushal (1982)
Celastrus paniculatus Willd.	Celastraceae	Tumors, dysentery, and diarrhea.	Fruit, seed	Singh et al. (2010)
Commelina benghalensis L.	Commelinaceae	Honeybee biting	Leaf	Joshi & Pant (2012)
Commelina paludosa Blume	Commelinaceae	Treatment of sexual impotency.	Whole plant	Joshi & Pant (2012)
Cordia myxa L.	Boraginaceae	Cold and cough	Fruit	Joshi & Pant (2012)
Corylus colurna L.	Betulaceae	Its tonic use as diuretic, aphrodisiac.	Nut	Ozturk et al. (2013)
Cydonia oblonga Mill.	Rosaceae	Diarrhea, dysentery, cough, sore throat, bronchitis, constipation, allergic rhinitis, and asthma.	Leaf, seed	Dwivedi et al. (2019)
Datura metel L.	Solanaceae	Antibiotics, antibacterial, antimicrobial activity.	Leaf	Sakthi et al. (2011)
Datura stramonium L.	Solanaceae	Analgesic, anti-asthmatic activities.	Flower	Soni et al. (2012)
Dicliptera bupleuroides Nees	Acanthaceae	Sexual problems, blood vomiting, respiratory issues, and urinary infection.	Seed, leaf	Shah et al. (2013)

TABLE 10.1 *(Continued)*

Plant Species	Family	Medicinal Values with Multipurpose Uses	Parts Used	References
Drimia indica (Roxb.) Jessop	Asparagaceae	Cancer, skin diseases, and ligament related issues.	Bulb	Bozorgia et al. (2017)
Euonymus tingens Wall.	Celastraceae	Kidney stone, eye diseases.	Whole plant	Pala et al. (2010)
Ficus auriculata Lour.	Moraceae	Gastrointestinal problems, hypoglycemic, and anthelmintic activity, anti-tumor activity.	Whole plant	Rout et al. (2009)
Ficus benghalensis L.	Moraceae	Treatment of sexual impotency.	Fruit, bark	Joshi & Pant (2012)
Ficus palmata Roxb.	Moraceae	Treatment of lung and bladder diseases, wound healing.	Fruit	Uniyal et al. (2006)
Glycosmis pentaphylla (Retz.) DC.	Rutaceae	Vomiting	Seed	Joshi & Pant (2012)
Grevillea robusta A.Cunn.ex R.Br.	Proteaceae	Burning and burning sensation.	Leaf	Joshi & Pant (2012)
Gymnema sylvestre (Retz.) R.Br. ex Sm.	Apocynaceae	Antidiabetic properties and eye problems.	Leaf	Kanetkar et al. (2007)
Hedychium spicatum G.Lodd.	Zingiberaceae	Neuromuscular disorders, body pain, anticancer, and antimicrobial activities.	Root	Giri et al. (2010)
Helicteres isora L.	Malvaceae	Excessive appetite, empyema, antioxidant, and anticancer properties.	Root	Venkatesh et al. (2014)
Holarrhena pubescens Wall. ex G.Don	Apocynaceae	Constipation, colic, and diarrhea.	Latex	Gilani et al. (2010)
Hydrocotyle sibthorpioides Colenso	Araliaceae	Dysentery, diarrhea, piles, rheumatism, digestive, diuretic, vermifuge, and menstrual problem.	Leaf	Das & Pathak (2013)
Hygrophila auriculata (Schumach.) Heine	Acanthaceae	Rheumatism, inflammation, jaundice, hepatic obstruction, gout, and pain.	Seed	Hussain et al. (2009)
Hypericum japonicum Thunb.	Hypericaceae	Hepatoprotective activity, protecting liver injury, and hepatotoxicity.	Whole plant	Wang et al. (2008)
Jacobaea nudicauli (Buch.-Ham. ex D.Don)	Asteraceae	Eye related and skin related diseases.	Whole plant	Pala et al. (2010)

TABLE 10.1 *(Continued)*

Plant Species	Family	Medicinal Values with Multipurpose Uses	Parts Used	References
Juniperus communis L.	Cupressaceae	Tuberculosis and respiratory disease, antioxidant, and antimicrobial activities.	Green wood	Gordien et al. (2009)
Justicia adhatoda Mart. ex Nees	Acanthaceae	Cough, fever, coryza, respiratory problems.	Leaf, flower	Biswas et al. (2011)
Lablab purpureus (L.) Sweet	Fabaceae	Anti-inflammatory, cytotoxic, and antiallergic.	Whole plant	Al-Snafi (2017)
Lannea coromandelica (Houtt.) Merr.	Anacardiaceae	Jaundice and liver pain	Bark	Anisuzzaman et al. (2007)
Leucas cephalotes (Roth) Spreng.	Lamiaceae	Asthma and cough	Whole plant	Rahman et al. (2014)
Lilium polyphyllum D.Don	Liliaceae	Galactagog, refrigerant, diuretic, and tonic.	Root	Dhyani (2007)
Litsea glutinosa (Lour.) C.B.Rob.	Lauraceae	Pain, arouse sexual power, bind fractured limbs.	Bark	Joshi & Pant (2012)
Melia azedarach (Spreng.) Balle	Meliaceae	Stomachache, cancer treatment, and malaria treatment.	Root	Sen & Batra (2012)
Mentha longifolia (L.) L.	Lamiaceae	Fever, coryza, rheumatism, dysentery, dyspepsia.	Leaf	Naghibi et al. (2005)
Micromeria biflora (Buch.-Ham. ex D.Don) Benth.	Lamiaceae	Fever, Sciatica, arthritis, skin disease.	Whole plant	Tiwari et al. (2010)
Morella esculenta Buch.-Ham. ex D.Don	Myricacea	Cardiac debility, edema, hemoptysis, ulcer healing, anti-inflammatory activity and antitumor activities.	Fruit, bark	Gusain & Khanduri (2016); Nitha et al. (2007)
Morina longifolia Wall. ex DC	Caprifoliaceae	Antimicrobial, cytotoxic, antiulcer, hypoglycemic, and hypotensive.	Whole plant	Katkar et al. (2010)
Nyctanthes arbor-tristis L.	Oleaceae	Immunotoxin and ulcerogenic activity.	Leaf	Deshmukh et al. (2007)
Origanum vulgare L.	Lamiaceae	Antifungal, anti-hyperglycemic activity.	Whole plant	Khosravi et al. (2011)
Oxalis corniculata L.	Oxalidaceae	Bacterial diseases, dysentery, diarrhea, skin disease.	Whole plant	Raghavendra et al. (2006)
Phyllanthus emblica L.	Phyllanthaceae	Gonorrhea, analgesic, and skin fairness, nausea, and vomiting, antitumor, and hepatoprotective activity.	Whole plant	Sharma et al. (2017)
Piper longum L.	Piperaceae	Muscular pains, carminative, and bronchitis.	Fruit	Manoj et al. (2004)

TABLE 10.1 *(Continued)*

Plant Species	Family	Medicinal Values with Multipurpose Uses	Parts Used	References
Plantago major L.	Plantaginaceae	Dysentery, wound healing, anti-inflammation, analgesic, antioxidant, antiulcerogenic activity.	Seed	Samuelsen (2000)
Plumbago zeylanica L.	Plumbaginaceae	Anti-atherogenic, cardiotonic, hepatoprotective, and neuroprotective properties.	Root, fruit	Tilak et al. (2004)
Polygonatum cirrhifolium (Wall.) Royle	Asparagaceae	Antioxidant and anti-inflammatory.	Root	Bhatt et al. (2014)
Polygonatum verticillatum (L.) All.	Asparagaceae	Aphrodisiac, emollient, cardio tonic, carminative, sialagog, stimulant.	Rhizome	Khan et al. (2010)
Prunus armeniaca L.	Rosaceae	Treatment for cancer	Fruit	Sharma et al. (2014)
Punica granatum L.	Lythraceae	Wound healing, anticancer, and antioxidant activities.	Fruit	Maria et al. (2010)
Pyracantha crenulata (D.Don) M.Roem.	Rosaceae	Used in cardiac failure and myocardial weakness.	Fruit, bark	Sharma et al. (2017)
Ribes nigrum L.	Grossulariaceae	Rheumatoid arthritis, antimicrobial, and anticancer.	Leaf	Sharma et al. (2017)
Ricinus communis L.	Euphorbiaceae	Toxicological, wound healing, and antimicrobial activity.	Leaf	Oyewole et al. (2010)
Rosa canina L.	Rosaceae	Treat urinary infection and kidney related infection.	Leaf, fruit	Sharma et al. (2017)
Rosa moschata Herrm.	Rosaceae	Stomach disorder	Leaf	Haq et al. (2011)
Roylea cinerea (D.Don) Baill.	Fabaceae	Skin infection	Leaf	Prakash & Aggarwal (2010)
Rumex hastatus D.Don	Polygonaceae	Antioxidant, anticholinesterase, and antituberculosis properties.	Root, leaf	Ahmad et al. (2015)
Rumex nepalensis Spreng.	Polygonaceae	Wound healing and anti-allergic.	Leaf	Uniyal et al. (2006)
Rumex vesicarius L.	Polygonaceae	Antibacterial properties	Leaf	Kambhar (2014)
Satyrium nepalense D.Don	Orchidaceae	Fever	Root	Mishra & Saklani (2012)
Sida cordifolia L.	Malvaceae	Blennorrhagia and nasal congestion.	Bark, root	Silva et al. (2006)

TABLE 10.1 *(Continued)*

Plant Species	Family	Medicinal Values with Multipurpose Uses	Parts Used	References
Smilax aspera L.	Smilacaceae	Diarrhea	Root	Longo & Vasapollo (2006)
Solanum nigrum L.	Solanaceae	Itching, burn relief, and pain.	Whole plant	Sharma et al. (2017)
Solanum violaceum R.Br.	Solanaceae	Hypertension, poisonous, insect bites.	Fruit	Rahmatullah et al. (2011)
Solena heterophylla Lour.	Cucurbitaceae	Gastrointestinal, respiratory, and vascular related issues.	Tuber	Janbaz et al. (2015)
Swertia chirayita (Roxb.) H.Karst.	Gentianaceae	Control fever and skin disease.	Whole plant	Kumar et al. (2010)
Syzygium cumini (L.) Skeels	Myrtaceae	Diabetes	Bark	Joshi & Pant (2012)
Trapa natans L.	Lythraceae	Blood pressure, cytotoxic activities, and cardiac problems.	Nut	Rahman et al. (2000)
Urena lobata L.	Malvaceae	Urinary problems and sexual transmitted diseases.	Whole plant	Hossan et al. (2010)
Urtica ardens Link	Urticaceae	Fever, bone fractures, digestion related issues, and skin infection.	Leaf	Shrestha et al. (2003)
Vachellia nilotica (L.) P.J.H. Hurter & Mabb.	Fabaceae	Treatment of cancers, tuberculosis, indurations of liver and spleen.	Whole plant	Shukla et al. (2010)
Vigna mungo (L.) Hepper	Fabaceae	Urinary tract infections, sexually transmitted diseases.	Fruit	Rahmatullah et al. (2011)
Vitex negundo L.	Lamiaceae	Bronchial issues and anti-inflammatory activities.	Leaf	Tandon (2005)
Vitis vulpina L.	Vitaceae	Treatment of sore throats, smallpox, infections in eye or skin, and tuberculosis.	Leaf, fruit	Sharma et al. (2017)
Zanthoxylum armatum (Thunb.) Druce	Rutaceae	Control allergy with carminative, stomachic, anthelmintic activities.	Leaf	Singh & Singh (2011)
Ziziphus jujuba Mill.	Ramanaceae	Anticancer treatment, blood purifier, insomnia, nervous exhaustion, treatment of fever, wound, ulcer treatment.	Whole plant	Sharma et al. (2017)

10.5 CONSTRAINTS AND FUTURE RESEARCH PROSPECTS

Variable results, inconsistent effects, inadequate research grants, and small sample size are some of the potential reasons for the low acceptance rate of TMs (Nahin & Straus, 2001). Traditional medicine works best in a specific environment in the specialized residential setting under experts' supervision, which is considered one of the other key problems with TMs (Telles et al., 2014). The interaction between healers and the patient can have a potential control effect and even possibly affect the outcome and duration of the disease (Brody, 2000). Furthermore, the development of TMs and natural products necessitates the convergence of current methodologies and integrated approaches connected to their evidence-based research in a variety of scientific domains, which demands national coordination and cooperation (Kunle et al., 2012). This integrated approach would result in cost and time savings, as well as a higher success rate in medication development (Katiyar et al., 2012). Scientific validation, quality control, and assurance of TMs are critical for ensuring therapeutic efficacy and safety, as well as rationalizing their usage in health care so that the required quantity of quality medicines is available (Sharma et al., 2010). Traditional formulations such as churns, bhasmas, liquid orals, Lehas, and others require quality assurance (Mukherjee et al., 2008). The use of fingerprinting and genetic marker-based compound analysis to standardize TM formulations is gaining popularity these days, as it aids in determining the correct botanical identity and regulating the chemical purity of the plants (Mukherjee et al., 2011). Attention should also be paid to intellectual property problems in order to prevent bio-piracy of historic indigenous knowledge and natural resources used in TMs (WHO, 2002). The improper, unscientific collecting of medical herbs depletes the richness of medicinal plants and puts genetic stocks and medicinal plant variety at risk (Chen et al., 2016). Over 70% of medicinal plant collection is detrimental, involving the removal of roots, bark, wood, stems, or the entire plant, especially in herbs that harm the plant (Sharma et al., 2010). As a result, sustainable harvesting and long-term use of natural products are required to protect these important plant resources from further loss and extinction (Liu et al., 2012).

Further research prospects of TMs can be triggered by many potential options:

- Policymaking for sustainable harvesting of plant resources and standardization of formulation.

- Training of researchers in traditional medicines with the inclusion of conventional and innovative research methods as well as with necessary information on human anatomy and physiology to equip them to work on designating with traditional medicine.
- Promoting research opportunities in traditional medicines by granting funds for research in traditional medicines.
- Further research in traditional medicines can potentially lead to the formulation of many life-saving miracle drugs which can not only be used by the local people but by the entire society.
- Traditional medicines are encountering significant obstacles in demonstrating their safety, mechanisms of action, quality, and efficacy, which necessitated extensive research to establish evidence-based efficacy, uncover mechanisms of action, define pharmacological and toxicological profiles, assess safety, and ensure quality.
- Moreover, the specific chemical ingredient(s) and mechanism(s) of action by which medicinal plants provide therapeutic advantages are still unknown.

10.6 CONCLUSION

The use of medicinal plants is an old therapeutic strategy by traditional healers, and following the emergence of modern medicines, it is still useful in treating ailments. Medicinal plants are compatible with the physiology of the human body, which has adapted over the centuries. Today's scenario is that scientists have to focus on determining the essential bioactive plant components responsible for a cure, mitigation, and disease healing. Folk medicines have several challenges, such as regulatory aspects, consumer acceptance, and market competition, to deliver safe and effective natural products. Furthermore, many known medicinal plants still need to be investigated and screened empirically for active chemical components, resulting in the identification of new innovative medications. Moreover, several endangered medicinal plants, whose use is scientifically proven, are over-exploited by medicinal plant traders, thus threatening the species' survival. There is an urgent need to come up with cultivation techniques to fulfill the demand as well as maintain the environment's health. Therefore, an urgent comprehensive study is needed using biotechnological tools to authenticate and produce new novel drugs from the region's rich bio-resources.

KEYWORDS

- **constraints**
- **medicinal plants**
- **natural forests**
- **prospects**
- **socio-economic significance**
- **traditional homegardens**
- **traditional medicines**

REFERENCES

Abdullahi, A. A., (2011). Trends and challenges of traditional medicine in Africa. *African Journal of Traditional, Complementary and Alternative Medicines, 8*(5S), 115–123.

Ahmad, S., Ullah, F., Ayaz, M., Sadiq, A., & Imran, M., (2015). Antioxidant and anticholinesterase investigations of *Rumex hastatus* D. Don: Potential effectiveness in oxidative stress and neurological disorders. *Biological Research, 48*(1), 1–8.

Akshay, K. R., Sudharani, N., Anjali, K. B., & Deepak, T. M., (2014). Biodiversity and strategies for conservation of rare, endangered and threatened medicinal plants. *Research and Reviews: Journal of Pharmacognosy and Phytochemistry, 2*(3), 12–20.

Al-Snafi, A. E., (2017). The pharmacology and medical importance of *Dolichos lablab* (*Lablab purpureus*): A review. *IOSR Journal of Pharmacy, 7*(2), 22–30.

Anisuzzaman, M., Rahman, A. H. M. M., Rashid, M., Harun-OrRashid, M., Naderuzzaman, A. T. M., & Islam, A. K. M. R., (2007). An ethnobotanical study of Madhupur, Tangail. *Journal of Applied Sciences Research, 3*(7), 519–530.

Balakumbahan, R., Rajamani, K., & Kumanan, K., (2010). *Acorus calamus*: An overview. *Journal of Medicinal Plants Research, 4*(25), 2740–2745.

Batish, D. R., Kaur, M., Harminder, P., Singh, R., & Kohli, K., (2007). Phytotoxicity of a medicinal plant, *Anisomeles indica*, against *Phalaris minor* and its potential use as natural herbicide in wheat fields. *Crop Protection, 26*(7), 948–952.

Bent, S., (2008). Herbal medicine in the United States: Review of efficacy, safety, and regulation. *Journal of General Internal Medicine, 23*(6), 854–859.

Bhatt, D., Kumar, R., Tewari, L. M., & Joshi, G. C., (2014). *Polygonatum cirrhifolium* Royle and *Polygonatum verticillatum* (L.) Allioni, Status assessment and medicinal uses in Uttarakhand, India. *Journal of Medicinal Plants Research, 8*(5), 253–259.

Biswas, A., Haq, W. M., Akber, M., Ferdausi, D., Seraj, S., Jahan, F. I., Chowdhury, A. R., & Rahmatullah, M., (2011). A survey of medicinal plants used by folk medicinal practitioners of Paschim Shawra and Palordi villages of Gaurnadi Upazila in Barisal district, Bangladesh. *Journal of Sustainable Agriculture, 5*, 15–22.

Bozorgia, M., Amin, G., Shekarchi, M., & Rahimia, R., (2017). Traditional medical uses of Drimia species in terms of phytochemistry, pharmacology and toxicology. *Journal of Traditional Chinese Medicine, 37*(1), 124–139.
Brody, H., (2000). The placebo response. *Journal of Family Practice, 49*(7), p. 649. ISSN-0094-3509.
Chen, S. L., Yu, H., Luo, H. M., Wu, Q., Li, C. F., & Steinmetz, A., (2016). Conservation and sustainable use of medicinal plants: Problems, progress, and prospects. *Chinese Medicine, 11*(1), 1–10.
Das, R. J., & Pathak, K., (2013). Use of indigenous plants in traditional health care systems by Mishing tribe of Dikhowmukh, Sivasagar District, Assam. *International Journal of Health and Medicine, 1*(3), 50–57.
DeFilipps, R. A., & Krupnick, G. A., (2018). The medicinal plants of Myanmar. *PhytoKeys, 102*, 1–341.
Deshmukh, R. D., Pokharkar, R. D., Takate, S. B., & Gite, V. N., (2007). Amelioration of CCl_4-induced hepatosuppression by *Nyctanthes arbortristis* Linn leaves in Wistar albino rats. *Pharmacology Online, 1*, 203–208.
Dhyani, A., (2007). Exploring *Lilium polyphyllum* in Uttarakhand, India. *The Lily Yearbook of North American Lily Society, 60,* 79–82.
Dwivedi, T., Kanta, C., Singh, L. R., & Prakash, I., (2019). A list of some important medicinal plants with their medicinal uses from Himalayan State Uttarakhand, India. *Journal of Medicinal Plants, 7*(2), 106–116.
Ekka, N. R., & Dixit, V. K., (2007). Ethno-pharmacognostical studies of medicinal plants of Jashpur District (Chhattisgarh). *International Journal of Green Pharmacy, 1*(1), 2–4.
Ekor, M., (2014). The growing use of herbal medicines: Issues relating to adverse reactions and challenges in monitoring safety. *Frontiers in Pharmacology, 4*, 1–10.
Eric, M., (1986). Handbook of American folklore. In: Dorson, R. M., (ed.), *Folklore Forum* (Vol. 19, No. 1, pp. 82–91). Indiana University Press.
Fabricant, D. S., & Farnsworth, N. R., (2001). The value of plants used in traditional medicine for drug discovery. *Environmental Health Perspectives, 109*, 69–75.
Gilani, A. H., Khan, A., Khan, A., Bashir, S., Rahman, N., & Mandukhail, S. R., (2010). Pharmacological basis for the medicinal use of *Holarrhena antidysenterica* in gut motility disorders. *Journal of Pharmaceutical Biology, 48*(11), 1240–1246.
Giri, D., Tamta, S., & Pandey, A., (2010). A review account on medicinal value of *Hedychium spicatum* Buch-Ham ex Sm: Vulnerable medicinal plant. *Journal of Medicinal Plants Research, 4*(25), 2773–2777.
GOI, Planning Commission, (2000). *Report of the Task Force on Conservation & Sustainable use of Medicinal Plants.* Available at URL. https://ruralindiaonline.org/en/library/resource/report-of-the-task-force-on-conservation-and-sustainable-use-of-medicinal-plants/ (accessed on 6 May 2024).
Gordien, Y. A., Gray, I. A., Franzblau, G. S., & Seidel, V., (2009). Antimycobacterial terpenoids from *Juniperus communis* L. (*Cupressaceae*). *Journal of Ethnopharmacology, 126*(3), 500–505.
Goyal, R. K., Singh, J., & Lal, H., (2003). *Asparagus racemosus* – an update. *Indian Journal of Medical Sciences, 57*(9), 408–414.
Gurib-Fakim, A., (2011). Traditional roles and future prospects for medicinal plants in health care. *Asian Biotechnology and Development Review, 13*(3), 77–83.

Gusain, Y. S., & Khanduri, V. P., (2016). *Myrica esculenta* wild edible fruit of Indian Himalaya: Need a sustainable approach for indigenous utilization. *Ecology, Environment and Conservation, 22*, 267–270.

Hand, W. D., (1985). Magical medicine: An alternative to "alternative medicine." *Western Folklore, 44*(3), 240–251.

Haq, F., Ahmad, H., & Alam, M., (2011). Traditional uses of medicinal plants of Nandiar Khuwarr catchment (District Battagram), Pakistan. *Journal of Medicinal Plants Research, 5*(1), 39–48.

Haque, M., Chowdhury, A. B. M., Shahjahan, M., Harun, M., & Dostogir, G., (2018). Traditional healing practices in rural Bangladesh: A qualitative investigation. *BioMed Central Complementary and Alternative Medicine, 18*(1), 1–15.

Hossain, K., & Ballardini, R. M., (2021). Protecting indigenous traditional knowledge through a holistic principle-based approach. *Nordic Journal of Human Rights, 39*(1), 51–72.

Hossan, M. S., Hanif, A., Khan, M., Bari, S., Jahan, R., & Rahmatullah, M., (2009). Ethnobotanical survey of the Tripura tribe of Bangladesh. *American-Eurasian Journal of Sustainable Agriculture, 3*(2), 253–261.

Hossan, S., Agarwala, B., Sarwar, S., Karim, M., Jahan, R., & Rahmatullah, M., (2010). Traditional use of medicinal plants in Bangladesh to treat urinary tract infections and sexually transmitted diseases. *Ethnobotany Research and Applications, 8*, 61–74.

Hussain, S., Ahmed, N., & Ansari, Z., (2009). Preliminary studies on diuretic effect of *Hygrophila auriculata* (Schum) Heine in rats. *International Journal of Health Sciences Education, 2*(1), 59–64.

Jain, J. B., Kumane, S. C., & Bhattacharya, S., (2006). Medicinal flora of Madhya Pradesh and Chattisgarh: A review. *Indian Journal of Traditional Knowledge, 5*(2), 237–242.

Janbaz, K. H., Akhtar, T., Saqib, F., Imran, I., Zia-Ul-Haq, M., Jansakul, C., De Feo, V., & Moga, M., (2015). Pharmacological justification of use of *Solena heterophylla* Lour, in gastrointestinal, respiratory and vascular disorders. *Journal of Translational Medicine, 13*, 2–8.

Joshi, B., & Pant, S. C., (2012). Ethnobotanical study of some common plants used among the tribal communities of Kashipur, Uttarakhand. *Indian Journal of Natural Products and Resources, 3*(2), 262–266.

Kala, C. P., (2006). Ethnobotany and ethno conservation of *Aegle marmelos* (L), correa. *Indian Journal of Traditional Knowledge, 5*(4)537–540.

Kambhar, S. V., (2014). *Rumex vesicarius* L. (*Polygonaceae*): An overview. *Journal of Global Ecology and Environment, 1*(1), 11–14.

Kamboj, A., & Saluja, A. K., (2008). *Ageratum conyzoides* L.: A review on its phytochemical and pharmacological profile. *International Journal of Green Pharmacy, 2*(2), 59–68.

Kanetkar, P., Singhal, R., & Kamat, M., (2007). *Gymnema sylvestre*: A Memoir. *Journal of Clinical Biochemistry and Nutrition, 41*(2), 77–81.

Katiyar, C., Gupta, A., Kanjilal, S., & Katiyar, S., (2012). Drug discovery from plant sources: An integrated approach. *Ayu, 33*(1), 10–19.

Katkar, K. V., Suthar, A. C., & Chauhan, V. S., (2010). The chemistry, pharmacologic, and therapeutic applications of *Polyalthia longifolia. Pharmacognosy Reviews, 4*(7), 62–68.

Kayne, S. B., (2010). Traditional medicine: A global perspective. In: *Books and Electronic Media* (pp. 88–953). London: Pharmaceutical Press, Bull World Health Organ. Available at URL. https://www.ncbi.nlm.nih.gov/pmc/articles/PMC2995194/ (accessed on 22 April 2024).

Khan, B., Abdukadir, A., Qureshi, R., & Mustafa, G., (2011). Medicinal uses of plants by the inhabitants of Khunjerab National Park, Gilgit, Pakistan. *Pakistan Journal of Botany, 43*(5), 2301–2310.

Khan, H., Saeed, M., Gilani, A. H., Khan, M. A., Dar, A., & Khan, I., (2010). The antinociceptive activity of *Polygonatum verticillatum* rhizomes in pain models. *Journal of Ethnopharmacology, 127*, 521–527.

Khatib, C., Nattouf, A., & Hasan, A. M. I., (2021). Traditional medicines and their common uses in central region of Syria: Hama and Homs–an ethnomedicinal survey. *Pharmaceutical Biology, 59*(1), 778–788.

Khosravi, A. R., Shokri, H., Kermani, S., Dakhili, M., Madani, M., & Parsa, S., (2011). Antifungal properties of *Artemisia sieberi* and *Origanum vulgare* essential oils against *Candida glabrata* isolates obtained from patients with vulvovaginal candidiasis. *Journal de Mycologie Médicale, 21*(2), 93–99.

Khumbongmayum, A. D., Khan, M. L., & Tripathi, R. S., (2005). Ethnomedicinal plants in the sacred groves of Manipur. *Indian Journal of Traditional Knowledge, 4*(1), 21–32.

Kumar, G., Karthik, L., & Rao, K. V. B., (2010). In vitro anti-candida activity of *Calotropis gigantea*. *Journal of Pharmacy Research, 3*(3), 539–542.

Kumar, P. P. N. V., Pammi, S. V. N., Kollu, P., Satyanarayanad, K. V., & Shameema, U., (2014). Green synthesis and characterization of silver nanoparticles using *Boerhaavia diffusa* plant extract and their anti-bacterial activity. *Industrial Crops and Products, 52*, 562–566.

Kunle, O. F., Egharevba, H. O., & Ahmadu, P. O., (2012). Standardization of herbal medicines-A review. *International Journal of Biodiversity and Conservation, 4*(3), 101–112.

Lakshman, C. D., (2016). Bio-Diversity and conservation of medicinal and aromatic plants. *Advances in Plants and Agriculture Research, 5*(4), 561–566.

Liu, Y., Shi, X. W., Liu, E. H., Sheng, L. S., Qi, L. W., & Li, P., (2012). More accurate matrix-matched quantification using standard superposition method for herbal medicines. *Journal of Chromatography A., 1254*, 43–50.

Longo, L., & Vasapollo, G., (2006). Extraction and identification of anthocyanins from *Smilax aspera* L. berries. *Food Chemistry, 94*, 226–231.

Maiti, S., (2004). Inventory, documentation and status of medicinal plants research in India. In: Batugal, P. A., Kanniah, J., Sy, L., & Oliver, J. T., (eds.), *Medicinal Plants Research in Asia-: The Framework and Project Workplans, Biodiversity International* (Vol. I, pp. 79–91).

Manoj, P., Soniya, E. V., Banerjee, N. S., & Ravichandran, P., (2004). Recent studies on well-known spice, *Piper longum* Linn. *Natural Product Radiance, 3*(4), 222–227.

Mao, A. A., & Hynniewta, T. M., (2000). Floristic diversity of North East India. *Journal Assam Science Society, 41*(4), 255–266.

Maria, G. M., Maria, A. N., & Maria, D. A., (2010). Pomegranate (*Punica granatum* L) A medicinal plant with myriad biological properties: A short review. *Journal of Medicinal Plants Research, 4*(25), 2836–2847.

Mensah, A. Y., Bonsu, A. S., & Fleischer, T. C., (2011). Investigation of the bronchodilator activity of *Abrus precatorius*. *International Journal of Pharmaceutical Sciences Review and Research, 6*(2), 9–13.

Mishra, A. P., & Saklani, S., (2012). *Satyrium nepalense*: A rare medicinal orchid of western Himalaya (India); phytochemical screening, antimicrobial evaluation and conservation studies. *Indonesian Journal of Pharmacy, 23*(3), 162–170.

Moreira, D. D. L., Teixeira, S. S., Monteiro, M. H. D., De-Oliveira, A. C. A., & Paumgartten, F. J., (2014). Traditional use and safety of herbal medicines. *Revista Brasileira de Farmacognosia, 24*, 248–257.

Mukherjee, P. K., & Wahile, A., (2006). Integrated approaches towards drug development from Ayurveda and other Indian system of medicines. *Journal of Ethnopharmacology, 103*(1);25–35.

Mukherjee, P. K., Ponnusankar, S., Pandit, S., Hazam, P. K., Ahmmed, M., & Mukherjee, K., (2011). Botanicals as medicinal food and their effects on drug-metabolizing enzymes. *Food and Chemical Toxicology, 49*(12), 3142–3153.

Mukherjee, P. K., Rai, S., Bhattacharya, S., Wahile, A., & Saha, B. P., (2008). Marker analysis of polyherbal formulation, triphala–A well-known Indian traditional medicine. *Council of Scientific and Industrial Research, 7*(3), 379–383.

Mukherjee, P. K., Shiv, B., Harwansh, R. K., & Chaudhary, S. K., (2014). Shifting paradigm for validation of medicinal plants in Indian traditional medicine. *Indian Drugs, 51*, 5–14.

Naghibi, F., Mosaddegh, M., Mohammadi, M. M., & Ghorbani, A., (2005). Labiatae family in folk medicine in Iran. *Journal of Ethnobiology and Ethnomedicine, 4*(2), 63–79.

Nahin, R. L., & Straus, S. E., (2001). Research into complementary and alternative medicine: Problems and potential. *British Medical Journal, 322*(7279), 161–164.

Negi, J. S., Singh, P., Joshi, G. P., Rawat, M. S., & Bisht, V. K., (2010). Chemical constituents of *Asparagus*. *Pharmacognosy Reviews, 4*(8), 215–220.

Nema, N. K., Maity, N., Sarkar, B., & Mukherjee, P. K., (2011). *Cucumis sativus* fruit-potential antioxidant, anti-hyaluronidase, and anti-elastase agent. *Archives of Dermatological Research, 303*(4), 247–252.

Ngo, H. V. V., Martinetz, T., Born, J., & Mölle, M., (2013). Auditory closed-loop stimulation of the sleep slow oscillation enhances memory. *Neuron, 78*(3), 545–553.

Nitha, B., Meera, C. R., & Janardhanan, K. K., (2007). Anti-inflammatory and anti-tumor activities of cultured mycelium of morel mushroom, *Morchella esculenta*. *Current Science, 92*(2), 235–239.

Okusa, P. N., Penge, O., De vleeschouwer, M., & Duez, P., (2007). Direct and indirect antimicrobial effects and antioxidant activity of *Cordia gilletii* de wild (*Boraginaceae*). *Journal of Ethnopharmacology, 112*(3), 476–481.

Oliveira, M., Simoes, M., & Sassi, C., (2006). Fitoterapia no sistema de saúde pública (SUS) no estado de São Paulo, Brasil. *Revista Brasileira de Plantas Medicinais, 8*(2), 39–41.

Oyewole, O. I., Owoseni, A. A., & Faboro, E. O., (2010). Studies on medicinal and toxicological properties of *Cajanus cajan, Ricinus communis,* and *Thymus vulgaris* leaf extracts. *Journal of Medicinal Plants Research, 4*(19), 2004–2006.

Ozioma, E. O. J., & Chinwe, O. A. N., (2019). Herbal medicines in African traditional medicine. *Herbal Medicine, 10*, 191–214.

Ozturk, M., Uysal, I., Gucel, S., Altundag, E., Dogan, Y., & Baslar, S., (2013). Medicinal Uses of natural dye-yielding plants in Turkey. *Research Journal of Textile and Apparel, 17*(2), 69–80.

Pala, N. A., Negi, A. K., & Todaria, N. P., (2010). Traditional uses of medicinal plants of Pauri Garhwal, Uttarakhand. *Nature and Science, 8*(6), 57–61.

Pan, S. Y., Litscher, G., Gao, S. H., Zhou, S. F., Yu, Z. L., Chen, H. Q., Zhang, S. F., et al., (2014). Historical perspective of traditional indigenous medical practices: The current renaissance and conservation of herbal resources. *Evidence-Based Complementary and*

Alternative Medicine (pp. 1–20). Available at URL. https://www.hindawi.com/journals/ecam/2014/525340/ (accessed on 22 April 2024).

Parisius, L. M., Stock-Schröer, B., Berger, S., Hermann, K., & Joos, S., (2014). Use of home remedies: A cross-sectional survey of patients in Germany. *BioMed Central Family Practice, 15*(1), 1–8.

Parmar, C., & Kaushal, M. K., (1982). *Wild Fruits* (pp. 78–80). Kalyani Publishers, New Delhi, India.

Patil, H. M., & Bhaskar, V. V., (2006). Medicinal uses of plants by tribal medicine men of Nandurbar district in Maharashtra Explorer. *Indian Journal of Natural Products and Resources, 5*(2), 125–130.

Patwardhan, B., (2014). Bridging Ayurveda with evidence-based scientific approaches in medicine. *European Association for Predictive, Preventive, and Personalized Medicine Journal, 5*(1), 1–7.

Payyappallimana, U., (2010). Role of traditional medicine in primary health care: An overview of perspectives and challenges. *Yokohama Journal of Social Sciences, 14*(6), 57–77.

Prakash, V., & Aggarwal, A., (2010). Traditional uses of ethnomedicinal plants of lower foothills of Himachal Pradesh-I. *Indian Journal of Traditional Knowledge, 9*(3), 519–521.

Prasad, A. G. D., Shyma, T. B., & Raghavendra, M. P., (2013). Plants used by the tribes for the treatment of digestive system disorders in Wayanad district, Kerala. *Journal of Applied Pharmaceutical Science, 3*(8), 171–175.

Raghavendra, M. P., Satish, S., & Raveesha, K. A., (2006). Phytochemical analysis and antibacterial activity of *Oxalis corniculata*; a known medicinal plant. *My Science, 1*(1), 72–78.

Rahman, A. H. M. M., Hossain, M. M., & Islam, A. K. M. R., (2014). Taxonomy and medicinal uses of angiosperm weeds in the wheat field of Rajshahi, Bangladesh. *Frontiers of Biological and Life Sciences, 2*(1), 8–11.

Rahman, M. M., Mosadik, M. A., Wahed, M. I., & Haque, M. E., (2000). Antimicrobial activity and cytotoxicity of *Trapa bispinosa*. *Fitoterapia, 71*, 704–706.

Rahmatullah, M., Azam, M. N. K., Rahman, M. M., Seraj, S., Mahal, M. J., Mou, S. M., Nasrin, D., et al., (2011). A survey of medicinal plants used by Garo and non-Garo traditional medicinal practitioners in two villages of Tangail District, Bangladesh. *American Eurasian Journal of Sustainable Agriculture, 5*(3), 350–357.

Rai, A. S., & Rahman, C. H., (2018). Pharmacognostic standardization and phytochemical investigation of *Cajanus scarabaeoides* (L.) thouars. *Research Journal of Pharmacognosy and Phytochemistry, 10*(1), 120–131.

Rajasekaran, A., & Singh, J., (2009). Ethnobotany of Indian horse chestnut (*Aesculus indica*) in Mandi district, Himachal Pradesh. *Indian Journal of Traditional Knowledge, 8*(2), 285–286.

Rajkumar, M. H., Sringeswara, A. N., & Rajanna, M. D., (2011). Ex-situ conservation of medicinal plants at university of agricultural sciences, Bangalore, Karnataka. *Recent Research in Science and Technology, 3*(4), 21–27.

Ramawat, K. G., Goyal, S., Kumar, N., & Sonie, K. C., (2009). Natural products of plant origin with therapeutic potentiality and their biotechnological production. In: Kumar, A., & Shekhawat, N. S., (eds.), *Plant Tissue Culture and Molecular Markers* (pp. 247–280). I. K. International Publishing House Pvt. Ltd.

Raskin, I., Ribnicky, D. M., Komarnytsky, S., Ilic, N., Poulev, A., Borisjuk, N., Brinker, A., et al., (2002). Plants and human health in the twenty-first century. *Trends in Biotechnology, 20*(12), 522–531.

Ravishankar, B., & Shukla, V. J., (2007). Indian systems of medicine: A brief profile. *African Journal of Traditional, Complementary and Alternative Medicines, 4*(3), 319–337.

Rout, S. D., Panda, T., & Mishra, N., (2009). Ethnomedicinal plants used to cure different diseases by tribal's of Mayurbhanj district of North Orissa. *Ethno-Medicine, 3*(1), 27–32.

Sakthi, S. S., Geetha, M., & Saranraj, P., (2011). Pharmacological screening of *Datura metel* and *Acalypha indica* for its antifungal activity against pathogenic fungi. *International Journal of Pharmaceutical Sciences Review and Research, 2*(1), 15–29.

Samuelsen, A. B., (2000). The traditional uses, chemical constituents and biological activities of *Plantago major* L.: A review. *Journal of Ethnopharmacology, 71*(1, 2), 1–21.

Sen, A., & Batra, A., (2012). Evaluation of antimicrobial activity of different solvent extracts of medicinal plant: *Melia azedarach* L. *International Journal of Current Pharmaceutical Research, 4*(2), 67–73.

Sen, S., & Chakraborty, R., (2017). Revival, modernization and integration of Indian traditional herbal medicine in clinical practice: Importance, challenges and future. *Journal of Traditional and Complementary Medicine, 7*(2), 234–244.

Sen, S., Chakraborty, R., & De, B., (2011). Challenges and opportunities in the advancement of herbal medicine: India's position and role in a global context. *Journal of Herbal Medicine, 1*(3, 4), 67–75.

Sever, M., (2015). *Folk Medicine, Folk Healing* (Vol. 9, No. 17, pp. 181–192). Gazi Akademik Bakış.

Shah, A., Marwat, S. K., Gohar, F., Khan, A., Bhatti, K. H., Amin, M., Din, N. U., et al., (2013). Ethnobotanical study of medicinal plants of semi-tribal area of Makerwal & Gulla Khel (lying between Khyber Pakhtunkhwa and Punjab provinces). *American Journal of Plant Sciences, 4*, 98–116.

Shankar, R., Deb, S., & Sharma, B. K., (2012b). Antimalarial plants of northeast India: An overview. *Journal of Ayurveda and Integrative Medicine, 3*(1), 10–16.

Shankar, R., Lavekar, G. S., Deb, S., & Sharma, B. K., (2012a). Traditional healing practice and folk medicines used by Mishing community of North East India. *Journal of Ayurveda and Integrative Medicine, 3*(3), 124–129.

Sharma, I. P., Kanta, C., Semwal, S. C., & Goswami, N., (2017). Wild fruits of Uttarakhand (India): Ethnobotanical and medicinal uses. *International Journal of Complementary & Alternative Medicine, 8*(3), 1–8.

Sharma, M., & Das, B., (2018). Medicinal plants of Northeast Region of India: A small review. *International Journal of Current Pharmaceutical Research, 10*(4), 11–12.

Sharma, M., & Kumar, A., (2013). Traditional medicinal plants of Rajasthan used in tribal medicine: A review. *International Journal of Life Science and Pharma Research, 3*(2), 38–42.

Sharma, S., Rathi, N., Kamal, B., Pundir, D., Kaur, B., & Arya, S., (2010). Conservation of biodiversity of highly important medicinal plants of India through tissue culture technology-a review. *Agriculture and Biology Journal of North America, 1*(5), 827–833.

Sharma, S., Satpathy, G., & Gupta, R. K., (2014). Nutritional, phytochemical, antioxidant and antimicrobial activity of *Prunus armenicus*. *Journal of Pharmacognosy and Phytochemistry, 3*(3), 23–28.

Sheri, K., Too, J. Y. J., Chuah, S. E. L., Toh, Y. P., Mason, S., & Radha, K. L. K., (2019). A scoping review of mentor training programs in medicine between 1990 and 2017. *Medical Education Online, 24*(1), 1–16.

Shi, Y., Zhang, C., & Li, X., (2020). Traditional Indian medicine, part I about traditional Indian medicine. *Journal of Traditional Chinese Medical Sciences, 21*, 1–6.

Shrestha, P., Gautam, R., Rana, R. B., & Sthapit, B., (2003). Home gardens and in situ conservation of plant genetic resources in farming systems. In: *Proceedings of the Second International Home Gardens Workshop* (pp. 105–108). Witzenhausen, Federal Republic of Germany.

Shukla, A. N., Srivastava, S., & Rawat, A. K. S., (2010). An ethnobotanical study of medicinal plants Rewa District, Madhya Pradesh. *Indian Journal of Traditional Knowledge, 9*(1), 191–202.

Siddiqui, M. Z., (2011). *Boswellia Serrata*, A potential anti-inflammatory agent: An overview. *Indian Journal of Pharmaceutical Sciences, 73*(3), 255–261.

Silva, R. L., Melo, G. B., Melo, V. A., Antoniolli, A. R., Michellone, P. R., Zucoloto, S., Picinato, M. A., & Franco, C. F., (2006). Effect of the aqueous extract of *Sida cordifolia* on liver regeneration after partial hepatectomy. *Acta Cirúrgica Brasileira, 21*(1), 37–39.

Singh, S., Srivastava, R., & Choudhary, S., (2010). Antifungal and HPLC analysis of the crude extracts of *Acorus calamus*, *Tinospora cordifolia*, and *Celestrus paniculatus*. *Journal of Agricultural Technology, 6*(1), 149–158.

Singh, T. P., & Singh, O. M., (2011). Phytochemical and pharmacological profile of *Zanthoxylum armatum* DC: An overview. *Indian Journal of Natural Products and Resources, 2*(3), 275–285.

Sofowora, A., Ogunbodede, E., & Onayade, A., (2013). The role and place of medicinal plants in the strategies for disease prevention. *African Journal of Traditional, Complementary, and Alternative Medicines, 10*(5), 210–229.

Soni, P., Siddiqui, A. A., Dwivedi, J., & Soni, V., (2012). Pharmacological properties of *Datura stramonium* L. as a potential medicinal tree: An overview. *Asian Pacific Journal of Tropical Biomedicine, 2*(12), 1002–1008.

Srivastava, R. C., Singh, R. K., & Mukherjee, T. K., (2010). Apatani community indigenous biodiversity of Apatani plateau: Learning on biocultural knowledge of Apatani tribes of Arunachal Pradesh for sustainable livelihoods. *Indian Journal of Traditional Knowledge, 9*(3), 432–442.

Tandon, V. R., (2005). Medicinal uses and biological activities of *Vitex negundo*. *Indian Journal of Natural Products and Resources, 4*(3), 162–165.

Telles, S., Pathak, S., Singh, N., & Balkrishna, A., (2014). Research on traditional medicine: What has been done, the difficulties, and possible solutions. *Evidence-Based Complementary and Alternative Medicine, 2014*, 1–5. Available at URL: https://doi.org/10.1155/2014/495635.

Tilak, J. C., Adhikari, S., & Devasagayam, T. P., (2004). Antioxidant properties of *Plumbago zeylanica* an Indian medicinal plant and its active ingredient, plumbagin. *Redox Report, 9*(4), 219–227.

Tiwari, G. B. G. P. K., Tadele, K., Aramde, F., & Tiwari, S. C., (2010). Community structure and regeneration potential of *Shorea robusta* forest in subtropical submontane zone of Garhwal Himalaya, India. *Nature and Science, 8*(1), 68–74.

Uniyal, S. K., Singh, K. N., Jamwal, P., & Lal, B., (2006). Traditional use of medicinal plants among the tribal communities of Chhota Bhangal, Western Himalaya. *Journal of Ethnobiology and Ethnomedicine, 2*(14), 1–8.

Venkatesh, S., Reddy, G. D., & Reddy, B. M., (2004). Antihyperglycemic activity of *Helicteres isora* roots in alloxan-induced diabetic rats. *Pharmaceutical Biology, 41*(5), 347–350.

Wang, R., Kong, J., Wang, D., Lien, L. L. M., & Lien, E. J. C., (2007). A survey of Chinese herbal ingredients with liver protection activities. *Chinese Medicine, 2*(1), 1–8.

Wang, Y. C., Chuang, Y. C., & Hsu, H. W., (2008). The flavonoid, carotenoid and pectin content in peels of citrus cultivated in Taiwan. *Food Chemistry, 106*, 277–284.

WHO, (2002). *WHO Traditional Medicine Strategy 2002–2005.* Geneva, Switzerland: World Health Organization. Available at URL. https://www.beovita.eu/pdf/WHO_EDM_TRM_2004.pdf (accessed on 22 April 2024).

WHO, (2013). *WHO Traditional Medicine Strategy: 2014–2023*. Available at URL. https://apps.who.int/iris/bitstream/handle/10665/92455/9789241506090_eng.pdf (accessed on 22 April 2024).

WHO, (2019). *WHO Global Report on Traditional and Complementary Medicine 2019*. WHO, Geneva. Available at URL. https://books.google.co.in/books?hl=en&lr =&id=WHOyDwAAQBAJ&oi=fnd&pg=PP1&dq=WHO+global+report+on+traditional+medicine&ots=h2gsx87PkE&sig=2X6iBLDh9NIIJKlPFTHawoc9cSI&redir_esc=y#v=onepage&q=WHO%20global%20report%20on%20traditional%20medicine&f=false (accessed on 22 April 2024).

WHO, (2022). *Traditional, Complementary and Integrative Medicine*. Available at URL. https://www.who.int/health-topics/traditional-complementary-and-integrative-medicine#tab=tab_1 (accessed on 22 April 2024).

Yakoot, M., (2013). Bridging the gap between alternative medicine and evidence-based medicine. *Journal of Pharmacology & Pharmacotherapeutics, 4*(2), 83–85.

Yuan, H., Ma, Q., Ye, L., & Piao, G., (2016). The traditional medicine and modern medicine from natural products. *Molecules, 21*(5), 1–18.

CHAPTER 11

Post-COVID-19 Opportunities in Himalayan Medicinal Plants

SAJITHA SIRIL,[1] K. ABHA MANOHAR,[2] N. N. SHAHINA,[3] GOPAL SHUKLA,[4] and SUMIT CHAKRAVARTY[1]

[1]*Department of Forestry, Uttar Banga Krishi Viswavidyalaya, Pundibari, Cooch Behar, West Bengal, India*

[2]*Department of Agronomy and Agroforestry, Centurion University, Parlakhamundi, Odisha, India*

[3]*Siviculture and Forest Management Division, ICFRE–Rain Forest Research Institute, Jorhat, Assam, India*

[4]*Department of Forestry, North Eastern Hill University, Tura Campus, Tura, Meghalaya, India*

ABSTRACT

The coronavirus has adversely impacted populations worldwide. The world is yet to endure the disturbance in its socio-economy caused by the pandemic. Traditional medicines are always a boon in such scenarios as evidenced from past experiences. All over the world, the utilization of medicinal plants has surged during the pandemic. In India, along with allelopathy, especially in the rural areas to prevent the occurrence of COVID-19, traditional medicinal practices like Ayurveda, Siddha, Unani, Yoga, and Homoeopathy are practiced. The Himalayas are a treasure trove of medicinal plants. Many of the utilizations and other therapeutic properties are still unknown to the public and the scientific community. This pandemic has opened up the various therapeutic values of Himalayan medicinal plants. Though many of them were not in common use previously among rural as well as urban populations.

Sustainable Forest Resources Management: Issues and Implications. Jahangeer A. Bhat, Gopal Shukla, Manmohan J. Dobriyal, Sumit Chakravarty, A. Arunachalam, and Rainer W. Bussmann (Eds.)

Most of the studies on Himalayan medicinal plants are restricted to inventory only. It's actually the need of the hour to explore the utility of the medicinal value of such plants. Proper validation from the authorities is inevitable in such circumstances to avoid any health consequences. Hence, this chapter opens up the opportunities for the reliance on Himalayan medicinal plants in the present scenario of COVID-19.

11.1 INTRODUCTION

The COVID-19 disease promptly surged to a pandemic scale, and it paralyzed societal and economic conditions all over the world. COVID-19 is caused by a virus called SARS-CoV-2, which causes severe acute respiratory syndrome and it is one of the life-threatening diseases for which no definite preventive or curative treatment has been devised so far (Babich et al., 2020). High fever along with cough, sore throat, headache, fatigue, and breathlessness are the usually identified symptoms in persons diagnosed with COVID-19 (Ge et al., 2020). In certain cases, during the advanced stages of the disease, it may result in loss of speech or movement, chest pain or pressure, severe cardiac injury, and even the death of the patient due to acute respiratory distress syndrome (Natesh et al., 2021). In India also, COVID is spreading too rapidly, and about 3.07 crores of people have been infected. Even though there is a 99% recovery rate, the death rate also shows a linear increase. The current situation necessitates improvised treatment modalities using traditional medicine. The application of traditional medicinal practices has had successful episodes during previous epidemic outbreaks such as SARS and H1N1 influenza (Luo et al., 2020). The role of medicinal plants in the treatment of existing as well as newly born diseases is highly important, particularly through traditional healers and the supply of plant-based raw materials to the pharmaceutical industries (Kala, 2004). The bioactive products derived from medicinal herbs are a potential source of New Chemical Entities (NCE) like anti-COVID-19 drugs. Such compounds play a significant role in the creation of many important drug molecules (Khan & Al-Balushi, 2021). During the pandemic, countries like India, South Korea, and China issued multiple guidelines for COVID-19 prevention and management that particularly focused on traditional medical practices (Ang et al., 2020).

The Indian System of Traditional Medicine is one of the world's oldest medical practices. Since the dawn of human civilization, traditional medicinal systems in India played a pertinent role in healthcare service. Traditional medicinal practices in India encompass six medicinal systems, namely

Ayurveda, Yoga & Naturopathy, Unani, Siddha, and Homeopathy (AYUSH) (Adhikari & Paul, 2018). About 25,000 formulations have been used so far in folk remedies in the Indian medicinal system (Pundarikakshudu & Kanaki, 2019). 'Ayurveda,' the most popular traditional medicinal practice in India, offers treatments for multiple diseases and health conditions. This medicinal system primarily relies on plants since the extracts from many of the medicinal plants contain compounds that have inhibitory effects on different strains of pathogenic viruses like HIV, smallpox virus, and hepatitis B virus (HBV) (Mukhtarad et al., 2008; Babich et al., 2020). The Ministry of AYUSH, New Delhi, has also provided an advisory on practices to boost immunity by using medicinal plants (Ayush, 2021; Table 11.1).

TABLE 11.1 Recommended Immunity Enhancers by the AYUSH Ministry to Ayurvedic Practitioners

Sl. No.	Common Name	Botanical Name	Dosage
1.	Guduchi	*Tinospora cordifolia*	500 to 1,000 mg of aqueous extract
2.	Amla	*Emblica officinalis*	Not specified
3.	Turmeric (Haldi)	*Curcuma longa*	Warm water gargle
4.	Tulsi	*Ocimum sanctum*	Sipping of Tulsi water
5.	Ashwagandha	*Withania somnifera*	Two times a day after mixing with warm water or milk.

Source: Adapted from: Prajapati & Kumar (2020).

The Himalayas are one of the largest mountain chains in the world with rich floral diversity. The studies on Himalayan medicinal plants are, however, limited (Badola & Aitken, 2013). The information on medicinal herbs in this region, such as their distribution, uses, and mode of consumption, is confined only to the indigenous people (Khan et al., 2010). Therefore, the present chapter discusses the importance and opportunities of the traditional system of plant-based healthcare in the Himalayan region of India and its milestones in formulating a potential therapy for the COVID-19 crisis. Further, the chapter also collaborates documented COVID-curing medicinal plants from the Himalayas, their phytochemistry, as well as their uses.

11.2 AN ANALYSIS OF POTENTIAL OF MEDICINAL PLANTS TO TREAT COVID

There are several data that support the therapeutic efficacy of herbs in the treatment of viral infections. For example, during the outbreak of SARS in 2003, traditional Chinese medicines reported positive responses in the treatment as well as prevention of the virus (Zhang et al., 2020; Yang et al., 2020). Ahmad et al. (2021) did an extensive study on the therapeutic properties of 83 medicinal plants. The potential of the medicinal herbs ranged from anti-inflammatory, immunomodulatory, antiviral, and anti-allergic. A categorization was also carried out in the study on the basis of priority, viz., most promising drugs (21 plants), equally promising drugs (44 plants), and possibly promising drugs (8 plants). They also reported that all the medicinal plants documented can also be included in the treatment of coronavirus disease.

Many researchers have also recently studied medicinal plants that are particularly used to prevent coronavirus infections. For instance, Babich et al. (2020) identified 26 plants with antiviral (SARS), immunoregulatory, and anti-inflammatory effects. In this study, out of the 26 species reported, 25 are recommended for treatment, and the remaining one species, namely the *Fortunes bossfern*, is used for preventing coronavirus disease. Around 20 easily accessible species of functional food plants that are being used as immune modulators and possess therapeutic properties in the respiratory tract and other viral diseases were documented by Yang et al. (2020). A total of 63 species of plants in 42 families from the Nepal region of the Himalayas were recorded by Khadka et al. (2020), who observed that the utilization of medicinal herbs has increased during the pandemic. The most commonly used species in the region was ginger (*Zingiber officinale*). Velingiri et al. (2020) have compiled a list of 25 medicinally significant plants that can be effective against Corona and various other viruses. Shinwari et al. (2020) also recorded around 38 important species with antiviral activity. Khan et al. (2021) provided a list of 49 species of plants with bioactivity for treating viral infections and 30 species for COVID-19 therapy.

11.3 HIMALAYA AS A STOREHOUSE OF ANTIVIRAL MEDICINES

The Himalayan ranges, the Global Centre of vascular plant diversity, situated in the northern part of India, are a hub of a diverse variety of medicinal plants (Barthlott et al., 2005; Bhat et al., 2013). The Indian Himalayas are home to more than 8,000, 44, and 600 species of angiosperms, gymnosperms, and pteridophytes, respectively. There are approximately 1,748 species reported from this region with therapeutic properties (Singh et al., 1996; Samant et al., 1998). The highest number of medicinal plant species is reported from the state of Uttarakhand (Singh et al., 2005) followed by Sikkim and North Bengal. Due to the distinct ecological and geographic conditions (Bhat et al., 2013), the diversity of medicinal plants in Trans-Himalayas is less compared to other Himalayan regions. In the Trans-Himalayan regions, 337 medicinal species are only reported (Kala, 2006). The higher plants also play a significant role in the livelihood of indigenous people through the provision of many forest products in the form of food as well as medicine (Joshi et al., 2016). Therefore, an extensive survey of the ethnomedicinal literature in the Himalayan region and consultation with elderly people of indigenous communities might reveal some valuable solutions for the treatment of Coronavirus disease.

Ministry of Ayush has recommended a list of 20 plant species with details of the botanical name, family, vernacular names, chemical constituents, properties, and action, pharmacological principle, parts used, therapeutic uses, dosage, and important formulations to increase immunity against the coronavirus (Ministry of Ayush, 2021). Out of the 20 medicinal plants recommended by AYUSH, about 14 species are present in the Himalayan region, indicating the importance of this region in traditional medical care during the pandemic situation (Table 11.2).

TABLE 11.2 List of AYUSH Recommended Medicinal Plants in the Himalayas Capable of Combating COVID-19 Infection

Sl. No.	The Scientific Name of the AYUSH Recommended Medicinal Plant	References Showing Plant's Presence in the Himalayan Region
1.	*Andrographis paniculata*	Sanwal et al. (2016)
2.	*Cinnamomum verum*	Rani et al. (2017); Joshi et al. (2016)
3.	*Clerodendrum serratum*	Sangle et al. (2016)
4.	*Curcuma longa*	Raina et al. (2005); Joshi et al. (2016)
5.	*Cymbopogon jwarancusa*	Joshi et al. (2016)
6.	*Hedychium spicatum*	Chhetri (2014); Kala (2006); Dwivedi et al. (2019); Joshi et al. (2016)
7.	*Justicia adhatoda*	Dwivedi et al. (2019)
8.	*Ocimum basilicum*	Chhetri (2014); Joshi et al. (2016)
9.	*Phyllanthus emblica*	Kala (2006); Dwivedi et al. (2019)
10.	*Pichrorhiza kurroa*	Chhetri (2014); Kala (2006); Bhat et al. (2013); Sharma et al. (2012)
11.	*Swertia chirata*	Chhetri (2014); Kala (2006); Dwivedi et al. (2019); Bhat et al. (2013)
12.	*Tinospora cordifolia*	Rana et al. (2012)
13.	*Withania somnifera*	Kala (2006)
14.	*Zingiber officinale*	Rai (2008)

11.4 PHYTOCHEMISTRY

The use of natural compounds in the treatment of infectious diseases originated from time immemorial. During the last few episodes of various virus outbreaks like influenza, HCV, and HIV, the efficacy of natural substances has already been proved. Nature-based drugs and their associated phytoconstituents have the capacity to develop as potential drugs against SARS-CoV-2 (Patel et al., 2021). Figure 11.1 illustrates the different antiviral mechanisms exhibited by phytochemicals of medicinal plants (Shinwari et al., 2020).

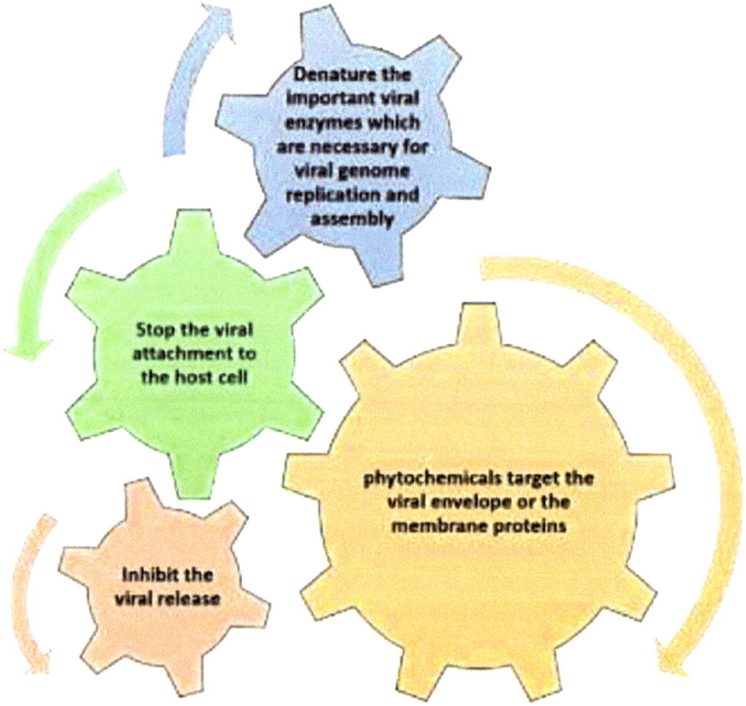

FIGURE 11.1 Antiviral mechanism by photochemical.

Source: Adapted from: Shinwari et al. (2020).

An *in-silico* study was done by Srivastava et al. (2020) on the potential of the extracts of Indian herbs in inhibiting COVID-19 protease. They elucidated that the inhibition potential of plant extracts (*Tinospora cordifolia, Curcuma longa, Zingiber officinale, Withania somnifera*) is higher than those of two anti-malarial drug compounds (chloroquine and hydroxychloroquine), which were already found to have inhibitory effects on COVID-19 protease *in vitro*. Plant materials are a rich source of bioactive compounds/phytochemicals, which have the potential to kill disease-causing pathogens and strengthen the immune system (Shinwari et al., 2020; Khan et al., 2021). A study by Natesh et al. (2021) using computational investigations revealed the potentially superior capacity of bio-actives in the medicinal plants in the Himalayas against COVID-19. Table 11.3 depicts the phytochemical constituents of the 14 antiviral medicinal plants.

TABLE 11.3 Major Phytochemical Constituents of 14 Himalayan Medicinal Plants

Scientific Name	Major Chemical Constituents	References
Andrographis paniculata	Andrographolide, deoxyandrographolide, neoandrographolide, andrographiside, homoandrographolide, andrographonand stigmasterol	Mishra et al. (2007); Sagadevan et al. (2015)
Cinnamomum verum	Eugenol, transcinnamaldehyde, cinnamaldehyde, benzaldehyde, caryophyllene, methyl eugenol, cinnazeylanine, cinnacassiol, cinnamyl acetate, linalool, and cineole	EMA (2011); Batihan et al. (2020)
Clerodendrum serratum	Hispidulin, scutellarein, uncinatone, pectolinarigenin, dmannitol, serratumin A, serratoside A, serratoside B, 7-O-P-coumaroyloxyugandoside, cleroserroside A, cleroserroside B, serratumoside A, myricoside, stigmasterol, serratin, bis(2-ethylhexyl) phthalate, and cleroflavone	Poornima et al. (2015); Wang et al. (2018)
Curcuma longa	Curcumene, curcuminoids, curcumenone, cineole, eugenol, and procurcumenol	Srivastava et al. (2011); Yang et al. (2020); Pardhe (2021)
Cymbopogon jwarancusa	Piperitone, camphene, geraniol, citral, alpha, and beta pinene, agarospirol, citronellol, and citronellal	Prasad et al. (2014)
Hedychium spicatum	Hedychenone, 7-hydroxy hedychenone, linalool, β-slinene, ethyl P-methoxy-cinnamate, ethyl cinnamate and eucalyptol	Rasool et al. (2019); Arumugam et al. (2021)
Justicia adhatoda	Vasicine, vasicinol, vitamin C, carotene, phenols, tannins, saponins, flavonoids, and alkaloids	Shinwari et al. (2020); Dhankhar et al. (2011)
Ocimum basilicum	Linalool, methyl cinnamate, quercetin, kaempferol, caryophyllene, phenolic acid, rosemarinic acid, flavonoids, and beta-carotene	Khan et al. (2021); Shinwari et al. (2020); Amber et al. (2017)

TABLE 11.3 *(Continued)*

Scientific Name	Major Chemical Constituents	References
Phyllanthus emblica	Phyllantidine, phyllan-tine, vitamin C, Ascorbic acids, linolic acid, corilangine, tannins, terpenoids, flavonoids, and saponins	Khan et al. (2021); Patel et al. (2021)
Pichrorhiza kurroa	Kutkiol, kutkisterol, phenol glucosides, kutkin, picrorhizin, iridoids, cucurbitacins, pikuroside, and veronicoside	Sharma et al. (2012); Masood et al. (2015)
Swertia chirata	Amarogentin, chiratol, gentianine, kairatenol, swertenol, mangiferin, swertiamarin, amaroswerin, ursolic acid, Gentianine, swertanone, and 1-hydroxy-3,5,8-trimethoxyxanthone	Aleem & Kabir (2018)
Tinospora cordifolia	Berberine, sitosterol, tinosporin, tinosporide, cordifolide, phenyl prophyl glycoside, palmatine, tinocordiside, cordifoliside	Srivastava et al. (2020); Choudhary et al. (2013)
Withania somnifera	Withaferin A, withanone, isopelletierine, anaferine, sitoindosideflavonoids, alkaloids, phenolic compounds and fatty acid	Khan et al. (2021); Patel et al. (2020); Srivastava et al. (2020)
Zingiber officinale	Zingerone, zingiberol, Ar-curcumene, A-zingiberene flavan, 6-shojail, Citral, Citronellol, paradols, gingerdiones, and gingerdiols	Patel et al. (2020); Srivastava et al. (2020); Shinwari et al. (2020); Chang et al. (2013); Yang et al. (2020); Li et al. (2021)

11.5 CONSERVATION OF HIMALAYAN MEDICINAL PLANTS DURING POST-COVID SCENARIO

Wild herbal plants serve about 90% of the raw materials for various herbal-based industries in India. The major portion of these herbs is collected from the alpine and sub-alpine regions of the Himalayas. Despite the presence of several laws and regulations existing for the prevention of illegal harvesting, Himalayan medicinal plants are under threat of extinction due to unscientific collection and over-harvesting (Kala, 2006). The coronavirus is proving to be a boon so far from the biodiversity conservation point of view because it has enriched biodiversity worldwide due to a reduction in human activities (Corlett et al., 2020). The Himalayan region is also relieving the natural environment from anthropogenic pressure. The well-organized participation of Himalayan hill indigenous people through the village stakeholder communities contributes to biodiversity conservation and the sustainability of production (Kala, 2006). The urgent need for medicinal plants compels public participation to continue using modern technologies in the conservation and management of the Himalayas.

The ethnic knowledge regarding the utilization, value, and importance of medicinally important plants is passed from the older to the younger generation orally; but, the difficulty in access, undulating terrain, and slow development in the Himalayan region hinder the treasure of knowledge from spreading worldwide (Farooquee et al., 2004; Kala, 2002). Therefore, the documentation of indigenous knowledge and ethnobotany supplemented by the latest scientific insights is urgently needed. The traditional and modern uses of 14 medicinal plants are given in Table 11.4. Plant-based compounds account for approximately 25% of allopathic drugs (Bhat et al., 2013), which indicates that the solutions for the current threatening COVID virus may be existing in nature itself. A wide range of diseases can be cured by allopathic medicines; however, the high costs and sometimes the associated side effects of modern medicines drag people to depend on herbal medicines, which are believed to have relatively fewer after-effects (Kala et al., 2005).

However, preclinical and clinical scientific studies as well as validation of medicinal plants are needed. In the present pandemic situation, Ayush scientists highly recommend plant-based medicine to boost immunity in order to prevent and overcome COVID and its related diseases. Recently, research and development programs associated with herbal medicinal systems are also highly encouraged by WHO (World Health Organization) (Ahmad et al., 2021). Generally, medicinal herbs are a rich source of vitamins, proteins, antioxidants, and minerals with multiple antiviral and antibacterial properties. For example, Ashwagandha (*Withania somnifera*), Kalamegha, Tulsi (*Ocimum tenuiflorum*), and Adrak (*Zingiber officinale*) help in enhancing the immune system and fighting against the invaded viruses (Ahmad et al., 2021). The Himalayas are a storehouse of large varieties of medicinal plants for respiratory ailments; however, only a few studies related to *in-vitro* validation have been done in the region. Amber et al. (2017) listed 55 ethnomedicinal plants for curing bronchitis from different Himalayan countries, but *in-vitro* validation has been done only for 6 species. Medicinal plants contain different chemical substances providing the basis for novel chemotherapeutic agents against various human pathogens (Amber et al., 2017). Therefore, it is essential to conduct detailed studies on ethnomedicinal aspects of medicinal herbs to support the researchers and enhance the authenticity of traditional medicines among the people. The unexplored and under-explored medicinal plants need thorough screening for their details like phytochemistry, toxicology, action mechanisms, and their activities *in-vitro* and *in-vivo*.

TABLE 11.4 Traditional and Modern Uses of Reported Medicinal Plants

Scientific Name	Part Used	Traditional Uses	Modern and Pharmacological Use
Andrographis paniculata	Whole plant	Treat liver disorders, common cough, cold (Mishra et al., 2007), antibacterial, antiviral, hypo glycemic, blood purifying activity, gonorrhea, skin eruptions, and fever (Akbar, 2011) respiratory infections, GIT infection, sore throat.	Antithrombotic actions, cardiovascular disease, cancer (Sagadevan et al., 2015) and anti-viral HIV (Akbar, 2011), HSV, HBV, HCV, SRV, H5N1, chikungunya virus, HPV (Enmozhi et al., 2021; Babich, 2020; Sytar et al., 2021; Gupta et al., 2016; Khan & Balushi, 2021; Sornpet et al., 2017; Singh et al., 2021); hepatoprotective, vermicide, hepatoprotective, antiviral, anti-inflammatory, and immune stimulant activity, antioxidant (Nagajothi et al., 2018).
Cinnamomum verum	Stem, bark	Spices (flavoring in food), treating indigestion, diabetes, acne, cold, headache, cough, microbial infections, cardiac ailments, urinary and respiratory disorders (Wang et al., 2020).	Anti-viral-H7N3 influenza (Brochot et al., 2017), anti-complement activity, anti-allergic activity, several biological activities such as peripheral, vasodilatory, antitumor, antifungal, cytotoxic and cardiovascular disease, anti-mutagenic, activities (Maruthamuthu & Ramanathan, 2016)
Clerodendrumserratum	Root	Pesticide properties, cure cold, fever, rhinitis, cough, and other inflammatory disease (Hazekamp et al., 2001). The roots of the plant have been claimed to be used in dyspepsia. Aqueous extracts of leaves of *C. serratum* possess bronchodilator property (Singh et al., 2012)	Anti-viral (Taraxerol have Inhibitory candidate against the SARS-CoV-2 spike, Mpro, and RdRp proteins (Kar et al., 2020)), anti-pyretic, antioxidant activity (Bhujbal et al., 2019), anti-inflammatory (Tiwari et al., 2021) and antinociceptive activity, anticancer activity, hepatoprotective activity (Wang et al., 2018), analgesic activity, anti-allergic activity, antifertility, antibacterial, antifungal, antimicrobial, antihistaminic activity, anti-asthmatic activity, mast cell stabilization (Joshi et al., 2016)
Curcuma longa	Rhizome	Antiseptic, cure gastritis, used ascarminative, anti-inflammatory, anthelmintic, and externally used to heal wounds, inflamed joints, and sprains (Joshi et al., 2016).	Antiviral (respiratory syncytial virus, inhibit viral replication, HSV-1, HIV, HCV) (Khan & Balushi, 2021; Babich, 2020) respiratory illness, tropical eosinophilia, cancer, jaundice, relieves toxicity, biliary disorders (Patel et al., 2020; Pardhe, 2021; Mehdi et al., 2021)

TABLE 11.4 *(Continued)*

Scientific Name	Part Used	Traditional Uses	Modern and Pharmacological Use
Cymbopogon jwarancusa	Whole plant	Flavoring agent, gout, rheumatism, skin disease, abdominal tumors, unconsciousness, and Fever, vomiting, anti-microbial, insecticidal, and insect repellent (Prasad et al., 2014)	Diaphoretic, anti-rheumatic, carminative, diuretic, emmenogog, flukicidal activity, antifungal, antibacterial, anti-filarial, anti-amoebic, and larvicidal activity (Avoseh et al., 2015; Shafiq et al., 2015)
Hedychium spicatum	Rhizome	Cough, colds, relieves paraxysmol attack of dyspnoea (Joshi et al., 2016) antimicrobial, antifungal, pediculicidal, and cytotoxic activities (Rasool et al., 2019) treatment of swelling, asthma, pain liver complaints, expectorant, and urinary disorders (Sravani & Paarakh, 2011)	Anti-malarial activity, anticancer, cytotoxic activity, hepato-protective activity, hypocholesterolemic activity, pediculicidal activity, antihistaminic, ulcer protective activity, antidiabetic activity, tranquilizing activity, anthelmintic activity, hypogly-cemic, vasodilator, bronchitis, eye disease and inflammations, stomachic, carminative (Rasool et al., 2019)
Justicia adhatoda	Leaf, root, flower	Cough, fever, and treats rheumatic pain (Joshi et al., 2016) dyspnoea, consumption, anemia, diarrhea, skin diseases, asthma, and dysentery (Singh & Huidrom, 2013)	Brochodilator activity, haemostatic, antiviral, abortifacient and easy child birth delivery, antimicrobial, and anti-inflammatory activity (Joshi et al., 2016; Sharma & Kumar, 2016; Chavan & Chowdhary, 2014; Dhankhar et al., 2011)
Ocimum basilicum	Leaf, flower	Headache, cough, diarrhea, feverish illness, indigestion, and insomnia, urinary disorders, skin diseases and fungal infections (Joshi et al., 2016; Shahrajabian et al., 2020; Bilal et al., 2012))	Anti-viral (Amber et al., 2016), anti-microbial, anti-inflamma-tory, and inhibitory activity in HIV-1, anti-allergenic, cytotoxic, and anti-spasmodic. antihyperlipidemic and anti-ulcerative activity, anti-cancer (Ch et al., 2015)
Phyllanthus emblica	Fruit pulp	Fruits used to treat diabetes, and jaundice bleeding disorders, dysuria, ulcer, leucorrhea, diarrhea, inflammation, cough, asthma, bronchitis, cephalalgia, ophthalmopathy, dyspepsia, diuretic, biliousness, colic, and hyperacidity.	Anti-bacterial, anti-oxidant, analgesic, anti-pyretic, hgypo-glycemic, scurvy, cancer, and heart diseases, antineutrophilic activity, anti-viral (Ahmad et al., 2021)
Pichrorhiza kurroa	Root, underground stem	Fever, diabetes, stomach disorder, jaundice dyspnoea, cough, skin diseases, stomachache, hepatitis, digestive disorders, diarrhea, and dysentery (Salma et al.,2017)	Anti-viral, anti-inflammatory, hepatoprotective, anticholestatic, antiulcerogenic, antidiabetic, and immune regulatory activities, purgative, brain tonic, stomachic, dyspepsia, and antiallergic, anticancer, enhancers of neuron growth and treatment of viral hepatitis (Sultan et al., 2016; Rathee et al., 2016).

TABLE 11.4 *(Continued)*

Scientific Name	Part Used	Traditional Uses	Modern and Pharmacological Use
Swertia chirata	Whole plant	Fever, ulcers, diabetes gouty arthritis, edema, chronic liver diseases, inflammation, digestive diseases, hepatitis, constipation, skin diseases, worms, epilepsy, scanty urine, melancholia, and blood purification (Kumar & Staden, 2016).	Anti-arthritic, anti-diabetic, hepatoprotective, anti-pyretic (Bhargava et al., 2009), anti-inflammatory (Kumari et al., 2021), hypoglycemic, hepatoprotective, antibacterial, antispasmodic, and antioxidant.
Tinospora cordifolia	Stem, leaf, aerial root	Diabetes, fever, cough, wheezing, brain tonic, hyperglycemia, stimulates bile secretion, enriches the blood, and cures jaundice, vermifuge, antidote to snake bite (Choudhary et al., 2013).	Anti-pyretic, immuno modulator, anti-rheumatic, hepatoprotective, hypoglycemic, antimicrobial, antidiarrheal, anticancer, antispasmodic (Sinha et al., 2004; Choudhary et al., 2013), inhibits the growth of HSV, antiviral potential against chikungunya virus (Ahmad et al., 2021).
Withania somnifera	Root, leaf, alkali.	Fatigue, skin problems, cough, weakness, emaciation, tumors, dyspnoea, insomnia, energy booster, increase vital fluids, muscle fat, blood, and cell production, counteract chronic fatigue, bone weakness, thirst, impotency, premature aging emaciation, debility, rejuvenating the reproductive organs, convalescence, and muscle tension.	Anti-viral, withaferin-A shows antiviral potential against SARS-CoV-2, anti-inflammatory, hepato protective, adaptogenic, anti-diabetic, treatment of nervous exhaustion, skin problems and coughing (Choudhary et al., 2013).
Zingiber officinale	Rhizome	Fever, cough, diarrhea, bleeding disease, enhancement of appetite, balancing circulation, enhance digestion, alleviate asthma, analgesic, anti-edematous, pain management, nerve stimulant, anti-pyretic (Dissanayake et al., 2020).	Anti-viral [Rhinovirus IB, human respiratory syncytial virus, inhibits viral attachment and penetration (Chang et al., 2013; Patel et al., 2021; Habeballa et al., 2020), avian influenza virus H9N2 (Khan & Balushi, 2021)], anti-bacterial, anti-inflammatory, bio-availability enhancer and hypolipidemic (Dissanayake et al., 2020).

The current scenario has created a sudden surge in the demand for plant-based products and drugs all over the world, resulting in the over-exploitation of these medicinal plants. Reports show that more than 150 species of medicinal plants have already become extinct in the wild due to habitat degradation, high demand, over-exploitation, and unsustainable harvesting (Singh & Rawat, 2011). Most of the selected herbal species only occur in the Himalayan regions (Table 11.1) and play a key role in curing COVID-19 as immunomodulators. Kala (2006) mentioned the urgent need for holistic development and sustainable management of the Himalayan medicinal plant sector with scientific cultivation methods, standardization of agro-techniques and propagation protocols, documentation of medicinal plant diversity, assessment of biochemical and genetic variability, value chain development, strengthening of medicinal plant trade and enterprises, capacity building of stakeholders, and legal reforms.

International organizations such as WWF and UNESCO as part of the ‘people and plants initiative’ are promoting research and developmental activities on ethnobotanical knowledge globally. Such initiatives are also aimed at integrating people’s perceptions and practices in resource management at the local and national levels (Bhat et al., 2013). The significance of ethnobiological knowledge on species-ecology can impart new dimensions in scientific research and conservation that have attained growing attention in resource management on a global scale.

11.6 CONCLUSION

Throughout the centuries, whenever an epidemic arises, humans turn to nature to prevent and cure diseases. The coronavirus, which has spread worldwide, has also led to such a scenario, where medicinal plants have gained increased importance regardless of their origin. The Himalayas are home to many unexplored medicinal plants that are unknown to the scientific community but are traditionally used. This study recommends some Himalayan medicinal plants that are believed to help prevent COVID-19. Due to the high demand, rural communities should be encouraged to cultivate these medicinal plants to generate additional income during this period. Furthermore, pharmacological and phytochemical studies are indeed necessary for the proper validation of dosages to be used by state and central authorities.

KEYWORDS

- **Himalayas**
- **immunity booster**
- **medicinal plants**
- **pandemic**
- **socio-economy**
- **traditional medicines**

REFERENCES

Adhikari, P. P., & Paul, S. B., (2018). History of Indian traditional medicine: A medical inheritance. *Asian J. Pharm. Clin. Res., 11*(1), 421.

Ahmad, S., Zahiruddin, S., Parveen, B., Basist, P., Parveen, A., Gaurav, Parveen, R., & Ahmad, M., (2021). Indian medicinal plants and formulations and their potential against COVID19–preclinical and clinical research. *Front. Pharmacol., 11*, 578970. doi:10.3389/fphar.2020.578970.

Akbar, S., (2011). Andrographis paniculata: A review of pharmacological activities and clinical effects. *Alternative Medicine Review, 16*(1), 66–77.

Aleem, A., & Kabir, H., (2018). Review on *Swertia chirata* as traditional uses to its pyhtochemistry and phrmacological activity. *Journal of Drug Delivery and Therapeutics, 8*(5-s), 73–78 DOI: Http://dx.doi.org/10.22270/jddt.v8i5-s.1957.

Amber, R., Adnan, M., Tariq, A., & Mussarat, S., (2017). A review on antiviral activity of the Himalayan medicinal plants traditionally used to treat bronchitis and related symptoms. *Royal Pharmaceutical Society, Journal of Pharmacy and Pharmacology, 69*, 109–122.

Ang, L., Lee, H. W., Choi, J. Y., Zhang, J., & Lee, M. S., (2020). Herbal medicine and pattern identification for treating COVID-19: A rapid review of guidelines. *Integr. Med. Res., 9*(2), 100407.

Arumugam, I., Krishnan, C., Ramachandran, S., Krishnan, D., Das, D., & Thamankar, V., (2021). Phytochemical investigation and *in-vitro* antimicrobial activity of the essential oil from rhizomes of Hedychium spicatum. *IJPSR. 12*(2), 853–858.

Avoseh, O., Oyedeji, O., Rungqu, P., Nkeh-Chungag, B., & Oyedeji, A., (2015). Cymbopogon species; ethnopharmacology, phytochemistry and the pharmacological importance. *Molecules, 20*, 7438–7453; doi:10.3390/molecules20057438.

AYUSH, (2021). *Ayurveda's Immunity-Boosting Measures for Self-Care During COVID 19 Crisis* | Central Government Health Scheme (CGHS Mumbai, India) [Internet]. Available from: https://www.mohfw.gov.in/pdf/ImmunityBoostingAYUSHAdvisory.pdf (accessed on 22 April 2024).

Babich, O., Sukhikh, S., Prosekov, A., Asyakina, L., & Ivanova, S., (2020). Medicinal plants to strengthen immunity during a pandemic. *Pharmaceuticals, 13*(10), 313. https://doi.org/10.3390/ph13100313.

Badola, H. K., & Aitken, S., (2003). The Himalayas of India: A treasury of medicinal plants under siege. *Biodiversity, 4*(3), 3–13.

Barthlott, W., Mutke, J., Rafiqpoor, D., Kier, G., & Kreft, H., (2005). Global centers of vascular plant diversity. *Nova Acta Leopoldina, 92*, 61–83.

Batiha, G. E., Beshbishy, A. M., Guswanto, A., Nugraha, A., Munkhjargal, T., Abdel-Daim, M. M., Mosqueda, J., & Igarashi, I., (2020). Phytochemical characterization and chemotherapeutic potential of *Cinnamomum verum* extracts on the multiplication of protozoan parasites in vitro and *in vivo*. *Molecules, 25*, 996; doi:10.3390/molecules25040996.

Bhargava, S., Rao, P. S., Bhargava, P., & Shukla, S., (2009). Antipyretic potential of *Swertia chirata* Buch ham. root extract. *Sci Pharm., 77*, 617–623.

Bhat, J. A., Kumar, M., & Bussmann, R. W., (2013). Ecological status and traditional knowledge of medicinal plants in Kedarnath Wildlife Sanctuary of Garhwal Himalaya, India. *J. Ethnobiology Ethnomedicine, 9*, 1. https://doi.org/10.1186/1746-4269-9-1.

Bhujbal, S. S., Kewatkar, S. M. K., More, L. S., & Patil, M. J., (2009). Antioxidant effects of roots of *Clerodendrum serratum* Linn. *Pharmacognosy Research, 1*(5), 294–298.

Bilal, A., Jahan, N., Ahmed, A., Bilal, S. N., Habib, S., & Hajra, S., (2012). Phytochemical and pharmacological studies on *Ocimum basilicum* Linn: A review. *IJCRR, 4*(23), 73–83.

Brochot, A., Guilbot, A., Haddioui, L., & Roques, C., (2017). Antibacterial, antifungal, and antiviral effects of three essential oil blends. *Microbiology Open, 6*(4), e00459. https://doi.org/10.1002/mbo3.459.

Ch, M., Naz, S., Sharif, A., Akram, M., & Saeed, M., (2015). Biological and pharmacological properties of the sweet basil (*Ocimum basilicum*). *Journal of Pharmaceutical Research International, 7*(5), 330–339. https://doi.org/10.9734/BJPR/2015/16505.

Chang, J. S., Wang, K. C., Yeh, C. F., Shieh, D. E., & Chiang, L. C., (2013). Fresh ginger (*Zingiber officinale*) has anti-viral activity against human respiratory syncytial virus in human respiratory tract cell lines. *J. Ethnopharmacol., 145*(1), 146–151. doi: 10.1016/j.jep.2012.10.043.

Chavan, R., & Chowdhary, A., (2014). *In vitro* inhibitory activity of *Justicia adhatoda* extracts against influenza virus infection and hemagglutination. *Int. J. Pharm. Sci. Rev. Res., 25*, 231–236.

Chhetri, D. R., (2014). *Medicinal Plants of the Himalaya: Production Technology and Utilization*. AGROBIOS, India.

Choudhary, N., Siddiqui, M. B., Azmat, S., & Khatoon, S., (2013). *Tinospora cordifolia:* Ethnobotany, phytopharmacology and phytochemistry aspects. *IJPSR, 4*(3), 891–899.

Corlett, R. T., Primackb, R. B., Devictorc, V., Maas, B., Goswamif, V. R., Batesg, A. E., Kohh, L. P., et al., (2020). Impacts of the coronavirus pandemic on biodiversity conservation. *Biological Conservation, 246*, 108571. https://doi.org/10.1016/j.biocon.2020.108571.

Dhankhar, S., Kaur, R., Ruhil, S., Balhara, M., Dhankhar, S., & Chhillar, A. K., (2011). A review on *Justicia adhatoda*: A potential source of natural medicine. *African Journal of Plant Science, 5*(11), 620–627.

Dissanayake, K. G. C., Waliwita, W. A. L. C., & Liyanage, R. P., (2020). A review on medicinal uses of *Zingiber officinale* (ginger). *International Journal of Health Sciences and Research, 10*(6), 142–148.

Dwivedi, T., Kanta, C., Singh, L., & Sharma, I. P., (2019). A list of some important medicinal plants with their medicinal uses from Himalayan State Uttarakhand, India. *Journal of Medicinal Plants Studies, 7*(2), 106–116.

Enmozhi, S. K., Raja, K., Sebastine, I., & Josep, J., (2021). Andrographolide as a potential inhibitor of SARS-CoV-2 main protease: An in-silico approach. *Journal of Biomolecular Structure and Dynamics, 39*(9), 3092–3098. http://dx.doi.org/10.1080/07391102.2020.1760136.

European Medicines Agency, (2011). Assessment report on Cinnamomum verum J. S. Presl, cortex and cortices aetheroleum EMA/HMPC/246773/2009.

Farooquee, N. A., Majila, B. S., & Kala, C. P., (2004). Indigenous knowledge systems and sustainable management of natural resources in a high-altitude society in Kumaun Himalaya, India. *J. Hum. Ecol., 16*, 33–42.

Ge, H., Wang, X., Yuan, X., Xiao, G., Wang, C., Deng, T., Yuan, Q., & Xiao, X., (2020). The epidemiology and clinical information about COVID-19. *Eur. J. Clin. Microbiol. Infect. Dis., 39*(6), 1011–1019.

Gupta, S., Mishra, K. P., & Ganju, L., (2016). Broad-spectrum antiviral properties of andrographolide. *Arch. Virol., 162*(3), 611–623.

Habeballa, R. S., Ahmedani, E. I., Awad, N. S., & Abdein, M. A., (2020). *In vitro* antiviral activity of *Illicium verum* and *Zingiber officinale* ethanolic extracts. *Medical Science, 24*, 3469–3480.

Hazekamp, A., & Verpoorte, R., (2001). Isolation of a bronchodilator flavonoids from the Thai medicinal plant *Clerodendrum petasites*. *J. Ethnopharmacology, 78*, 45–49.

Joshi, R. K., Satyal, P., & Setzer, W. N., (2016). Himalayan aromatic medicinal plants: A review of their ethnopharmacology, volatile phytochemistry, and biological activities. *Medicines, 3*(6), 1–55.

Kala, C. P., (2002). *Medicinal Plants of Indian Trans-Himalaya*. Dehradun, Bishen Singh Mahendra Pal Singh.

Kala, C. P., (2004). Current status of medicinal plants used by traditional Vaidyas in Uttaranchal state of India. *Ethnobot. Res. Appl., 3*, 267–278.

Kala, C. P., (2006). Problems and prospects in the conservation and development of the Himalayan medicinal plants sector. *International Journal of Sustainable Development, 9*(4), 370–389. 10.1504/IJSD.2006.014221.

Kala, C. P., Farooquee, N. A., & Dhar, U., (2005). Traditional uses and conservation of Timur (*Zanthoxylum armatum* DC.) through social institutions in Uttaranchal Himalaya, India, *Conservation and Society, 3*, 224–230.

Kar, P., Sharma, N., Singh, B., Sen, A., & Roy, A., (2020). Natural compounds from *Clerodendrum* spp. as possible therapeutic candidates against SARS-CoV-2: An in-silico investigation. *Journal of Biomolecular Structure & Dynamics*. https://doi.org/10.1080/07391102.2020.1780947.

Khadka, D., Dhamala, M. K., Li, F., Aryal, P. C., Magar, P. R., Bhatta, S., Thakur, M. S., et al., (2020). *The Use of Medicinal Plant to Prevent COVID-19 in Nepal*. PREPRINT (Version 1) available at Research Square. https://doi.org/10.21203/rs.3.rs-88908/v.

Khan, B. A., Akhtar, N., & Mahmood, T., (2010). A comprehensive review of a magic plant, *Hippophae rhamnoides*. *Pharmacognosy Journal, 2*, 58–61.

Khan, S. A., & Al-Balushi, K., (2021). Combating COVID-19: The role of drug repurposing and medicinal plants. *Journal of Infection and Public Health, 14*(4), 495–503. https://doi.org/10.1016/j.jiph.2020.10.012.

Khan, T., Khan, M. A., Mashwani, Z., Ullah, N., & Nadhman, A., (2021). Therapeutic potential of medicinal plants against COVID-19: The role of antiviral medicinal metabolites.

Biocatalysis and Agricultural Biotechnology, 31, 101890. https://doi.org/10.1016/j.bcab.2020.101890.

Kumar, V., & Staden, J. V., (2016). A review of *Swertia chirayita* (*Gentianaceae*) as a Traditional medicinal plant. *Front. Pharmacol., 6*, 308. doi: 10.3389/fphar.2015.00308.

Kumari, P., Rajoria, V., & Jain, A. P., (2021). *In Vivo* anti-inflammatory and antimicrobial effect of extract of herbal plant *Swertia chirata* L. *Asian Journal of Pharmaceutical Education and Research, 10*(1), 87–95. https://dx.doi.org/10.38164/AJPER/10.1.2021.87-95.

Li, X., Ao, M., Zhang, C., Fan, S., Chen, Z., & Yu, L., (2021). Zingiberis rhizoma recens: A review of its traditional uses, phytochemistry, pharmacology, and toxicology. *Evidence-Based Complementary and Alternative Medicine, 6668990*, 1–20. https://doi.org/10.1155/2021/6668990.

Luo, H., Tang, Q. L., Shang, Y. X., Liang, S. B., Yang, M., Robinson, N., & Liu, J. P., (2020). Can Chinese medicine be used for prevention of corona virus disease 2019 (COVID-19)? A review of historical classics, research evidence and current prevention programs. *Chin. J. Integr. Med., 26*(4), 243–250. doi: 10.1007/s11655-020-3192-6.

Maruthamuthu, R., & Ramanathan, K., (2016). Phytochemical analysis of bark extract of *Cinnamomum verum*: A medicinal herb used for the treatment of coronary heart disease in Malayali Tribes, Pachamalai Hills, Tamil Nadu, India. *International Journal of Pharmacognosy and Phytochemical Research, 8*(7), 1218–1222.

Masood, M., Arshad, M., Qureshi, R., Sabir, S., Amjad, M. S., Qureshi, H., & Tahir, Z., (2015). *Picrorhiza kurroa*: An ethnopharmacologically important plant species of Himalayan region. *Pure and Applied Biology, 4*(3), 407–417. http://dx.doi.org/10.19045/bspab.2015.43017.

Mehdi, S., Siddique, R., Mehdi, S., & Ali, M. M., (2021). Phytochemical evaluation of *Curcuma longa* and its beneficial effects. *J. Toxicol. Pharmaceut. Sci., 5*(1), 5–10.

Ministry of AYUSH, (2021). *20 Medicinal Plants of '21 for COVID-19 Care*. National Medicinal Plants Board, Ministry of AYUSH, Government of India.

Mishra, S. K., Sangwan, N. S., & Sangwan, R. S., (2007). Plant review *Andrographis paniculata* (Kalmegh): A review. *Pharmacognosy Reviews, 1*(2), 283–298.

Mukhtarad, M., Arshad, M., Ahmad, M., Pomerantz, R. J., Wigdahl, B., & Parveen, Z., (2008). Antiviral potentials of medicinal plants. *Virus Res., 131*, 111–120.

Nagajothi, S., Mekala, P., Raja, A., Raja, M. J., & Senthilkumar, P., (2018). Andrographis paniculata: Qualitative and quantitative phytochemical analysis. *Journal of Pharmacognosy and Phytochemistry, 7*(4), 1251–1253.

Natesh, J., Mondal, P., Kaur, B., Ajees, A., Salam, A., Kasilingam, S., & Meeran, S. M., (2021). Promising phytochemicals of traditional Himalayan medicinal plants against putative replication and transmission targets of SARS-CoV-2 by computational investigation. *Computers in Biology and Medicine, 133*, 104383. https://doi.org/10.1016/j.compbiomed.2021.104383.

Pardhe, D. D., (2021). An overview of medicinal and pharmacological properties of turmeric (*Curcuma longa*) in the treatment of various human ailments. *Epitome: International Journal on Multidisciplinary Research, 7*(4), ISSN: 2395-6968.

Patel, B., Sharma, S., Nair, N., Majeed, J., Goyal, R. K., & Dhobi, M., (2021). Therapeutic opportunities of edible antiviral plants for COVID-19. *Molecular and Cellular Biochemistry, 476*, 2345–2364. https://doi.org/10.1007/s11010-021-04084-7.

Poornima, B. S., Hegde, P. L., Pradeep, & Harini, A., (2015). Pharmacological review on *Clerodendrum serratum* Linn. Moon. *Journal of Pharmacognosy and Phytochemistry, 3*(5), 126–130.

Prajapati, S., & Kumar, N. G. V., (2020). SARS-CoV-2 pandemic: An opportunity for Indian traditional medicines (AYUSH). *Int. J. Complement Alt. Med., 13*, 103–105. doi: 10.15406/ijcam.2020.13.00502.

Prasad, C., Singh, D., Shukla, O., & Singh, U. B., (2014). *Cymbopogon jwarancusa* – An important medicinal plant: A review. *The Pharma. Innovation Journal, 3*(6), 13–19.

Pundarikakshudu, K., & Kanaki, N. S., (2019). Analysis and regulation of traditional Indian medicines (TIM). *Journal of AOAC International, 102*, 977, 978.

Rai, S., (2008). *Management of Ginger (Zingiber officinale Rosc.) Rhizome rot in Darjeeling and Sikkim Himalayan Region.* https://lib.icimod.org/record/13347 (accessed on 6 May 2024).

Raina, V. K., Srivastava, S. K., & Syamasundar, K. V., (2005). Rhizome and leaf oil composition of *Curcuma longa* from the lower Himalayan Region of Northern India. *Journal of Essential Oil Research, 17*, 556–559. 10.1080/10412905.2005.9698993.

Rana, V., Thakur, K., Sood, R., Sharma, V., & Sharma, T., (2012). Genetic diversity analysis of *Tinospora cordifolia* germplasm collected from North Western Himalayan region of India. *Journal of Genetics, 91*, 99–103. https://doi.org/10.1007/s12041-012-0137-7.

Rani, A., Pande, C., Tewari, G., & Patni, K., (2017). A review on aroma profile of *Cinnamomum* species in North and North East India. *World Journal of Pharmaceutical Research, 6*(11), 200–221.

Rasool, S., & Maqbool, M., (2019). An overview about *Hedychium spicatum*: A review. *Journal of Drug Delivery and Therapeutics, 9*(1-s), 476–480. http://dx.doi.org/10.22270/jddt.v9i1-s.2347.

Rathee, D., Rathee, P., Rathee, S., & Rathee, D., (2016). Phytochemical screening and antimicrobial activity of *Picrorrhiza kurroa*, an Indian traditional plant used to treat chronic diarrhea. *Arabian Journal of Chemistry, 9*, S1307–S1313. http://dx.doi.org/10.1016/j.arabjc.2012.02.009.

Sagadevan, P., Suresh, S. N., Ranjithkumar, R., Rathishkumar, S., Sathish, S., & Chandarshekar, B., (2015). Traditional use of *Andrographis paniculata*: Review and perspectives. *International Journal of Biosciences and Nanosciences, 2*(5), 123–131.

Salma, U., Kundu, S., & Gantait, S., (2017). Phytochemistry and pharmaceutical significance of *Picrorhiza kurroa* royle ex Benth. *Phytochem. and Pharm. of Med. Herbs.* (pp. 26–37). ISBN: 978-93-85995-46-0.

Samant, S. S., Dhar, U., & Palni, L. M. S., (1998). *Medicinal Plants of Indian Himalaya: Diversity Distribution Potential Values.* Almora: G.B. Pant Institute of Himalayan Environment and Development.

Sangle, A. D., Salahudiin, K. K., Ayalasomayajula, P. R., & Ansari, I., (2016). Phyto-chemical and pharmacological review of *Clerodendrum serratum* (bharangi). *International Journal of Current Research, 8*(7), 34338–34341.

Sanwal, C. S., Kumar, R., & Bhardwaj, S. D., (2016). Integration of *Andrographis paniculata* as potential medicinal plant in chir pine (*Pinus roxburghii* Sarg.) plantation of North-Western Himalaya. *Hindawi Publishing Corporation Scientifica* (p. 7). Article ID 2049532. http://dx.doi.org/10.1155/2016/2049532.

Shafiq, A., Kanwal, R., Qureshi, R. U., & Chaudhry, F. R., (2015). In vitro screening of *Cymbopogon jwarancusa* and *Conyza canadensis* against liver flukes. *Tropical Biomedicine, 32*(3), 407–412.

Shahrajabian, M. H., Sun, W., & Cheng, Q., (2020). Chemical components and pharmacological benefits of basil (*Ocimum basilicum*): A review. *International Journal of Food Properties, 23*(1), 1961–1970. https://doi.org/10.1080/10942912.2020.1828456.

Sharma, A., & Kumar, A., (2016). Pharmacognostic studies on medicinal plants: *Justicia adhatoda. World Journal of Pharmaceutical Research, 5*(7), 1674–1704.

Sharma, N., Pathania, V., Singh, B., & Gupta, R. C., (2012). Intraspecific variability of main phytochemical compounds in *Picrorhiza kurroa* Royle ex Benth. from North Indian higher altitude Himalayas using reversed phase high-performance liquid chromatography. *Journal of Medicinal Plants Research, 6*(16), 3181–3187.

Shinwari, Z. K., Qaiser, M., Nasar, M. Q., & Ali, A., (2020). Indigenous knowledge based herbal medicine for corona (COVID-19) treatment. *Pak. J. Bot., 52*(4), 1427–1434. http://dx.doi.org/10.30848/PJB2020-4(13).

Singh, D. K., & Hajra, P. K., (1996). Floristic diversity. In: Gujral, G. S., & Sharma, V., (eds.), *Changing Perspective of Biodiversity Status in the Himalaya* (pp. 23–38). New Delhi: British Council Division, British High Commission Publ. Wildlife Youth Services.

Singh, D., Srivastava, R. K., & Khanduri, V. P., (2005). Marketing strategies and trade of medicinal plants in Uttaranchal: Present and future prospects. *Indian Forester, 131*(3), 330–340.

Singh, G., & Rawat, G. S., (2011). Ethnomedicinal survey of Kedarnath wildlife sanctuary in western Himalaya, India. *Ind. J. Fundam. Appl. Life Sci., 1*, 35, 36.

Singh, K. J., & Huidrom, D., (2013). Ethnobotanical uses of medicinal plant, *Justicia adhatoda* L. by Meitei community of Manipur, India. *Journal of Coastal Life Medicine, 1*(4), 322–325.

Singh, M. K., Khare, G., Iyer, S. K., Sharwan, G., & Tripathi, D. K., (2012). *Clerodendrum serratum*: A clinical approach. *Journal of Applied Pharmaceutical Science, 2*(2), 11–15.

Singh, R. S., Singh, A., Kaur, H., Batra, G., Sarma, P., Kaur, H., Bhattacharyya, A., et al., (2021). Promising traditional Indian medicinal plants for the management of novel coronavirus disease: A systematic review. *Phytotherapy Research* (pp. 1–29). https://doi.org/10.1002/ptr.7150.

Sinha, K., Mishra, N. P., Singh, J., & Khanuja, S. P. S., (2004). *Tinospora cordifolia* (Guduchi), a reservoir plant for therapeutic applications: A review. *Indian Journal of Traditional Knowledge, 3*(3), 257–270.

Sornpet, B., Potha, T., Tragoolpua, Y., & Pringproa, K., (2017). Antiviral activity of five Asian medicinal pant crude extracts against highly pathogenic H5N1 avian influenza virus. *Asian Pac. J. Trop. Med., 10*, 871–876.

Sravani, T., & Paarakh, P. M., (2011). *Hedychium spicatum* Buch. Ham. – An overview *Pharmacologyonline, 2*, 633–642.

Srivastava, A. K., Kumar, A., & Misra, N., (2020). *On the Inhibition of COVID-19 Protease by Indian Herbal Plants: An in-Silico Investigation.* arXiv: Other Quantitative Biology.

Srivastava, R. M., Singh, S., Dubey, S. K., Misra, K., & Khar, A., (2011). Immunomodulatory and therapeutic activity of curcumin. *Int. Immunopharmacol., 11*, 331–341.

Sultan, P., Jan, A., & Pervaiz, Q., (2016). Phytochemical studies for quantitative estimation of iridoid glycosides in *Picrorhiza kurroa* royle. *Bot. Stud., 57*, 7. doi: 10.1186/s40529-016-0121-2.

Sytar, O., Brestic, M., Hajihashemi, S., Skalicky, M., Kubes, J., Lamilla-Tamayo, L., Ibrahimova, U., et al., (2021). COVID-19 prophylaxis efforts based on natural antiviral plant extracts and their compounds. *Molecules, 26*, 727. https://doi.org/10.3390/molecules26030727.

Tiwari, R. K., Chanda, S., M, U., Singh, M., & Agarwal, S., (2021). Anti-inflammatory and anti-arthritic potential of standardized extract of *Clerodendrum serratum* (L.) moon. *Front. Pharmacol., 12*, 629607. doi: 10.3389/fphar.2021.629607.

Vellingiri, B., Jayaramayya, K., Iyer, M., Narayanasamy, A., Govindasamy, V., Giridharan, B., Ganesan, S., et al., (2020). COVID-19: A promising cure for the global panic. *Science of the Total Environment, 725*. https://doi.org/10.1016/j.scitotenv.2020.138277.

Wang, J. H., Luan, F., He, X. D., Wang, Y., & Li, M. X., (2018). Traditional uses and pharmacological properties of *Clerodendrum* phytochemicals. *Journal of Traditional and Complementary Medicine, 8*, 24–38.

Wang, J., Su, B., Jiang, H., Cui, N., Yu, Z., Yang, Y., & Sun, Y., (2020). Traditional uses, phytochemistry and pharmacological activities of the genus *Cinnamomum* (*Lauraceae*): A review. *Fitoterapia, 146*, 104675. https://doi.org/10.1016/j.fitote.2020.104675.

Yang, F., Zhang, Y., Tariq, A., Jiang, X., Ahmed, Z., Zhihao, Z., Idrees, M., et al., (2020). Food as medicine: A possible preventive measure against coronavirus disease (COVID-19). *Phytotherapy Research, 34*, 3124–3136. https://doi.org/10.1002/ptr.6770.

Zhang, D. H., Wu, K. L., Zhang, X., Deng, S. Q., & Peng, B., (2020). In silico screening of Chinese herbal medicines with the potential to directly inhibit 2019 novel coronavirus. *J. Integr. Med., 18*(2), 152–158.

CHAPTER 12

Atmospheric N Deposition and Fertilization Impact on Forest Soil Organic Carbon and Plant Communities

KINGSHUK MODAK,[1] P. S. SRIKANTH,[2] AVIJIT GHOSH,[3] and SUKANYA MISRA[4]

[1]*Forest Ecology and Climate Change Division, Rain Forest Research Institute, Jorhat, Assam, India*

[2]*Chemistry and Bioprocessing Division, Rain Forest Research Institute, Jorhat, Assam, India*

[3]*ICAR–Indian Grassland and Fodder Research Institute, Jhansi, Uttar Pradesh, India*

[4]*Rani Lakshmi Bai Central Agriculture University, Jhansi, Uttar Pradesh, India*

ABSTRACT

Atmospheric N deposition (N) is a threat to global plant biodiversity and forest soil carbon. Increasing load of N deposition due to anthropogenic activities has had a negative impact on the environment, climate, and ecosystem functions. Plant communities and soil microorganisms retain a major portion of deposited N in a forest ecosystem, but when the soil reaches its N saturation level, loss of N occurs in the form of leaching and N oxide emission. Excess N deposition has been reported to deteriorate soil health by promoting soil acidification, which leads to a decline in forest productivity and loss of terrestrial biodiversity in Europe, North America, and tropical countries like China. Long-term studies have shown that N deposition has

Sustainable Forest Resources Management: Issues and Implications. Jahangeer A. Bhat, Gopal Shukla, Manmohan J. Dobriyal, Sumit Chakravarty, A. Arunachalam, and Rainer W. Bussmann (Eds.)

a positive effect on SOC accumulation in N-limited temperate forests but has a negative or neutral effect on N-rich tropical forests. Evidence also suggests a loss in biodiversity, a shift in species richness, and abundance in tropical forests due to elevated N deposition. However, the unrealistic dose and duration of added N in these field experiments in tropical forests compared to actual atmospheric N input are major limitations. Therefore, more simulated N fertilization modeling should be carried out across the global forest ecosystem with appropriate amounts and durations of N loads to quantify the critical load for sensitive plant communities and SOC. These studies will be useful in framing policies to alleviate the harmful impact of N deposition on the environment.

12.1 INTRODUCTION

Worldwide forest ecosystems are deficient in nitrogen (N), which limits net primary productivity (Harrison et al., 2005). In the pre-industrialization era, biological N fixation and lightning were the major sources of terrestrial N (Galloway et al., 2008). However, in the current world, atmospheric deposition is one of the dominant sources of N and a major driver of the global N cycle. Anthropogenic N is mainly released into the atmosphere by fossil fuel burning, application of fertilizer, and manures in the form of oxides of N (NO_x, NO_3^-, N_2O) and reduced forms (NH_4^+, NH_3). Generally, the forest receives a considerable amount of N either in the form of fertilizer or as atmospheric N deposition. Experiments suggest that anthropogenic N deposition is mainly responsible for carbon (C) sequestration in the forest ecosystem of the northern hemisphere (Erisman and de Vries, 2000). Thus, both biomass and forest soil are important sinks of C in the terrestrial ecosystem (Van Miegroet & Jandl, 2007). The increase in biomass C due to N fertilization has been demonstrated through various experiments (Spiecker et al., 2012), but the potential of forest soil to sequester C is less highlighted. In the last few decades, irrespective of land use, there has been a decline in C in forest soil due to increased temperature (Trumbore et al., 1996). The majority of experimental results on the impact of N fertilization on forest soil C pools are site-specific. Moreover, studying the direct effect of N fertilization on soil C pools requires extensive sampling to observe temporal changes at a particular site. Additionally, significant variation is also observed in soil C pools within different forest types. Therefore, the knowledge of fertilizer response on forest productivity and changes in C and nutrient dynamics in terms of soil chemistry is insufficient to predict the gain and loss of soil C. A comprehensive assessment is required to study the effect of N fertilization on soil C sequestration, taking into account aboveground biomass, C input in soil (litterfall and root biomass), soil solution phase (dissolved organic C and N), and the response of soil microorganisms.

12.2 PROCESSES OF SOIL N CYCLE

Soil organic matter (SOM) constitutes nearly 90% of the total N stored in the forest ecosystem. Many studies on litter decomposition show exponential loss of C and an increase in N over time (Heim & Frey, 2004; Finn et al., 2015). In order to understand the dynamics of N and its effect on the forest ecosystem, there is a need for a clear understanding of the processes, pools, and fluxes of N. The soil N cycle involves 3 processes: input, transformation, and output in the forest ecosystem (Figure 12.1). These processes include biological N fixation (BNF), atmospheric deposition, litter decomposition, N mineralization, ammonification, nitrification, denitrification, and N leaching. BNF is the primary source of N contributing about 90–130 Tg N ha^{-1} (Cleveland et al., 1999). There are three types of N fixers viz. symbiotic N fixer (*Rhizobium* sp.), non-symbiotic N fixer (free-living e.g., *Azotobacter* sp) and filamentous actinomycetes (*Frankia* sp). Based on 12 field studies, the BNF in tropical forest soils varied between 15 and 36 kg N hm^{-2} yr^{-1} (Zhu et al., 2015). On the other hand, the BNF in temperate forest ecosystem (7–27 kg N hm^{-2} $year^{-1}$) is much lower or equal to tropical forest. Houlton et al. (2008) reported that high rates of BNF have a positive effect on production of soil enzymes required for absorption of phosphorus (P) and other essential nutrients. Majority of N assimilated by the plant are derived from decomposition of litter and dead organic matter in forest environments. The rate of litter decomposition plays a pivotal role in soil N cycle and dynamics. Different climatic conditions around the globe also have a significant effect on decomposition rate in addition to litter quality. Therefore, stoichiometric ratios like C/N, N/P, lignin/N are often used to predict the soil N availability and dynamics effectively (Talbot et al., 2012).

Ammonification governs the mineralization of soil N from insoluble organic N to dissolved organic N (DON). The NH_4^+–N generated through this microbially mediated process is either assimilated by plants or has different fates like adsorption by clay, microbial immobilization, or conversion to NO_3^- by the nitrification process. Gilliam et al. (2001) reported that 90% of NH_4^+ is transformed into NO_3^-, which is utilized by plants or subjected to leaching and denitrification loss. The loss of N from forest ecosystems is mainly in the form of NO_2, NO, and N_2O. Among these greenhouse gases, N_2O contributes to ozone layer depletion and NO emission induces acid rain. Several studies have demonstrated that N-rich tropical forests have a higher rate of nitrification and emit more NO_x compared to N-limited temperate forests (Hall & Matson, 2003; Corre et al., 2010). Therefore, N deposition in these forests can further exacerbate the emission of N-oxides. NH_3 volatilization is another important avenue for the loss of N from ecosystems. Soil properties like pH and cation-exchange capacity can significantly stimulate gaseous loss of N from the soil (Li et al., 2019). The soil solution phase of forests mainly consists of DON and NO_3^-. Mature and undisturbed temperate forests have a low N deposition rate (Goodale et al., 2000). Hence, less leaching is observed as most of the N is either assimilated by plants or immobilized by microbes. On the contrary, highly weathered tropical rainforests are subjected to prominent leaching loss of NO_3^-, which can be aggravated by elevated N deposition (Zhu et al., 2015).

12.3 ATMOSPHERIC N DEPOSITION

Generally, in the absence of human interference, the major sources of forest N are BNF and litter decomposition, where N is transferred among SOM, plants, and microorganisms over time. In the last few decades, intensive N fertilization, industrialization, fossil fuel combustion, and cultivation of N-fixing plants have elevated N deposition. Atmospheric N depositions have a significant effect on carbon, nitrogen, and water cycling. It is reported that global atmospheric N deposition would be the main source of N in the forest ecosystem by 2050 with an estimated rate of 200 Tg N per year (Galloway et al., 2004). The major biotic and abiotic mechanisms for the retention or incorporation of deposited N in the forest are microbial immobilization, plant assimilation, soil cation exchange capacity, and SOM adsorption (Silver et al., 2005). However, when the forest ecosystem reaches its N saturation potential, plants and microbes cannot retain more N (Aber et al., 1998). At this point, N is lost from the ecosystem primarily in the form of nitrate leaching and NOx emissions. Positive feedback is observed between rates of N deposition and photosynthesis (Liu et al., 2018), tree growth (Etzold et al.,

2020), and water use efficiency (Hu et al., 2019) as nitrogen is a major driver of all these processes. Elevated N deposition also acts as a primary factor for the C sequestration status of the forest ecosystem, particularly in N-deprived environments in temperate and boreal forests (Fernández-Martínez et al., 2017). Hence, atmospheric N deposition can be considered a climate change mitigation strategy under such circumstances. Higher N input in the form of deposition can abruptly increase tree canopy cover, which indirectly increases the albedo and thereby has a positive feedback on climate. On the other hand, increased atmospheric N deposition can accelerate the rate of N cycling in forest ecosystems. This may lead to a cascade of negative phenomena such as eutrophication, nutrient imbalance, loss of N in the pathway (leaching, emission of greenhouse gases like NO_x, etc.), loss of biodiversity due to soil acidification, and even forest degradation under extreme scenarios (Bobbink et al., 2010; Penuelas et al., 2013). These environmental issues have raised serious concerns about N deposition among researchers and policymakers worldwide.

12.4 STUDY METHODS TO ESTIMATE N DEPOSITION AND ITS EFFECT ON N CYCLE IN FOREST

12.4.1 LONG-TERM SIMULATED N-DEPOSITION EXPERIMENTS

Long-term experiments are set up to observe the response of different forms and doses of inorganic fertilizers (urea, ammonium nitrate, sodium nitrate) in different forest types. These experiments help us to gather knowledge about the effect of N deposition on N cycle, SOC sequestration, changes in plant community structure by simulating the dose and duration of fertilization. EXMAN (Experimental Manipulation of Forest Ecosystems in Europe) and NITREX (Nitrogen saturation experiments) projects are supported by the European Union Commission of European Communities. The NITREX mainly focuses on the effect of atmospheric N deposition on factors and processes affecting N saturation in forest ecosystems. The EXMAN study the effect of atmospheric N deposition by manipulating the chemical composition/quantity of coniferous forests in Europe (Wright & Rasmussen, 1998). These projects study the dose-response relationship of N deposition and estimate the critical load which helps in the formulation of environmental policies. However, these projects have their own limitations as they are not useful in short-term studies of N deposition in forest ecosystems.

12.4.2 N DEPOSITION GRADIENT METHOD

In this method, different levels of background N deposition are selected to observe their response in the N cycle. In comparison to NITREX and EXMAX programs in Europe, this method is convenient to study the short-term changes in the forest ecosystem in response to atmospheric N deposition (Gundersen et al., 1998). However, the interferences observed due to heterogeneity in soil properties, topography, climate, plant composition, and land-use history sometimes make it difficult to interpret the results.

12.4.3 N INPUT MANIPULATION METHOD

A manipulation of atmospheric N deposition is done using a roof to check whether a N-saturated ecosystem can be reversed by limiting N addition. Through this method, N cycling in an ecosystem was assessed in the absence of N deposition in a Norway spruce stand (*Picea abies*), Solling, Germany (Corre & Lamersdorf, 2004). In this method, one roof was used to simulate N throughfall and acid input based on the reference of the Solling spruce stand. Another roof was treated as a control plot (ambient roof). To study the roof effect, an ambient plot without a roof was also selected of the same age as the adjacent spruce woodland.

12.4.4 ^{15}N ISOTOPE TRACING METHOD

The ^{15}N isotope tracing method is generally used to study the changes in N translocation in plants and microorganisms and soil N fluxes (NO_x emission, NO_3^- leaching, nitrification, and mineralization). Corre et al. (2010) studied the impact of elevated N addition on soil N cycling and its fluxes by the ^{15}N isotope tracing method in the semi-deciduous forest of Panama.

12.5 EFFECT OF NITROGEN FERTILIZATION ON SOIL ORGANIC CARBON IN FOREST SOIL

Globalization has led to anthropogenically elevated N deposition and increased N fertilization which stimulated ecosystem C sequestration (Reay et al., 2008). Except in tropical forests, N deficiency is widespread in most forests, limiting net primary productivity (NPP) (Matson et al., 1999).

However, previous studies on forest C sequestration emphasized more on plant productivity rather than soil as a C sink (Lu et al., 2021). Soil is the largest pool of C in terrestrial ecosystems, with more than half of its C stored in forest ecosystems (Schlesinger & Bernhardt, 2013). Hence, C sequestered in soil in the form of organic matter (OM) could be quantitatively more significant than vegetation in terms of the forest C budget (Hyvönen et al., 2007). Our knowledge on the effect of N fertilization on SOC sequestration has numerous limitations. Firstly, studies have shown N deposition promoted SOC sequestration by decreasing the decomposition of plant litter, OM, and suppressing soil respiration (Frey et al., 2014). Secondly, long-term experiments showed N deposition did not affect SOC sequestration (Lu et al., 2011). Thirdly, N deposition also depleted the SOC pool due to increased microbial activity (Van Miegroet & Jandl, 2007). Despite a few studies, questions still remain on how nitrogen fertilization affects SOC stabilization and sequestration by altering the SOC pools in terms of chemical quality and quantity. Geng et al. (2020) reported that C sequestration efficiency under added N varies from 0 to 70 kg C per kg N and 16 to 470 kg C kg^{-1} N for soil and vegetation, respectively. Long-term experimental application of N fertilizer at rates of 5 and 15 g N m^{-2} yr^{-1} in hardwood (*Quercus rubra*) and pine (*Pinus resinosa*) forest plots, respectively showed that dissolved organic C (DOC) flux is insensitive to N status (McDowell et al., 2004) in temperate forests. In the temperate Duke Forest, N fertilization induces the buildup of recalcitrant SOC in the form of non-hydrolysable alkyl C and preferential utilization of labile C by microorganisms due to increased leaf litter production (Feng et al., 2010). Lucander et al. (2021) used the SAFE biogeochemical model to simulate the response of Norway spruce forest growth and SOC to N fertilization. SAFE is a forest ecosystem model used to manage forests by considering decomposition, tree growth, soil processes, hydrology, and weathering. They observed that SOC increased by 1–2% in the fertilized site compared to the control site before tree harvesting. However, after 20 years of felling, the SOC was 4% higher in the N fertilized site than the control site. The simulated change in SOC after felling in the N fertilized site was attributed to an increase in wood litter and a slow wood decomposition rate due to suppressed production of lignin-degrading enzymes. Therefore, they concluded that N fertilization may contribute to SOC accumulation in the long term in N-limited forest soil but has virtually no effect on SOC build-up in the short term.

A study conducted in sub-tropical plantation forests in China showed that the application of sodium nitrate fertilizer for four years did not change total

SOC but promoted the formation of micro-aggregate-associated organic C and fine-particulate organic C by 83.4% and 133% respectively (Geng et al., 2020). The molecular composition also changed due to the exogenous application of N fertilizer with an increase in aliphatic/aromatic and furfural/pyrrole ratios in the mineral-associated SOM. This suggests a passive C pool decline due to decreased humification and mineralization of SOM, which could have negative consequences on SOC stabilization in subtropical plantation forests. In the Chir pine forest of sub-tropical South China, after 12 years of N addition, there was no significant change in SOC concentration. The probable explanation is that the activities of soil microorganisms were reduced by N addition, which hindered the mixing of litter C with the surface soil (Wu et al., 2013). Li et al. (2006) also reported that the application of N fertilizer marginally increased total SOC in sub-tropical forests. Cusack et al. (2010) reported that heterotrophic respiration declined due to N fertilizer application. As a result, SOC concentration was approximately 30% and 100% higher in the fertilized plot than in the control plot in lower and higher elevation tropical forests. This might be due to lower C input from aboveground net primary production (ANPP) to the soil, which suppressed microbial respiration, leading to an increase in bulk SOC stock (Cusack et al., 2009).

Long-term N fertilizer application is also reported to improve SOM stabilization in the light and heavy fractions (Swanston et al., 2004). On the contrary, a 15-year long-term field study of Pinus radiata plantation showed that in the N applied plots, the light and heavy fractions had a mean residence time of 23 and 67 years, respectively, which are lower than those in the control sub-plots (36 and 133 years in the light and heavy fractions, respectively) (Huang et al., 2011). The ^{14}C and lignin analysis of soil C showed that nitrogen addition led to a higher decomposition rate of labile and recalcitrant C. However, the N-fertilized plots still have an increased SOC concentration by 14.6% in the 0–5 cm layer, which can be attributed to an increase in C input in the form of litterfall and forest debris. Thus, long-term N addition can improve tree biomass and SOC content in the added plots in the 0–5 cm soil layer despite increased decomposition of SOM.

The studies related to the linkage between microbial communities and SOC dynamics under elevated N fertilization are limited. Experiments related to N fertilization suggested that microbial biomass decreased (Mo et al., 2008), increased (Li et al., 2006), and also remained unchanged (Davidson et al., 2004) in tropical forest soil. Cusack et al. (2011) reported that nitrogen fertilization increased the microbial biomass in tropical forest soil with a corresponding decline in low-density SOM pools due to an increase in DOC and mineral-associated C. In the lower elevation forest, the population of gram-negative bacteria was higher with a significant loss in labile SOC of O-alkyl functional group. On the contrary, fungal phospholipid-derived fatty acids (PLFA) were dominant in higher elevation tropical forest with a decline in aliphatic hydrophobic compounds from SOM. The quality of litter in different forest types also has a great impact on microbial biomass C (MBC). Evidence suggests that there is a decline in heterotrophic respiration and an increase in SOC concentration in tropical forests with high lignin content and C: N due to N addition (McGroddy & Silver, 2000). But the plant-derived lignin compound does not retain over a long time in SOM (Gleixner et al., 2002). In due course of time, the addition of N promotes the transformation of lignin to mineral-associated C, which is chemically recalcitrant via a condensation reaction (Moran et al., 2005). In the Alaskan boreal forest, N addition significantly decreased C:N ratio of microbial biomass from 16.0 $\pm$ 1.4 to 5.2 $\pm$ 0.3 due to an increase in the activity of cellulose-degrading enzymes (Allison et al., 2008). In the nutrient-limited soil ecosystem, growth of soil microbes is restricted due to inadequate C supply (Del Giorgio & Cole, 1998). N fertilization at this stage will exert a demand on C supply. To meet this demand, C will be mined from soil that entered via litterfall. This transformation process is governed by C:N ratio and C utilization efficiency (CUE) of soil microorganisms. Thus, for incorporation of supplied N, SOM will be converted to more labile microbial biomass C with concomitant loss of soil C through respiration (Van Miegroet & Jandl, 2007). Similar observations were also made by Mansson & Falkengren-Grerup (2003) in the deciduous forest soils dominated by *Quercus robur* L. Hence, N fertilization may be a poor strategy to sequester C, especially in marginally productive forest soil.

12.6 EFFECT OF ATMOSPHERIC N DEPOSITION ON PLANT COMMUNITIES IN FOREST SOIL

Several studies have documented that nitrogenous fertilizer application and atmospheric N deposition could affect plant community composition in forest ecosystems (Clark & Tilman, 2008; Maskell et al., 2010). External N fertilization increases plant productivity in terms of biomass by eliminating N deficiency. But excess N addition would hamper the carboxylation step of photosynthesis and hinder plant growth (Bauer et al., 2004). Initially,

North American and European ecologists showed interest in studying the response of temperate forest plant species to N deposition (Dirkse et al., 1991). Due to the insignificant effect of N deposition on forest biodiversity, the number of studies started to decline (Hurd et al., 1998; Gilliam et al., 2006). In 2003, while studying the understory vegetation of 531 boreal forest trees in Sweden, Strengbom et al. (2003) found that increased N deposition decreased the abundance of *Vaccinium myrtillus,* which was once dominant. Xia & Wan (2008) reported that N supply has a significant effect on plant functional type. Applications of fertilizer have a positive effect on tree biomass compared to shrubs and herbaceous plants rather than woody species. Long-term experiments simulating N deposition through nitrogenous fertilizer application have raised some concerns among researchers. Firstly, N accumulation causes a decline in biodiversity with an abrupt increase in nitrophilous plants and the elimination of native species as a result of steep competition (Phoenix et al., 2012). Secondly, excess N in the soil system makes the plant more succulent and susceptible to secondary stress like poor resistance to pest attacks, increased herbivory, and frost injury (Sheppard et al., 2008). Thirdly, changes in plant community composition and a decline in species richness occur due to soil acidification, loss of basic cations, metal toxicity (Al^{3+}, Fe^{3+}), and species invasion (Horswill et al., 2008, Bobbink et al., 2010).

In the Tundra ecosystem, fields are covered with permafrost and waterlogged in summer. Many researchers reported that several field manipulations of study sites with NPK fertilizer addition make it difficult to interpret the changes in plant communities in response to N addition. Soudzilovskaia et al. (2005) demonstrated that annual fertilization of N increased the population of vascular plants and decreased the lichen and bryophytes. Similarly, Madan et al. (2007) also found a slight increase in the vascular plant population. Several observations made by different researchers confirmed that 5–15 kg N ha^{-1} $year^{-1}$ was the minimum threshold value of N deposition which stimulated the decrease in bryophytes and lichens in the Arctic region (Gordon et al., 2001, Soudzilovskaia et al., 2005, Madan et al., 2007).

In the Boreal Forest ecosystem, plant growth is restricted due to low nutrient availability, cold climate, and a short growing season. In the current scenario, the load of N deposition in European boreal forests is <6 kg N ha^{-1} $year^{-1}$ (Bobbick et al., 2010). With an increase in N deposition rate to >10 kg N ha^{-1} $year^{-1}$, the population of native boreal forest species like *Hylocomium splendens* starts to decline (Mäkipää and Heikkinen, 2003) with a simultaneous increase in *Brachythecium* spp. and *Plagiothecium* spp., which are dominant in a nutrient-rich environment (Strengbom et al., 2001). The number of understory shrubs like *Vaccinium myrtillus* also starts to decrease due to pathogen attacks, and graminoids like *Deschampsia flexuosa* start to proliferate in response to N deposition (Strengbom et al., 2002). The chemical species of deposited N available to the plants also have a profound impact on growth in addition to quantity (Bobbink et al., 2010). Due to poor N mineralization rates in these forests, the availability of NH_4^+–N and NO_3^-–N is less in comparison to organic N. Hence, the steep competition exhibited by plant communities due to NH_4^+ and NO_3 scarcity may also limit their growth. It is concluded that a threshold value of 5–10 kg N deposited per ha per year has the possibility to decrease lichen, shrubs, and bryophyte species in boreal forests.

In temperate regions, high-altitude forests are mainly at risk of losing biodiversity due to wet N deposition caused by orographic rainfall. Herbs, being the most dominant type of vegetation in temperate forests, are mostly affected by N deposition. A decline in biomass and density of herb species, particularly *Maianthemum canadense*, was found after seven years of high N addition in red pine forest stands in Harvard Forest, Massachusetts. Initially, Gilliam & Roberts (2003) also reported that excess N increased herbaceous cover and decreased species richness. However, in 2006, Gilliam found in another study that high N deposition decreased soil N spatial heterogeneity, which helped maintain high herbaceous species diversity. Similarly, Schleppi et al. (1999) also observed no significant changes in the herb layer after the addition of 30 kg N ha^{-1} yr^{-1}. This suggests that high ambient N deposition has no effect on herbaceous species biodiversity by nullifying the spatial variability of soil N. In support of experimental results, evidence from monitoring and field surveys also suggested that N deposition in the range of 15–40 kg N ha^{-1} yr^{-1} increased the abundance of nitrophilous herb species in Dutch forests (Dirkse and van Dobben, 1989), German spruce/fir forests (Kraft et al., 2000), and deciduous forests of France (Thimonier et al., 1994). Another study on modeling of simulated N fertilization in Norway spruce (*Picea abies*) in Sweden indicated that the tree growth in the northern region is low where historically high N deposition rate was observed (Lucander et al., 2021). On the contrary, the southern region reported to have increased tree growth with low N deposition. This result was in agreement with N saturation theory where plant productivity is not hindered under N-saturated forest (Emmett, 2007).

In the Mediterranean forest, the change in lichen community is very pronounced in response to N deposition. Even at a low level of 3 kg N ha^{-1}

yr^{-1}, Fenn et al. (2008) observed a significant shift in lichen community composition from sensitive to N-tolerant types. In another study, the natural shift in native dominant acidophyte (N-sensitive species) was a matter of concern for its link to the food web (McCune et al., 2007). Allen et al. (2007) demonstrated a pronounced loss in biodiversity in the understory vegetation of mixed conifer forest in the San Bernardino Mountains in southern California. N deposition stimulated the growth of invasive species with high foliage and litter production. Evidence from long-term studies implied that N deposition of 3–5 kg N ha^{-1} yr^{-1} had an impact on the epiphytic lichen community of the Mediterranean ecosystem.

The tropical forests are storehouses of biodiversity (Mittermeier et al., 1998). These forests are rich in SOC, N, and often low in P. The shift in composition of plant communities in response to N deposition is still a key question of research in tropical forest ecosystems. In most of the field experiments, complexity arises in comprehending the effect of N fertilization on plant communities because the doses of fertilizer applied are clearly higher than natural atmospheric deposition. Besides this, the rich biodiversity observed among all the tropical forest species makes it hard to interpret the effect of N added on plant community composition since not all the species in the ecosystem are nutrient-limited. The tropical rainforests are dominant in broadleaf trees with multi-layer canopies and different leaf shedding cycles. The biodiversity of epiphytes and lianas is also considerably rich. The Amazon rainforest is a typical example in this category with high spatial variability in species richness. Although many researchers have documented that tropical forests are nutrient-limited (Tamm, 1990; Aber et al., 1995), low light availability to understory plants can serve as a hindrance to their growth even with increasing nutrient availability (Bobbink et al., 2010). Wu et al. (2013) reported that understory plant species richness declined due to a decrease in the population of arbuscular mycorrhizal fungi (AMF), an increase in the bacterial stress index, and a reduction in soil pH after eight years of simulated N deposition in a sub-tropical forest. Similarly, high N deposition (150 kg N ha^{-1} yr^{-1}) reduced the understory species richness in a 50-year-old forest, but the effect was less pronounced in rehabilitated and disturbed forests (Lu et al., 2010). This indicates that N deposition over young or disturbed forest ecosystems can have a positive effect on understory plant species richness compared to mature and undisturbed forests. A high dose of N also resulted in a decline in understory diversity, particularly in tree seedlings and fern species, due to enhanced competitive behavior of fast-growing species (Aerts & Chapin, 1999).

The tropical montane forests are characterized by low above-ground biomass, litterfall, and foliage in comparison to tropical rainforests. A positive correlation was observed between N-limited montane forest and N addition (LeBauer & Treseder, 2008). Osterlag & Verville (2002) reported a significant increase in invasive species after 10 years of N addition on young soil (N-limited, 200–400 years) in wet tropical montane forest.

12.7 LIMITATIONS OF THE STUDY

Significant variability in observations from long-term studies of atmospheric N deposition on SOC and composition of plant communities makes it often difficult to draw a logical conclusion. Limitations in research funding also restrict the use of more reliable methods like the ^{15}N isotope tracing method. It is also unfortunate that most of the long-term simulated experiments in tropical forests use high doses of N, which are quite unrealistic in comparison to natural atmospheric deposition. Hence, more experiments should be conducted in tropical and sub-tropical forests with appropriate doses and durations of load. Additionally, the impact on forest biodiversity and SOC pools should be extensively studied in the absence of anthropogenic N deposition before the atmospheric N load starts to increase in the coming years in tropical regions.

12.8 CONCLUSIONS

- Atmospheric N deposition may stimulate recalcitrant SOC buildup in N-limited temperate forests but have a negative or neutral effect on N-rich tropical forests.
- In tropical forest soil, suppression of microbial respiration with N fertilization has a positive impact on SOC accumulation in the long term.
- In marginally productive forest soil, N fertilization may increase heterotrophic respiration rate, which will convert SOM to more labile microbial biomass and prevent soil C sequestration.
- Elevated N deposition is associated with the loss of biodiversity in Mediterranean, boreal, and tropical montane forests.
- N deposition may lead to a shift in the abundance of plant species and understory diversity (particularly fern species) in tropical rainforests with a simultaneous increase in nitrophilous plants.

- Loss or decline in the richness of the lichen community can serve as a potential indicator of biodiversity loss in a particular forest ecosystem.
- Modeling on simulated N fertilization is a promising approach to quantify the critical N load of global forest ecosystems which are sensitive to biodiversity degradation in response to N deposition.

KEYWORDS

- **anthropogenic activities**
- **biodiversity**
- **ecosystem functions**
- **forest soil carbon**
- **global plant biodiversity**
- **N deposition**
- **plant communities**
- **soil microorganism**

REFERENCES

Aber, J. D., Magill, A., Mcnulty, S. G., Boone, R. D., Nadelhoffer, K. J., Downs, M., & Hallett, R., (1995). Forest biogeochemistry and primary production altered by nitrogen saturation. *Water, Air, and Soil Pollution, 85*(3), 1665–1670.

Aber, J., McDowell, W., Nadelhoffer, K., Magill, A., Berntson, G., Kamakea, M., & Fernandez, I., (1998). Nitrogen saturation in temperate forest ecosystems: Hypotheses revisited. *BioScience, 48*(11), 921–934.

Aerts, R., & Chapin, III. F. S., (1999). The mineral nutrition of wild plants revisited: A re-evaluation of processes and patterns. In: *Advances in Ecological Research* (Vol. 30, pp. 1–67). Academic Press.

Allen, E. B., Temple, P. J., Bytnerowicz, A., Arbaugh, M. J., Sirulnik, A. G., & Rao, L. E., (2007). Patterns of understory diversity in mixed coniferous forests of southern California impacted by air pollution. *The Scientific World Journal, 7*, 247–263.

Allison, S. D., Czimczik, C. I., & Treseder, K. K., (2008). Microbial activity and soil respiration under nitrogen addition in Alaskan boreal forest. *Global Change Biology, 14*(5), 1156–1168.

Bauer, G. A., Bazzaz, F. A., Minocha, R., Long, S., Magill, A., Aber, J., & Berntson, G. M., (2004). Effects of chronic N additions on tissue chemistry, photosynthetic capacity, and carbon sequestration potential of a red pine (*Pinus resinosa* Ait.) stand in the NE United States. *Forest Ecology and Management*, *196*(1), 173–186.

Bobbink, R., Hicks, K., Galloway, J., Spranger, T., Alkemade, R., Ashmore, M., Bustamante, M., et al., (2010). Global assessment of nitrogen deposition effects on terrestrial plant diversity: A synthesis. *Ecological Applications*, *20*(1), 30–59.

Clark, C. M., & Tilman, D., (2008). Loss of plant species after chronic low-level nitrogen deposition to prairie grasslands. *Nature*, *451*(7179), 712–715.

Cleveland, C. C., Townsend, A. R., Schimel, D. S., Fisher, H., Howarth, R. W., Hedin, L. O., & Wasson, M. F., (1999). Global patterns of terrestrial biological nitrogen (N_2) fixation in natural ecosystems. *Global Biogeochemical Cycles*, *13*(2), 623–645.

Corre, M. D., & Lamersdorf, N. P., (2004). Reversal of nitrogen saturation after long-term deposition reduction: Impact on soil nitrogen cycling. *Ecology*, *85*(11), 3090–3104.

Corre, M. D., Veldkamp, E., Arnold, J., & Wright, S. J., (2010). Impact of elevated N input on soil N cycling and losses in old-growth lowland and montane forests in Panama. *Ecology*, *91*(6), 1715–1729.

Cusack, D. F., Chou, W. W., Yang, W. H., Harmon, M. E., Silver, W. L., & Lidet Team, (2009). Controls on long-term root and leaf litter decomposition in neotropical forests. *Global Change Biology*, *15*(5), 1339–1355.

Cusack, D. F., Silver, W. L., Torn, M. S., Burton, S. D., & Firestone, M. K., (2011). Changes in microbial community characteristics and soil organic matter with nitrogen additions in two tropical forests. *Ecology*, *92*(3), 621–632.

Cusack, D. F., Torn, M. S., Mc Dowell, W. H., & Silver, W. L., (2010). The response of heterotrophic activity and carbon cycling to nitrogen additions and warming in two tropical soils. *Global Change Biology*, *16*(9), 2555–2572.

Davidson, E. A., Reis De, C. C. J., Vieira, I. C., Figueiredo, R. D. O., Moutinho, P., Yoko, I. F., & Tuma, S. R., (2004). Nitrogen and phosphorus limitation of biomass growth in a tropical secondary forest. *Ecological Applications*, *14*(sp4), 150–163.

Del Giorgio, P. A., & Cole, J. J., (1998). Bacterial growth efficiency in natural aquatic systems. *Annual Review of Ecology and Systematics*, 503–541.

Dirkse, G. M., & Van, D. H. F., (1989). Het effect van bemesting op de samenstelling van de kruidlaag van dennenbossen. *Natura*, *9*, 208–212.

Dirkse, G. M., Van, D. H. F., & Tamm, C. O., (1991). *Effects of Fertilization on Herb and Moss Layers of a Scots Pine Stand in Lisselbo (Sweden); A Multivariate Analysis* (No. 91/7). Rijksinstituut voor Natuurbeheer.

Emmett, B. A., (2007). Nitrogen saturation of terrestrial ecosystems: Some recent findings and their implications for our conceptual framework. *Acid rain-Deposition to Recovery*, 99–109.

Erisman, J. W., & Vries, W. D., (2000). Nitrogen deposition and effects on European forests. *Environmental Reviews*, *8*(2), 65–93.

Etzold, S., Ferretti, M., Reinds, G. J., Solberg, S., Gessler, A., Waldner, P., & De Vries, W., (2020). Nitrogen deposition is the most important environmental driver of growth of pure, even-aged and managed European forests. *Forest Ecology and Management*, *458*, 117762.

Feng, X., Simpson, A. J., Schlesinger, W. H., & Simpson, M. J., (2010). Altered microbial community structure and organic matter composition under elevated CO_2 and N fertilization in the duke forest. *Global Change Biology*, *16*(7), 2104–2116.

Fenn, M. E., Jovan, S., Yuan, F., Geiser, L., Meixner, T., & Gimeno, B. S., (2008). Empirical and simulated critical loads for nitrogen deposition in California mixed conifer forests. *Environmental Pollution, 155*(3), 492–511.
Fernández-Martínez, M., Vicca, S., Janssens, I. A., Ciais, P., Obersteiner, M., Bartrons, M., & Peñuelas, J., (2017). Atmospheric deposition, CO_2, and change in the land carbon sink. *Scientific Reports, 7*(1), 1–13.
Finn, D., Page, K., Catton, K., Strounina, E., Kienzle, M., Robertson, F., & Dalal, R., (2015). Effect of added nitrogen on plant litter decomposition depends on initial soil carbon and nitrogen stoichiometry. *Soil Biology and Biochemistry, 91*, 160–168.
Frey, S. D., Ollinger, S., Nadelhoffer, K. E., Bowden, R., Brzostek, E., Burton, A., & Wickings, K., (2014). Chronic nitrogen additions suppress decomposition and sequester soil carbon in temperate forests. *Biogeochemistry, 121*(2), 305–316.
Galloway, J. N., Dentener, F. J., Capone, D. G., Boyer, E. W., Howarth, R. W., Seitzinger, S. P., Asner, G. P., et al., (2004). Nitrogen cycles: Past, present, and future. *Biogeochemistry, 70*(2), 153–226.
Galloway, J. N., Townsend, A. R., Erisman, J. W., Bekunda, M., Cai, Z., Freney, J. R., & Sutton, M. A., (2008). Transformation of the nitrogen cycle: Recent trends, questions, and potential solutions. *Science, 320*(5878), 889–892.
Geng, J., Cheng, S., Fang, H., Pei, J., Xu, M., Lu, M., & Cao, Z., (2020). Nitrogen fertilization changes the molecular composition of soil organic matter in a subtropical plantation forest. *Soil Science Society of America Journal, 84*(1), 68–81.
Gilliam, F. G., Barry, J. J., Hermann, B. P., Meador, K. J., Vahle, V., & Kanner, A. M., (2006). Rapid detection of major depression in epilepsy: A multicenter study. *The Lancet Neurology, 5*(5), 399–405.
Gilliam, F. S., & Roberts, M. R., (2003). Interactions between the herbaceous layer and overstory canopy of eastern forests. *The Herbaceous Layer in Forests of Eastern North America*, 233–254.
Gilliam, F. S., (2006). Response of the herbaceous layer of forest ecosystems to excess nitrogen deposition. *Journal of Ecology, 94*(6), 1176–1191.
Gilliam, F. S., Yurish, B. M., & Adams, M. B., (2001). Temporal and spatial variation of nitrogen transformations in nitrogen-saturated soils of a central Appalachian hardwood forest. *Canadian Journal of Forest Research, 31*(10), 1768–1785.
Gleixner, G., Poirier, N., Bol, R., & Balesdent, J., (2002). Molecular dynamics of organic matter in a cultivated soil. *Organic Geochemistry, 33*(3), 357–366.
Goodale, C. L., Aber, J. D., & McDowell, W. H., (2000). The long-term effects of disturbance on organic and inorganic nitrogen export in the White Mountains, New Hampshire. *Ecosystems, 3*(5), 433–450.
Gordon, C., Wynn, J. M., & Woodin, S. J., (2001). Impacts of increased nitrogen supply on high Arctic heath: The importance of bryophytes and phosphorus availability. *New Phytologist, 149*(3), 461–471.
Gundersen, P., Emmett, B. A., Kjønaas, O. J., Koopmans, C. J., & Tietema, A., (1998). Impact of nitrogen deposition on nitrogen cycling in forests: A synthesis of NITREX data. *Forest Ecology and Management, 101*(1–3), 37–55.
Hall, S. J., & Matson, P. A., (2003). Nutrient status of tropical rain forests influences soil N dynamics after N additions. *Ecological Monographs, 73*(1), 107–129.
Harrison, R., Prietzel, J., & Jandl, R., (2005). Nutritional management of coniferous forests. *Coniferous Forests*, 485–501.

Heim, A., & Frey, B., (2004). Early stage litter decomposition rates for Swiss forests. *Biogeochemistry, 70*(3), 299–313.

Horswill, P., O'Sullivan, O., Phoenix, G. K., Lee, J. A., & Leake, J. R., (2008). Base cation depletion, eutrophication and acidification of species-rich grasslands in response to long-term simulated nitrogen deposition. *Environmental Pollution, 155*(2), 336–349.

Houlton, B. Z., Wang, Y. P., Vitousek, P. M., & Field, C. B., (2008). A unifying framework for dinitrogen fixation in the terrestrial biosphere. *Nature, 454*(7202), 327–330.

Hu, Y., Zhao, P., Zhu, L., Zhao, X., Ni, G., Ouyang, L., & Shen, W., (2019). Responses of sap flux and intrinsic water use efficiency to canopy and understory nitrogen addition in a temperate broadleaved deciduous forest. *Science of the Total Environment, 648*, 325–336.

Huang, Z., Clinton, P. W., Baisden, W. T., & Davis, M. R., (2011). Long-term nitrogen additions increased surface soil carbon concentration in a forest plantation despite elevated decomposition. *Soil Biology and Biochemistry, 43*(2), 302–307.

Hurd, T. M., Brach, A. R., & Raynal, D. J., (1998). Response of understory vegetation of Adirondack forests to nitrogen additions. *Canadian Journal of Forest Research, 28*(6), 799–807.

Hyvönen, R., Ågren, G. I., Linder, S., Persson, T., Cotrufo, M. F., Ekblad, A., & Wallin, G., (2007). The likely impact of elevated [CO_2], nitrogen deposition, increased temperature and management on carbon sequestration in temperate and boreal forest ecosystems: A literature review. *New Phytologist., 173*(3), 463–480.

Kraft, M., Schreiner, M., Reif, A., & Aldinger, E., (2000). Veranderung von bodenvegetation und humusauflage im nordschwarzwald. *Allgemeine Forst Zeitschrift, 55*, 222–224.

LeBauer, D. S., & Treseder, K. K., (2008). Nitrogen limitation of net primary productivity in terrestrial ecosystems is globally distributed. *Ecology, 89*(2), 371–379.

Li, S., Zheng, X., Zhang, W., Han, S., Deng, J., Wang, K., & Liu, C., (2019). Modeling ammonia volatilization following the application of synthetic fertilizers to cultivated uplands with calcareous soils using an improved DNDC biogeochemistry model. *Science of the Total Environment, 660*, 931–946.

Li, Y., Xu, M., & Zou, X., (2006). Effects of nutrient additions on ecosystem carbon cycle in a Puerto Rican tropical wet forest. *Global Change Biology, 12*(2), 284–293.

Liu, N., Wang, J., Guo, Q., Wu, S., Rao, X., Cai, X. A., & Lin, Z., (2018). Alterations in leaf nitrogen metabolism indicated the structural changes of subtropical forest by canopy addition of nitrogen. *Ecotoxicology and Environmental Safety, 160*, 134–143.

Lu, M., Zhou, X., Luo, Y., Yang, Y., Fang, C., Chen, J., & Li, B., (2011). Minor stimulation of soil carbon storage by nitrogen addition: A meta-analysis. *Agriculture, Ecosystems & Environment, 140*(1, 2), 234–244.

Lu, X., Mo, J., Gilliam, F. S., Zhou, G., & Fang, Y., (2010). Effects of experimental nitrogen additions on plant diversity in an old-growth tropical forest. *Global Change Biology, 16*(10), 2688–2700.

Lu, X., Vitousek, P. M., Mao, Q., Gilliam, F. S., Luo, Y., Turner, B. L., & Mo, J., (2021). Nitrogen deposition accelerates soil carbon sequestration in tropical forests. *Proceedings of the National Academy of Sciences, 118*(16), e2020790118.

Lucander, K., Zanchi, G., Akselsson, C., & Belyazid, S., (2021). The effect of nitrogen fertilization on tree growth, soil organic carbon and nitrogen leaching—A modeling study in a steep nitrogen deposition gradient in Sweden. *Forests, 12*(3), 298.

Madan, N. J., Deacon, L. J., & Robinson, C. H., (2007). Greater nitrogen and/or phosphorus availability increase plant species' cover and diversity at a high arctic polar semidesert. *Polar Biology, 30*(5), 559–570.

Mäkipää, R., & Heikkinen, J., (2003). Large-scale changes in abundance of terricolous bryophytes and macrolichens in Finland. *Journal of Vegetation Science, 14*(4), 497–508.

Månsson, K. F., & Falkengren-Grerup, U., (2003). The effect of nitrogen deposition on nitrification, carbon and nitrogen mineralization and litter C: N ratios in oak (*Quercus robur* L.) forests. *Forest Ecology and Management, 179*(1–3), 455–467.

Maskell, L. C., Smart, S. M., Bullock, J. M., Thompson, K. E. N., & Stevens, C. J., (2010). Nitrogen deposition causes widespread loss of species richness in British habitats. *Global Change Biology, 16*(2), 671–679.

Matson, P. A., Mc Dowell, W. H., Townsend, A. R., & Vitousek, P. M., (1999). The globalization of N deposition: Ecosystem consequences in tropical environments. *Biogeochemistry, 46*(1), 67–83.

McCune, B., Grenon, J., Mutch, L. S., & Martin, E. P., (2007). Lichens in relation to management issues in the Sierra Nevada national parks. *North American Fungi, 2*, 1–39.

McDowell, W. H., Magill, A. H., Aitkenhead-Peterson, J. A., Aber, J. D., Merriam, J. L., & Kaushal, S. S., (2004). Effects of chronic nitrogen amendment on dissolved organic matter and inorganic nitrogen in soil solution. *Forest Ecology and Management, 196*(1), 29–41.

McGroddy, M., & Silver, W. L., (2000). Variations in belowground carbon storage and soil CO_2 Flux rates along a wet tropical climate gradient 1. *Biotropica, 32*(4a), 614–624.

Mittermeier, R. A., Myers, N., Thomsen, J. B., Da Fonseca, G. A., & Olivieri, S., (1998). Biodiversity hotspots and major tropical wilderness areas: Approaches to setting conservation priorities. *Conservation Biology*, 516–520.

Mo, J., Zhang, W. E. I., Zhu, W., Gundersen, P. E. R., Fang, Y., Li, D., & Wang, H. U. I., (2008). Nitrogen addition reduces soil respiration in a mature tropical forest in southern China. *Global Change Biology, 14*(2), 403–412.

Moran, K. K., Six, J., Horwath, W. R., & Van, K. C., (2005). Role of mineral-nitrogen in residue decomposition and stable soil organic matter formation. *Soil Science Society of America Journal, 69*(6), 1730–1736.

Ostertag, R., & Verville, J. H., (2002). Fertilization with nitrogen and phosphorus increases abundance of non-native species in Hawaiian montane forests. *Plant Ecology, 162*(1), 77–90.

Penuelas, J., Poulter, B., Sardans, J., Ciais, P., Van, D. V. M., Bopp, L., Boucher, O., et al., (2013). Human-induced nitrogen–phosphorus imbalances alter natural and managed ecosystems across the globe. *Nature Communications, 4*(1), 1–10.

Phoenix, G. K., Emmett, B. A., Britton, A. J., Caporn, S. J., Dise, N. B., Helliwell, R., & Power, S. A. (2012). Impacts of atmospheric nitrogen deposition: Responses of multiple plant and soil parameters across contrasting ecosystems in long-term field experiments. *Global Change Biology, 18*(4), 1197–1215.

Reay, D. S., Dentener, F., Smith, P., Grace, J., & Feely, R. A., (2008). Global nitrogen deposition and carbon sinks, *Nature Geoscience, 1*, 430–437.

Schleppi, P., Bucher-Wallin, L., Siegwolf, R., Saurer, M., Muller, N., & Bucher, J. B., (1999). Simulation of increased nitrogen deposition to a montane forest ecosystem: Partitioning of the added 15N. *Water, Air, and Soil Pollution, 116*(1), 129–134.

Schlesinger, W. H., & Bernhardt, E. S., (2013). *Biogeochemistry: An Analysis of Global Change.* Waltham, MA.

Sheppard, L. J., Leith, I. D., Crossley, A., Van, D. N., Fowler, D., Sutton, M. A., & Woods, C., (2008). Stress responses of *Calluna vulgaris* to reduced and oxidized N applied under real world conditions. *Environmental Pollution, 154*(3), 404–413.

Silver, W. L., Thompson, A. W., Reich, A., Ewel, J. J., & Firestone, M. K., (2005). Nitrogen cycling in tropical plantation forests: Potential controls on nitrogen retention. *Ecological Applications, 15*(5), 1604–1614.

Soudzilovskaia, N. A., & Onipchenko, V. G., (2005). Experimental investigation of fertilization and irrigation effects on an alpine heath, northwestern Caucasus, Russia. *Arctic, Antarctic, and Alpine Research, 37*(4), 602–610.

Spiecker, H., Mielikäinen, K., Köhl, M., & Skovsgaard, J. P., (2012). *Growth Trends in European Forests: Studies from 12 Countries.* Springer Science & Business Media.

Strengbom, J., Nordin, A., Näsholm, T., & Ericson, L., (2001). Slow recovery of boreal forest ecosystem following decreased nitrogen input. *Functional Ecology, 15*(4), 451–457.

Strengbom, J., Nordin, A., Näsholm, T., & Ericson, L., (2002). Parasitic fungus mediates change in nitrogen-exposed boreal forest vegetation. *Journal of Ecology*, 61–67.

Strengbom, J., Olofsson, J., Witzell, J., & Dahlgren, J., (2003). Effects of repeated damage and fertilization on palatability of *Vaccinium myrtillus* to grey sided voles, *Clethrionomys rufocanus*. *Oikos, 103*(1), 133–141.

Swanston, C., Homann, P. S., Caldwell, B. A., Myrold, D. D., Ganio, L., & Sollins, P., (2004). Long-term effects of elevated nitrogen on forest soil organic matter stability. *Biogeochemistry, 70*(2), 229–252.

Talbot, J. M., & Treseder, K. K., (2012). Interactions among lignin, cellulose, and nitrogen drive litter chemistry–decay relationships. *Ecology, 93*(2), 345–354.

Tamm, C. O., (1990). *Nitrogen in Terrestrial Ecosystems: Questions of Productivity, Vegetational Change, and Ecological Stability.* Springer–Verlag, Berlin. doi: 10.1007/978-3-642-75168-4.

Thimonier, A., Dupouey, J. L., Bost, F., & Becker, M., (1994). Simultaneous eutrophication and acidification of a forest ecosystem in North-East France. *New Phytologist, 126*(3), 533–539.

Trumbore, S. E., Chadwick, O. A., & Amundson, R., (1996). Rapid exchange between soil carbon and atmospheric carbon dioxide driven by temperature change. *Science, 272*(5260), 393–396.

Van, M. H., & Jandl, R., (2007). Are nitrogen-fertilized forest soils sinks or sources of carbon? *Environmental Monitoring and Assessment, 128*(1), 121–131.

Wright, R. F., & Rasmussen, L., (1998). Introduction to the NITREX and EXMAN projects. *Forest Ecology and Management, 101*(1–3), 1–7.

Wu, J., Liu, W., Fan, H., Huang, G., Wan, S., Yuan, Y., & Ji, C., (2013). Asynchronous responses of soil microbial community and understory plant community to simulated nitrogen deposition in a subtropical forest. *Ecology and Evolution, 3*(11), 3895–3905.

Xia, J., & Wan, S., (2008). Global response patterns of terrestrial plant species to nitrogen addition. *New Phytologist, 179*(2), 428–439.

Zhu, X., Zhang, W., Chen, H., & Mo, J., (2015). Impacts of nitrogen deposition on soil nitrogen cycle in forest ecosystems: A review. *Acta Ecologica Sinica, 35*(3), 35–43.

CHAPTER 13

Waste to Wealth from Forests: Perspective and Constraints

M. S. SANKANUR,[1] R. P. GUNAGA,[2] T. R. AHLAWAT,[3] S. K. SINHA,[2] R. S. CHAUHAN,[1] A. D. CHAUDHARY,[4] and VRUTTI PATEL[4]

[1]*Department of Forest Biology and Tree Improvement, College of Forestry, Navsari Agricultural University, Navsari, Gujarat, India*

[2]*Department of Forest Products and Utilization, College of Forestry, Navsari Agricultural University, Navsari, Gujarat, India*

[3]*Navsari Agricultural University, Navsari, Gujarat, India*

[4]*CAAST Secretariat, Directorate of Research, College of Forestry, Navsari Agricultural University, Navsari, Gujarat, India*

ABSTRACT

Forests play a crucial role as a valuable resource; however, conventional methods often result in substantial waste. This waste from forestry operations has the potential to be transformed into various products, thereby adding extra value. The manufacturing of charcoal and briquettes from forest tree waste, including plant debris, plant parts, and tree bark, has a significant potential for adding value to wood fuel. Such biomass is used to make charcoal and briquettes, which are effective for cooking and ironing in urban and peri-urban areas. Additionally, forest waste can be used to make agarbatti, prepare bamboo and wooden decorates, and produce biofuel, essential oils, and other value-added products. Hence, natural forests would be preserved by using biomass waste as a substrate to produce value-added products from waste. Hence in this chapter, we discussed various possibilities to convert waste resources from forest and wood industries into value-added products *viz*.,

Sustainable Forest Resources Management: Issues and Implications. Jahangeer A. Bhat, Gopal Shukla, Manmohan J. Dobriyal, Sumit Chakravarty, A. Arunachalam, and Rainer W. Bussmann (Eds.)

agarbatti preparation and its utility, charcoal and its utility, wooden decorate and its utility, biofuel productions and its utility, essential oil extraction and its utility, fuel from bark and its utility, to generate wealth from waste.

13.1 INTRODUCTION

India produces around 350 million tons of waste from agriculture and allied sectors every year (ICAR Report, 2020). Among agriculture and allied sectors, the forestry sector also generates waste that has utilization potential as value-added products. Forest tree waste from plants and plant parts such as *Lantana camara* stem, Mango leaves, *Eucalyptus* leaves, Teak leaves, *Casuarina* leaf litter, Almond leaves, Coconut leaves, *Delonix regia* pods, *Prosopis* twigs, *Ipomoea* stem, Lemon leaves, *Parthenium* plants, and tree barks have a high potential for value-added wood fuel production in the form of charcoal and briquettes. The charcoal and briquettes prepared from such biomass are efficient for cooking and ironing in rural households and are useful in small tea shop boilers and in tandoor chulhas in small hotels. Though capital-intensive, biomass charcoal and briquette technology have long-term utility because of their eco-friendly nature with air quality improvement, economical upliftment of rural people, and cost-effectiveness. Forest waste can also be utilized for the production of agarbatti, wooden, and bamboo decorations preparation, biofuel, and essential oil production, etc. Biomass waste would be used as a substrate, thus saving natural forests.

13.2 WASTE MANAGEMENT

Waste management always emphasizes on disposing waste, treating waste, recovering energy, recycling waste, reusing waste, avoiding and reducing waste (Figure 13.1). Through these processes, waste to wealth may be generated in all sectors of agriculture and allied science in general, and forestry in particular.

By looking into the production potential of forestry wastes as value-added products, the whole chapter is divided into different parts. Details of each part are elaborated under the headings Agarbatti making, charcoal preparation, wooden, and bamboo decorations preparation, biofuel, and essential oil production, and fuel value index of waste bark.

FIGURE 13.1 Waste management least preferable to most preferable.

Source: Reprinted with permission from Lange et al., 2022. Copyright © 2022 by the authors. Licensee MDPI, Basel, Switzerland. https://creativecommons.org/licenses/by/4.0/

13.3 AGARBATTI MAKING

13.3.1 AGARBATTI AND ITS PROCESSING TECHNIQUES

Since ancient times, India has been practicing the burning of Agarbatti for religious and social functions. Agarbatti is also known as Udubattis and is similar to joss sticks made of Dhup. It is an aromatic powder or paste burnt in homes as a fragrant fumigant. These are available in diverse colors and aromas. Moreover, agarbatties are available in other forms including dashang, which may be powder (cones and Dhup, tablets) or sticky paste. Based on size and quality, agarbatti burns around 15 to 30 minutes. The majority (about 75%) of agarbatties produced are of poor quality, either comprising simply charcoal powder or low-grade sandalwood powder mixed with a combination of 50% wood gum powder. They are given a top note using inexpensive fragrances. Natural fixatives like amber, musk, and civet are combined with synthetic aromatics in higher-quality types, along with essential oils, refined resins, and other substances.

India is a large nation, and the people who live there practice diverse religious ceremonies, speak different dialects, and adhere to unique customs and traditions. Despite these differences, everyone uses agarbatti (incense sticks) at weddings, religious worships, and celebrations. Moreover, it serves as an air freshener. The agarbatti industry is a real source of employment that may take advantage of the weaker groups in both urban and rural communities. Women constitute the majority of those who make agarbattis. An important aspect of the agarbatti industry is the provision of raw materials by agarbatti units to artisans who roll agarbatti in their homes while also managing their families and children. They depend greatly on this to complement their current family income. Due to its popularity in foreign markets, agarbatti has grown to be a substantial foreign exchange earner (Nuthan et al., 2009 & Ramana et al., 2010).

13.3.2 PREPARATION OF AGARBATTI

Agarbaties are often made in two different sorts: (i) dipped varieties or perfumed agarbatties; and (ii) masala agarbatties.

1. **Dipped Varieties or Perfumed Agarbatties:** A semi-solid paste is made by combining water with powders such as charcoal, gigatu, white chips, etc. This mixture is applied to sticks by hand rolling after being taken the composition on a wooden plank. Finally, white oil or other solvents, such as diethyl phthalate (D.E.P.), are used to dilute a suitable fragrance ingredient, after which the raw sticks are dried and packaged.
2. **Masala Agarbatties:** The needed amount of water is added to the powdered ingredients (charcoal, jigatu, white chips, local herbs, resin, etc.) together with the perfumery compound (essential oils, aromatic compounds, purified resins, and natural fixatives like civet, musk, etc.). Hand rolling and drying are used to apply this substance on bamboo sticks. The final agarbatti sticks are packaged in a printed carton that is again coated in clear cellophane paper and bundled either according to weight or according to number. They are also wrapped in moisture-proof cellophane paper or butter paper.

Masala Agarbatti includes ingredients like white chips 40%, gigatu 20%, charcoal 20%, aromatic chemicals, essential oil, and other ingredients 20%. The compositions of agarbatti raw materials can be modified as per the requirements. The process of manufacturing Agarbatti is simple; mostly all the process of agarbatti preparation are similar among different manufacturers; however, only formulation of jigat and ingredient may change; however, it depends upon the manufacturer.

13.3.3 SIGNIFICANT CHARACTERISTICS OF THE AGARBATTI INDUSTRY

- Technical-know-how is simple;
- No need of sophisticated equipments;
- Implements/tools used are cost effective, simple, and portable;
- Large capital investment is not required;
- Initially the industry can be started with a small investment even at the residence of the artisans;
- Eco-friendly;
- Highly labor-intensive industry;
- No dependence on power supply;
- No work-related hazards.

Raw materials required for the preparation of 10,000 sticks of Agarbatti include charcoal six kg, gigatu five kg, white chips powder two kg, sandalwood powder one kg, bamboo sticks six kg, kuppam dust two kg, perfumes one and a half kg, and DEP oil five kg.

The plant and machinery for manufacturing Agarbathi are as follows: aluminum trays 5′×5′ (4 nos), wooden planks 5′×5′ (4 nos), heat sealing machine (2 nos), sprayer (4 nos), weighing balance (1 no), plastic buckets (4 nos), jug (4 nos), bag stitching machine (1 no), hdpe container for storing and mixing of chemicals (4 nos), working table for packing (2 nos), agarbatti dryer (1 no), agarbatti machine (1 no), high-speed auto feeder (1 no), masala mixer (1 no), masala filter (1 no), pouch sealing machine/band sealer (1 no), and agarbatti counting machine (1 no).

13.3.4 EMPLOYMENT OPPORTUNITY IN AGARBATTI INDUSTRY

About the quantity of agarbatti workers in the nation, no trustworthy government figures are available. Nearly 12 lakh people are directly or indirectly employed in the agarbatti industry (NCAER, 1990). Out of this, about 20% of them are employees in the organized sector (factory-based) and the remaining 80% of them are working in the unorganized sector (home-based). Around 10,934 agarbatti workers were counted in the Census of Handicrafts, with the majority living in Bengal, Delhi, Uttar Pradesh, Orissa, and Maharashtra. Hence, relying on field observations and imprecise projections from industry sources is necessary. There are around 500,000 agarbatti workers in the nation, according to these estimates. Of these, 90% are women and about 80% are home-based family laborers. Of these workers, 60,000 or so are from Gujarat. Ahmedabad city alone had about 20,000 agarbatti rollers. About 70% are home-based workers. Agarbatti rolling is the primary occupation of practically every home, especially in the regions of Bapunagar, Odhav, Rakhial, Amraiwadi, Meghaninagar, and Naroda. There are very few homes in Ahmedabad's Bapunagar district where this work is not the primary source of income (Hazarika et al., 2018).

13.3.5 RECENT DEVELOPMENTS IN AGARBATTI PREPARATION

On the basis of a conservative assumption of 20% industry growth annually, the market for agarbatti in India is projected to be worth 2,000 crores. According to a market study carried out by the National Council of Applied Economic Research (NCAER, 1990), 147 billion agarbattis were produced nationwide in 1990, with an estimated value of Rs. 7 billion. ITC hopes that by investing in the company and contributing its complementary R&D-based product development and distribution, it can increase the small- and medium-scale industries' competitiveness. The company keeps up its social commitment through advancing quality and procedure standards inside small and medium-sized businesses. With the assistance of the company's technical and process contributions, six agarbatti manufacturing facilities have achieved ISO 9001–2000 certification. The company continues to work with numerous NGOs in Bihar, Karnataka, Pondicherry, and Tamil Nadu to give vocational opportunities as part of its commitment to rural youth and economically disadvantaged women. The Vyakti Vikas Kendra, a charitable organization supported by renowned spiritual leader Sri Sri Ravishankar and situated close to Bangalore, is also receiving assistance from ITC for its "Agarbatti Community Participation Project." Through the program, more than 100 village women are learning how to prepare and roll agarbattis (Hazarika et al., 2018).

India's Agarbatti industry is mainly in the form of a cottage industry and is also labor-intensive. The modern era of Agarbatti manufacturing in India was initiated for the first time in Thanjavur, Tamil Nadu. Later, it has gradually expanded to other parts of neighboring states. Karnataka, Andhra Pradesh, Kerala, Orissa, Tamil Nadu, Gujarat, Dadra, and Nagar Haveli, Bihar, Tripura, and Assam are presently the leading states in the agarbatti sector. Karnataka is the country's leading production state, featuring major towns such as Mysore and Bangalore. Nearly 1,000 units are reported to exist in Karnataka. The Agarbatti industry is increasingly expanding (Hazarika et al., 2018). According to Bordoloi & Sarmah (2007), approximately Rs 7.1 billion comes from the total domestic sales, with the contribution of South

India at 35%, West at 30%, North at 18%, and East at 17%. In addition to Karnataka, Agarbatti industries are also expanding in Bihar, Dadra, and Nagar Haveli, Gujarat, Tamil Nadu, Tripura, and Assam.

13.4 CHARCOAL PREPARATION

13.4.1 CHARCOAL AND ITS PROCESSING TECHNOLOGY

Charcoal is a renewable fuel with several appealing characteristics. It has a low nitrogen and ash content, almost no sulfur, and is highly reactive while being simple to store and handle. As carbonized charcoal possesses a semimetal with a higher electrical resistivity than graphite, it can be a more effective adsorbent with a larger surface area. The growing use of charcoal as a renewable fuel, reductant, adsorbent, and soil amendment is foreshadowed by recent breakthroughs in our understanding of its manufacture and qualities. In India, Coimbatore, Hyderabad, Bangalore, Ahmedabad, Rajkot, Kolkata, Delhi, and Mumbai, which are located in various agro-climate zones, are known for producing charcoal on a huge scale.

13.4.2 PRODUCTION OF CHARCOAL IN INDIA

Major charcoal producing states of India are Tamil Nadu (2,34,000 MT per annum), Andhra Pradesh (2,08,000 MT per annum), Gujarat (1,14,000 MT per annum), Rajasthan (49,200 MT per annum), and Karnataka (30,000 MT per annum). Adopted from www.indiastat.com.In the industrial sectors, silicon and cooking industries are major charcoal users. However, there are many industrial sectors which require charcoal in small quantities, such as Mosquito Coil, Agarbatti, Automobile batteries, Carbon disulfide Manufacturers, Pharma, and Water purification (http://seiaa.gujarat.gov.in).

According to estimates from the IEA (International Energy Agency), more than 2.5 billion people, or 52% of the population in developing nations, rely on biomass as their main cooking fuel. More than half of these people live in India, China, and Indonesia. Nevertheless, biomass is the most important source of energy in sub-Saharan Africa. More than 90% of the rural population in certain areas of this region rely on fuelwood and charcoal. The share is smaller in China, where a large proportion of households use coal instead.

The biomass resource is abundant in many countries and is based primarily on organic waste products, which include forest waste. Utilization of pine needles is crucial in preventing forest fires. Utilizing briquette technology is a first step. Compared to biomass briquettes coupled with clay, starch-based briquettes are far more effective. The efficiency is determined by its capacity to ignite readily and safely, produce less smoke, have a high calorific value, produce less ash as this will be a nuisance during cooking, and be robust enough for safe shipping and storage. The technique has the potential to significantly improve residential fuel quality while reducing costs, improving efficiency, and protecting the environment.

There are two methods of charcoal preparation (Traditional kilns viz., Earth Pit Kilns and Earth Mound Kiln) and Modern Methods (Improved Kilns viz., Brick Kiln, Casamance Kiln, Steel Kiln, Adam Retort). Numerous agencies in India provide this.

13.4.3 THE ADVANTAGES AND DISADVANTAGES (FAO REPORT) OF USING CHARCOAL

- **Advantages:**
  - More energy content than other biomass;
  - Simple transportation and storage without insect or fungus threat;
  - Flavor in grilled food;
  - Burns quickly, constantly, and easy extinguishable;
  - Smokeless and sulfur free.
- **Disadvantages:**
  - Pyrolysis releases Carbon monoxide;
  - Half of the energy lost in the process of pyrolysis;
  - Efficient charcoal production methods are usually unaffordable;
  - Producers target specific wood.

Biochar is the name given to charcoal that is produced from non-wood materials such as coconut shells, bamboo, etc. The advantages of biochar or charcoal are as follows:

- The general health of the soil ecosystem is enhanced by the habitats that biochar provides for soil bacteria.
- It delivers a variety of nutrients, including potassium, which would otherwise be highly expensive to add with full-spectrum chemical fertilizers.

- With the use of charcoal, the soil is able to store more nutrients and experience less nutrient losses due to water. Due to their very low natural fertility, sandy soils are especially affected by this.
- Compost frequently loses the carbon it adds to the soil fairly fast. But charcoal's carbon may linger in the soil for a very long time, and it also helps the soil hold onto the carbon that compost, plant roots, and other biomass add.
- The soil's ability to store moisture is improved by charcoal. This strengthens the soil's structure and lessens soil cracking. It can lower the expense of irrigating the fields with regular use (adopted from: http://www.ecopro.in/outreach-biochar.html).

13.4.4 CERTAIN OF INDUSTRIAL USES OF CHARCOAL

Use of charcoal by different industries are briefly summarized in Table 13.1.

TABLE 13.1 Major Uses of Charcoal

Major Uses of Charcoal	Type of Industries
Outdoor grilling, backyards, and on camping trips, everyday cooking.	Cooking fuel
Smelting iron.	Heavy metal
Charcoal filters, removes coloring material from solutions, glue, and gelatin industries, avoid fluoride, sugar refining.	Purification/filtration
Art charcoal pencils, sticks.	Art
To improve soils, compost pile to increase its carbon content.	Horticulture
Dietary supplement for gastric problems, charcoal biscuits, teeth whitening, treat alcohol poisoning, anti-aging, reduces high cholesterol.	Pharmaceuticals
Hookah.	Smoking
Syngas ($CO + H_2 + CO_2 + N_2$ mixtures)/wood gas/producer gas/ suction gas.	Automotive, chemical, fuel
Rubber vulcanization, number of dyes containing sulfur.	Carbon di-sulfide
Charcoal battery	Automotive

13.4.5 EMPLOYMENT OPPORTUNITY IN CHARCOAL PRODUCTION

Up to 6% to 10% of the rural inhabitants in India, Pakistan, the Philippines, Nepal, and Burma are employed by the industry that produces charcoal. By giving residents of rural areas part-time or alternative work opportunities, biomass charcoal briquette production technology can not only help supplement the fuel needs in those areas but also serve as an example for effective biomass waste management and utilization (Antal & Gronli, 2003).

13.4.6 RECENT DEVELOPMENTS IN CHARCOAL PRODUCTION

Our review of the literature offers the following remarks:

1. The highest yield of carbon that may be generated from a specific biomass feedstock is meaningfully constrained by thermochemical equilibrium calculations, which are known as "theoretical yields." The measured fixed-carbon yield of charcoal derived from a feedstock can be compared to the calculated theoretical yield of carbon for the feedstock. This comparison serves as a gauge for the carbonization process' effectiveness.
2. Increased pressure during carbonization increases the output of charcoal from biomass particles. The specific volume of the tarry pyrolytic vapors decreases under pressure; as a result, their residence duration inside the particle and in its immediate surroundings increases. Moreover, there is a larger partial pressure of the tarry vapor inside and around the particle. These effects are enhanced when there is less gas flow across the particle bed. The tarry vapor is made up of a complex reactive mixture of organic substances, including lignin moieties that are extremely unstable at high temperatures, anhydrosugars, and associated oligomers, sugar fragments, and vapor-phase sugars. On top of the charcoal, these tarry vapors quickly decompose, creating secondary charcoal and a gas predominantly made of water, carbon dioxide, methane, hydrogen, and carbon monoxide.
3. The yield of charcoal from the dry portion of the feedstock can change depending on whether there is water present. Naturally, more feedstock must be consumed during carbonization in an air-blown kiln in order to provide heat to dry the wet feed before pyrolysis can start. On the other hand, the yield of charcoal produced by an externally heated retort increases at high pressure as the moisture content of the feed increases. There is some

evidence to support the idea that moisture serves as a catalyst in the production of charcoal.

4. The amount of charcoal produced by the primary and secondary carbonization reactions determines how much heat is released during pyrolysis. The chemistry of pyrolysis is significantly exothermic when the output of charcoal is high. If the feedstock is dry and the yield is high, the exothermic evolution of heat can elevate the temperature of incoming feedstock sufficiently to initiate the carbonization reactions.
5. Natural minerals found in biomass, such as potassium, act as "built-in" catalysts for the creation of charcoal, oxygen chemisorption on charcoal, and charcoal combustion.
6. Charcoal is a renewable, highly reactive, truly "clean coal." Relative to fossil fuels, charcoal contains virtually no sulfur or mercury and very little nitrogen and ash. Carbonized charcoal has a high surface area and can conduct electricity almost as well as a metal. Charcoal is the carbon of choice for applications ranging from potting media to the fabrication of pure silicon for the semiconductor industry because of its unique mix of qualities.

13.5 WOODEN AND BAMBOO DECORATES PREPARATION

Forest waste is also one of the major parts of Indian handicrafts, and various beautiful and esthetic things are crafted out of them. The art and craft of wood and bamboo is a source of employment for thousands of individuals in India. Most of the offcuts of wood and bamboo, such as branches, roots, and stumps left from the forests, plantations, and/or sawmills, are considered as forest waste materials. These may be utilized to create value-added decorative products, turning waste into wealth. Important value-added products of handicrafts like human figures, statues, hornbills, kitchen utilities, table lamps, frames, boxes, sofas, chairs, teapots, table tops, etc., are prepared from waste wood and bamboo resources.

Handicrafts are prepared using bamboo, wooden resources, non-wood timber forest resources, and other inputs with the help of small hand tools. Forest waste resources mainly used are branches, stumps, wood and bamboo chips, sawdust, wooden blocks, etc. Non-wood timber forest resources like climbers, inflorescence, leaves, fruits, kernels, roots, etc., are also used in the preparation of handicrafts. Apart from raw materials, other inputs are needed such as paints, varnishes, nails, hooks, sandpapers of different grits, Fevicol, and clay powder required for the preparation of bamboo/wood handicrafts. In addition to this, small tools like a hammer, circular saw, hand saw, gas burner, cutter, planer, wood carving tools, etc., are also required, which cost less than Rs. 5,000/–. Wooden and bamboo decorative entrepreneurship is one such program of the project that has yielded excellent results. The preparation of handicrafts does not require sophisticated tools and equipment. It is rather an art that requires skills that can be honed with time and practice. There is ample opportunity for entrepreneurship in the field of handicrafts. The handicraft market is characterized by novelty, and to flourish in a demand-based environment, it becomes imperative to manufacture innovative products. The CAAST sub-project at NAU, Navsari, worked on one of the innovative ideas of this project on the utilization and value addition of forest waste resources. Under this, wooden decorative items have been prepared using some forest waste resources. This activity also finds a place in the Vocal for Local concept of the Government of India, wherein local products are promoted at the global level through various government interventions.

13.5.1 SUCCESS STORY OF PREPARATION VALUE-ADDED PRODUCTS FROM WASTE BAMBOO AND WOOD

Under the banner of ICAR-NAHEP-CAAST sub-project entitled "Establishment of Secondary Agriculture unit for skill development in students and farmers."

13.5.1.1 PREPARATION OF BAMBOO CANDLE

- **Description:** This Bamboo candle (Figure 13.2(a)) is prepared by collecting waste wood and bamboo internodes coated with eco-friendly varnish and painted with the help of trained students, faculties, and artisans.
- **Raw Materials Used:** Clay or chalk powder, wooden block or sample, sandpaper, adhesive, varnish, bamboo internodes.

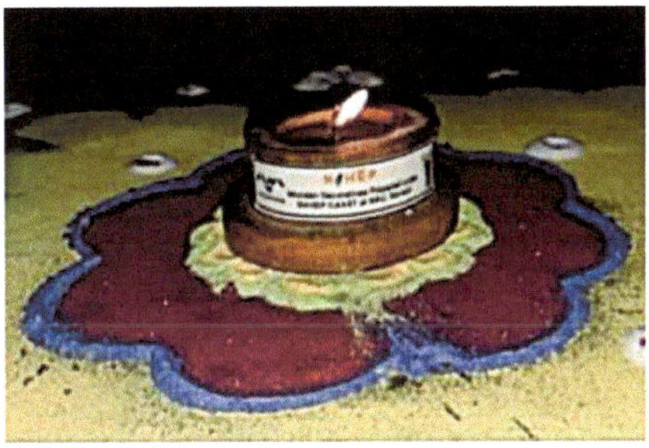

(a) Bamboo candle

(b) Eco-friendly fancy Rakhi

(c) Wooden bamboo tree

(d) Decorative table

(e) Wooden mirror.

FIGURE 13.2 Value added products from the waste material.

- **Procedure:** First, collected wood is debarked and dried. After that, sanding is carried out to smooth the wooden surface and china clay is applied to fill the pores on wood followed by repeated sanding. The wooden block is cut into a circular shape. After that, the internode of bamboo is cross-cut at 2 inches above and another cut just beneath the node. Then bamboo is fixed with the help of wood glue to the circular shaped cut wooden piece of thickness 5 mm and diameter 3 inches. Later, sanding is done to the bamboo piece. Then a coat of eco-friendly wood preservative is applied to prevent the entire structure from decay or insect attack. Apply a coat of varnish on it and leave it to dry. Measure the wax. Make sure you have a clean, flat place to work on before you start creating candles. Melt the wax along with a little Crayons color stick. Attach a wick and secure it during pouring the wax. Add more wax if required. Cut the wick. Different color Crayons sticks may be used to impart color to the candles. The candle will burn about 5–6 hours when lit continuously.
- **Product Dimension:** Length: $2^{1/2}$″ Width: 3″: The dimensions of the final product may vary with the availability of raw materials, and its cost is INR 15.00.
- **Utility:** This is one of the best eye-catching and creative products prepared from waste wood and bamboo materials. It can be used as a decorative bamboo candle on table tops in offices, for home decoration, and as gift items for various functions. Many visitors appreciated this product, and many units of this product were sold during the Deepavali festival (Sankanur et al., 2022a).

13.5.1.2 PREPARATION OF ECO-FRIENDLY FANCY RAKHI

- **Description:** This eco-friendly fancy Rakhi is prepared using wooden pieces (Figure 13.2(b)), a few NTFPs, and varnish with the help of trained students, faculties, and artisans.
- **Raw Materials Used:** Wooden pieces, seeds, beads, thread, bamboo pieces, coconut holder, sandpaper, and varnish.
- **Procedure:** Firstly, collected solid branch wood is debarked and dried. Then it is converted into various shapes of 5 mm thickness. After that, sanding is carried out to smoothen the wood and bamboo piece surfaces, and varnish is applied. Later, a design is made on the wooden/bamboo

pieces if required, and holes are drilled using a drilling machine to tie thread along with a combination of colorful beads.

- **Product Dimensions:** Length: 6″ including thread, Width: 1 to $1^{1/2}$." The dimensions of the final product may vary with the availability of raw materials, and its cost per piece is INR 12.00.
- **Utility/Purpose:** This is a creative product prepared from waste wood, NTFPs, and other raw materials. It can be used as an attractive gift item during *Raksha Bandhan*, and many visitors appreciated this product. Many such products were sold during the *Raksha Bandhan* festival (Sankanur et al., 2022b).

13.5.1.3 PREPARATION OF WOODEN TREE SCULPTURE

- **Description:** This wooden-bamboo tree is prepared by collecting the waste branches of trees and bamboo internodes coated with eco-friendly varnish and paint (Figure 13.2(c)) under the banner of ICAR-NAHEP-CAAST sub-project entitled "Establishment of Secondary Agriculture unit for skill development in students and farmers" with the help of trained students, faculties, and artisans.
- **Raw Materials Used:** China clay/chalk powder, wooden block, sandpaper, nails, adhesive, varnish, bamboo internodes, stump of wooden branches, paint, and paintbrush.
- **Procedure:** Firstly, collected branch wood stump is debarked and dried. After sanding to smoothen the wooden surface, China clay is applied to fill the pores on wood, followed by repeated sanding. Holes are made randomly at an angle of 45° on the wooden stump by a drilling machine. A hole of suitable diameter is made in the wooden block to hold the stump properly. After this, small size leaf structures prepared from bamboo internode are pasted on the wooden branched structure. An eco-friendly wood preservative coat is applied to prevent decay or insect attack on the entire structure. Black and green paints are applied to color the stump and leaves, respectively. Finally, after drying the wooden stump, it is fixed on the wooden block with the help of nails and gaps are filled with a mixture of sawdust and adhesive.
- **Product Dimensions:** Length: 15″ Width: 10″ Perimeter: 10″ (Approximately). The dimensions of the final product may vary with availability of raw materials and its cost is about INR 170.00.

- **Utility:** This is one of the best eye-catching and creative products prepared from waste wood and bamboo materials. It can be used as a decorative article on table tops in offices, home decor, and as gift items for various functions. Many visitors have appreciated this product, and many samples have been sold as decorative articles in various offices (Sankanur et al., 2022c).

13.5.1.4 PREPARATION OF DECORATIVE TABLE LAMP

- **Description:** This decorative table lamp is prepared by collecting waste wood and bamboo internodes (Figure 13.2(d)) from sawmills with the help of trained students, faculty, and artisans.
- **Raw Materials Used:** Wooden block, sandpaper, nails, adhesive, varnish, bamboo internode, paint, paintbrush, electric wire, electric bulb holder, and electric bulb.
- **Procedure:** Firstly, dried rectangular wooden blocks and bamboo internodes are collected from sawmills. Then, sanding is carried out to smoothen the wooden blocks. A hole is made in the center of the wooden block with a drilling machine to fit a 15 cm long cylindrical bamboo internode of 4 cm diameter to hold an electric bulb. A cover on the top of the electric bulb holder is fitted using a 15 cm long internode of bamboo having an 8 cm diameter, and circular holes of 2 cm and 4 cm diameters are made on the internode to allow sufficient light to pass through. The base and top of the cover are decorated using bamboo sticks and rectangular and semicircular bamboo pieces. A coat of eco-friendly wood preservative is applied to protect the structure from decay and insects. Finally, the lamp is coated with varnish and black paint at desired places.
- **Product Dimensions:** Length: 15″ Width: 4″ Perimeter: 6″ (Approximately). The dimensions of the final product may vary with availability of raw materials and its production cost is INR 250.00.
- **Utility:** This is one of our attractive and popular creative products prepared from waste wood and bamboo internodes. It can be used as an eco-friendly decorative table lamp at night for home. More than 70 samples of this product have been sold to staff at various events within and outside the university. Scientists from different places visited NAU during different CAAST events and appreciated and encouraged the sale of this product in the market (Sankanur et al., 2022d).

13.5.1.5 PREPARATION OF WOODEN MIRROR

- **Description:** This Wooden mirror (Figure 13.2(e)) is prepared by collecting waste wood blocks and stems.
- **Raw Materials Used:** Wooden blocks, sandpaper, nails, adhesive, varnish, bamboo internode, paint, paintbrush, and hook for hanging on the wall with the help of trained students, faculty, and artisans.
- **Procedure:** Firstly, dried wooden blocks and bamboo internodes will be collected from construction sites and/or roadside plantations. After that, sanding will be carried out to smoothen the wooden blocks. A mirror of the desired/chosen shape will be brought from the hardware shop. Accordingly, wooden blocks will be cut into wooden pieces. Then adhesive will be applied in order to fix the mirror onto the wooden block. Furthermore, wooden stems will be cut into small disc-shaped structures in order to fix these around the mirror after sanding on the sides and top. Later, a coat of eco-friendly wood preservative will be applied to prevent the entire structure from decay or insect attack. Finally, the mirror will be coated with varnish and various paints at desired places.
- **Product Dimensions:** Length: 30″ Width: 30″ (Approximately). The dimensions of the final product may vary with the availability of raw materials, and its cost of production is INR 150.00.
- **Utility:** This is one of our attractive and well-liked artistic products prepared from waste wood, small stems, and bamboo, and it can be used as an eco-friendly decorative mirror in households (Sankanur et al., 2022).

13.6 SUCCESS STORY OF PREPARATION VALUE ADDED PRODUCTS FROM PINE NEEDLES AND GENERATION OF ELECTRICITY FROM PINE NEEDLES: WASTE TO WEALTH

Malti Devi once walked miles to gather firewood from her village in Pithoragarh, Berinag. Today, she has one less burden to bear because she was married early. In addition, she has her husband living in a different city, working for a different company than her own. In Berinag village, a green organization generates electricity without creating any environmental hazards. Coniferous pine trees lose their needles during the summer (April to June), creating a dense carpet on the forest floor. The high resin concentration in the needles can start forest fires if the smallest embers, such as a partially burned bidi, are scattered. Rajnish Jain founded the non-profit company Avani Bio-Energy after realizing that these pine needles may be used to meet the rural areas' energy demands. The organization employs local people, mostly women, to gather vast quantities of pine needles. After being cut to maximize their density, the collected needles are fed into a gasifier facility. These needles are burned by the plant under controlled circumstances, at high temperatures, and with a limited amount of oxygen.

The end product is charred biomass that is further separated into producer gas, a mixture of carbon monoxide, hydrogen, and methane gases. After being cleaned and cooled, the gas is utilized to power electricity generators. In order to meet household demands, the solid residual charcoal remains after the producer gas has been separated. In 2009, the first gasifier facility was built at Avani in Berinag. It was a little, nine-kilowatt facility that produced energy for the group, according to Jain. The first plant's construction cost around Rs 6 lakh. Soon after, a 120-kilowatt second plant was built in Pithoragarh's Chachreda hamlet with a budget of Rs 1.2 crore crowdsourced funds.

In Almora, the organization plans to construct two more 120-kilowatt plants. According to Jain, these plants will provide clean energy for 125,000 people, 200 families with cooking fuel, 200 jobs, and help 400 acres of forested land regain its ecological equilibrium. Small gasifiers are being installed right now in surrounding villages to guarantee access to electricity and offer locals stable employment. Cash payments are made to those who collect pine needles in addition to the charcoal that is made. To recuperate the running costs of the facilities, they must purchase power at a nominal cost. Uttarakhand Power Company Ltd (UPCL), inspired by their success, has a deal with them to buy electricity. To provide bioelectricity, the organization has negotiated a 20-year contract with local power firms. Also, they are aiming to collaborate with businesses in order to sell electricity and reinvest the money made into the neighborhood. Leading national bodies and media sources have recognized the organization's efforts. Local residents are being forced to relocate and hunt for work in major cities as a result of declining ecosystem services and a lack of job prospects. They may be able to find jobs in their villages as a result of initiatives like these. Using environmental trash, such as pine needles, reduces its destructiveness and aids in the management of forest fires. According to environmentalist Suresh Bhai, the charcoal created discourages communities from cutting down trees for firewood (India's The Times, 2014: This story was published on December 14, 2014, by Seema Sharma).

13.7 BIOFUEL PRODUCTION

India is the ninth-largest economy in the world, with real GDP growth of 8.7% in the last five years. To deliver a sustained growth rate of 9% and to meet the energy needs of the country, India needs to increase its primary energy supply by 3 to 4 times. Presently, India is the fourth-largest consumer of crude oil and natural gas in the world, and out of the total consumption of petroleum products, more than 75% of the requirement is met through imports for which a maximum (37%) share of foreign currency reserve is being paid. The transportation sector is the major consumer of petroleum energy (70%) and a contributor to air pollution. Biofuels are being promoted widely as a solution to rising fuel prices, growing energy demands, and cutting greenhouse gases, but the supply of raw materials is still a question. To address these issues, the use of biofuel or at least using them blended with fossil fuels is recommended. Looking to the future demand for biofuel (increasing 7.5% annually) and wasteland availability (43.15 M ha) in India, agroforestry may be a potential land use system to integrate perennial woody biofuel crops with traditional agriculture crops to satisfy energy demands without sacrificing food and nutritional security issues.

Biofuels may offer an effective solution for both energy security and the environmental crisis. Until alternative fuels based on domestically produced renewable feedstocks are created to replace or supplement petroleum-based fuels, India's energy security will remain at risk. Biofuels are fuels derived from various biological materials like plants and animals. All the resources are obtained from either plants or animals, and on most occasions, agricultural waste and wood are also used. The use of available non-edible oil seeds is the ideal and replenishing source, including bioethanol from sugar and cellulose, and biodiesel from plant and animal fat and algae. Biodiesel is less harmful than table salt, produces less skin irritation than soap, and degrades faster than sugar (McMillen et al., 2005).

13.7.1 WHY TO PROMOTE BIOFUELS?

- Renewable and sustainable source of energy;
- Energy security and decreased dependence on oil imports;
- Near carbon neutral and low greenhouse gas emission;
- Employment generation in rural sector.

The comparative GHG emission of biodiesel is very low compared to other fuels. This indicates that they are safer for human health and the environment.

13.7.2 BIOFUELS CLASSIFICATION

1. **First-Generation Biofuel:** Made from sugar, starch, and vegetable oil.
2. **Second-Generation Biofuel:** Made from non-edible plant materials.
3. **Third-Generation Biofuel:** Made from algae and microbes.

13.7.3 TYPES OF BIOFUELS

13.7.3.1 BIOETHANOL

Bioethanol is an alcohol created through fermentation, primarily from carbohydrates found in crops that produce sugar or starch. Cellulosic biomass derived from non-food sources such as trees and grasses is also used as feedstock for bioethanol production. Sugarcane, sugar beet, sweet sorghum, maize, straw, grass, and other biomass serve as raw material sources for bioethanol production.

13.7.3.2 BIODIESEL

Biodiesel is a mono alkyl ester made from oils (plant or animal) that is similar to diesel oil made from petroleum. Surprisingly, in 1897, Rudolf Diesel demonstrated his first diesel engine running on peanut oil.

The most extensively used method for producing biodiesel is called transesterification, in which triglycerides react with alcohol (often methanol) in the presence of chemical (acid or alkali) or biological (enzyme) catalysts to create mono alkyl esters. There have been some commercialized batch alkali-catalyzed procedures. Bioethanol and biodiesel are becoming more popular as alternative fuels to petrol and diesel, respectively. It is possible to make bioethanol from a variety of leftovers and wastes in addition to the way it is typically made from cane molasses by yeast fermentation. The cost-effective synthesis of bioethanol may result from effective process optimization and integration that combines production and recovery operations. Switchgrass is mostly grown in the United States in extreme climatic circumstances and has a high concentration of cellulose and hemicelluloses, which has sparked considerable interest in the bioethanol industry. Biodiesel, on the other hand, is generally produced from vegetable oils.

Biofuels are widely promoted as a solution to rising fuel prices, growing energy demands, and reducing greenhouse gases, but the supply of raw materials remains uncertain. In India, demand for diesel is five times more than that for gasoline. But while the ethanol industry is mature, the biodiesel industry is still in its infancy. Indian biodiesel plant construction efforts by the private sector have stalled due to a lack of reliable feedstock supplies (Table 13.2). To achieve 20% mandatory blending targets by 2017 as advocated in National Biofuel Policy, 2009, there will be a need for nearly 19 MT of biodiesel, for which nearly 20 M ha area would be required under biofuel plantations against the 0.5 M ha at present (Raju et al., 2012).

13.7.3.2.1 Production of Biodiesel From Agricultural Wastes

Biodiesel is mainly manufactured from vegetable oils or animal fats. Rapeseed oil, rice bran oil, soybean oil, palm oil, and sunflower oil are just a few of the oils that are employed. The choice of vegetable oil relies on how readily it is accessible in the nation where biodiesel is made. Nevertheless, waste-derived bioethanol can be used to transesterify vegetable oils and create biodiesel from monoethyl esters of fatty acids. Certain residues can be successfully used as carbon sources for the production of single-cell oil, which lowers the cost of fermentation. Whey concentrates and tomato waste hydrolysate, both of which contain more than 1 g/l total organic nitrogen and yield 14.3% and 39.6% lipid, respectively, can be utilized to create gamma-linolenic acid. As a result of the 14.1% and 11.5% gamma-linolenic acid yielded from these wastes, whey concentrate and tomato waste hydrolysate are two excellent sources of raw materials for the synthesis of biodiesel (Fakas et al., 2008) (Table 13.2).

TABLE 13.2 Projected Demand and Area for Biodiesel at Different Blending Rates

Year	5% Blending		10% Blending		20% Blending	
	Biodiesel Demand (Mt)	Area (Mha)	Biodiesel Demand (Mt)	Area (Mha)	Biodiesel Demand (Mt)	Area (Mha)
2011–2012	3.21	3.42	6.42	6.85	12.84	13.69
2016–2017	4.61	4.91	9.21	9.83	18.43	19.66
2020–2021	6.15	6.56	12.31	13.13	24.61	26.25

Source: Adapted from: Raju et al. (2012).

India is home to more than 200 tree species that naturally grow in a variety of agroclimatic and agroecological environments and produce oil from their seeds in the range of 25–70%. In Table 13.3, a few of the promising species are included.

TABLE 13.3 Promising Biodiesel Trees Species

Species	Seed Yield (kg/tree)	Oil Percentage
Azadirachta indica	20–35	30–40
Calophyllum inophyllum	25–50	45–70
Garcinia indica	30–130	40–52
Jatropha curcas	0.5–2.0	30–51
Madhuca longifolia	30–200	30–35
Mimusops elengi	40–60	25–30
Mesua farrea	40–100	35–40
Pongamia pinnata	30–150	27–42
Simarouba glauca	10–20	60–75
Scleichera oleosa	40–60	30–35
Simmondsia chinensis	2–5	30–50
Sterculia gattata	30–40	30–45

Source: Adapted from: Dhyani et al. (2015).

13.8.3 PROMINENT FOREST-BASED TREE BORNE OIL SEED (TBOS)

There are several promising tree species that can be used for biodiesel production, each with its own advantages and disadvantages. A few of the most notable ones are given in Tables 13.4–13.9.

TABLE 13.4 Karanj

Common Name	Karanj and Karanja
Scientific Name	*Milletti pinnata* (L) Panigrhahi (Syn: Pongamia pinnata (L) Pierre)
Family	Fabaceae (Leguminaceae)
Habit	Medium sized tree
Flowering Season	March to May
Harvest Season	January to March
Yield starts at	5 years
Seed Yield	15–40 kg per tree
Kernel Oil	27–39%
Uses	Biodiesel, cake as manure green manure soap production, lather industry and biogas production.

Source: Adapted from: Dhyani et al. (2015).

TABLE 13.5 Mahuda

Common Name	Mahuda and Mahua
Scientific Name	*Madhuca longifolia* (J. Kaonig) J.F.Macbr
Family	Sapotaceae
Habit	Tree
Flowering Season	March to April
Harvest Season	June to August
Yield Starts at	8–10 years
Seed Yield	10–40 kg per tree
Kernel Oil	33–43%
Uses	Biodiesel, cake as manure agricultural equipments, soap industry and biogas production.

Source: Adapted from: Dhyani et al. (2015).

TABLE 13.6 Calophyllum

Common Name	Suttan Champa and Undi
Scientific Name	*Calophyllum inophyllum* L.
Family	Clusiaceae
Habit	Medium to large trees grown along the seacoast (littoral tree)
Flowering Season	March to April
Harvest Season	June to August
Yield Starts at	5–8 years
Seed Yield	30–80 kg per tree
Kernel Oil	30 to 80%
Uses	Biodiesel, Seed oil for lighting lamp, as varnish wood for construction of boats and medicine

Source: Adapted from: Dhyani et al. (2015).

TABLE 13.7 Simaroba

Common Name	Simaroba and Paradise Tree
Scientific Name	*Simarouba glauca* DC
Family	Simaroubaceae
Habit	Medium sized trees
Flowering Season	December to February
Harvest Season	February to April
Yield Starts at	4 years
Seed Yield	10–25 kg per tree
Kernel Oil	60–70%
Uses	Biodiesel, medicine, cake as manure, insecticide, biogas production and edible oil

Source: Adapted from: Dhyani et al. (2015).

TABLE 13.8 Neem

Common Name	Neem
Scientific Name	*Azadirachta indica* A Juss.
Family	Meliaceae
Habit	Tree
Flowering Season	March to May
Harvest Season	June to November
Yield Starts at	8 years
Seed Yield	20–40 kg per tree
Kernel Oil	30%
Uses	Biodiesel, medicine, cake as manure, agricultural equipment, insecticide, soap industry and biogas production.

Source: Adapted from: Dhyani et al. (2015).

TABLE 13.9 Kokum

Common Name	Kokum
Scientific Name	*Garcinia indica* Choisy
Family	Clusiaceae
Habit	Tree
Flowering Season	October to February
Harvest Season	March to June
Yield Starts at	7–8 years
Seed Yield	30–173 kg per tree
Kernel Oil	22–25%
Uses	Seed oil as biofuel, fruits in culinary purpose confectionery, medicines, and cosmetics.

Source: Adapted from: Dhyani et al. (2015).

13.8.4 CONSTRAINTS FOR BIOFUEL PROGRAM IN INDIA

- Lack of confidence in farmers due to profitability;
- Lack of availability of improved propagation material for high productivity;
- No subsidy and benefits for the farmers proposed by government;
- No exemption on state taxes and subsidy on production and marketing of biofuels.

13.8.5 FUTURE CHALLENGES

1. Non-edible oilseeds, mostly of forest origin, are potential candidates for biofuel, but these crops are growing wild. Therefore, systematic efforts for their domestication should be made.
2. Systematic efforts towards tree improvement programs and propagation techniques should be undertaken so that high-yielding genotypes can be developed for further plantation programs.
3. The negative and positive interactions in biofuel plant-based agroforestry systems should be evaluated to develop a sound agroforestry system suitable for different geographic areas.
4. A sound seed collection mechanism from fields as well as the wild should be developed by involving JFM, *Panchayat Samiti*, SHG, NGOs, and Cooperatives at the village level.
5. At the village and community levels, small to medium capacity biodiesel units must be developed and commissioned to ensure accessibility to energy security.

13.9 ESSENTIAL OIL PRODUCTION

Essential oils are volatile oils having a strong aromatic odor and a pleasant taste that evaporate quickly when they come into contact with air. They are also referred to as aetherolea, volatile oils, and ethereal oils. Essential oils possess the characteristic fragrance of the plant from which they are derived (Anonymous, 2014). The physiological importance of essential oils in the plant body is not completely known. It is, however, understood that they are byproducts of metabolic processes rather than reserve foods. A pleasant odor and aroma which they impart attract insects and animals for pollination or the dispersal of the fruits and seeds, while unpleasant odors ward off enemies (Krishnamurthy, 1993).

Several consumer products, including detergents, soaps, cosmetics, pharmaceuticals, perfumes, confectionery products, food, soft drinks, distilled alcoholic beverages, and insecticides, contain essential oils. Essential oils are not only confined to natural products, but nowadays more pleasant and long-lasting synthetic perfumes and essences are prepared. It is often said that synthetic essential oils are replacing the natural ones. However, it is not possible to say that synthetic oils will replace the natural ones in general. In some cases, synthetic ones are inferior to the natural ones because small traces of highly odorous substances (esters, etc.) occur in the natural ones, without which the products are deficient in quality. The synthesis of compounds artificially is highly complex and expensive and may not compete with natural products. However, in the current era where science and technology are progressing so rapidly, it is difficult to foresee the future position. But according to the current trend, there is still a great demand for natural oil, and global production and consumption of essential oils and perfumes are rapidly expanding.

13.9.1 EXTRACTION METHODS OF ESSENTIAL OILS

The quality and production of essential oils are greatly improved by the extraction processes. Traditional techniques for processing essential oils are still widely utilized around the world and are of great significance. Essential oils are extracted from the plant in different ways depending on the nature of the plant and its oil. There are principally three methods used widely for extraction of essential oils, namely distillation, expression, and extraction by solvents.

13.9.2 UTILIZATION

Essential oils are utilized for various purposes. They are utilized in perfumes, cosmetics, soaps, as flavoring agents or essences in culinary preparations, chewing gum, tobacco, toothpaste, liquors, and beverages, and as disinfectants. Some have medicinal properties and are used therapeutically (Eucalyptus oil), as antiseptics (clove oil), antispasmodics (oils of Ginger, Lemon, Peppermint, Fennel, etc.), and in aromatherapy (Cinnamon leaf oil). However, others are used as solvents in the varnish and paint industries like turpentine oil. This oil is also used as a starting material in the manufacturing of synthetic camphor.

13.9.3 PROMINENT FOREST-BASED ESSENTIAL OIL PRODUCING PLANTS

Forests hold a vast collection of fragrant plants that yield a diverse range of essential oils. These oils are potent liquids comprising volatile aromatic compounds that encapsulate the plant's fragrance and essence. A few notable plants found in forests that are known for their production of essential oils are given in Tables 13.10–13.19.

TABLE 13.10 Peppermint

Common Name	Peppermint
Scientific Name	*Mentha balsamea* Wild.
Habit	Herbaceous perennial plant
Gujarati Name	Fudino
Family	Lamiaceae
Part Used	Whole plant above ground (just before flowering)
Plant is Native to	Various countries like Eurasia, North America, Southern Africa and Australia
Commonly Grown in	Europe, Africa, Asia, Australia, and North America
Major Producers	USA, France, Brazil, and India
Chief Importing Countries are	Germany
Oil Price per kg (INR)	2200/-
Uses	• Great mood enhancer;
	• Calms mind, bad breath;
	• Treats cold, flu, muscle joint plains.

Source: Adapted from: Babu et al. (2016); and Rohini & Smitha (2021).

TABLE 13.11 Basil

Common Name	Basil
Scientific Name	*Ocimum basilicum*
Habit	Herbaceous annual plant
Gujarati Name	Damro
Family	Lamiaceae
Part Used	Leaves and the flowering tops
Plant is Native to	Tropical Regions
Commonly Grown in	India, Pakistan, Iran, and Thailand
Major Producers	France, Egypt, Indonesia, US, Greece, and Israel
Chief Importing Countries are	China, Hong Kong, USA, and Germany
Oil Price per kg (INR)	950/-
	• Promote mental health;
Uses	• Improves breathing;
	• Promotes great sleep.

Source: Adapted from: Babu et al. (2016); and Rohini & Smitha (2021).

TABLE 13.12 Lemon Grass

Common Name	Lemon grass
Scientific Name	*Cymbopogon citratus*
Habit	Perennial grass
Gujarati Name	Lemongrass
Family	Poaceae
Part Used	Leaves
Plant is Native to	Combodia, India, and Sri Lanka
Commonly Grown in	Vietnam, Laos, Burma, and Thailand
Major Producers	India
Chief Importing Countries are	US, China, and Germany
Oil Price per kg (INR)	1,750/–
Uses	• Will purify skin, enhances skin tone; • Fever, digestion, anti-bacterial; • Fungal and antiseptic.

Source: Adapted from: Babu et al. (2016); and Rohini & Smitha (2021).

TABLE 13.13 Ylangylang

Common Name	Ylangylang
Scientific Name	*Cananga odorata*
Habit	Tree
Family	Annonaceae
Part Used	Flowers
Plant is Native to	India to Papua New Guinea and to Queensland, Australia (Temperate climate)
Commonly Grown in	Micronesia, Madagascar, Melanesia, the Comoros islands, and Polynesia
Major Producers	Comoros
Chief Importing Countries are	France
Oil Price per kg (INR)	2,250/–
Uses	• Strengthen hairs; • Keep skin imperfections at bay; • Act as a natural room freshener.

Source: Adapted from: Babu et al. (2016); and Rohini & Smitha (2021).

TABLE 13.14 Eucalyptus

Common Name	Eucalyptus
Scientific Name	*Eucalyptus globule* sLabill.
Habit	Tree
Gujarati Name	Nilgiri
Family	Myrtaceae
Part Used	Leaves
Plant is Native to	Australia
Commonly Grown in	Most extensively planted eucalypt species in the world
Major Producers of Oil	Brazil, India
Chief Oil Importing Countries are	People's Republic of China
Oil Price per kg (INR)	875/–
Uses	• Improves the respiratory health; • Combat cold and flu; • Use as a home cleaner.

Source: Adapted from: Babu et al. (2016); and Rohini & Smitha (2021).

TABLE 13.15 Citronella

Common Name	Citronella
Scientific Name	*Cymbopogon winterianus* Jowitt
Habit	Herb
Gujarati Name	Lemongrass
Family	Poaceae
Part Used	Leaves and stems
Plant is Native to	Tropical and semitropical areas of Asia, India, Indonesia
Commonly Grown in	South and Central America
Major Producers	Taiwan, China, Indonesia, and Guatemala
Chief Importing Countries are	USA, UK, West Germany, Japan, and Hong Kong
Oil Price per kg (INR)	1,025/–
Uses	• For glowing skin, rid of toxics; • Orthopedic issues, fungal; • Antiseptic, natural dewormer; • Useful in perfumes, insect repellents, soaps, and candles industries.

Source: Adapted from: Babu et al. (2016); and Rohini & Smitha (2021).

TABLE 13.16 Cinnamon

Common Name	Cinnamon
Scientific Name	*Cinnamomum zeylanicum* Breyn
Habit	Tree
Gujarati Name	Tamal
Family	Lauraceae
Part Used	Bark and leaves
Plant is Native to	Bangladesh, India, Myanmar, and Sri Lanka
Commonly Grown in	Sri Lanka, Indonesia, and China
Major Producers	Indonesia, China, and Vietnam
Chief Importing Countries are	US, Mexico, India, and Vietnam
Oil Price per kg (INR)	2,250/-
Uses	• For good blood circulation; • Act as room freshener; • Arthritis, digestion, stress reliever.

Source: Adapted from: Babu et al. (2016); and Rohini & Smitha (2021).

TABLE 13.17 Clove

Common Name	Clove
Scientific Name	*Syzygium aromaticum*
Gujarati Name	Taj
Family	Myrtaceae
Part Used	Flower and Leaves
Plant is Native to	Molucca Island
Commonly Grown in	Indonesia
Major Producers	Indonesia
Chief Importing Countries are	India
Oil Price per kg (INR)	1,700/–to 3,000/–
Uses	• Blood circulation; • Room fresheners; • Secretion of breast milk.

Source: Adapted from: Babu et al. (2016); and Rohini & Smitha (2021).

TABLE 13.18 Davana

Common Name	Davana
Scientific Name	*Artemisia pallens*
Gujarati Name	Damnak
Family	Asteraceae
Part Used	Leaves and flowers
Plant is Native to	Southern India
Commonly Grown in	Southern India
Major Producers	India
Chief Importing Countries are	France
Price Oil per kg (INR)	2,300/–
Uses	• For all types of skin infections; • Best relaxant, antiseptic; • Treats wounds.

TABLE 13.19 Vetiver

Common Name	Vetiver
Scientific Name	*Chrysopogon zizanioides*
Gujarati Name	Khas
Family	Poaceae
Part Used	Roots
Plant is Native to	Tropical Asia
Commonly Grown in	India
Major Producers	India and Indonesia
Chief Importing Countries are	Romania, Hungary, and Poland
Price Oil per kg (INR)	30,000/–
Uses	• Treats acne, scars on skin; • Anxiety problems, anti-inflammatory; • Nerviness, sedative, and antiseptic.

Source: Adapted from: Babu et al. (2016); and Rohini & Smitha (2021).

13.10 FUEL VALUE INDEX OF WASTE BARK

Fuelwood accounts for around 90% of the total wood production in India. It is still the main source of energy in rural India, which indicates the scarcity of alternatives and creates a lot of pressure on forest resources in the countryside (Shrivastava & Saxena, 2017). The bark is one of the few renewable energy resources after wood and has negligible sulfur content when compared to coal (Corder, 1976). Furthermore, due to the increase in the price of fossil fuels and its less availability, bark will become a more important energy source in the future.

Currently, a lump of bark after removal from wood is creating a major residue problem in many wood conversion industries such as sawmills, pulp mills, and composite wood industries. The actual production of bark residue by these wood industries depends on the use of different tree species. Bark occupies about 10% to 22% of the total log volume on a dry weight basis depending on the size and type of tree species used in wood industries (FAO, 1990). This bark residue creates a serious waste disposal problem unless it can be converted into high-energy solid fuel value like charcoal or briquettes. The proximate chemical composition of bark is much more extensive than wood. Tree bark contains a higher amount of lignin, extractives, ash, and moisture and a lesser amount of polysaccharides compared to wood (Harkin & Rowe, 1971). The high amount of lignin and extractives increases the calorific value of wood, whereas the high amount of moisture and ash content decreases the calorific value. Rathod et al., 2023 studied the fuel value index of 10 commercially important tree barks and revealed that out of 10 tree species, the bark of *Casuarina equisetifolia* showed the highest fuel value index (FVI). However, *Acacia nilotica, Adina cordifolia, Acacia auriculiformis,* and *Mangifera indica* showed moderately high FVI. They concluded that the bark of these five species can be further utilized for making briquettes.

13.11 CONCLUSION

Agarbattis have huge potential as a manufacturing business because their demand is at an all-time high and increases even more during festivals. India needs to undergo more R&D work for the formulation of quality agarbatti from natural products and upgrade skills, tools, and appliances involved in this industry. It is seen that no patent has been made so far for any brand of agarbatti product. Additionally, producers have a monopoly on products with a formula or a specific composition. Government initiatives and policies are required to bring the industry out of this monopoly. On the other hand, charcoal is a renewable and highly reactive form of "clean coal." It contains lower levels of nitrogen, ash, and sulfur compared to fossil fuels. Carbonized charcoal can conduct electricity nearly as well as metal, and it can have a high surface area. This extraordinary combination of properties makes charcoal the carbon of choice for applications ranging from potting media to the production of pure silicon for the semiconductor industry. The waste bark of commercially important trees being used in furniture industries cannot be ignored, and it can be utilized for briquette making after evaluating its fuel value index because bark contains more ash content than wood. Off-cut materials of wood and bamboo from sawmills can also be converted into novel value-added products. Forest-based TBOs may be a viable option for biofuel supplies in sufficient volume to meet current and future demands as well as generate green and clean energy. Essential oils are extracted from a lot of plant material to produce a small amount of oil and they also have a variety of uses in routine life, including medicinal uses. Extraction of essential oils by distillation, expression, and solvent methods is widely used on a commercial scale. Modern extraction techniques, with technological improvements, are useful in some circumstances, such as the creation of expensive essential oils in their natural condition without affecting their thermosensitive components or the extraction of essential oils for microanalysis, but they are not yet commonly used on a commercial basis. Hence, there are a lot of scopes to convert waste materials from forest and wood industries into surplus value-added products.

ACKNOWLEDGMENT

We are grateful to funding agency the ICAR (GoI), New Delhi and World Bank for sponsoring NAHEP-CAAST sub-project at NAU, Navsari, India.

KEYWORDS

- **agarabatti**
- **forestry**
- **Parthenium plants**
- **recover energy**
- **recycle**
- **value-added products**
- **waste management**

REFERENCES

Anonymous, (2014). *Essential Oil*. Oxford English Dictionary (online, American English ed.). Retrieved from: https://web.archive.org/web/20140809030553/http:/www.oxforddictionaries.com/us/definition/american_english/essential-oil?q=essential+oil/ (accessed on 22 April 2024).

Antal, M. J., & Gronli, M., (2003). The art, science, and technology of charcoal production. *Ind. Eng. Chem. Res., 42*(8), 1619–1640. https://doi.org/10.1021/ie0207919.

Babu, N., Srivastava, S. K., Prusty, M., & Sahoo, T., (2016). Medicinal and aromatic plant production technologies: A step towards farm women prosperity (p. 42). ICAR- Central Institute for Women in Agriculture, Bhubaneswar, 751 003 Odisha, Technical Bulletin No. 28. Retrieved from: https://krishi.icar.gov.in/jspui/bitstream/123456789/12920/2/ICAR-CIWA-TB%2828%29%202017.pdf (accessed on 22 April 2024).

Bordoloi, B., & Sarmah, E., (2007). Feasibility Study for Establishing Agarbatti Manufacturing Units in North East India and Developing a Comprehensive Marketing Strategy for CBTC-BASIX Brand of Agarbatti (p. 112). A project report on post graduate program in agri-business management national institute of agricultural extension management, Hyderabad cane and bamboo technology center (CBTC), Guwahati & BASIX, Hyderabad. Retrieved from: https://www.scribd.com/document/334472262/Feasibility-Study-for-CBTC-BASIX-MANAGE-2008/ (accessed on 22 April 2024).

Corder, S. E., (1976). Properties and Uses of Bark as an Energy Source (pp. 1–21). Research paper 31 Forest Research Laboratory, School of Forestry, Forest Research Laboratory, Oregon State University, USA. Retrieved from: https://ir.library.oregonstate.edu/downloads/6682x523q?locale=en/ (accessed on 22 April 2024).

Dhyani, S. K., Vimala, D. S., & Handa, A. K., (2015). Tree Borne Oilseeds for Oil and Biofuel (p. 50). Technical Bulletin 2/2015. ICAR-CAFRI, Jhansi. Retrieved from: https://krishi.icar.gov.in/jspui/handle/123456789/4871/ (accessed on 22 April 2024).

Fakas, S., Čertik, M., Papanikolaou, S., Aggelis, G., Komaitis, M., & Galiotou-Panayotou, M., (2008). γ-Linolenic acid production by *Cunninghamella echinulata*, growing on complex organic nitrogen sources. *Bioresour. Technol.*, *99*(13), 5986–5990. https://doi.org/10.1016/j.biortech.2007.10.016.

FAO, (1990). The potential use of wood residues for energy generation. In: Energy Conservation in the Mechanical Forest Industry. Rome: Food and Agriculture Organization of the United Nations. FAO Forestry Paper 93, ISBN 92-5-102912. Retrieved from: https://www.fao.org/4/T0269E/t0269e08.htm#6.%20the%20potential%20use%20of%20wood%20residues%20for%20energy%20generation (accessed on 22 April 2024).

Harkin, J. M., & Rowe, J. W., (1971). Bark and Its Possible Uses (Vol. 60, p. 56). Research note FPL: 091, U.S. Department of Agriculture, Forest Service and Forest Product Laboratory: Madison, WI, USA. Retrieved from: https://www.fpl.fs.usda.gov/documnts/fplrn/fplrn091.pdf/ (accessed on 22 April 2024).

Hazarika, P., Dutta, N. B., Biswas, S. C., Dutta, R. C., & Jayaraj, R. S. C., (2018). Status of agarbatti industry in India with special reference to Northeast. *Int. J. Adv. Res. Biol. Sci.*, *5*(1), 173–186. http://dx.doi.org/10.22192/ijarbs.2018.05.01.024.

http://carbon.atomistry.com/carbon_disulphide.html (accessed on 22 April 2024).

http://seiaa.gujarat.gov.in Retrieved from: https://seiaa.gujarat.gov.in/Grasim_Cellulosic_EC.pdf (accessed on 22 April 2024).

http://www.ecopro.in/outreach-biochar.html (accessed on 22 April 2024).

http://www.fao.org/3/t0269e/t0269e08.htm (accessed on 22 April 2024).
http://www.fao.org/docrep/X5328E/X5328E00.htm (accessed on 22 April 2024).
http://www.gsfdcltd.co.in (accessed on 22 April 2024).
http://www.newworldencyclopedia.org/entry/Charcoal#Uses (accessed on 22 April 2024).
http://www.nordicinnovation.org (accessed on 22 April 2024).
https://carbon.atomistry.com (accessed on 22 April 2024).
https://docplayer.net/52040245-The-feasibility-of-wood-based-charcoal-in-industrial-processes.html (accessed on 22 April 2024).
https://en.wikipedia.org/wiki/Charcoal#Metallurgical_fuel (accessed on 22 April 2024).
https://en.wikipedia.org/wiki/Producergas (accessed on 22 April 2024).
https://www.newworldencyclopedia.org//entry/Charcoal#Uses (accessed on 22 April 2024).
https://www.scribd.com/document/216542816/Antal-2003 (accessed on 22 April 2024).
ICAR Report, (2020). Creating Wealth from Agricultural Waste (p. 172). Indian Council of Agricultural Research, New Delhi. Retrieved from: https://icar.org.in/sites/default/files/Circulars/Creating-Wealth-From-Agricultural-Waste.pdf (accessed on 22 April 2024).
Lange, S.L.; Senekane, M.F.; Naicker, N. Understanding the Socio-Demographic Profile of Waste Re-Users in a Suburban Setting in South Africa. Resources 2022, 11, 45. https://doi.org/10.3390/resources11050045
Nuthan, D., Sankanur, M., Chandrashekara, K. M., Sunil, K. P., & Shashidhar, R., (2009). *Exposure Visit of Selected User Groups* (pp. 1–45). Village Forest Committee members, SHG members and Farmers to witness value addition to Bamboo and Agarbathi making in Adilabad District, Andhra Pradesh, RC, NAEB, Bangalore.
Pandey, S., Chaudhary, P., & Vadivelu, G. A., (2017). *The Report on the Feasibility of Wood Based Charcoal in Industrial Processes* (p. 77). by submitted by Indian Institute of Forest Management, Bhopal. Retrieved from: https://iifm.ac.in/uploads/files/REPORTS_of_2017_-_18.pdf (accessed on 22 April 2024).
Raju, S. S., Parappurathu, S., Chand, R., Joshi, P. K., Kumar, P., & Msangi, S., (2012). *Biofuels in India: Potential, Policy, and Emerging Paradigms* (p. 27). Policy Paper – National Centre for Agricultural Economics and Policy Research. Retrieved from: https://krishi.icar.gov.in/jspui/bitstream/123456789/733/1/pp27.pdf (accessed on 22 April 2024).
Ramana, P., Sankanur, M., & Patil, S. K., (2010). Prospects of agarbathi industry: A case study from Sirsi Area of Uttarakannada District, Karnataka. In: *National Seminar on Processing Technologies for Value Addition of Minor Forest Produce in Tribal Areas: A Step in Rural Development* (pp. 70–72).
Rathod, K. R., Sinha, S. K., Gunaga, R. P., & Sankanur, M. S., (2023). Fuel bark quality evaluation of commercial tropical tree species: An approach to waste utilization. In: Thakur, N. S., et. al., (eds.), Proceeding of NAU-IES-IURFO Conference on Tree Based Diversified Land-use System: Augmenting Livelihood Security and Industrial Growth (pp. 225, 226). N.A.U., Navsari.
Rohini, M. R., & Smitha, G. R., (2021). Glossary on use and cultivation of common medicinal herbs, ICAR-IIHR. Extension Bulletin No. 118 (p. 27). Bengaluru. https://doi.org/10.13140/RG.2.2.33186.96965.
Sankanur, M. S., Gunaga, R. P., Ahlawat, T. R., & Ahir, B., (2022b). *Preparation of Fancy Rakhi, Pamphlet* (p. 2). Published by CAAST Secretariat, Director of Research, Navsari Agricultural University, Navsari under NAHEP-CAAST sub-project. University Publication No. 39/2021-22.

Sankanur, M. S., Gunaga, R. P., Ahlawat, T. R., & Ahir, B., (2022a). *Preparation of Bamboo Candle, Pamphlet* (p. 2). Published by CAAST Secretariat, Director of Research, Navsari Agricultural University, Navsari under NAHEP-CAAST sub-project. *Univ*ersity Publication No. 38/2021-22.

Sankanur, M. S., Gunaga, R. P., Ahlawat, T. R., & Ahir, B., (2022c). *Preparation of Tree Sculpture, Pamphlet* (p. 2). Published by CAAST Secretariat, Director of Research, Navsari Agricultural University, Navsari under NAHEP-CAAST sub-project. University Publication No. 41/2021-22.

Sankanur, M. S., Gunaga, R. P., Ahlawat, T. R., & Ahir, B., (2022d). *Preparation of Decorative Table Lamp, Pamphlet* (p. 2). Published by CAAST Secretariat, Director of Research, Navsari Agricultural University, Navsari under NAHEP-CAAST sub-project, University Publication No. 40/2021-22.

Sankanur, M. S., Gunaga, R. P., Ahlawat, T. R., & Ahir, B., (2022e). Preparation of fancy mirrors, Pamphlet (p. 2). Published by CAAST Secretariat, Director of Research, Navsari Agricultural University, Navsari under NAHEP-CAAST sub-project, University Publication No. 42/2021-22.

Sewa, A., (2000). *Fragrance of Hard Work: Women Agarbatti Rollers of Gujarat* (p. 10). SEWA Academy, Self Employed Women's Association (SEWA), Ahmedabad, India. Retrieved from: http://www.sewaresearch.org/pdf/researches/fragrance.pdf (accessed on 22 April 2024).

Shrivastava, S., & Saxena, A. K., (2017). Wood is Good: But, is India Doing Enough to Meet Its Present and Future Needs? (p. 45). A status report Published by Centre for Science and Environment New Delhi.

CHAPTER 14

Waste-to-Energy Generation: Perspectives and Constraints

MUDASIR YOUSSOUF,[1] AADIL YOUSUF,[2] MOHD. HUSSAIN NAIK,[3] and ZUBAIR A. MALIK[4]

[1]*Department of Environmental Sciences, Government HSS Chittergul, Anantnag, Jammu and Kashmir, India*

[2]*Department of Civil Engineering, National Institute of Technology, Patna, Bihar, India*

[3]*Department of Geography, Central University of Punjab, Bathinda, Punjab, India*

[4]*Department of Biology, Government HSS Hardutooru, Anantnag, Jammu and Kashmir, India*

ABSTRACT

The increase in global municipal solid waste (MSW) generation from the previous four decades and the expected doubling of waste generation by 2025 demand attention for its scientific and sustainable management. In view of this huge waste generation and rampant fossil fuel exploitation, Waste to Energy (WtE) management practices are a perfect solution for sustainable waste management for clean energy generation. The priorities of WtE technologies are different for both developed and developing countries due to differences in waste composition, waste segregation and collection practices, government policies regarding waste management, economical and technological issues. Whereas landfill-derived gas (LFG) and anaerobic digestion are practiced in both developed and developing

Sustainable Forest Resources Management: Issues and Implications. Jahangeer A. Bhat, Gopal Shukla, Manmohan J. Dobriyal, Sumit Chakravarty, A. Arunachalam, and Rainer W. Bussmann (Eds.)

nations, these practices are preferred in low-income countries with high ratios of organic wastes and easily available land resources. Incineration is preferred for mass burning of wastes for energy derivation and is gaining momentum, but the high moisture content in wastes and generation of air pollutants pose challenges to this method. Gasification and pyrolysis are more sophisticated waste treatment technologies that produce clean and energy-efficient fuels like char, syngas, and bio-oil. These technologies are still in the initial stages in developing nations due to their high cost, complex energy requirements, and the need for homogeneous and high calorific waste.

14.1 INTRODUCTION

The rapid increase in economic development and population growth has greatly increased the consumption of natural resources and generation of different waste streams, like municipal solid wastes, industrial, agricultural, commercial wastes, etc. (Chen, 2018). Mismanagement of wastes will impart adverse effects on local ecology, human health, as well as on the environment. Millions of tons of wastes are generated worldwide every day by population increase and intense economic activities. Lack of modern waste management training both at government and personal levels, limited availability of land resources, and unprecedented industrialization further worsens the situation for safe disposal of wastes (Korai, 2017; Varjani, 2020). Unscientific handling of wastes has led to adverse effects on air, both groundwater and surface water, and soil surfaces, also posing public health risks (Triassi et al., 2015).

Different treatment methods are used to alter the physical and chemical properties of wastes for their safe disposal. Encompassed by different waste treatment methods, waste-to-energy recovery treatment methods are gaining momentum with rising energy deficiency across the globe (Ramachandra et al., 2018). Waste-to-energy treatment methods are based on the principle of sustainable development which exploit the wastes for nutrients, organic compounds, and energy resources (Mubarok et al., 2017; Banu et al., 2018). The three basic waste-to-energy technologies include (i) thermal conversion methods (incineration, pyrolysis, and gasification); (ii) biochemical conversion (anaerobic digestion); and (iii) landfill (Tozlu et al., 2016).

While developed countries utilize different technologies and treatments for the recovery of energy, biofuels, and compost from wastes, in developing countries even the collection, segregation, and transport of wastes is still a large challenge (Guerrero et al., 2013). Waste-to-energy technologies are considered as sustainable opportunities to address the waste management challenge and the creation of decentralized energy systems in developing countries (Moya et al., 2017). Globally, 1.4 billion people lack basic electricity and 85% of this population lives in rural areas (Yadav et al., 2020). Waste-to-energy technologies can fulfill the basic energy needs (like cooking, lighting, cooling, and heating) of these rural areas. On a large scale, this energy can be used to run businesses, industries, on-grid electricity, etc. (Chakraborty et al., 2016).

Waste management process is not static, but rather more challenging and should be continuously changed according to the type of waste generated, geographical and climatic conditions, and the social and economic status of residents. Slight variation in any one of these factors demands a separate management plan, and the same strategy cannot be applied in all situations. The failure of various state-of-the-art waste management projects in developing countries is because these projects were just followed without considering the local factors (Kumar & Agrawal, 2020). In this chapter, we will briefly discuss the current status of available WTE technologies, their advantages, and shortcomings.

14.2 SOLID WASTE GENERATION, MANAGEMENT, AND ENERGY POTENTIAL

Waste quantity and composition greatly vary with income levels and urbanization of a region. High-income countries contribute most of the global waste generation ratio with higher per capita waste generation as compared to the low-income countries. Due to rapid economic growth and urbanization in low and middle-income countries, the amount of waste generation is rapidly increasing and will replace the high-income countries in coming decades (Yang et al., 2018). Moreover, low and middle-income countries generate higher quantities of organic and inert wastes and with economic prosperity the generation of inorganic wastes (paper, plastics, metals, etc.) increases (Wilson et al., 2015). MSW management is one of the costliest services that is mostly practiced in high-income countries as compared to low-income ones. In high-income countries, about 96% of MSW is collected while it is only 39% in low-income countries and in both cases the waste collection is efficient in urban areas as compared to rural areas (WBG, 2018).

Global solid waste generation is around 2.01 billion tons per year and will increase to 4 billion tons by the end of this century. Of the waste generated, around 37% of global MSW is dumped in landfills, 33% is openly dumped, 13.5% is recycled, 5.5% is composted, and the remaining 11% is treated by different WTE treatments to recover energy as shown in Figure 14.1 (Wilson et al., 2015; World Bank Group, 2018). Approximately 2 billion tons of agricultural waste, 160 million tons of forest residues, and 9.1 billion tons of industrial waste are generated annually worldwide (Millati et al., 2019).

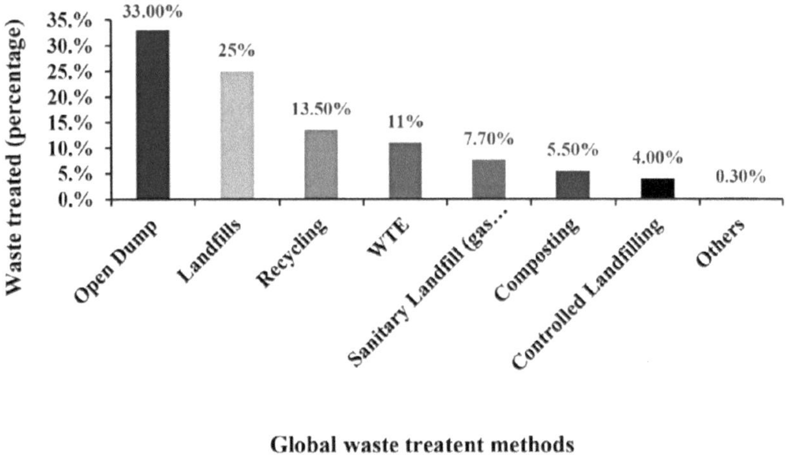

FIGURE 14.1 Status of current global waste treatment methods.

Source: Adapted from Kaza et al., 2018. http://creativecommons.org/licenses/by/3.0/igo.

Low-income countries mostly practice open dumping (93%) whereas high-income countries follow the waste treatment methods of sanitary landfilling (39%), recycling (29%), and incineration (22%).

Along with municipal solid waste (MSW), there are other considerable wastes generated, like industrial, agricultural, hazardous, medical, and e-waste (Figures 14.2 and 14.3). Industrial and agricultural waste are, respectively 18 and 4 times higher compared to MSW generation rate. Hazardous, medical, and e-waste are only a fraction of MSW generation and are treated in special facilities (Ferronato & Torretta, 2019).

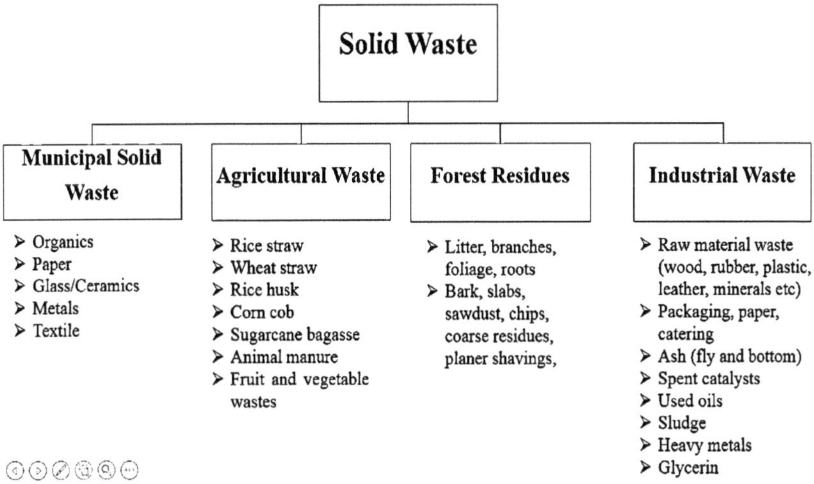

FIGURE 14.2 Common examples of solid wastes generated.

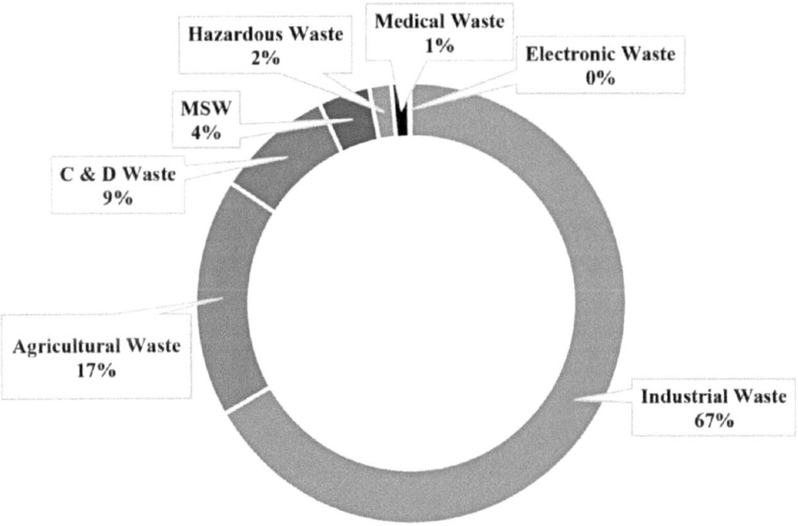

FIGURE 14.3 Global average waste generation by type.

Source: Adapted from: World Bank Report (2018).

14.3 ENERGY POTENTIAL

Different techniques are used in the conversion of carbon-rich biomass as a source of renewable energy. For sustainable management, solid waste is treated as a biomass source for energy derivation by using various waste-to-energy (WtE) methods (Cutz et al., 2016). To calculate the bioenergy potential of various waste-biomass sources at a global level, different scientists have used different methods and projected different results. Kim & Dale (2004) projected the annual bioenergy potential of global waste crops and crop residues of major crops (corn, barley, oat, rice, wheat, sorghum, and sugarcane) as bioethanol production potential. They projected that 1.6 Pg of dry wasted crops and crop residues could potentially produce 491 GL of bioethanol per year (Kim & Dale, 2004). Fischer & Schrattenholzer (2001) assumed the bioenergy potential of five biomass sources (crop residues, energy crops, wood from forests and forest residues, animal wastes, and MSW) through 2050. The assumed bioenergy potential from all these five sources in 2050 was estimated to increase between 370 and 450 exajoules (8.8 and 10.8 Gtoe) as compared to the base year, 1990, which was only 46 exajoules (Fischer & Schrattenholzer, 2001). Similarly, the biomass energy potential of bioenergy crops, agriculture, and forestry residues and wastes in 2050 was studied by Smeets et al. (2007) using the Quickscan model. The bioenergy potential of these three biomass sources was calculated to be 74 EJ yr^{-1} in the year 2050 (Smeets et al., 2007). In another study, the gross bioenergy potential of global agricultural areas was projected to be between 64 and 161 EJ per year in 2050 (Haberl et al., 2011). The bioenergy potential of global MSW was also calculated in terms of electricity generation and biogas production. In terms of electricity generation, the global bioenergy potential of MSW is 352 GW per day whereas it can be predicted that the biogas potential of the world MSW is 3.3×10^8 m^3 per day (Bölükbaş and Akıncı, 2018; Sharma & Jain, 2020).

14.4 WASTE TO ENERGY (WTE) TECHNOLOGIES

For sustainable management of waste, the world is rapidly adapting and inventing new WtE technologies because of their economic and environmental benefits. WtE technologies harness the stored energy of waste in the form of heat, combustible gases, or electricity, thus decreasing the use of fossil fuels and providing energy security (Vaish et al., 2016; Srivastava et al., 2020). With a global value of 35.1 billion U.S. dollars in 2019, the WtE market is expected to grow up to 50 billion U.S. dollars in 2027 (Statista, 2020). Countries like Japan, Scandinavia, Switzerland, USA, France, etc., are leading in WtE generation (incineration), whereas developing countries like India, Brazil, Indonesia, etc. (Figure 14.4) are slowly taking the lead due to financial constraints and poor SWM strategies (Sharma & Jain, 2020; EIA, 2020).

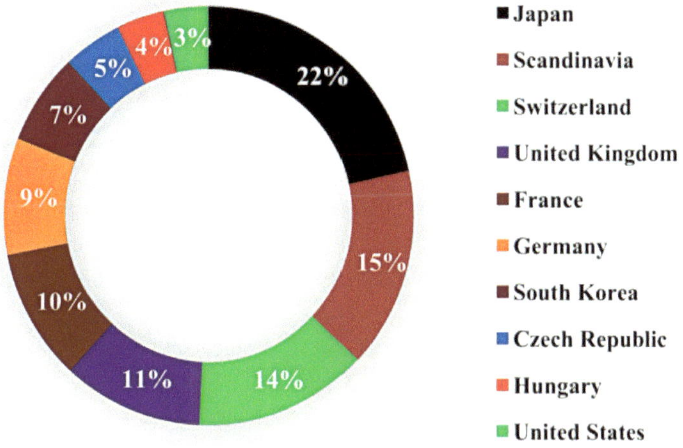

FIGURE 14.4 Percent of total MSW that is burned with energy recovery in selected countries.

The commonly practiced WtE technologies can be differentiated into three types, Thermal conversion methods (incineration), Thermochemical conversion methods (pyrolysis, gasification, plasma treatment) and Biochemical conversion methods (anaerobic digestion, landfills, fermentation as shown in Figure 14.5 (Kumar & Samadder, 2017). These techniques can be used in combinations to derive different recovery products as fuels like direct heat, gases (H_2, CH_4), liquids (Bio-oil) and solids (biochar). These technologies are not so simple and there are various technical, financial, and decisional difficulties with these technologies (Agaton et al., 2020; Khan & Kabir, 2020).

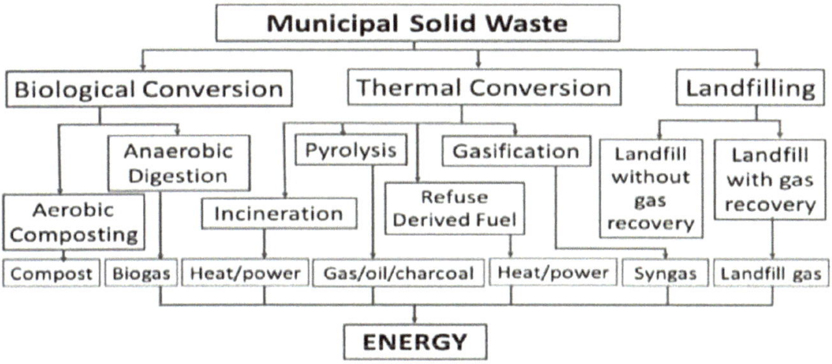

FIGURE 14.5 Different WtE techniques and their products.

Source: Adapted from: Kumar & Samadder (2017).

14.5 BIOCHEMICAL CONVERSION

Biochemical conversion processes are unique eco-friendly methods that involve the bacterial, fungal, and other microbial breakdown of waste biomass to derive the energy in various forms such as biogas, bioethanol, and bio-hydrogen. By this method variety of wastes like MSW, agricultural, and industrial wastes can be treated. Anaerobic digestion and fermentation are the commonly used biochemical methods that involve microorganisms for the conversion process (Beyene et al., 2018; Bader et al., 2020).

14.5.1 ANAEROBIC DIGESTION (AD)

Anaerobic digestion (AD) is a process that involves breaking down a substrate without oxygen to produce various products such as CH_4, CO_2, and other minor gases, along with a solid nutrient-rich leftover called digestate. This process is mostly suitable for waste materials that have a higher percentage of organic matter and moisture, such as food waste, agricultural waste, bagasse, and animal waste, among others (Beyene et al., 2020).

The AD process is completed in three steps: hydrolysis, acidogenesis, and methanogenesis. In the hydrolysis step, the substrate is broken down from a complex state into simpler and more soluble units such as glucose, amino acids, glycerol, and long-chain carboxylic acid. These readily available units are further oxidized by fermentative bacteria. Acidogenesis is considered the fastest step in anaerobic digestion, generating organic products such as organic acids, alcohols, lactates, among others. In the methanogenesis step, two groups of methanogenic bacteria produce methane from acetate or hydrogen and carbon dioxide (Liu et al., 2019; Koyama et al., 2017).

Biogas generation is carried out on different scales, from micro-digesters used for biogas generation to large-scale digesters for electricity generation and production of biomethane (Jain et al., 2019). Micro-digesters are particularly popular in developing countries and are mainly used for heating and cooking purposes. There are approximately 50 million micro-digesters in operation worldwide, with 42 million operating in China alone, producing 13 million m^3 of biogas. India has 4.9 million micro-digesters producing 2 million m^3 of biogas, and around 700,000 biogas plants are installed in the rest of Asia, Africa, and South America (IRENA, 2017; Yang et al., 2019). The global installed biogas capacity and global electricity generation from biogas are depicted in Figures 14.6 and 14.7.

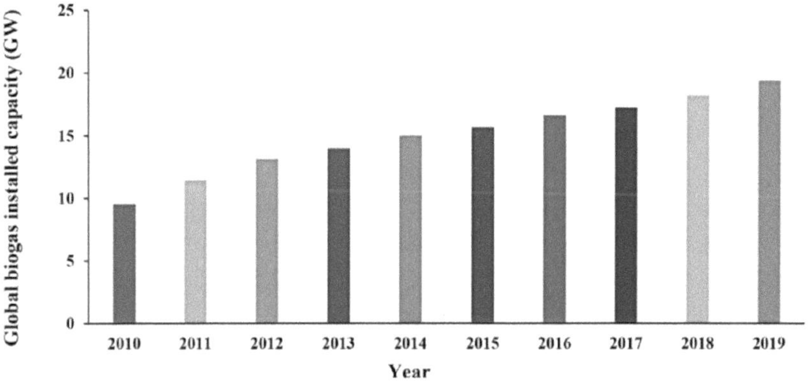

FIGURE 14.6 Total biogas installed capacity of world.

Source: Adapted from: IRENA (2020).

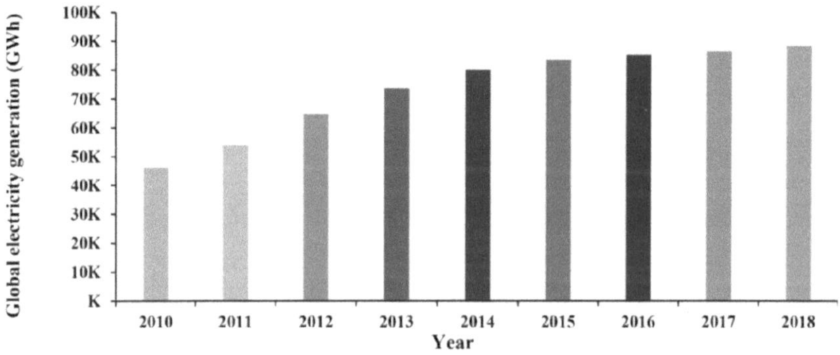

FIGURE 14.7 Total global electricity generation from biogas.

Source: Adapted from: IRENA (2020).

Electricity generation from large-scale digesters is a well-established technology worldwide. In terms of electricity generation from biogas, Europe is currently leading with an installed capacity of 11 GW from 18,202 units. On the other hand, China's total installed capacity of biogas power generation is only 0.6 GW (IEA, 2018). India has a current installed capacity of 9 MW from 389 medium-sized biogas power plants (MNRE, 2019). Among European countries, Germany is leading with 10,971 biogas plants, followed by Italy, France, Switzerland, and the UK with 1655, 742, 632, and 613 installed biogas plants, respectively. In the USA, there are 2200 biogas plants with an installed capacity of 977 MW, whereas Canada has

180 plants with an installed capacity of 196 MW (EBA, 2018; WBA, 2017; CBA, 2020).

14.5.1.1 BENEFITS

Biogas generation has a number of environmental benefits including sustainable waste management, renewable energy source, reduction in the generation of greenhouse gases and availability of biofertilizer. To prevent the ammonia poisoning from the digestion of nitrogen-rich wastes like food waste, co-digesting these wastes with high carbon wastes like agricultural residues are practiced for better biogas yield and management of different wastes in the same digester (Khan et al., 2017). AD is a unique method in treating putrescible wastes that can generate GHG's in atmosphere and leachate into groundwater (Kumar et al., 2019). Using 1 m^3 of biogas 2.04 kW of electricity could be generated also it can be used for heating, cooking, and in vehicles as fuel (Sharma et al., 2020).

14.5.1.2 CONSTRAINTS

Based on microbial activity, the productivity and maintenance of anaerobic digestion is greatly influenced by several factors like reactor types, substrate source, process variables (pH, temperature, alkalinity, C/N ratio, etc.), and microbial consortium (Raheem et al., 2016). Temperature is the main factor that affects biogas yield, which can be observed through very low biogas production in psychrophilic temperatures at higher altitudes (Jang et al., 2018). During the initial phase of digestion, the formation of volatile fatty acids, hydrogen sulfide, and ammonia lowers the pH of digesters and hinders the AD process (Chen et al., 2008).

Operating and maintaining costs of biogas-based plants are very high when compared to other renewable energy sources like solar, hydro, and wind power due to government support like feed-in tariffs (Schmidt & Dabur, 2013). Availability of low-cost electricity as compared to electricity produced from biogas plants in most parts of the world is also hindering its popularity. The free availability of feedstock for AD is a positive factor, but the cost of handling (cutting, pre-treatment), and transporting the wastes is negatively affecting the plant economics (Mittal, 2018). Lack of proper legal standards or guidelines by local governments is also affecting the utilization of biogas as bottled fuel for transportation or injection into the natural gas grid (Vijay et al., 2015). Due to poor segregation of wastes (organic and inorganic), inert material, toxic wastes, and unavailability of committed quantity and quality of wastes, it results in the failure of biogas plants (Planning Commission, 2014).

14.6 BIOETHANOL FERMENTATION

Bioethanol fermentation is a biochemical process that involves the breakdown of sucrose into glucose and fructose, followed by their conversion into ethanol through enzymatic reactions. The physiochemical properties of bioethanol are listed in Table 14.1. To extract bioethanol from a substrate, the process involves continuous hydrolysis and fermentation, followed by distillation and dehydration (Raheem, 2016).

TABLE 14.1 Physicochemical Properties of Bioethanol

Fuel Property	Unit	Bio-Ethanol [C2H5OH]
Density at 15°C	Kg/m^3	790
Kinetics viscosity at 40°C	mm^2/s	1.13
Oxygen	Mass Percentage	34.7
Cetane number	–	5, 8
Octane number	–	110
Latent heat of vaporization	MJ/kg	0.91
Lower calorific value	MJ/kg	25.22,26.70
Flash point	°C	13
Auto-ignition temperature	°C	332.8, 366.0
Water content	mg/kg	2024
Stoichiometric fuel/air ratio	–	1/9.01

Source: Adapted from: Khuong et al. (2016).

First-generation biofuels, made from sugarcane, corn, wheat, and other crops, have raised environmental and economic concerns as they compete with food and land resources. This has led to the creation of second-generation biofuels, which are derived from waste materials such as municipal solid waste, crop residues, sludge, and food waste. Now, third-generation biofuels are being made from algal biomass, including micro- and macroalgae (Nair et al., 2017).

Wastes that contain carbon-rich lignocellulosic matter, such as MSW, fruit waste, vegetable waste, and food waste, can be used to produce bioethanol. Different types of pretreatments, such as thermal, enzymatic, acidic, and alkali, are applied to lignocellulosic substrates to increase yield and improve cellulose solubility (Vavouraki et al., 2013; Matsakas et al., 2014).

Of all the biofuels, bioethanol is the most successfully and widely used in transportation. Brazil began using bioethanol in 1925, and in 1984, bioethanol was used as a fuel in internal combustion engines in Germany and France (Ahar et al., 2017). Although bioethanol was initially neglected because of its expensive production cost when compared to cheaply available petroleum fuels, the oil crisis of the 1970s led to its renewal as an alternative fuel in several countries (Demirbas, 2009).

The global production of bioethanol has increased significantly over the years. In 2007, the production was 13.12 billion gallons, while in 2019 it rose to 29.03 billion gallons. The United States is currently the largest producer of bioethanol, with a yearly production of 15.78 billion gallons in 2019. Brazil comes a close second, with a production of 8.57 billion gallons of bioethanol in the same year (as shown in Figure 14.8 by AFDC, 2020).

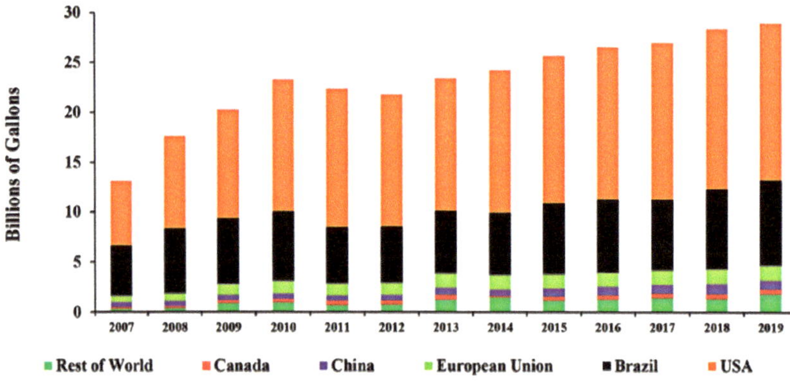

FIGURE 14.8 Global ethanol production by country from 2007 to 2019 (AFDC, 2020).

Source: Adapted from AFDC, 2020

14.6.1 ADVANTAGES

Bioethanol is a biofuel that is commonly used as a direct fuel or blended with gasoline to be used in vehicles without any mechanical modifications. When blended with gasoline, bioethanol improves the engine efficiency or octane rating, while also reducing the emission of harmful exhaust gases in vehicles (Staniszewski et al., 2007). Bioethanol is the preferred choice for vehicular fuel because it has a higher heat of vaporization and flame speeds compared to petroleum fuel. Additionally, it is less toxic, easily biodegradable, and generates fewer primary air pollutants, which in turn reduces the formation of secondary air pollutants (Balat & Balat, 2009).

14.6.2 CONSTRAINTS

The use of bioethanol in vehicles is limited due to its low energy density and lower heating value when compared to gasoline. At temperatures below 20°C, pure bioethanol can cause engine start problems, as it is more difficult to vaporize and has low Reid vapor pressure (Khuong et al., 2016). Moreover, bioethanol is corrosive to engine components since it is miscible with water, unlike gasoline which is immiscible. The bioethanol blended fuel may contain water, which could cause more corrosion damage to engine parts. Additionally, bioethanol can absorb high levels of moisture, which can decrease its octane rating when it reaches its saturation point and phase separation occurs (Chen et al., 2019). Although the use of bioethanol considerably reduces greenhouse gas emissions, it increases the generation of other air pollutants such as acetaldehyde, formaldehyde, and acetone (Chao et al., 2000).

14.7 LANDFILL PROCESSES

Landfilling is a common and cost-effective method of managing municipal solid waste (MSW) (Figure 14.9). Despite the availability of more advanced waste-to-energy (WtE) methods, landfilling remains popular in developing nations due to its low cost. In developed or industrialized nations, landfilling is an essential part of solid waste management infrastructure. In these nations, waste residues treated by WtE methods are dumped in landfills, and their volume is reduced by 90% (Agamuthu, 2013; Pressley et al., 2014). When the organic fraction of solid waste is dumped in landfills, it undergoes anaerobic decomposition, which releases landfill gases (LFG). This process is similar to the generation of biogas in anaerobic digestion, except that the initial decomposition in landfills is aerobic until the available oxygen is depleted. LFG is a combination of different gases, with methane (50 to 75%) and carbon dioxide (25 to 50%) being the primary gases, along with minor gases such as ammonia, nitrogen, and non-methane volatile organic compounds (Zuberi & Ali, 2015).

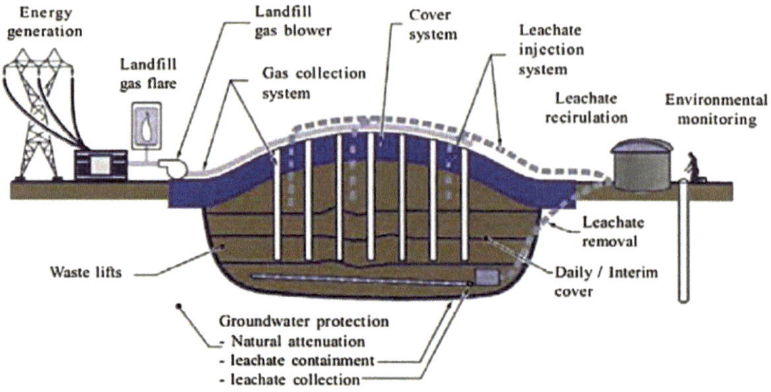

FIGURE 14.9 A typical engineered landfill with methane recovery system.

Source: Reprinted with permission from Roy and Tarafdar, 2017. Copyright © 2022, The Author(s), under exclusive license to Springer Nature Switzerland AG

Methane gas generated from landfills, known as landfill gas (LFG), is harmful to the environment. Methane gas (CH_4) has a 28 times higher global warming potential than CO_2 over 100 years. It is considered the third-largest anthropogenic source and contributes 12% of the total man-made methane emissions. The global methane generation from landfills is estimated to be 50 million tons per year, of which only 5 million tons are captured for proper utilization (Themelis & Ulloa, 2007; Beyene et al., 2018). LFG gases released from landfills pose serious environmental threats as they contribute to global warming. Moreover, methane being a flammable gas is also a safety risk. Therefore, it becomes necessary to manage and collect methane from landfills for energy recovery (Gautam & Kumar, 2019) (Figure 14.10).

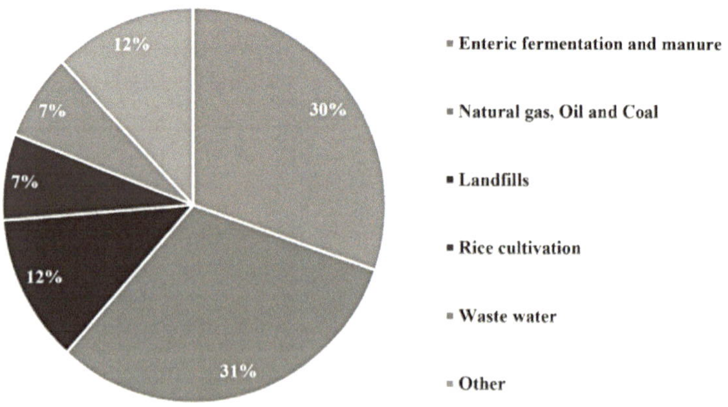

FIGURE 14.10 Sources of global anthropogenic methane emissions (USEPA, 2012).

Source: Adapted from USEPA, 2012

The United States generated the highest amount of landfill methane, with 1.03 × 108 metric tons of carbon equivalent in 2011. This accounted for approximately 17.7% of the total methane emissions in the country. In China, around 13% of methane generation came from 580 landfills in 2013. Meanwhile, in Europe, landfill methane generation contributed to 22% of total methane generation. The top landfill methane generating countries worldwide are listed in Table 14.2, and their emissions account for 58.4% of the total world landfill methane emission, according to Zhao et al. (2019).

TABLE 14.2 Countries with Highest Landfill Methane Emissions, 1970–2017 (*Mt* CO_2 *e*)

Country	1970	1980	1990	2000	2010	2012	2017
USA	151.6	180.5	176.2	133.4	130.4	121.9	127.9
China	9.4	14.1	24.5	40.4	63.7	69.2	79.6
Russia	28.3	31.5	33.7	41.5	58.1	63.6	64.7
Turkey	2.6	4.1	7.8	34.4	39.2	41.4	46.5
India	7.3	12.4	17.6	23.7	31.8	33.6	37.9
UK	52.6	58.9	56.7	54.4	28.6	24.4	25.7
Canada	18.1	22.5	20.8	22.5	22.1	22.6	24
Iran	2.4	4.5	7.9	11.2	14.6	15.3	16.9
Brazil	4.4	8.5	12.5	17	14.5	14.9	15.8
Indonesia	1.3	2.7	5.1	9.2	13.3	14.3	16.1

Landfill gas (LFG) can have hazardous effects on the environment, but it can be used as a fuel if it is captured and collected by a gas recovery system. However, LFG contains moisture and other trace gases that must be separated to improve its calorific value and storage options and minimize damage to engines or gas burners during fuel oxidation. LFG can be used as a nonconventional energy source for direct use (e.g., cooking, heating, lighting), generation of electricity for distribution through the power grid, and injecting the methane gas into the gas distribution grid (Terraza & Grajales, 2006).

There are more than 1000 LFG plants worldwide, with most of these plants located in Europe and the USA. In developing countries, the technology is still growing (Urase et al., 2008). The European Union has the highest number of LFG utilization plants in the world, with 136 in Croatia alone. Germany is the largest producer of LFG-generated electricity, and in 2009, about 3.5 million homes were powered by it. In Sweden, trains run on biogas (Njoku et al., 2018). China is the leading country in Asia in landfill methane recovery, with 44 LFG utilization plants. Thailand and Indonesia have 5, while Malaysia, the Philippines, and Korea have 4. India has 2, and Bangladesh has 1 LFG utilization Plant (Menikpura et al., 2013).

14.7.1 ADVANTAGES

Landfill gas (LFG) is the third largest human-made source of methane. However, it can also be a renewable energy source, just like biogas, and a clean fuel. Utilizing LFG can help improve air quality in local areas around landfill sites by capturing LFGs and reducing the use of fossil fuels. The collected LFG can be used for heating, cooking, electricity generation, and can also be supplied to the natural gas grid, providing countries with new sources of revenue. Moreover, LFG utilization can create job opportunities for the construction and maintenance of landfills.

14.7.2 CONSTRAINTS

The utilization and implementation of LFG (landfill gas) face several barriers, including technical, institutional, political, and financial problems. Developing countries often lack the necessary technology for LFG utilization and cannot afford the available technology of international markets due to high costs. Additionally, there is a lack of skilled and qualified personnel for designing, constructing, and maintaining LFG recovery plants, which further hinders adoption. The absence of political will and national action plans to promote LFG recovery as a sustainable strategy for waste management and energy derivation also contributes to the problem. However, local governments can identify LFG recovery projects for carbon credits under the Clean Development Mechanism (CDM) by developing a framework (Rajaram et al., 2011).

14.8 MICROBIAL FUEL CELLS (MFCS) AND MICROBIAL ELECTROLYSIS CELLS (MECS)

MFC technology is a promising bioenergy solution that can convert low-grade organic waste from municipal and wastewater into electricity (as shown in Figure 14.11(A)). On the other hand, MECs utilize exoelectrogenic microbes to convert the organic matter present in wastewater into electrical current and produce protons (H^+). These protons are then further reduced at the cathode to produce hydrogen gas (as illustrated in Figure 11(B)) (Zhou et al., 2013).

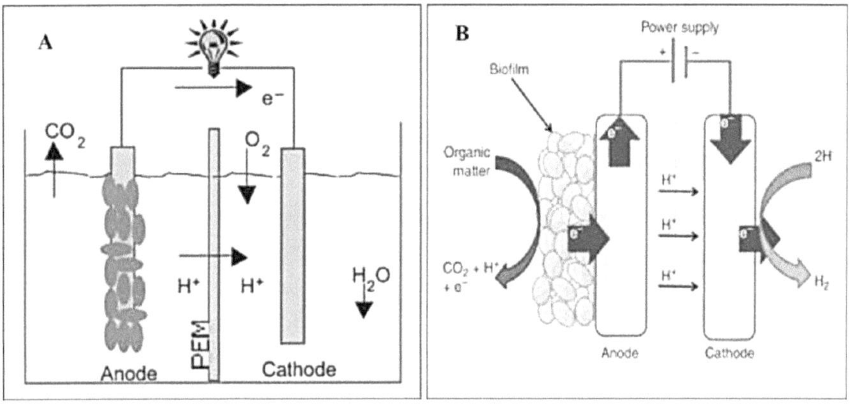

FIGURE 14.11 Schematic representation and operation of typical (A) MFC; and (B) MEC.
Source: (A) Adapted from Zhou et al., 2013 (B) Adapted from Wang et al., 2020

A typical microbial fuel cell or a microbial electrolysis cell consists of two chambers – anode and cathode – separated by a cationic membrane. In the anode chamber, wastewater containing a carbon source and a microbial consortium is present, which oxidizes organic substrates and produces electrons and protons (Zhou et al., 2013). The electrons generated are first transferred to the anode surface and then flow through an electrical circuit to reach the cathode. At the same time, protons migrate through the cationic membrane and enter the cathode chamber to maintain electrical neutrality (You et al., 2006).

In the presence of an electron acceptor such as oxygen, both the electrons and protons are reduced in the cathode chamber. The electricity produced can be stored or used to power an energy storage or consumption device that is connected between the electrodes (Rabaey et al., 2004). These MFCs can produce electricity and biofuel, such as hydrogen, which can be stored and used to power chemical fuel cells or internal combustion engines to address the low output voltage of MFCs (Zhou et al., 2012).

14.8.1 ADVANTAGES

Both microbial fuel cells (MFCs) and microbial electrolysis cells (MECs) are capable of treating waste and generating electricity and biofuels simultaneously. These cells can effectively remove various contaminants such as nitrate, nitrite, ammonium, and COD to produce treated wastewater with lower organic and nutrient loads. To measure the efficiency and scalability of MFCs/MECs, researchers have used a stack of MFCs to charge phone batteries, run a small DC motor, power kitchen clocks and LEDs, and even test ecological robots (Khan et al., 2017).

14.8.2 CONSTRAINTS

There is ongoing work being done to improve the optimization, applicability, scale, and cost of microbial fuel cells and microbial electrolysis cells (MFCs/MECs). However, their power output still lags behind that of conventional chemical fuel cells by three to four times. Additionally, the high cost of electrodes and membranes, as well as lower power production, are major setbacks for this technology. Scalability is also limited due to thermodynamic limitations and voltage losses (Flimban et al., 2018).

14.9 THERMO-CHEMICAL CONVERSIONS

Thermo-chemical conversion technologies involve breaking down and reforming organic matter at high temperatures to produce biochar, syngas, and bio-oil. Gasification, pyrolysis, and liquefaction are the three main thermo-chemical conversion processes available, depending on the type and quantity of substrate and energy demand (Lee et al., 2019). Among the waste-to-energy (WtE) techniques, thermal conversion technologies have become popular due to their well-established processes and equipment, environmental independence, low water usage, short processing time, and ability to process substrates that are difficult for microbes to break down, such as plastics (Uzoejinwa et al., 2018). Therefore, it is important to focus on exploring the future energy potential of thermo-chemical technologies among the available WtE techniques.

14.9.1 INCINERATION

Incineration is a waste management technique that involves burning waste at a very high temperature (750 to 1100°C) in the presence of oxygen. This technique is used to reduce the mass and volume of MSW (municipal solid waste) by 70% and 90% respectively, while converting the reduced mass into heat and energy (Tozlu et al., 2016). The main parts of an incineration process are the feed system, combustion chamber, exhaust system, and residual dispose system (Figure 14.12). Incineration not only minimizes waste but also reduces the infectious and hazardous properties of medical and hazardous waste. The efficiency of energy generated in the form of heat, steam, and electricity depends on various factors, such as the calorific value of the waste and the efficiency of the incineration plant. The average efficiency of energy generated lies in the range of 80% for heat, 20 to 30% for cogeneration, and 20% for pure electricity (Beylot et al., 2018; Makarichi et al., 2018).

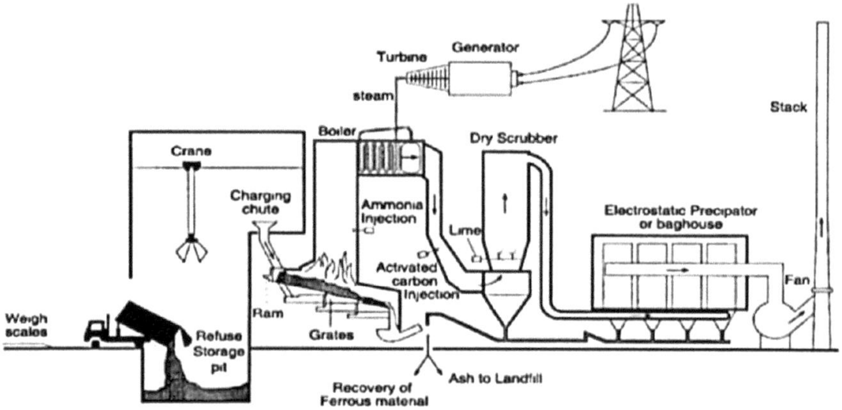

FIGURE 14.12 Schematic view of waste-to-energy incinerator.

Source: Reprinted with permission from: Tammemagi (1999).

The process of incineration produces flue gas, bottom ash, and fly ash. Flue gas is a mixture of harmful gases like Carbon Oxides (CO_x), Nitrogen Oxides (NO_x), Sulfur Oxides (SO_x), Polyaromatic Hydrocarbons (PAH), and Particulate Matter (PM). On the other hand, bottom ash and fly ash contain toxic organic and inorganic substances such as heavy metals (Wang et al., 2019; Khan & Kabir, 2020). Before being released into the atmosphere, flue gas needs to be treated through various air pollution control methods and equipment (Tozlu et al., 2016; Lino et al., 2017).

Incineration is considered as an effective method for treating MSW (Municipal Solid Waste) because of its capacity to clean flue gas. The fly ash and bottom ash produced during the process can be used in various construction sectors (Kan et al., 2020). To ensure the process is self-sustaining and economically feasible, non-combustible wastes such as glass, ceramics, and metals should be separated from the combustible wastes before feeding the incinerator (Defra 2013). The minimum calorific value of MSW required for incineration should be an average of 7 to 8 MJ/kg, with a minimum annual availability of 50,000 tons per year and additional fuel (Qazi et al., 2018).

The process of energy recovery from Municipal Solid Waste (MSW) started in the USA in 1885 with the installation of the world's first waste incinerator. As of 2016, there were 1618 Waste-to-Energy (WtE) facilities for MSW treatment worldwide, including 512 in Europe, 822 in Japan, 166 in China, and 88 in the USA (Scarlat et al., 2019).

In the European Union, MSW incineration increased from 32 million tons in 1995 to 70 million tons in 2018. During the same period, landfilling decreased by 61%, and material recycling increased by 200% (Levaggi et al., 2020). Among European nations, Germany, France, and the UK have the highest number of WtE treatment facilities, whereas 18 countries incinerate less than 25% of the total wastes generated, and 10 countries lack any waste incineration facility.

Incineration is prevalent in countries like Japan and Germany due to land resource constraints. In contrast, in the USA, landfilling is still considered a more viable option due to its low cost (Scarlat et al., 2019).

Japan is at the forefront of waste incineration in Asia due to limited land resources for landfilling and strict environmental regulations. In India, a few incineration plants have been installed, but some have been shut down. There are currently two waste incinerators in Hyderabad with the capacity to handle 500 TPD of MSW and a power generation capacity of 6 MW, and another plant with a capacity of 700 TPD of MSW and a power generation capacity of 6.6 MW (Nixon et al., 2017).

However, for many developing countries, incineration is not practical due to poor waste collection and separation, high capital, operating, and maintenance costs, lack of technical manpower, and the availability of cheap landfilling options (Kumar & Samadder, 2017).

14.9.1.1 ADVANTAGES

Incineration has several advantages, such as being suitable for all types of waste in rural or urban areas. The process is efficient and continuously fed with less noise and odor generation. The efficiency of incineration plants can be improved by installing air pollution control devices and safely disposing of fly and bottom ash. Compared to other thermochemical WtE techniques, incineration requires low levels of highly skilled manpower and capital while having a high energy-producing capacity (Qazi et al., 2018; dos Santos et al., 2020).

14.9.1.2 CONSTRAINTS

The emission of toxic air pollutants such as dioxins, furans, and particulate matter does not stop even after modifying or installing pollution control devices. These modifications or installations increase construction and maintenance costs. The fly ash generated is loaded with toxic heavy metals and must be handled and disposed of properly to prevent air, land, and water pollution. Poor waste segregation, variability, and high moisture content can cause operational problems in incineration plants (Qazi et al., 2018; dos Santos et al., 2020).

14.9.2 PYROLYSIS

Pyrolysis is a thermochemical process in which organic biomass is decomposed at high temperatures (300 to 1300°C) in the absence of oxygen. The process includes dehydration, fragmentation, and product formation, resulting in gaseous products, pyrolytic fluids, char, and ash. The temperature applied depends on the product, material, and process type. Conventional pyrolysis (300 to 600°C), fast pyrolysis (600 to 1000°C), and flash pyrolysis (800 to 1200°C) are the three types of pyrolysis, each with a different temperature range. The composition of MSW, heating rate, time, pressure, and reactor type also affect the pyrolysis process and products. Lower temperatures produce more solid products (biochar), whereas higher temperatures produce more liquid products (bio-oil), and minor gaseous products called syngas, like CO, H_2, CO_2, and CH_4, are also produced (Figure 14.13). Pyrolysis is considered the best practice in terms of waste reduction and energy recovery, with bio-oil having a calorific value of 38 MJ/kg and biochar being useful (Nizami et al., 2016).

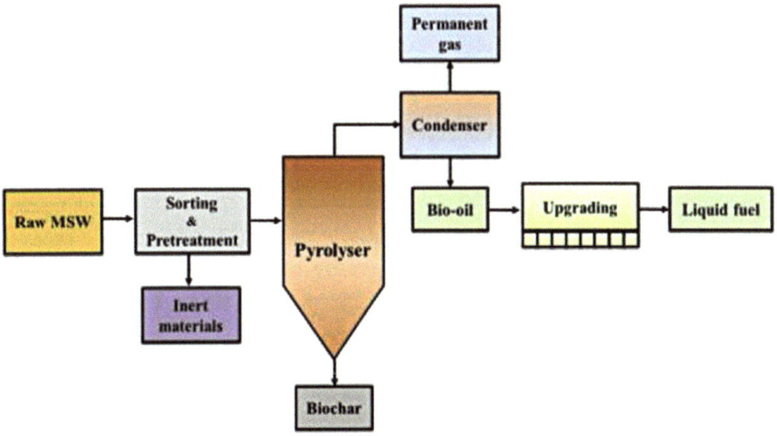

FIGURE 14.13 Schematic representation of MSW flow chart.

Source: Adapted from: Malav et al. (2020).

Pyrolysis oil, also known as bio-oil, can be used to replace petroleum products in stationary power stations, and can also be upgraded to replace gasoline in vehicles. Waste pyrolysis facilities are becoming increasingly popular in developed nations, particularly for treating plastic waste into oil. However, in developing countries, this treatment is still lagging due to its high cost, high organic content, and non-segregation of MSW, as well as advanced technology (Garcia-Nunez et al., 2017).

In the European Union, there are several installed working plants, with some under construction. For example, there are three operational plants, one each in the Netherlands, Finland, and Germany, with a waste treatment capacity of 24,000, 50,000, and 24,000 tons per year, respectively. Two more pyrolysis facilities are currently under construction in Finland and Sweden, each with a total waste treatment capacity of 24,000 tons per year. Additionally, two other facilities will be installed in Norway and Germany (ETIP Bioenergy, 2020).

In the USA, several pyrolysis facilities are currently treating plastic waste. Oregon has a daily bio-oil production of 60 barrels from 10 tons of waste, while in Maryland, a small pyrolysis plant produces 3–5 barrels of bio-oil from the treatment of one ton of plastic waste. Large-scale deployment of pyrolysis technology in the USA could produce bio-oil from plastics worth $9 billion and create 40,000 jobs (ACC, 2014).

Similarly, in other countries like Georgia and the UK, small-scale pyrolysis facilities are treating plastic waste and producing high-quality bio-oil (Mockler Martens, 2018). In China, waste pyrolysis is slowly emerging as an alternative to landfilling, composting, and incineration. Kingtiger, Henan Doing, Huayin, and Niutech are four pyrolytic processes by which futuristic pyrolysis facilities will be built in China (Ma et al., 2018).

There are currently no incineration plants in India. However, a pyrolysis/gasification plant was installed in Pune back in 2012. This plant has a waste treatment capacity of 700 TPS but has never operated at full capacity (Gupta et al., 2019).

14.9.2.1 ADVANTAGES

The hazardous air pollutants like furans, dioxins, NO_x, SO_x, etc., are not produced in an inert atmosphere by the installation of air pollution control devices in pyrolysis plants. The products of pyrolysis like biochar, bio-oil, or syngas can be used as fuel in power plants, industries, or vehicles. Like incineration, pyrolysis also reduces the volume (70 to 90%) of waste for landfilling.

Except for plasma arc gasification, pyrolysis is considered an eco-friendly method for waste treatment. The initial low temperature of pyrolysis also provides an option for metal recovery (Chen et al., 2014; Chen et al., 2015).

14.9.2.2 CONSTRAINTS

The pyrolysis technique has high capital, operational, and maintenance costs. The production of tar poses operational challenges; it causes inefficiencies both at commercial and pilot-scale plants. Pyrolysis is extremely sensitive to the presence of moisture and air, so wastes should be pretreated for moisture and inert matter, and an airtight reaction chamber should be maintained. The crude products produced from pyrolysis contain low energy which needs further upgrading, and the global waste consumption rate is very low due to a limited number of facilities. Due to the complexity of the pyrolysis process and high economic costs, the technology remained limited to developed nations (Chen et al., 2014; Chen et al., 2015).

14.9.3 GASIFICATION

Gasification is a process of partial oxidation that converts biomass into combustible gases at a temperature ranging between 750 and 1100°C. The gases produced are called synthesis gas or syngas, which is a mixture of H_2, CO, CH_4, and N_2 (Figure 14.14) (Shahbaz et al., 2016). The gas may

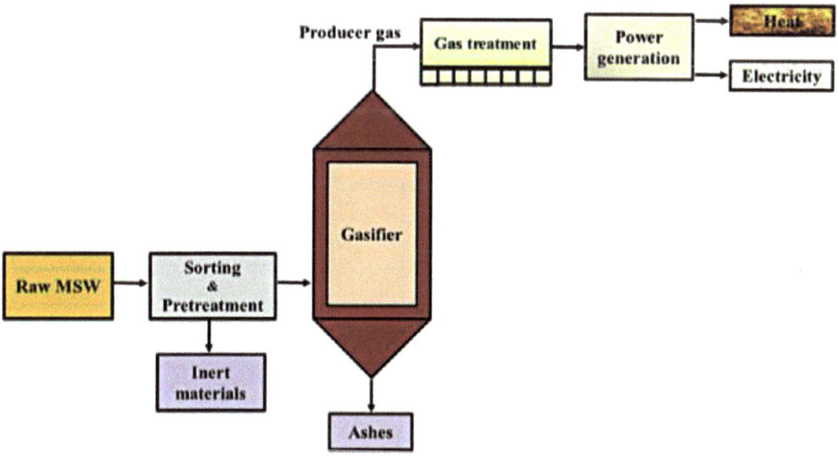

FIGURE 14.14 Schematic diagram of a MSW gasification process.

Source: Adapted from: Malav et al. (2020).

also contain contaminants, such as minute char particles, which need to be removed for efficient energy generation. To ensure a successful gasification process, it is essential to sort MSW to remove inert materials, such as glass and metals. The high calorific value of syngas (4–10 MJ/Nm^3) makes it suitable for power generation (Watson et al., 2018).

Gasification is considered an endothermic reaction as external energy is applied for the completion of the process (Bosmans et al., 2013). The sorted and dried MSW enters the reactor, called a gasifier, either in the absence or presence of O_2 at very high temperature and pressure. The gasification process generates only bottom ash, which needs to be detached and managed properly (Klinghoffer & Castaldi, 2013). Commonly, steam, air, carbon dioxide, and oxygen are used as agents of gasification, but steam and air are preferred. Air, being the cheaper agent, is not preferred as high N_2 concentration lowers the calorific value of the product gas by diluting it. Steam, however, is preferred due to its higher H_2 content, the higher calorific value of syngas, and easy removal of unreacted steam (Hosseini & Wahid, 2014).

Japan is leading with 112 waste gasification plant installations in 2012, whereas South Korea has only 7 gasification plants with 325,000 tons of waste treatment per year. Among European Union nations, Germany has installed a gasification plant with a waste treatment capacity of 608 tons per year, and two gasification plants are under construction, one each in the UK and France. In countries like Spain, the UK, and Italy, at least one gasification plant is under construction, whereas in the Netherlands and Sweden, two gasification plants in each state are under construction. In developing countries like India, Vietnam, and China, WtE projects like pyrolysis and gasification are being installed. However, due to poor waste management, the low calorific value of waste, and the availability of low-cost landfilling, thermochemical waste treatment is being blocked (Waldheim, 2018; ETIP-Bioenergy, 2020). Even developed countries like the USA have only 3 operational gasification plants at the experimental and trial phase. The operation and maintenance costs of gasification, upgrading of syngas, and the complexity of technology are factors affecting the production of technology in both developed and developing nations (Mukherjee et al., 2020).

14.9.3.1 ADVANTAGES

Gasification is considered a cleaner option than incineration as it reduces the formation of hazardous byproducts like dioxins and furans. The syngas produced requires less expensive and smaller gas cleaning equipment and after the cleaning and scrubbing produces superior quality combustible gas and energy. The syngas produced can be used in internal combustion engines and combined cycle turbines. The treatment reduces the waste volume by 90% with only 8 to 12% of inert and vitrified ash which is less than incineration (15 to 20%) (Consonni and Viganò, 2012; Dong et al., 2018).

14.9.3.2 CONSTRAINTS

The gasification process produces byproducts like tar, char, particulates, and halogens within the syngas. These products can pose operational and efficiency challenges, and the pretreatment methods required for syngas cleaning increase the production cost. Not only the pretreatment methods but also the use of catalysts for the up-gradation of syngas purity also increases the process cost. Ceramic filters, sulfur removal systems, and catalysts are ruined by the produced tar and high molecular weight organic gases. Another important problem with gasification is that it is suited for homogeneous wastes, and the heterogeneous wastes reduce the efficiency of syngas cleaning and production. In order to increase the calorific value of wastes, the process may require pretreatment of wastes by shredding and using methods like torrefaction and pelletization for densification (Pieta et al., 2018; Mukherjee et al., 2020).

14.10 CONCLUSION

Solid waste generation is one of the major environmental problems the world is facing. For the proper management of waste, the emphasis is given on reducing, recycling, reusing, and recovering energy and materials. Due to the carbonaceous nature of MSW, different treatments are used for energy recovery, such as direct utilization of MSW as fuel in incineration, converting MSW into more refined fuel (syngas, char, bio-oil) in pyrolysis/gasification, or microbial conversion of MSW into gaseous fuels (anaerobic digestion, landfilling) and liquid fuels (fermentation). Technologies like landfilling and incineration are popular and cost-effective, but both have environmental constraints such as causing groundwater pollution and air pollution. Gasification and pyrolysis are far more advanced technologies that are restricted to trial phases in certain developed nations. These technologies are unattractive in developing nations due to high capital, operational, and maintenance costs, technological dependence, and unscientific waste management. In order to increase the scale of waste management by WtE conversion technologies in developing countries, the government should emphasize framing policies and regulations, providing financial support, and developing the latest technologies.

KEYWORDS

- **anaerobic digestion**
- **fermentation**
- **gasification**
- **incineration**
- **landfilling**
- **municipal solid waste**
- **pyrolysis**
- **waste-to-energy**

REFERENCES

Agamuthu, P., (2013). Landfilling in developing countries. *Waste Management & Research, 31*(1), 1, 2. https://doi.org/10.1177/0734242X12469169.

Agaton, C. B., Guno, C. S., Villanueva, R. O., & Villanueva, R. O., (2020). Economic analysis of waste-to-energy investment in the Philippines: A real options approach. *Applied Energy, 275*(11526), 5. https://doi.org/10.1016/j.apenergy.2020.115265.

Alternative Fuels Data Center, (2020). American Chemistry Council. https://www.americanchemistry.com/Media/PressReleasesTranscripts/ACC-newsreleases/Converting-Non-Recycled-Plastics-Into-Oil-Could-Add-9-billion-to-US-Economy.html (accessed on 6 May 2024).

American Chemistry Council. https://www.americanchemistry.com/Media/PressReleasesTranscripts/ACC-newsreleases/Converting-Non-Recycled-Plastics-Into-Oil-Could-Add-9-billion-to-US-Economy.html (accessed on 6 May 2024).

Azhar, S. H. M., Abdulla, R., Jambo, S. A., Marbawi, H., Gansau, J. A., Faik, A. A. M., & Rodrigues, K. F., (2017). Yeasts in sustainable bioethanol production: A review. *Biochemistry and Biophysics Reports, 10*, 52–61. https://doi.org/10.1016/j.bbrep.2017.03.003.

Bader, N. B., Germec, M., & Turhan, I., (2021). *Scheffersomyces stipitis* biofilm reactor for ethanol production from acid-pretreated/detoxified and glucose- or xylose enriched rice husk hydrolysate under a continuous process. *Biomass Conversion and Biorefinery, 11*, 2909–2921. https://doi.org/10.1007/s13399-020-00611-6.

Balat, M., & Balat, H., (2009). Recent trends in global production and utilization of bio-ethanol fuel. *Applied Energy, 86*(11), 2273–2282. https://doi.org/10.1016/j.apenergy.2009.03.015.

Banu, J. R., Yukesh, K. R., Dinesh, K. M., Gunasekaran, M., Sivagurunathan, P., Park, J. H., & Kumar, G., (2018). Recent advances on biogranules formation in dark hydrogen fermentation system: Mechanism of formation and microbial characteristics. *Bioresour Technol, 268*, 787–796. https://doi.org/10.1016/j.biortech.2018.07.034.

Beyene, H. D., Werkneh, A. A., & Ambaye, T. G., (2018). Current updates on waste to energy (WtE) technologies: A review. *Renewable Energy Focus, 24*, 1–11. https://doi.org/10.1016/j.ref.2017.11.001.

Beylot, A., Hochar, A., Michel, P., Descat, M., Ménard, Y., & Villeneuve, J., (2018). Municipal solid waste incineration in France: An overview of air pollution control techniques, emissions, and energy efficiency. *Journal of Industrial Ecology, 22*(5), 1016–1026. https://doi.org/10.1111/jiec.12701.

Bolan, N. S., Thangarajan, R., Seshadri, B., Jena, U., Das, K. C., Wang, H., & Naidu, R., (2013). Landfills as a biorefinery to produce biomass and capture biogas. *Bioresource Technology, 135*, 578–587. https://doi.org/10.1016/j.biortech.2012.08.135.

Bölükbaş, A., & Akıncı, G., (2018). Solid waste composition and the properties of biodegradable fractions in Izmir City, Turkey: An investigation on the influencing factors. *Journal of Environmental Health Science and Engineering, 16*(2), 299–311. https://doi.org/10.1007/s40201-018-0318-2.

Bosmans, A., Vanderreydt, I., Geysen, D., & Helsen, L., (2013). The crucial role of waste-to-energy technologies in enhanced landfill mining: A technology review. *Journal of Cleaner Production, 55*, 10–23. https://doi.org/10.1016/j.jclepro.2012.05.032.

Canadian Biogas Association (CBA). https://biogasassociation.ca/about_biogas/projects_canada (accessed on 22 April 2024).

Chakraborty, S., Sadhu, P., & Goswami, U., (2016). Barriers in the advancement of solar energy in developing countries like India. *Problemy Ekorozwoju–Problems of Sustainable Development, 11*(2), 75–80.

Chen, D., Yin, L., Wang, H., & He, P., (2014). Pyrolysis technologies for municipal solid waste: A review. *Waste Management, 34*(12), 2466–2486. https://doi.org/10.1016/j.wasman.2014.08.004.

Chen, L., Gao, K., Zhang, C., & Lang, W., (2019). *Alternative Fuels for IC Engines and Jet Engines and Comparison of Their Gaseous and Particulate Matter Emissions*; Woodhead Publishing. https://doi.org/10.1016/B978-0-08-102791-2.00002-7.

Chen, Y., (2018). Evaluating greenhouse gas emissions and energy recovery from municipal and industrial solid waste using waste-to-energy technology. *Journal of Cleaner Production, 192*, 262–269. https://doi.org/10.1016/j.jclepro.2018.04.260.

Chen, Y., Cheng, J. J., & Creamer, K. S., (2008). Inhibition of anaerobic digestion process: A review. *Bioresource Technology, 99*(10), 4044–4064. https://doi.org/10.1016/j.biortech.2007.01.057.

Consonni, S., & Viganò, F., (2012). Waste gasification vs. conventional waste-to-energy: A comparative evaluation of two commercial technologies. *Waste Management, 32*(4), 653–666. https://doi.org/10.1016/j.wasman.2011.12.019.

Cutz, L., Haro, P., Santana, D., & Johnsson, F., (2016). Assessment of biomass energy sources and technologies: The case of central America. Renewable and sustainable energy reviews. *Renewable and Sustainable Energy Reviews, 58*, 1411–1431. https://doi.org/10.1016/j.rser.2015.12.322.

Demirbas, A., (2009). Biofuels securing the planet's future energy needs. *Energy Conversion and Management*, *50*(9), 2239–2249. https://doi.org/10.1016/j.enconman.2009.05.010.

Dong, J., Tang, Y., Nzihou, A., Chi, Y., Weiss-Hortala, E., & Ni, M., (2018). Life cycle assessment of pyrolysis, gasification and incineration waste-to-energy technologies: Theoretical analysis and case study of commercial plants. *Science of the Total Environment*, *626*, 744–753. https://doi.org/10.1016/j.scitotenv.2018.01.151.

dos Santos, I. F. S., Mensah, J. H. R., Gonçalves, A. T. T., & Barros, R. M., (2020). Incineration of municipal solid waste in Brazil: An analysis of the economically viable energy potential. *Renewable Energy*, *149*, 1386–1394. https://doi.org/10.1016/j.renene.2019.10.134.

EIA. https://www.eia.gov/energyexplained/biomass/waste-to-energy.php (accessed on 22 April 2024).

ETIP-Bioenergy. https://www.etipbioenergy.eu/images/ETIP-B-SABS2_WG2_Current_Status_of_Adv_Biofuels_Demonstrations_in_Europe_Mar2020_final.pdf (accessed on 22 April 2024).

Ferronato, N., & Torretta, V., (2019). Waste mismanagement in developing countries: A review of global issues. *International Journal of Environmental Research and Public Health*, *16*(6), 1060. https://doi.org/10.3390/ijerph16061060.

Fischer, G., & Schrattenholzer, L., (2001). Global bioenergy potential through 2050. *Biomass and Bioenergy*, *20*(3), 151–159. https://doi.org/10.1016/S0961-9534(00)00074-X.

Flimban, S. G., Kim, T., Ismail, I. M. I., & Oh, S. E., (2018). Overview of microbial fuel cell (MFC) recent advancement from fundamentals to applications: MFC designs, major elements, and scalability. *Energy & Fuel Technology*. https://doi.org/10.20944/preprints201810.0763.v1.

Garcia-Nunez, J. A., Pelaez-Samaniego, M. R., Garcia-Perez, M. E., Fonts, I., Abrego, J., Westerhof, R. J. M., & Garcia-Perez, M., (2017). Historical developments of pyrolysis reactors: A review. *Energy & Fuels*, *31*(6), 5751–5775. https://doi.org/10.1021/acs.energyfuels.7b00641.

Gautam, P., & Kumar, S., (2019). Landfill gas as an energy source. In: *Current Developments in Biotechnology and Bioengineering* (pp. 93–117). Elsevier. https://doi.org/10.1016/B978-0-444-64083-3.00006-3.

Guerrero, L. A., Maas, G., & Hogland, W., (2013). Solid waste management challenges for cities in developing Countries. *Waste Management*, *33*(1), 220–232. https://doi.org/10.1016/j.wasman.2012.09.008.

Gupta, M., Srivastava, M., & Agrahari, K., (2018). Waste to energy technologies in India: A review. *Journal of Energy and Environmental Sustainability*, *6*, 29–35.

Haberl, H., Erb, K. H., Krausmann, F., Bondeau, A., Lauk, C., Müller, C., et al., (2011). Global bioenergy potentials from agricultural land in 2050: Sensitivity to climate change, diets and yields. *Biomass and Bioenergy*, *10*(12), 4753–4769. https://doi.org/10.1016/j.biombioe.2011.04.035.

Hosseini, S. E., & Wahid, M. A., (2014). Utilization of palm solid residue as a source of renewable and sustainable energy in Malaysia. *Renewable and Sustainable Energy Reviews*, *40*, 621–632. https://doi.org/10.1016/j.rser.2014.07.214.

IEA. *Biogas Installed Power Generation Capacity, 2010–2018*. https://www.iea.org/data-and-statistics/charts/biogas-installed-power-generation-capacity-2010-2018 (accessed on 22 April 2024).

International Renewable Energy Agency, (2020). IRENA. International Renewable Energy Agency. www.irena.org/-/media/Files/IRENA/Agency/Publication/2017/Dec/IRENA_Biogas_for_domestic_cooking_20 (accessed on 6 May 2024).

IRENA. International Renewable Energy Agency. www.irena.org/-/media/Files/IRENA/Agency/Publication/2017/Dec/IRENA_Biogas_for_domestic_cooking_2017.pdf (accessed on 22 April 2024).

Jain, S., Newman, D., Nzihou, A., Dekker, H., Feuvre, L., Richter, P., Gobe, H., & Morton, F., (2019). *Global Potential of Biogas*. Doctoral dissertation, The World Biogas Association.

Jang, H. M., Choi, Y. K., & Kan, E., (2018). Effects of dairy manure-derived biochar on psychrophilic, mesophilic and thermophilic anaerobic digestions of dairy manure. *Bioresource Technology*, *250*, 927–931. https://doi.org/10.1016/j.biortech.2017.11.074.

Kan, L., Shi, R., Zhao, Y., Duan, X., & Wu, M., (2020). Feasibility study on using incineration fly ash from municipal solid waste to develop high ductile alkali-activated composites. *Journal of Cleaner Production*, *254*(12016), 8. https://doi.org/10.1016/j.jclepro.2020.120168.

Kaza, Silpa, Lisa Yao, Perinaz Bhada-Tata, and Frank Van Woerden. 2018. *What a Waste 2.0: A Global Snapshot of Solid Waste Management to 2050.* Urban Development Series. Washington, DC: World Bank. doi:10.1596/978-1-4648-1329-0.

Khan, I. U., Othman, H. D., Hashim, M., Matsuura, H., Ismail, T., Rezaei-DashtArzhandi, A. F., Wan, M., & Azelee, I., (2017). Biogas as a renewable energy fuel – a review of biogas upgrading, utilization and storage. *Energy Conversion and Management*, *150*, 277–294. https://doi.org/10.1016/j.enconman.2017.08.035.

Khan, I., & Kabir, Z., (2020). Waste-to-energy generation technologies and the developing economies: A multi-criteria analysis for sustainability assessment. *Renewable Energy*, *150*, 320–333. https://doi.org/10.1016/j.renene.2019.12.132.

Khan, M. D., Khan, N., Sultana, S., Joshi, R., Ahmed, S., & Yu, E., (2017). Bioelectrochemical conversion of waste to energy using microbial fuel cell technology. *Process Biochemistry*, *57*, 141–158. https://doi.org/10.1016/j.procbio.2017.04.001.

Khuong, L. S., Zulkifli, N. W. M., Masjuki, H. H., Mohamad, E. N., Arslan, A., Mosarof, M. H., & Azham, A., (2016). A review on the effect of bioethanol dilution on the properties and performance of automotive lubricants in gasoline engines. *RSC Advances*, *6*(71), 66847–66869.

Kim, S., & Dale, B. E., (2004). Global potential bioethanol production from wasted crops and crop residues. *Biomass and Bioenergy*, *26*(4), 361–375. https://doi.org/10.1016/j.biombioe.2003.08.002.

Klinghoffer, N. B., & Castaldi, M. J., (2013). Gasification and pyrolysis of municipal solid waste (MSW). In: *Waste to Energy Conversion Technology* (pp. 146–176). Woodhead Publishing Series in Energy. https://doi.org/10.1533/9780857096364.2.146.

Korai, M. S., Mahar, R. B., & Uqaili, M. A., (2017). The feasibility of municipal solid waste for energy generation and its existing management practices in Pakistan. *Renewable and Sustainable Energy Reviews*, *72*, 338–353. https://doi.org/10.1016/j.rser.2017.01.051.

Koyama, M., Yamamoto, S., Ishikawa, K., Ban, S., & Toda, T., (2017). Inhibition of anaerobic digestion by dissolved lignin derived from alkaline pre-treatment of an aquatic macrophyte. *Chemical Engineering Journal*, *311*, 55–62. https://doi.org/10.1016/j.cej.2016.11.076.

Kumar, A., & Agrawal, A., (2020). Recent trends in solid waste management status, challenges and potential for the future Indian cities–A Review. *Current Research in Environmental Sustainability*, *2*, 10001. https://doi.org/10.1016/j.crsust.2020.100011.

Kumar, A., & Samadder, S. R., (2017). A review on technological options of waste to energy for effective management of municipal solid waste. *Waste Management*, *69*, 407–422. https://doi.org/10.1016/j.wasman.2017.08.046.

Kumar, S., Gandhi, P., Yadav, M., Paritosh, K., Pareek, N., & Vivekanand, V., (2019). Weak alkaline treatment of wheat and pearl millet straw for enhanced biogas production and its economic analysis. *Renewable Energy*, *139*, 753–764. https://doi.org/10.1016/j.renene.2019.02.133.

Lee, S. Y., Sankaran, R., Chew, K. W., Tan, C. H., Krishnamoorthy, R., Chu, D. T., & Show, P. L., (2019). Waste to bioenergy: A review on the recent conversion technologies. *BMC Energy*, *1*(1), 1–22. https://doi.org/10.1186/s42500-019-0004-7.

Levaggi, L., Levaggi, R., Marchiori, C., & Trecroci, C., (2020). Waste-to-ENERGY IN the EU: The effects of plant ownership, waste mobility, and decentralization on environmental outcomes and welfare. *Sustainability*, *12* (14), 5743. https://doi.org/10.3390/su12145743.

Lino, F. A. M., & Ismail, K. A. R., (2017). Incineration and recycling for MSW treatment: Case study of Campinas, Brazil. *Sustainable Cities and Society*, *35*, 752–757. https://doi.org/10.1016/j.scs.2017.09.028.

Liu, J., Li, Q., Gu, W., & Wang, C., (2019). The impact of consumption patterns on the generation of municipal solid waste in China: Evidences from provincial data. *International Journal of Environmental Research and Public Health*, *16*(10), 1–19. https://doi.org/10.3390/ijerph16101717.

Loy, A. C. M., Yusup, S., Chin, B. L. F., Gan, D. K. W., Shahbaz, M., & Acda, M. N., (2018). Comparative study of in-situ catalytic pyrolysis of rice husk for syngas production: Kinetics modeling and product gas analysis. *Journal of Cleaner Production*, *197*(1), 1231–1243. https://doi.org/10.1016/j.jclepro.2018.06.245.

Ma, W., He, L., Dan, Z., Chen, G., & Lu, X., (2018). Municipal solid waste properties in China: A comparative study between Beijing, Guangzhou and Lhasa. In: Crocker, R., Saint, C., Chen, G., & Tong, Y., (eds.), *Unmaking Waste in Production and Consumption: Towards the Circular Economy*. Emerald Publishing Limited. https://doi.org/10.1108/978-1-78714-619-820181019.

Mahajan, D., Tonjes, D. J., Mamalis, S., Boudreaux, R., Hasty, J. K., & Danhui, X., (2015). Effective landfill gas management strategies for methane control and reuse technology. *Journal of Renewable and Sustainable Energy*, *7*(4), 41511. https://doi.org/10.1063/1.4929383.

Makarichi, L., Jutidamrongphan, W., & Techato, K. A., (2018). The evolution of waste-to-energy incineration: A review. *Renewable and Sustainable Energy Reviews*, *91*, 812–821. https://doi.org/10.1016/j.rser.2018.04.088.

Malav, L. C., Yadav, K. K., Gupta, N., Kumar, S., Sharma, K. G., & Krishnan, S., (2020). A review on municipal solid waste as a renewable source for waste-to-energy project in India: Current practices, challenges, and future opportunities. *Journal of Cleaner Production*, *277*, 12322. https://doi.org/0.1016/j.jclepro.2020.123227.

Matsakas, L., Kekos, D., Loizidou, M., & Christakopoulos, P., (2014). Utilization of household food waste for the production of ethanol at high dry material content. *Biotechnology for Biofuels and Bioproducts*, *7*(4), 1–9. https://doi.org/10.1186/1754-6834-7-4.

Menikpura, S. N. M., Sang-Arun, J., & Bengtsson, M., (2013). Climate co-benefits of energy recovery from landfill gas in developing Asian cities: A case study in Bangkok. *Waste Management & Research*, *31*(10), 1002–1011. https://doi.org/10.1177/0734242X13492004.

Millati, R., Cahyono, R. B., Ariyanto, T., Azzahrani, I. N., Putri, R. U., & Taherzadeh, M. J., (2019). Agricultural, industrial, municipal, and forest wastes: An overview. In: *Sustainable Resource Recovery and Zero Waste Approaches* (pp. 1–22). Elsevier. https://doi.org/10.1016/B978-0-444-64200-4.00001-3.

Mittal, S., Ahlgren, E. O., & Shukla, P. R., (2018). Barriers to biogas dissemination in India: A Review. *Energy Policy*, *112*, 361–370. https://doi.org/10.1016/j.enpol.2017.10.027.

MNRE. *Ministry of New and Renewable Energy*. Bio Energy Overview, Ministry of New and Renewable Energy, India (mnre.gov.in)

Mockler, M. K., (2018). *Implementing Plastic Pyrolysis into Chile's Plastic Recycling Infrastructure*. Oregon State University.

Moya, D., Aldás, C., Jaramillo, D., Játiva, E., & Kaparaju, P., (2017). Waste-to-energy technologies: An opportunity of energy recovery from municipal solid waste, using Quito-Ecuador as case study. *Energy Procedia*, *134*, 327–336. https://doi.org/10.1016/j.egypro.2017.09.537.

Mubarok, M. Z., Winarko, R., Chaerun, S. K., Rizki, I. N., & Ichlas, Z. T., (2017). Improving gold recovery from refractory gold ores through bio-oxidation using iron-sulfur-Oxidizing/Sulfur-oxidizing mixotrophic bacteria. *Hydrometallurgy*, *168*, 69–75. https://doi.org/10.1016/j.hydromet.2016.10.018.

Mukherjee, C., Denney, J., Mbonimpa, E. G., Slagley, J., & Bhowmik, R., (2020). A review on municipal solid waste-to-energy trends in the USA. *Renewable and Sustainable Energy Reviews*, *119*(10951). https://doi.org/10.1016/j.rser.2019.109512.

Nair, R. B., Lennartsson, P. R., & Taherzadeh, M. J., (2017). Bioethanol production from agricultural and municipal wastes. In: *Current Developments in Biotechnology and Bioengineering: Solid Waste Management* (pp. 157–190). Elsevier. https://doi.org/10.1016/B978-0-444-63664-5.00008-3.

Nixon, J. D., Dey, P. K., & Ghosh, S. K., (2017). Energy recovery from waste in India: An evidence-based analysis. *Sustainable Energy Technologies and Assessments*, *21*, 23–32. https://doi.org/10.1016/j.seta.2017.04.003.

Nizami, A. S., Ouda, O. K. M., Rehan, M., El-Maghraby, A. M. O., Gardy, J., Hassanpour, A., Kumar, S., & Ismail, I. M. I., (2016). The potential of Saudi Arabian natural zeolites in energy recovery technologies. *Energy*, *108*, 162–171. https://doi.org/10.1016/j.energy.2015.07.030.

Njoku, P. O., Odiyo, J. O., Durowoju, O. S., & Edokpayi, J. N., (2018). A review of landfill gas generation and utilization in Africa. *Open Environmental Sciences*, *10*, 1–15. https://doi.org/10.2174/1876325101810010001.

Pieta, I. S., Epling, W. S., Kazmierczuk, A., Lisowski, P., Nowakowski, R., & Serwicka, E. M., (2018). Waste into fuel—Catalyst and process development for MSW valorization. *Catalysts*, *8*(3), 113. https://doi.org/10.3390/catal8030113.

Pressley, P. N., Aziz, T. N., DeCarolis, J. F., Barlaz, M. A., He, F., Li, F., & Damgaard, A., (2014). Municipal solid waste conversion to transportation fuels: A life-cycle estimation of global warming potential and energy consumption. *Journal of Cleaner Production*, *70*, 145–153. https://doi.org/10.1016/j.jclepro.2014.02.041.

Qazi, W. A., Abushammala, M. F., & Younes, M. K., (2018). Waste-to-energy technologies: A literature review. *The Journal of Solid Waste Technology and Management*, *44*(4), 387–409. https://doi.org/10.5276/JSWTM.2018.387.

Raheem, A., Hassan, M. Y., & Shakoor, R., (2016). Bioenergy from anaerobic digestion in Pakistan: Potential, development and prospects. *Renewable and Sustainable Energy Reviews*, *59*, 264–275. https://doi.org/10.1016/j.rser.2016.01.010.

Rajaram, V., Siddiqui, F. Z., & Khan, M. E., (2011). *From Landfill Gas to Energy: Technologies and Challenges*. CRC Press.

Ramachandra, T. V., Bharath, H. A., Kulkarni, G., & Han, S. S., (2018). Municipal solid waste: Generation, composition and GHG emissions in Bangalore, India. *Renewable and Sustainable Energy Reviews*, *82*(Part 1), 1122–1136. https://doi.org/10.1016/j.rser.2017.09.085.

Report of the Task Force on Waste to Energy, (2014). In: *Planning Commission* (Vol. 1). New Delhi.

Scarlat, N., Fahl, F., & Dallemand, J. F., (2019). Status and opportunities for energy recovery from municipal solid waste in Europe. *Waste and Biomass Valorization*, *10*(9), 2425–2444. https://doi.org/10.1007/s12649-018-0297-7.

Schmidt, T. S., & Dabur, S., (2014). Explaining the diffusion of biogas in India: A new functional approach considering national borders and technology transfer. *Environmental Economics and Policy Studies*, *16*(2), 171–199. https://doi.org/10.1007/s10018-013-0058-6.

Shahbaz, M., Yusup, S., Inayat, A., Patrick, D. O., & Partama, A., (2016). System analysis of poly-generation of SNG, power and district heating from biomass gasification system. *Chemical Engineering Transactions*, *52*, 781–786.

Sharma, K. D., & Jain, S., (2020). Municipal solid waste generation, composition, and management: The global scenario. *Social Responsibility Journal*, *16*(6), 917–948. https://doi.org/10.1108/SRJ-06-2019-0210.

Sharma, S., Basu, S., Shetti, N. P., Kamali, M., Walvekar, P., & Aminabhavi, T. M., (2020). Waste-to-energy nexus: A sustainable development. *Environmental Pollution*, *267*, 115501. https://doi.org/10.1016/j.envpol.2020.115501.

Singh, C. K., Kumar, A., & Roy, S. S., (2018). Quantitative analysis of the methane gas emissions from municipal solid waste in India. *Scientific Reports*, *8*(1), 2913. https://doi.org/10.1038/s41598-018-21326-9.

Smeets, E. M. W., Faaij, A. P. C., Lewandowski, I. M., & Turkenburg, W. C., (2007). A bottom-up assessment and review of global bio-energy potentials to 2050. *Progress in Energy and Combustion Science*, *33*(1), 56–106. https://doi.org/10.1016/j.pecs.2006.08.001.

Staniszewski, M., Kujawski, W., & Lewandowska, M., (2007). Ethanol production from whey in bioreactor with co-immobilized enzyme and yeast cells followed by pervaporative recovery of product–kinetic model predictions. *Journal of Food Engineering*, *82*(4), 618–625. https://doi.org/10.1016/j.jfoodeng.2007.03.031.

Statistical, E. *EBA Statistical Report*. https://www.europeanbiogas.eu/wp-content/uploads/2019/11/EBA_report2018_abriged_A4_vers12_220519_RZweb.pdf (accessed on 22 April 2024).

Summary for Policymakers, (2013). In: Stocker, T. F., Qin, G. K. D., Plattner, M., Tignor, S. K., Allen, J., Boschung, A., Nauels, Y., et al., (eds.), *Climate Change 2013: The Physical Science Basis*. Contribution of Working Group I to the Fifth Assessment Report of the Intergovernmental Panel on Climate Change; Cambridge University Press: United Kingdom.

Tammemagi, H. Y., (1999). *The Waste Crisis: Landfills, Incinerators, and the Search for a Sustainable Future*. Oxford University Press: New York.

Terraza, H., & Grajales, F., (2006). *The Landfill Gas-to-Energy Initiative for Latin America and the Caribbean*. The World Bank.

Themelis, N. J., & Ulloa, P. A., (2007). Methane generation in landfills. *Renewable Energy*, *32*(7), 1243–1257. https://doi.org/10.1016/j.renene.2006.04.020.

Tozlu, A., Özahi, E., & Abuşoğlu, A., (2016). Waste to energy technologies for municipal solid waste management in Gaziantep. *Renewable and Sustainable Energy Reviews*, *54*, 809–815. https://doi.org/10.1016/j.rser.2015.10.097.

Triassi, M., Alfano, R., Illario, M., Nardone, A., Caporale, O., & Montuori, P., (2015). Environmental pollution from illegal waste disposal and health effects: A review on the "triangle of death." *International Journal of Environmental Research and Public Health*, *12*(2), 1216–1236. https://doi.org/10.3390/ijerph120201216.

Urase, T., Okumura, H., Panyosaranya, S., & Inamura, A., (2008). Emission of volatile organic compounds from solid waste disposal sites and importance of heat management. *Waste Management & Research: The Journal for Sustainable Circular Economy*, *26*(6), 534–538. https://doi.org/10.1177/0734242X07084321.

USEPA, (2012). *Global Anthropogenic Emissions of Non-CO_2 Greenhouse Gases: 1990–2030* (pp. 1–22). Available online on: Summary Report – Global Non-CO_2 Greenhouse Gas Emissions Projections (epa.gov).

Uzoejinwa, B. B., He, X., Wang, S., Abomohra, A. E. F., Hu, Y., & Wang, Q., (2018). Co-pyrolysis of biomass and waste plastics as a thermochemical conversion technology for high-grade biofuel production: Recent progress and future directions elsewhere worldwide. *Energy Conversion and Management*, *163*, 468–492. https://doi.org/10.1016/j.enconman.2018.02.004.

Varjani, S., Pandey, A., Gnansounou, E., Khanal, S., & Sindhu, R., (2020). Resource recovery from waste: An introduction. In: Otoo, M., & Drechse, P., (eds.), *Current Developments in Biotechnology and Bioengineering* (pp. 3–11). Elsevier.

Vavouraki, A. I., Angelis, E. M., & Kornaros, M., (2013). Optimization of thermo-chemical hydrolysis of kitchen wastes. *Waste Management*, *33*(3), 740–745. https://doi.org/10.1016/j.wasman.2012.07.012.

Vijay, V. K., Kapoor, R., Trivedi, A., & Vijay, V., (2015). Biogas as clean fuel for cooking and transportation needs in India. In: Ravindra, P., (ed.), *Advances in Bioprocess Technology* (pp. 257–275). Springer, Cham. https://doi.org/10.1007/978-3-319-17915-5_14.

Waldheim, L., (2018). *Gasification of Waste Forenergy Carriers: A Review*; IEA Bioenergy Task.

Wang, A., Liu, W., Zhang, B., & Cai, W., (2020). *Bioelectrosynthesis: Principles and Technologies for Value-Added Products*. Wiley-VCH.

Waste to Energy Global Market Size Outlook 2027. Statista. https://www.statista.com/statistics/480452/market-value-of-waste-to-energy-globally-projection/ (accessed on 22 April 2024).

Watson, J., Zhang, Y., Si, B., Chen, W. T., & De Souza, R., (2018). Gasification of biowaste: A critical review and outlooks. *Renewable and Sustainable Energy Reviews*, *83*, 1–17. https://doi.org/10.1016/j.rser.2017.10.003.

Wilson, D. C., Rodic, L., Modak, P., Soos, R., Carpintero, A., Velis, C., Iyer, M., & Simonett, O., (2015). *United Nations Environment Program; International Solid Waste Association.* Global Waste Management Outlook.

World Biogas Association Market Report – USA, (2017).

Yadav, A., Pal, N., Patra, J., & Yadav, M., (2020). Strategic planning and challenges to the deployment of renewable energy technologies in the world scenario: Its impact on global sustainable development. *Environment, Development and Sustainability*, *22*(1), 297–315. https://doi.org/10.1007/s10668-018-0202-3.

Yang, H., Ma, M., Thompson, J. R., & Flower, R. J., (2018). Waste management, informal recycling, environmental pollution and public health. *Journal of Epidemiol Community Health*, *72*(3), 237–243. https://doi.org/10.1136/jech-2016-208597.

Yang, Y., Ni, J. Q., Zhu, W., & Xie, G., (2019). Life cycle assessment of large-scale compressed bio-natural gas production in China: A case study on manure co-digestion with corn stover. *Energies*, *12* (3), 429. https://doi.org/10.3390/en12030429.

Yousuf, A., Manzoor, S. O., Youssouf, M., Malik, Z. A., & Khawaja, K. S., (2020). Fly ash: Production and utilization in India-an overview. *Journal of Materials and Environmental Science*, *11*(6), 911–921.

Zhao, H., (2019). *Methane Emissions from Landfills*. Earth and Environmental Engineering, Columbia University. https://doi.org/10.13140/RG. 2.2. 17326.77120.

Zhou, M., Jin, T., Wu, Z., Chi, M., & Gu, T., (2012). Microbial fuel cells for bioenergy and bioproducts. In: Opalakrishnan, K., & Van, L. J. B. R., (eds.), *Sustainable Bioenergy and Bioproducts*(pp. 131–171). Springer: London. https://doi.org/10.1007/978-1-4471-2324-8_8.

Zhou, M., Yang, J., Wang, H., Jin, T., Xu, D., & Gu, T., (2013). Microbial fuel cells and microbial electrolysis cells for the production of bioelectricity and biomaterials. *Environmental Technology*, *34*(13, 14), 1915–1928. https://doi.org/10.1080/09593330.2013.813951.

Zuberi, M. J. S., & Ali, S. F., (2015). Greenhouse effect reduction by recovering energy from waste landfills in Pakistan. *Renewable and Sustainable Energy Reviews*, *44*, 117–131. https://doi.org/10.1016/j.rser.2014.12.028.

CHAPTER 15

Ecosystem Services of Agroforestry

YASHMITA-ULMAN[1] and MANOJ SINGH[2]

[1]*Department of Silviculture and Agroforestry, College of Horticulture and Forestry, ANDUAT, Kumarganj, Ayodhya, Uttar Pradesh, India*

[2]*Department of Zoology, Kalinga University, Naya Raipur, Raipur, Chhattisgarh, India*

ABSTRACT

With limited land resources at hand, agricultural landscapes are continuously being managed so that multiple benefits are derived and provisioning services are enhanced, leading to sustainability in production. Agroforestry is a practice that has been commonly implemented since ancient times and comprises integrated natural resource management. Although agroforestry systems are assumed to provide numerous ecosystem services, the synthesis of evidence and literature supporting these perceptions is limited. Therefore, this chapter attempts to condense the existing data on ecosystem and provisioning services provided by agroforestry systems by aggregating a number of papers from around the world. Based on scientific papers published over time and space, it can be concluded that agroforestry systems contribute to providing: (i) provisioning services including food, nutrition, medicinal plants, and energy; (ii) regulatory services such as carbon sequestration, pest and erosion control; and (iii) supporting services like improvement of soil fertility, air and water quality, biodiversity conservation, and pollination. It is evident from the literature review that in the current era of scarce resources and climate change, agroforestry systems are a viable land-use option that can alleviate poverty and provide ecosystem services. Therefore, these systems should be replicated and recommended in landscapes worldwide by policymakers.

Sustainable Forest Resources Management: Issues and Implications. Jahangeer A. Bhat, Gopal Shukla, Manmohan J. Dobriyal, Sumit Chakravarty, A. Arunachalam, and Rainer W. Bussmann (Eds.)

15.1 INTRODUCTION

In the tropical regions, agroforestry systems have been widely advocated as a natural resource management strategy. Agroforestry, being a traditional land use system, provides solutions to many current and future environment-related problems. There is evidence that agroforestry not only meets agricultural goals but also helps in soil conservation, maintaining watershed functions, carbon sequestration, climate amelioration, and biodiversity conservation (Schroth et al., 2004). So, agroforestry provides various environmental benefits along with ecosystem services which are provisioning, regulating, cultural, and supporting in nature. Agroforestry is a sustainable land use practice that purposefully incorporates woody vegetation along with crops and/or animals and obtains benefits from the possible ecological and economic interaction between these components. The agroforestry systems, by retaining native species, prove to be a substitute habitat for the species, giving them favorable conditions for establishment, especially when these agroforestry systems are adjacent to deforested, fragmented, and degraded forests (Hartley, 2002; Lindenmayer & Hobbs, 2004). These native species interact with each other and their environment, keeping alive the functioning and services of the ecosystem within the agroforestry systems (Simonetti et al., 2006).

The traditional agroforestry systems, by incorporating different tree species, help in conserving plant and animal diversity and their habitats and reduce pressure on the adjacent forests (McNeely & Schroth, 2006). The trees in these systems provide cover, protection, fruits, fodder, fuelwood, medicines, etc. (Ouinsavi et al., 2005). Wild fruits are highly nutritious and supplement the daily diet, act as a substitute for exotic fruits, provide food security, and income (Djossa et al., 2008). The numerous functions of agroforestry have been appreciated by the International Assessment of Agricultural Science and Technology for Development (2008) and the Millennium Ecosystem Assessment (2005). This chapter aims to bring together the findings of different research teams working on environmental and ecosystem services supported by agroforestry systems. The identification and knowledge of agroforestry and its benefits might help pave the way for new effective land use policies and assist land managers in making decisions leading to sustainable management.

15.2 AGROFORESTRY FOR BIODIVERSITY CONSERVATION

With the passage of time, many plants and animals are nearing extinction, and many more are threatened. Out of 147,500 species on the IUCN Red

List, species threatened with extinction are more than 41,000 (IUCN Red List, 2022). So, there is an urgent need for effective plans to conserve biodiversity. The occurrence of high biological diversity in agroforestry systems is reported by various researchers (Schroth et al., 2004; McNeely, 2004; Harvey et al., 2006a). Agroforestry helps in germplasm conservation of sensitive species, provides habitat for species, acts as corridors between forest patches and adjacent vegetation, ameliorates the environmental conditions by preventing soil erosion, enhances water recharge, thus maintaining the regulating and provisional services of the ecosystem, and reducing the rate of conversion of forests into agricultural fields. Thus, scientists and policymakers recommend agroforestry as one of the land use options.

Numerous studies support the role of agroforestry in the conservation of global biodiversity. Different agroforestry systems support diverse taxa as shown in Table 15.1. When compared with traditional agricultural practices, shade-grown coffee agroforestry shows good potential for biodiversity enhancement (Perfecto et al., 1996; Mogl & Toledo, 1999). Many authors have focused on the habitat features and management practices in agroforestry systems which contribute to bird species richness and diversity (Godinho & Rabaca, 2011; Leal et al., 2013; Catarino et al., 2014), ant diversity (Philpott, 2005), and beetle diversity (Bos et al., 2007). Studies in shaded agroforestry systems have reported high levels of diversity of insect-eating birds (Perfecto et al., 1996; Rice & Greenberg, 2000; Van Bael et al., 2008), and bird species that are forest specialists (Schulze et al., 2004; Williams-Guillen et al., 2006; Bos et al., 2007; Philpott et al., 2008a; Francesconi et al., 2013). The forest reserves surrounding the agroforestry systems increased their conservation value (Munyuli, 2012). For example, in Brazil, the shaded cacao plantations contiguous to the forests showed rich bat communities (Faria & Baumgarten, 2007). Similarly, in Mexico, ant species richness is higher in coffee plantations near forests (Perfecto & Vandermeer, 2002; Armbrecht & Perfecto, 2003). The forest remnants have influenced the bird diversity and composition in cacao plantations of Sulawesi, Indonesia (Abrahamczyk et al., 2008), bat diversity in shaded coffee agroforests of Quindio, Colombia (Numa et al., 2005) and Chiapas, Mexico (Williams-Guillen & Perfecto, 2010). But there are a few studies which highlight that the agroforestry systems are no surrogates of forests as the source population and forest specialists are restricted only to forests. For example, the study carried out in Brazil showed that the bat, birds (Faria et al., 2006), ferns, frogs, and lizards' assemblages (Faria et al., 2007) were species-rich in cacao agroforests when forests were nearby.

TABLE 15.1 Inventory of Diverse Species Found in Different Agroforestry Systems

Types of Agroforestry Systems	Taxa (Species Richness)	Study Area	Author (Year)
Agrisilviculture	Birds (71)	Maharashtra, India	Mandan et al. (2014)
Forests, shade-grown coffee, live fence, pastures, riparian forests	Trees (145), bird (139), butterfly (119)	Honduras	Milder et al. (2010)
Coffee agroforests	Amphibians (26)	Mexico	Murrieta-Galindo et al. (2013)
Tea and coffee plantations	Bats (19)	Western ghats, India	Wordley et al. (2014)
Agrisilviculture, Tea gardens, homegardens	Mammals (14), birds (185), butterflies (133), reptiles (18), amphibians (14)	Assam, India	Yashmita-Ulman et al. (2021a)

Structurally complex vegetation supports higher wildlife diversity than simple vegetation structures. For example, forests that are structurally more complex protect more herpetofaunal species than agroforests that are structurally simple (Palacios et al., 2013). Canopy cover and leaf litter thickness (Wanger et al., 2010; Pineda et al., 2005), native vegetation (Murrieta-Galindo et al., 2013) have been shown to affect the species richness of amphibians, while canopy heterogeneity and cover influence the species richness of reptiles (Wanger et al., 2010; Godinho et al., 2011). Agroforestry practices that increase the complexity of plant architecture provide habitat heterogeneity and niche diversity. This feature has been found to support a high diversity of arthropods in alley cropping systems in Missouri, USA (Stamps et al., 2002), herpetofauna in Sulawesi, Indonesia (Wanger et al., 2009), and spiders in cacao agroforests in Indonesia (Stenchly et al., 2012). Canopy cover, tree species, and flowering species diversity influence bee abundance and species richness in shade coffee plantations in Chiapas, Mexico (Jha & Vandermeer, 2010). Agroforestry has been found to support similar levels of mobile taxa such as bats, birds, and mammals compared to their adjacent forests (Bhagwat et al., 2008). Studies have revealed six major reasons why agroforestry systems are more valuable for the conservation of biodiversity.

First, many agroforestry systems have been reported to conserve biodiversity and support species movement between them and adjacent forest reserves. For example, Williams-Guillen et al. (2006) reported that in Nicaragua, the mantled howling monkeys (*Alouatta palliata* Gray) used shade coffee plantations as an alternative habitat as it acted as a corridor between the forest fragments. A similar case is observed with the Golden-headed lion tamarin (*Leontopithecus chrysomelas* Kuhi) in cabruca cacao agroforests of Brazil (Rice & Greenberg, 2000). Vaughan et al. (2007) reported two species of sloth bear actively using and moving between riparian forests through cacao agroforests and live fences in Costa Rica. The coffee plantations and tea gardens acted as corridors connecting the forests for frugivorous and nectarivorous bats (Estrada et al., 2006; Mexico) and birds (Yashmita-Ulman et al., 2018; India), respectively.

Second, agroforestry systems cannot support the forest specialist species, but they have been found to support the generalists. For example, in a study conducted in Talamanca, Costa Rica, Harvey, and Villobos (2007) reported that the bird species composition in cacao and banana agroforestry systems contained fewer forest-dependent species and more non-forest species compared to adjacent forests. Similarly, Greenler & Ebersole (2015) observed in Costa Rica that 31 bird species were restricted only to rainforest habitats, and 106 bird species occurred only in live fences, shade-grown cacao, and riparian forest buffers. In southern Costa Rica, Daily et al. (2001) observed that 149 (55%) of the bird species were restricted only to forest areas, 60 (22%) occurred in both forests and agricultural areas, and 66 (24%) occurred in agricultural areas only. In Talamanca, Costa Rica, the species richness and diversity of dung beetles were highest in the forests, followed by agroforestry systems (Harvey et al., 2006b). Wal et al. (2012) found no specific relationship between structural complexity and bird species richness in agroforestry systems in Tabasco, Mexico.

Third, agroforestry systems protect species and habitat outside the protected areas. For example, Round et al. (2006) found that in Thailand, the mixed fruit orchards harbored 38%–48% bird species as found in the adjacent forest areas. Tylianakis et al. (2005) reported similar Hymenopteran communities in coffee agroforests as well as adjacent forest fragments in Ecuador, and Sakchoowong et al. (2008) reported high ant species richness in jungle tea plantations compared to forests in Thailand. The woody elements in riparian strips were observed to have higher bird density and diversity compared to that of maize (*Zea mays* L.) monoculture (Gillespie et al., 1995). In homegardens of southwestern Bangladesh, Kabir & Webb (2009) observed 419 plant species, out of which 59% were indigenous and six were species of conservation importance. Homegardens of Assam were found to support good populations of *Gekko gecko* Linnaeus (Yashmita-Ulman & Singh, 2021). In Sweden, the species richness of birds in pastureland increased with an increase in the shrub and tree species (Soderstrom et al., 2001) as it also increased the diversity and populations of invertebrates. In Sulawesi, Indonesia, agroforestry supported high numbers of butterfly species compared to other land uses (Schulze et al., 2004).

Fourth, agroforestry systems provide good foraging and nesting habitats for native and threatened wild animals alike. In shaded coffee, the tree canopies provide food resources such as fruits, flowers, nectar, arthropods (Parrish & Petit, 1995; Greenberg et al., 1997; Wunderle & Latta, 1998) and the understory provides bark insects (Roberts et al., 2000a) to the resident as well as migratory birds. The field margins with seed-rich habitats, perennial forbs provide diversified plants and insects as foraging options for birds (Vickery et al., 2009) in summer as well as winter when managed sympathetically (Vickery et al., 2002). The native tree species in shaded coffee plantations in Mexico provide roosting sites and food for bats (Pineda et al., 2005). The agrisilvicultural systems in Rajasthan and Uttaranchal support foraging birds and the tree components support cavity-nesting birds (Pandey & Mohan, 1993). Ciconiiformes species were found nesting on trees of cereal farms in lowland Nepal (Koju et al., 2019). Nesting of birds has been observed in various agroforestry systems in Sulawesi, Indonesia (Sodhi et al., 2005) and other systems such as windbreaks and hedgerows in Quebec, Canada (Jobin et al., 2000), hedgerows in Mexico (Zuria & Gates, 2013), field margins in England (Marshall et al., 2006), and homegardens in India (Yashmita-Ulman et al., 2017). The agroforestry systems have been documented to support nesting of various insects. For example, the shaded coffee provided ample woody debris for *Eciton burchelli* Westwood to nest (Roberts et al., 2000b). Some researchers also focused on the factors affecting the choice of nesting. For example, Jha & Vandermeer (2010) concluded that the amount of canopy cover affected the bee abundance of cavity-nesters and wood-nesters, and the number of flowering tree species affected the bee abundance of ground nesters in shaded coffee agroforests.

Fifth, agroforestry systems reduce the pressure on the adjoining protected forests. In Sumatra, Indonesia, Murniati et al. (2001) observed that households owning mixed gardens were less dependent on adjoining protected forests compared to households that did not own mixed gardens. Similar observations were made by Masozera & Alavalapati (2004) in Rwanda, where households that owned diversified agricultural fields were less dependent on adjacent forests. This reduction in dependence is due to the floral diversity of agroforestry systems. In Karnataka, Shastri et al. (2002) found 93 species with 952 individuals in a 1.7 ha area of agricultural landscapes, and homegardens had 20 to 40 tree species per unit area. Many researchers considered homegardens as a miniature forest (e.g., Ewel, 1999). In homegardens, the diversity of plant species varied from the lowest, being 27 species in Sri Lanka, to the highest, being 602 species in West Java (Kumar & Nair, 2004). Yashmita-Ulman et al. (2021b) recorded 516 species of plants in different agroforestry systems such as tea gardens, homegardens, and agrisilvicultural systems of Assam in which trees were retained due to their timber, medicinal, economic, or fodder value.

Sixth, agroforestry systems provide a congenial microclimate and microhabitats. The microclimatic conditions such as shade and moisture in agroforestry systems affect insect abundance and species diversity. Harvey et al. (2006b) observed that the diversity and densities of dung beetles and butterflies in southwestern Nicaragua were positively correlated with the microclimatic conditions. These conditions provided varied habitat factors for the different life cycle stages and numerous sources of nectar, pollen, insects, and hosts. Brandle et al. (2004) highlighted higher insect densities in windbreaks. Da Silva et al. (2009) found greater diversity of soil fauna in shaded cacao agroforests in Brazil due to diverse plants providing diverse microhabitats. The landscape heterogeneity and riparian buffers affected mammal species richness positively and grazing negatively. The occurrence of small mammals in agrisilviculture systems was influenced by the presence of trees and undisturbed understory (Klaa et al., 2005; Gonçalves et al., 2012). Migratory bird species have been observed to prefer agroforestry systems over rainforests due to the structure and resources in agroforestry (Greenler & Ebersole, 2015). McDermott et al. (2015) suggested that 25 to 40% canopy cover and emergent shade trees with a basal area of >5 m^2/ha were important for making shade-coffee and silvopastoral systems suitable for migratory birds. Most of the migratory bird species found in the cacao plantations were insectivorous and fed on the arthropods which were found in abundance (Greenberg et al., 2000). The floristic and structural diversity of tree cover influenced bird and bat communities (Harvey et al., 2006), birds (Reitsma et al., 2001; Clough et al., 2009), soil invertebrates (Rahman et al., 2012), and soil macrofauna (Mendes et al., 2011). Thus, these studies highlighted that the agroforestry systems supported biodiversity conservation to a large extent.

15.3 AGROFORESTRY FOR FOOD SECURITY AND NUTRITIONAL NEEDS

Harvesting and selling of Non-Timber Forest Produce (NTFP) is a traditional activity and main source of income for tribals (Affandi et al., 2017).

In Indonesia, many agroforestry systems that comprise NTFP cultivation can be witnessed in the form of Repong, Parak, mixed rubber gardens, Kemenyan gardens, and Reba Juma (Affandi et al., 2017). Reba Juma is an agrisilvicultural system which consists of tree crops such as candlenut, mangosteen, bitter bean, durian, garcinia, areca nut, and annual crops such as turmeric, papaya, chili, ginger, and banana. Agroforestry raises and stabilizes farm income. It leads to a reduction in farmers' vulnerability to shocks. In western Kenya, food insecurity during droughts and floods was reduced by 25% due to the presence of agroforestry and its associated income and livelihood strategies (Thorlakson & Neufeldt, 2012). Muliawati et al. (2018) found that 27% of the cultivated plants in home gardens met subsistence needs. In Java, Rahman et al. (2016) reported that the establishment of fruit trees could address the issue of temporary reduction in the availability of nutrient-rich foods. The levels of crop diversification increased in farmers who took agroforestry training (Pratiwi & Suzuki, 2019). A positive correlation was found between agroforestry systems and the increased consumption of Vitamin A-rich fruits, legumes, green leafy vegetables, and meat in children under the age of five at the national and regional level, respectively (Ickowitz et al., 2016). Homegardens were also reported to be repositories of medicinal plants. Prihatini et al. (2018) reported that homegardens are important sources of diverse nutritional and medicinal plant species. Non-Timber Forest Products (NTFPs) such as honey, firewood, and medicinal plants found in agroforestry systems also increased dietary diversity (Belcher et al., 2004) and were also a source of livelihood (Njurumana & Octavia, 2020). Thus, it is clear that agroforestry systems have immensely contributed to supplying food and nutritional supplements.

15.4 AGROFORESTRY FOR ENERGY PRODUCTION

Fuelwood production in agroforestry systems is a sustainable alternative to the collection of fuelwood from adjacent forests and provides multiple benefits to people. Agroforestry systems with energy-rich tree species such as *Acacia* sp., *Prosopis* sp., *Leucaena* sp., etc., can play a very important role in charcoal production. For the production of charcoal, trees with diameters greater than 4 cm are preferred. For fuel in kilns and kiln spacers, trees with diameters between 4 cm and 2 cm are used, and for firewood purposes, trees with diameters less than 2 cm are preferred (Okello et al., 2001). There have been studies on multipurpose agroforestry systems to test their potential for fuelwood production (Jama & Getahun, 1991; Jama et al., 2008; Siri & Raussen, 2003). Many researchers have studied agroforestry systems and tree species preferred for energy plantations. For example, in Tanzania, Kimarao et al. (2007) observed that in five years, the rotational woodlot system with N_2-fixing fast-growing tree species had the potential of producing 20–50 t/ha of wood. In Rwanda, Biggelaar (1996) documented 60 tree species commonly planted in agroforestry systems as scattered trees or in blocks and preferred for fuelwood use by local farmers. *Eucalyptus* sp. and *Grevillea robusta* A. Cunn. Ex R. Br. were commonly used in woodlots, and *Grevillea robusta* and *Toona sinensis* (A. Juss.) M. Roem. as hedgerows for fuelwood production (Balasubramanian & Sekayange, 1992). The average wood production through trees incorporated by farmers in farming systems such as alley cropping, boundary planting, short-term improved fallows with fast-growing trees in Rwanda was nearly 1.5 m^3 ha^{-1} $year^{-1}$ (Samyn, 1993). This also reduced the burden on women and children from carrying fuelwood loads from distant forests. The desirable characteristics for fuelwood production in agroforestry systems included the ability to fix N_2, coppice, grow on degraded lands, and to grow rapidly (Mead, 2005). Kerkhof (1990), through his research in woodlots of Rwanda, indicated that fast-growing and coppicing trees in woodlots have a high potential for fuelwood production.

In India, multipurpose trees are preferred for firewood use (Sharma & Kapoor, 2005). In Kerala, the fuelwood requirements of households are met from homesteads. Due to policies in India, people are forced to increase fuelwood production in areas such as home gardens, wastelands, agro-silviculture, along canals, and highways, i.e., areas outside forests. Very few studies have focused on agroforestry systems with respect to their potential in fuelwood supply and its valuation. Nearly 50% of firewood is collected from agro-silvicultural lands owned by the people themselves, 33% from highway plantations, and 17% from forests (Natrajan, 1996). Trees outside forests, e.g., home gardens, agro-silviculture, roadsides, canals, railways, and other common lands, are the source of 75% of fuelwood in India (Singh et al., 2021). In the year 2011–2012, agroforestry systems (64%) provided the highest percentage of fuelwood supply, followed by forests (23.8%), and the least from common property resources (12%) (Singh et al., 2021). Nearly 1116.6 million can be saved if firewood is used from agroforestry systems (Singh et al., 2021). From the above, it can be concluded that agroforestry systems contributed to fuelwood supply, foreign exchange savings, self-reliance, carbon neutrality, and energy security.

15.5 AGROFORESTRY FOR EROSION CONTROL AND SOIL FERTILITY IMPROVEMENT

The trees and shrubs present in agroforestry systems provide litter that prevents the dislodging of soil particles caused by the impact of raindrops. The factors affecting water erosion are erosivity of rainfall, degree of slope, length of slope, and vegetative cover (FAO & ITPS, 2015). In the United States, a reduction in sediment loss has been observed in watershed areas where corn and soybean are grown rotationally along with buffers planted on contours (Udawata et al., 2011a, b). Contour buffers also reduce soil erosion by decreasing the length and degree of slope. In the United States, there has been an improvement in water holding capacity, rates of infiltration, bulk density, and porosity through the deployment of alley cropping systems and grazing management (Kumar et al., 2008). Corn-soybean rotation with agroforestry buffer had infiltration rates 40 times (Seobi et al., 2005), 80 times (Kumar et al., 2008), and five times (Bharati et al., 2002) greater compared to that of claypan soils, pasture land, and corn and soybean fields alone, respectively. The agroforestry buffers in North and South America have lower water content and higher soil porosity, increasing their water storage capacity and infiltration rates, thereby reducing runoff (Udawatta et al., 2011a).

Agroforestry systems with perennial trees help in soil aggregate stability and carbon sequestration (Paudel et al., 2011). Trees in agroforestry fields help in organic matter and nutrient accumulation in biomass and runoff reduction from the croplands due to high root density. Reduction in soil erodibility is caused by aggregate stability brought about by an increase in soil organic matter (Al-Kaisi et al., 2014). The floral diversity of agroforestry systems enhances exudation from roots and creates a microhabitat favorable for diverse microbes and soil fauna. The root exudates, detritus, and excreta of fauna increase soil organic matter and bring greater stability in soil aggregates (Kremer & Kussman, 2011). Riparian buffers with a buffer strip between 4 and 7.5 m have been proven to be the most effective in filtering nutrients and sediments, thus reducing water erosion (Blanco-Canqui et al., 2006). The width of the buffer strip depends on the landscape, objectives, management practices, and weather. Dissolved nutrients are retained by wider buffers, and the amount of sediment loss is governed by the strip width of grass and shrubs.

The traditional method followed in agroforestry is to incorporate nitrogen-fixing trees and crops. By addition of the above-ground and below-ground organic manure and nutrient recycling, trees improve the physical, chemical, and biological properties of soil. Lee & Jose (2003) compared the soil organic matter and microbes in an alley cropping system involving cotton and pecan and monoculture cotton and found higher soil organic matter and microbes in the former. The soil pores under grass were three times and under agroforestry buffer strips were five times greater as compared to that of maize-soybean rotation (Udawatta et al., 2008a). Udawatta et al. (2008b) further showed that the soil under agroforestry buffers showed higher stability of soil aggregates, carbon, nitrogen, and enzymatic activity as compared to those under row crops. Litterfall is the main source (80%) of total nitrogen input for two years in Poplar plantation (Meiresonne et al., 2007). In alley cropping systems, the roots of trees uptake the nitrogen fertilizer from deep soil layers and through litterfall return it back to the soil surface (Zamora et al., 2009). Over five years, the phosphorus (P) content in zero to 20 cm soil depth of agricultural land under gray alder increased from 22.53 mg kg^{-1} to 47.20 mg kg^{-1} (Uri et al., 2002), and the P content increased with an increase in the age of mixed pine/oak trees (Slazak et al., 2010). This increase in P content can be attributed to the mycorrhizal activities in soil and P drawn deep down from soil layers (Diemont et al., 2006). In Germany, soil Ca under beech increased with stand age (Rumberger et al., 2004) due to the nutrient pump effect, which means capturing nutrients from deep soil layers and recycling them (van Noordwijk et al., 1996). The combined effect of N-fixing and non-N-fixing species is advantageous for the crops (Grunewald et al., 2007). Similarly, the intercropping of trees and crops is productive. Rivest et al. (2010) reported that the soybean (*Glycine max* (L.) Merr.) intercropped with hybrid poplar (*Populus nigra* L. x *P. maximowiczii* A. Henry) trees increased turnover of nutrients and their supply to poplar trees.

Agroforestry systems such as silvopasture, alley cropping, riparian buffers, and windbreaks have been reported to improve the physical, chemical (Sauer & Hernandez-Ramirez, 2011), and biological soil properties (Paudel et al., 2011). Microbial and macrofaunal activity, enzymatic activity are greater in agroforestry systems compared to grazing lands (Sauer & Hernandez-Ramirez, 2011). Agroforestry reduces the loss of antibiotics from surface and subsurface waters as the soils under agroforestry bond strongly with antibiotics due to the presence of different microbes, exudates, soil nutrients, and chemical bonding compared to grazing and agricultural lands (Chu et al., 2010). Yang et al. (2012) reported shortening of half-life, faster decomposition of herbicides and antibiotics in multispecies agroforestry systems due to the presence of enzymes, root exudates, and physical parameters. Lin et al. (2011) have concluded that the pollutants are metabolized by the multispecies microbial community present in the root area, which converts the pollutants to less harmful compounds by biochemical processes. These studies suggest that agroforestry improves nutrient uptake, nutrient cycling, and the efficiency of nutrient use in trees and crops.

15.6 AGROFORESTRY FOR CARBON SEQUESTRATION

The Clean Development Mechanism of the Kyoto Protocol approved afforestation and reforestation programs as an approach for sequestration of soil carbon (IPCC, 2007). The trees, shrubs, and grasses growing in agroforestry systems were seen capable of storing carbon in soil, aboveground and belowground biomass, living, and dead organisms (Pinho et al., 2012) through long growth periods (Morgan et al., 2010), thus rendering agroforestry systems more efficient in carbon sequestration as compared to that of monocultural agricultural systems or pasture (Sharrow & Ismail, 2004; Kirby & Potvin, 2007). It has been observed that light energy is utilized very efficiently in pine-pasture alley cropping systems, as compared to that of monoculture system. The photosynthesis begins earlier than the cash crop and continues even after its harvest due to the vegetation associated with agroforestry in a corn-soybean alley cropping system. The carbon sequestration potential differs according to the age of individuals, species composition, type of the system, geographic factors, environmental factors, and management practices (Pinho et al., 2012). Nair et al. (2009) observed that the ability to sequester carbon of the aboveground and belowground vegetation varied system-wise and biogeography-wise, for example, fodder bank agroforestry system in West African Sahel (0.29 Mg ha^{-1} yr^{-1}), mixed species stand in Puerto Rico (15.21 Mg ha^{-1} yr^{-1}), alley cropping system in Canada (1.25 Mg ha^{-1}), and silvopastoral system in Costa Rica (173 Mg ha^{-1}). Soil carbon is not only beneficial for agricultural productivity but also important for improving environmental services. Carbon can be stored in trees for extended periods of time in two ways, i.e., *in-situ* method (store carbon in biomass by increasing the rotation age of trees) and *ex-situ* method (store carbon in products; Montagnini & Nair, 2004). Restoring degraded cropland, pasturelands (Albrecht & Kandji, 2003), and riparian buffers (Udawatta & Jose, 2012) can help in effectively storing carbon and provide other ecosystem services.

The deep roots of trees, the quantity and quality of litterfall and pruned material in agroforestry systems not only recycle the mineral nutrients but also increase the soil carbon. The carbon sequestration of alley cropping systems is enhanced due to cycling of nutrients, diverse microbial communities, high soil carbon and nitrogen (Weerasekera et al., 2016). In countries such as Kenya, Madagascar, and Sumatra, agroforestry has helped in pressure reduction on forests (Silcshi et al., 2007), the largest sink of carbon on lands (Montagnini & Nair, 2004). According to IPCC (2000) report, by the year 2010 and 2040, 12,000 Mg C yr^{-1} and 17,000 Mg C yr^{-1} additional carbon can be sequestered by improving the current management practices in agroforestry. In addition to this, 630 million ha of unfruitful croplands and grasslands can be brought under agroforestry systems, resulting in an increased rate of carbon sequestration, i.e., from 3,91,000 Mg C yr^{-1} in 2010 to 5,86,000 Mg C yr^{-1} in 2040. This carbon can then be sold in carbon credit markets. Agroforestry reduces the carbon footprint by reducing the use of synthetic fertilizers, shade induced due to microclimate, reduced stress on trees, and improved soil aggregate stability (Udawatta et al., 2009).

15.7 AGROFORESTRY FOR IMPROVED AIR QUALITY AND WATER MANAGEMENT

Certain agroforestry systems, namely windbreaks and shelterbelts, protect buildings and roads from snow, reduce wind chills for livestock, protect crops, provide habitat for wildlife, produce oxygen, reduce carbon dioxide, reduce wind velocity and erosion, reduce noise pollution, particulate matter, and odor, etc. Tyndall & Colletti (2007) observed that when the shelterbelts are well planned and designed strategically, they can effectively reduce the odor released due to livestock operations.

Riparian buffers in agricultural fields help in the reduction of pollution from non-point sources (Udawatta et al., 2002; Lee et al., 2003; Anderson et al., 2009). The runoff water is cleaned by reducing the runoff velocity, promoting infiltration, deposition of sediments, and retention of nutrients. Lee et al. (2003) reported that an additional 20% of nutrients were removed when woody stems were incorporated with switchgrass (*Panicum virgatum* L.) in a riparian buffer strip. As trees have long growth periods, they utilize the nutrients prior to the cropping season and succeeding cropping season, thus increasing the efficiency of nutrient use in agroforestry systems. In northwest Florida, at 0.9 m depth, a reduction of 72% in nitrate-nitrogen was observed in pecan-cotton alley cropping system as compared to monoculture cotton by Allen et al. (2004). Nair et al. (2007) compared soil phosphorus concentration in grassland and slash pine (*Pinus elliottii* Engelm.) silvipastoral system, which was 20 years old, and found that the soil nutrient retention was higher in the silvipastoral system.

Using isotopic labeling, Guo & Zhao (2021) proved that different plants in alley cropping systems use water from different soil depths. In this study, the Poplar trees absorbed water from the deeper layers of soil, and the Walnut trees acted as windbreaks as per the observations made by Yukhnovskyi et al. (2021). In temperate regions, water availability is a major problem in agricultural systems. Alley cropping systems help in the uniform distribution of snowfall, and the snow melts in early spring and becomes available to the crops (Scholten, 1988). These systems obstruct wind and alter the wind flow both windward and leeward (Brandle et al., 2004), leading to reduced wind speed and evaporation from the surface of soil on the leeward side, thus helping in soil moisture conservation (McNaughton, 1988). In Germany, Grunewald et al. (2009) observed soil moisture content nearly 3% higher in the topsoil (0–10 cm) layer at a distance of 4–8 m away from the hedgerow in alley cropping systems using poplar (*Populus nigra* L. x *Populus suaveolens* Fisch) after one year. Anderson et al. (2009) reported improved water storage in an alley cropping system using oak in hedgerows. The soil water use is efficient in intercropping in comparison to monocropping (Fernandez et al., 2008) due to the process called hydraulic lift in which the deep-growing tree roots draw the water from underground and release it in upper soil layers, thus making water available to annual intercrops (Burgess et al., 2001). Perry et al. (2001) reported that short rotation coppices of poplar could more effectively reduce the water yielded per year, amount of peak flow discharged from precipitation per year, and floods as compared to annual crops. Linderson et al. (2007) observed that certain clones of willow increased the efficiency of water use but decreased the level of available water. Busch (2009) suggested the use of short rotation coppice in areas having less precipitation and low water availability. Studies must be encouraged to understand the impact of trees on the availability of water so that species selection and species combination can be recommended according to the site conditions.

15.8 AGROFORESTRY FOR CONTROL OF PESTS

Agroforestry systems with diverse trees create an environment that helps in the natural control of pests. Many studies have reported that ants predate upon arthropods, thus regulating the pest attack in coffee agroforests (Way & Khoo, 1992; Schmitz et al., 2000; Philpott & Armbrecht, 2006). Various ant species predate upon pests such as the coffee berry borer (*Hypothenemus hampei* Ferrari) (Philpott & Armbrecht, 2006), and pupae of *Ceratitis capitata* Wiedemann (Armbrecht & Perfecto, 2003). Ants predate upon canopy arthropods (Philpott et al., 2004) and herbivores occupying the floor of coffee agroforests (Philpott et al., 2008a). Ants (*Azecta* sp.) found in shaded coffee of Mexico limit spider populations and serve as biological control agents (Vandemeer et al., 2002). Farmers promote ants in citrus and coconut gardens, banana groves, sweet potato, corn, bean, and squash fields because they act as natural predators for pests that attack these crops (Perfecto & Castineiras, 1998). The effects of ants on several arthropod orders depend on the species and seasons (Philpott et al., 2004). A few studies have been done on the effect of management intensity on ant predation (Philpott et al., 2008b), and ant nesting sites (Philpott & Foster, 2005).

Out of 1293 bat species currently known in the world, two-thirds are insectivorous. Insectivorous bats control the populations of herbivorous and disease-carrying insects (Wordley et al., 2014). In coffee plantations of Mexico, Williams-Guillen et al. (2008) reported an 84% increase in insect pests if the bats were excluded in the wet season. Kalka et al. (2008) found a growth of 65% in the insect population and a 68% increase in leaf damage if bats were excluded from coffee plantations. Bats have been reported to forage on mosquitoes in coffee plantations (Wordley et al., 2014). Karp et al. (2013) reported that birds predate upon the berry borer (*Hypothenemus hampei* Ferrari). When biological predators such as birds and bats are present in cacao plantations, the yield is higher by 31% (Mas et al., 2013). The above literature review confirms the potential of agroforestry systems in increasing the populations of biological predators and providing habitats for them to survive.

15.9 AGROFORESTRY FOR POLLINATION

The pollinators depend on high floral diversity with flowering plants to meet their nutritional requirements (IPBES, 2016). Many agroforestry systems such as hedgerows, riparian buffers, windbreaks, etc., are used by bees (Morandin et al., 2013; M'Gonigle et al., 2015; Cole et al., 2017; Garratt et al., 2017; Ponisio et al., 2017; Sutter et al., 2018; Kremen et al., 2018; Timberlake et al., 2019), butterflies (Cole et al., 2017; Dainese et al., 2017;

Luppi et al., 2018; Rosas-Ramos et al., 2019), moths (Rosas-Ramos et al., 2019; Froidevaux et al., 2019), and flies (Morandin et al., 2013; Haenke et al., 2014; M'Gonigle et al., 2015; Cole et al., 2017; Schirmel et al., 2018). The use of agroforestry systems by pollinators basically depends on the presence of pollens and flowering plants having nectar and sugar-rich sources in this system. Biological indicators of good agri-environment schemes are the pollinating agents (Sepp et al., 2004). Marshall et al. (2006) observed a high number of bees in agricultural fields with broad marginal strips. This indicated that the agricultural fields with boundary strips have good floral resources. The fruit bats disperse the seeds of forests and shade trees, e.g., white cedar (*Melia* sp.), wild figs (*Ficus* sp.), and wild date (*Phoenix sylvestris* (L.) Roxb.) (Wordley et al., 2014). Studies have been conducted on how changing precipitation affects coffee flowering phenology and coffee plant-pollinator relationship (Peters & Carroll, 2012). Wild bees were found foraging on native plants planted in hedgerows (Morandin et al., 2013) and on cherry trees in orchards (Kay et al., 2020). The trees in agroforestry systems also offer nesting opportunities for cavity nesters. Such studies emphasize the importance of indigenous flowering fruit trees in agroforestry systems for maintaining wild bee populations and associated pollination services. Timberlake et al. (2019) reported that hedgerows produced twice and four times the amount of nectar per unit of land compared to that of forests and grasslands, respectively. Studies conducted by Staley et al. (2016, 2018) investigated the effect of management on pollinators and found that trimming of hedgerows had a negative effect on pollinators.

Few studies show that agroforestry can contribute to pollination services that can lead to an increase in crop yield and quality of produce. Sutter et al. (2018) observed that the presence of hedgerows increased the yields of winter oilseed rape (*Brassica napus* L.). Smith & Lewis (1972) found that artificial windbreaks adjacent to apple orchards increased pollinator abundance, leading to an increase in fruit set by 20% to 30% with no reduction in fruit size. Thus, suggesting that planted windbreaks may increase yield due to an increase in pollinator activity. Factors influencing the potential of agroforestry practices in pollination services include attributes of the pollinator, field area, crop type, floristic composition, presence of forests, and landscape. In Canada, Kremen et al. (2004) demonstrated that in a total of 1.2 km of field area, 30% of natural vegetation is enough to provide effective pollination services by native bees for watermelon (*Citrullus lanatus* Thunb.). Castle et al. (2019) observed an increase in the quantity and quality of strawberries cultivated in hedgerows close to forested areas compared to strawberries grown on isolated hedgerows, which might be due to better pollination services in areas near the forest. Based on yields estimated from pollination services and pest control, Morandin et al. (2016) concluded that in hedgerows farmers require seven years to recover the initial input costs. Another study conducted by Sabino et al. (2022) in the Amazon showed that agroforestry area increased in number (3.27%) and size (23.18%) between the years 2006 and 2017. In the same duration, a 45.61% increase was recorded in the crop production of forested areas, where a high number of crops were dependent on pollinators such as bees and beetles for pollination. Studies on pollination in agroforestry are few, and thus studies focusing on documenting and mapping pollinators of essential crops are suggested for higher production and conservation.

15.10 CONCLUSION

It is evident from the above that the agroforestry systems support numerous ecosystem services and environmental benefits which cannot be overlooked by policymakers, farmers, and practitioners. In the era of climate change and environmental sustainability, the role of agroforestry in being an alternative to monocropping, being convergent with the traditional farming practices, and its contribution towards ecosystem services needs to be fully explored. Agroforestry not only alleviates poverty but also helps in carbon sequestration, biodiversity conservation, improvement in soil and water quality for the society as a whole. Therefore, these systems must be promoted all over the world.

KEYWORDS

- **biodiversity conservation**
- **carbon sequestration**
- **energy production**
- **food security**
- **pest control**
- **pollination**

REFERENCES

Abrahamczyk, S., Kessler, M., Putra, D. D., Waltert, M., & Tscharntke, T., (2008). The value of differently managed cacao plantations for forest bird conservation in Sulawesi, Indonesia. *Bird Conserv. Int.*, *18*, 349–362.

Affandi, O., Zaitunah, A., & Batubara, R., (2017). Potential economic and development prospects of non timber forest products in community agroforestry land around Sibolangit Tourism Park. *For. Soc.*, *1*, 68–77.

Albrecht, A., & Kandji, S. T., (2003). Carbon sequestration in tropical agroforestry systems. *Agric. Ecosyst. Environ.*, *99*, 15–27.

Al-Kaisi, M. M., Douelle, A., & Kwaw-Mensah, D., (2014). Soil microaggregate and macroaggregate decay over time and soil carbon change as influenced by different tillage systems. *J. Soil Water Conserv.*, *69*, 574–580.

Allen, S., Jose, S., Nair, P. K. R., Brecke, B. J., Nkedi-Kizza, P., & Ramsey, C. L., (2004). Safety net role of tree roots: Experimental evidence from an alley cropping system. *For. Ecol. Manag.*, *192*, 395–407.

Anderson, S. H., Udawatta, R. P., Seobi, T., & Garrett, H. E., (2009). Soil water content and infiltration in agroforestry buffer strips. *Agrofor. Syst.*, *75*, 5–16.

Armbrecht, I., & Perfecto, I., (2003). Litter-twig dwelling ant species richness and predation potential within a forest fragment and neighboring coffee plantations of contrasting habitat quality in Mexico. *Agric. Ecosyst. Environ.*, *97*, 107–115.

Balasubramanian, V., & Sekayange, L., (1992). Effet de la culture en couloir sur les proprie´te´s du sol et les performances des arbustes et des cultures vivrie`resdans un environnement semi-aride au Rwanda. *Bull. Reseau. Erosion*, *12*, 180–190.

Belcher, B., Imang, N., & Achdiawan, R., (2004). Rattan, rubber, or oil palm: Cultural and financial considerations for farmers in Kalimantan. *Econ. Bot.*, *58*, 77–87.

Bhagwat, S. A., Willis, K. J., Birks, J. B., & Whittaker, R. J., (2008). Agroforestry: A refuge for tropical biodiversity? *Trends Ecol. Evol.*, *23*, 261–267.

Bharati, L., Lee, K. H., Isenhart, T. M., & Schultz, R. C., (2002). Soil-water infiltration under crops, pasture, and established riparian buffer in Midwestern USA. *Agroforest. Syst.*, *56*, 249–257.

Biggelaar, D. C., (1996). *Farmer Experimentation and Innovation: A Case Study of Knowledge Generation Processes in Agroforestry Systems in Rwanda*. Community Forestry Case Study Series 12.

Blanco-Canqui, H., Gantzer, C. J., & Anderson, S. H., (2006). Performance of grass barriers and filter strips under interrill and concentrated flow. *J. Environ. Qual.*, *35*, 1969–1974.

Bos, M. M., Steffan-Dewenter, I., & Tscharntke, T., (2007). The contribution of cacao agroforests to the conservation of lower canopy ant and beetle diversity in Indonesia. *Biodivers. Conserv.*, *16*, 2429–2444.

Brandle, J. R., Hodges, L., & Zhou, X. H., (2004). Windbreaks in North American agricultural systems. *Agrofor. Syst.*, *61, 62*, 65–78.

Burgess, S. S. O., Adams, M. A., Turner, N. C., White, D. A., & Ong, C. K., (2001). Tree roots: Conduits for deep recharge of soil water. *Oecologia*, *126*, 158–165.

Busch, G., (2009). The impact of short rotation coppice cultivation on groundwater recharge—A spatial (planning) perspective. Landbauforschung-vTI. *Agric. For. Res.*, *3*, 207–222.

Castle, D., Grass, I., & Westphal, C., (2019). Fruit quantity and quality of strawberries benefit from enhanced pollinator abundance at hedgerows in agricultural landscapes. *Agric. Ecosyst. Environ.*, *275*, 14–22.

Catarino, L., Godinho, C., Pereira, P., Luis, A., & Rabaca, J. E., (2014). Can birds play a role as high nature value indicators of montado system? *Agrofor. Syst.*, *90*, 45–56.

Chu, B., Goyne, K. W., Anderson, S. H., Lin, C. H., & Udawatta, R. P., (2010). Veterinary antibiotic sorption to agroforestry buffer, grass buffer and cropland soils. *Agrofor. Syst.*, *79*, 67–80.

Clough, Y., Putra, D. D., Pitopang, R., & Tscharntke, T., (2009). Local and landscape factors determine functional bird diversity in Indonesian cacao agoforestry. *Biol. Conserv.*, *142*, 1032–1041.

Cole, L. J., Brocklehurst, S., Robertson, D., Harrison, W., & McCracken, D. I., (2017). Exploring the interactions between resource availability and the utilization of semi-natural habitats by insect pollinators in an intensive agricultural landscape. *Agric. Ecosyst. Environ.*, *246*, 157–167.

Da Silva, M. K. M., Da Gama-Rodrigues, E. F., Da Gama-Rodrigues, A. C., Machado, R. C. R., & Baligar, V. C., (2009). Soil and litter fauna of cacao agroforestry systems in Bahia, Brazil. *Agrofor. Syst.*, *76*, 127–138.

Daily, G. C., Ehrlich, P. R., & Sanchez-Azofeifa, G. A., (2001). Countryside biogeography: Use of human-dominated habitats by the avifauna of Southern Costa Rica. *Ecol. Appl.*, *11*, 1–13.

Dainese, M., Montecchiari, S., Sitzia, T., Sigura, M., & Marini, L., (2017). High cover of hedgerows in the landscape supports multiple ecosystem services in Mediterranean cereal fields. *J. Appl. Ecol.*, *54*, 380–388.

Diemont, S. A. W., Martin, J. F., Levy-Tacher, S. I., Nigh, R. B., Lopez, P. R., & Golicher, J. D., (2006). Lacandon maya forest management: Restoration of soil fertility using native tree species. *Ecol. Eng.*, *28*, 205–212.

Djossa, B. A., Fahr, J., Wiegand, T., Ayihouenou, B. E., Kalko, E. K., & Sinsin, B. A., (2008). Land use impact on *Vitellaria paradoxa* C.F. Gaerten. Stand structure and distribution patterns: A comparison of biosphere reserve of Pendjari in Atacora District in Benin. *Agrofor. Syst.*, *72*, 205–220.

Estrada, C. G., Damon, A., Hernandez, C. S., Pinto, L. S., & Nunez, G. I., (2006). Bat diversity in montane rainforest and shaded coffee under different management regimes in Southeastern Chiapas, Mexico. *Biol. Conserv.*, *132*, 351–361.

Ewel, J. J., (1999). Natural systems as models for the design of suitable systems of land use. *Agrofor. Syst.*, *45*, 1–21.

FAO; ITPS, (2015). *Status of the World's Soil Resources (SWSR) – Main Report.* Food and Agriculture Organization of the United Nations and Intergovernmental Technical Panel on Soils, Rome, Italy.

Faria, D., & Baumgarten, J., (2007). Shaded cacao plantations (*Theobroma cacao*) and bat conservation in southern Bahia, Brazil. *Biodivers. Conserv.*, *16*, 291–312.

Faria, D., Laps, R. R., Baumgarten, J., & Cetra, M., (2006). Bat and bird assemblages from forests and shade cacao plantations in two contrasting landscapes in the Atlantic Forest of Southern Bahia, Brazil. *Biodivers. Conserv.*, *15*, 587–612.

Faria, D., Paciencia, M. L. B., Dixo, M., Laps, R. R., & Baumgarten, J., (2007). Ferns, frogs, lizards, birds and bats in forest fragments and shade cacao plantations in two contrasting landscapes in the Atlantic Forest, Brazil. *Biodivers. Conserv.*, *16*, 2335–2357.

Fernandez, M., Gyenge, J., Licata, J., Schlichter, T., & Bond, B., (2008). Belowground interactions for water between trees and grasses in a temperate semiarid agroforestry system. *Agrofor. Syst., 74*, 185–197.
Francesconi, W., Nair, P. K. R., Levey, D. J., Daniels, J., & Cullen, L. Jr., (2013). Butterfly distribution in fragmented landscapes containing agroforestry practices in southeastern Brazil. *Agrofor. Syst., 87*, 1321–1338.
Froidevaux, J. S. P., Broyles, M., & Jones, G., (2019). Moth responses to sympathetic hedgerow management in temperate farmland. *Agric. Ecosyst. Environ., 270*, 55–64.
Garratt, M. P. D., Senapathi, D., Coston, D. J., Mortimer, S. R., & Potts, S. G., (2017). The benefits of hedgerows for pollinators and natural enemies depends on hedge quality and landscape context. *Agric. Ecosyst. Environ., 247*, 363–370.
Gillespie, A. R., Miller, B. K., & Johnson, K. D., (1995). Effects of ground cover on tree survival and growth in filter strips of the Cornbelt Region of the midwestern US. *Agric. Ecosyst. Environ., 53*, 263–270.
Godinho, C., & Rabaca, J. E., (2011). Birds like it corky: The influence of habitat features and management of 'montados' in breeding bird communities. *Agrofor. Syst., 82*, 183–195.
Godinho, S., Santos, A. P., & Sa-Sousa, P., (2011). Montado management effects on the abundance and conservation of reptiles in Alentejo, Sothern Portugal. *Agrofor. Syst., 82*, 197–207.
Gonclaves, P., Alcobia, S., Simoes, L., & Santos-Reis, M., (2012). Effects of management options on mammal richness in a Mediterranean agro-silvo-pastoral system. *Agrofor. Syst., 85*, 383–395.
Greenberg, R., Bichier, P., & Angon, A. C., (2000). The conservation value for birds of cacao plantations with diverse planted shade in Tabaco, Mexico. *Anim. Conserv., 3*, 105–112.
Greenberg, R., Bichier, P., & Sterling, J., (1997). Bird populations in rustic and planted shade coffee plantations of eastern Chiapas, Mexico. *Biol. Conserv., 80*, 235–247.
Greenler, S. M., & Ebersole, J. J., (2015). Bird communities in tropical agroforestry ecosystems: An underappreciated conservation resource. *Agrofor. Syst., 89*, 691–704.
Grunewald, H., Bohm, C., Barwolff, M., Wollecke, J., Quinkenstein, A., Schwarz, K. U., & Hoffmann, J., (2009). *Okologische Aspekte von Agroforstsystemen 2* (pp. 233–263). Symposium Energiepflanzen 2009 abgehalten in Berlin am 17.–18. November, Fachagenturfu¨r Nachwachsende Rohstoffe (FNR).
Guo, H., & Zhao, Y., (2021). Using isotopic labelling to investigate root water uptake in an alley cropping system within Taklimakan Desert Oasis, China. *Agrofor. Syst., 95*, 907–918.
Haenke, S., Kovacs-Hostyanszki, A., Frund, J., Batary, P., Jauker, B., Tscharntke, T., & Holzschuh, A., (2014). Landscape configuration of crops and hedgerows drives local syrphid fly abundance. *J. Appl. Ecol., 51*, 505–513.
Hartley, M., (2002). Rationale and methods for conserving biodiversity in plantation forests. *For. Ecol. Manag., 155*, 81–95.
Harvey, C. A., & Villalobos, J. A. G., (2007). Agroforestry systems conserve species-rich but modified assemblages of tropical birds and bats. *Biodivers. Conserv., 16*, 2257–2292.
Harvey, C. A., Gonzalez, J., & Somarriba, E., (2006b). Dung beetle and terrestrial mammal diversity in forests, indigenous agroforestry systems and plantain monocultures in Talamanca, Costa Rica. *Biodivers. Conserv., 15*, 555–585.
Harvey, C. A., Medina, A., Sanchez, D. M., Vilchez, S., Hernandez, B., Saenz, J. C., Maes, J. M., et al., (2006a). Patterns of animal diversity in different forms of tree cover in agricultural landscapes. *Ecol. Appl., 16*, 1986–1999.

Ickowitz, A., Rowland, D., Powell, B., Salim, M. A., & Sunderland, T., (2016). Forests, trees, and micronutrient-rich food consumption in Indonesia. *PLoS One, 11*, 1–15.

International Assessment of Agricultural Science and Technology for Development Executive summary of the synthesis report, (2008). http://www.agassessment.org/docs/SR_ExecSum280508English.htm (accessed on 22 April 2024).

IPBES, (2016). *Assessment Report on Pollinators, Pollination and Food Production.* Intergovernmental Science-Policy Platform on Biodiversity and Ecosystem Services: Bonn, Germany.

IPCC, (2000). *Land Use, Land-Use Change, and Forestry* (p. 375). A special report of the IPCC. Cambridge University Press, Cambridge.

IPCC., (2007). *Intergovernmental Panel on Climate Change 2007.* Synthesis Report, http://www.ipcc.ch/pdf/assessment-report/ar4/syr/ar4_syr.pdf (accessed on 22 April 2024).

IUCN, (2022). *The IUCN Red List of Threatened Species.* https://www.iucnredlist.org/about/background-history (accessed on 22 April 2024).

Jama, B. A., & Getahun, A., (1991). Fuelwood production from *Leuceana leucocephala* established in fodder crops at Mtwapa, Coast Province Kenya. *Agrofor. Syst., 16*, 119–128.

Jama, B. A., Mutegi, J. K., & Njui, A. N., (2008). Potential of improved fallows to increase household and regional fuelwood supply: Evidence from western Kenya. *Agrofor. Syst., 73*, 155–166.

Jha, S., & Vandermeer, J. H., (2010). Impacts of coffee agroforestry management on tropical bee communities. *Biol. Conserv., 143*, 1423–1431.

Jobin, B., Choiniere, L., & Belanger, L., (2000). Bird use of three types of field margins in relation to intensive agriculture in Quebec, Canada. *Agric. Ecosyst. Environ., 1664*, 1–13.

Kabir, E. M., & Webb, E. L., (2008). Can homegardens conserve biodiversity in Bangladesh? *Biotropica, 40*, 95–103.

Kalka, M. B., Smith, A. R., & Kalko, E. K. V., (2008). Bats limit arthropods and herbivory in a tropical forest. *Science, 320*, 71.

Karp, D. S., Mendenhall, C. D., Sandí, R. F., Chaumont, N., Ehrlich, P. R., Hadly, E. A., & Daily, G. C., (2013). Forest bolsters bird abundance, pest control and coffee yield. *Ecol. Lett., 16*, 1339–1347.

Kay, S., Kuhn, E., Albrecht, M., Sutter, L., Szerencsits, E., & Herzog, F., (2020). Agroforestry can enhance foraging and nesting resources for pollinators with focus on solitary bees at the landscape scale. *Agrofor. Syst., 94*, 379–387.

Kerkhof, P., (1990). *Agroforestry in Africa: A Survey of Project Experience*. Panos Institute, London.

Kimaro, A. A., Timmer, V. R., Mugasha, A. G., Chamshama, S. A. O., & Kimaro, D. A., (2007). Nutrient use efficiency and biomass production of tree species for rotational woodlot systems in semi-arid Morogoro, Tanzania. *Agrofor. Syst., 71*, 175–184.

Kirby, K. R., & Potvin, C., (2007). Variation in carbon storage among tree species: Implications for the management of a smallscale carbon sink project. *For. Ecol. Manag., 246*, 208–221.

Klaa, K., Mill, P. J., & Incoll, L. D., (2005). Distribution of small mammals in a silvoarable agroforestry system in northern England. *Agrofor. Syst., 63*, 101–110.

Koju, R., Maharajan, B., Gosai, K. R., Kittur, S., & Sundar, K. S. G., (2019). Ciconiiformes nesting on trees in cereal-dominated farmlands: Importance of scattered trees for heronries in lowland Nepal. *Waterbirds, 42*, 355–453.

Kremen, C., MGonigle, L. K., & Ponisio, L. C., (2018). Pollinator community assembly tracks changes in floral resources as restored hedgerows mature in agricultural landscapes. *Front. Ecol. Evol.*, *6*, 170.
Kremen, C., Williams, N. M., Bugg, R. L., Fay, J. P., & Thorp, R. W., (2004). The area requirements of an ecosystem service: Crop pollination by native bee communities in California. *Ecol. Lett.*, *7*, 1109–1119.
Kremer, R. J., & Kussman, R. D., (2011). Soil quality in a pecan – Kura clover alley cropping system in the Midwestern USA. *Agrofor. Syst.*, *83*, 213–223.
Kumar, B. M., & Nair, P. K. R., (2004). Tropical homegardens: A time-tested example of sustainable agroforestry. *Advances in Agroforestry* (Vol. 3). Springer, Dordrecht.
Kumar, S., Anderson, S. H., Bricknell, L. G., & Udawatta, R. P., (2008). Soil hydraulic properties influenced by agroforestry and grass buffers for grazed pasture systems. *J. Soil Water Conserv.*, *63*, 224–232.
Leal, A. I., Correia, R. A., Palmeirim, J. M., & Granadeiro, J. P., (2013). Does canopy pruning affect foliage-gleaning birds in managed cork oak woodlands? *Agrofor. Syst.*, *87*, 355–363.
Lee, K. H., & Jose, S., (2003). Soil respiration and microbial biomass in a pecan–cotton alley cropping system in southern USA. *Agrofor. Syst.*, *58*, 45–54.
Lee, K. H., Isenhart, T. M., & Schultz, R. C., (2003). Sediment and nutrient removal in an established multi-species riparian buffer. *J. Soil Water Conserv.*, *58*, 1–8.
Lin, C. H., Lerch, R. N., Goyne, K. W., & Garrett, H. E., (2011). Reducing herbicides and veterinary antibiotic losses from agroecosystems using vegetative buffers. *J. Environ. Qual.*, *40*, 791–799.
Lindenmayer, D. B., & Hobbs, R. J., (2004). Fauna conservation in Australian plantation forests—A review. *Biol. Conserv.*, *119*, 151–168.
Linderson, M., Iritz, Z., & Lindroth, A., (2007). The effect of water availability on stand level productivity, transpiration, water use efficiency and radiation use efficiency of field-grown willow clones. *Biomass Bioenerg.*, *31*, 460–468.
Luppi, M., Dondina, O., Orioli, V., & Bani, L., (2018). Local and landscape drivers of butterfly richness and abundance in a human-dominated area. *Agric. Ecosyst. Environ.*, *254*, 138–148.
M'Gonigle, L. K., Ponisio, L. C., Cutler, K., & Kremen, C., (2015). Habitat restoration promotes pollinator persistence and colonization in intensively managed agriculture. *Ecol. Appl.*, *25*, 1557–1565.
Mandan, J. K., Patil, V., Narkhede, S. S., Gunaga, R. P., & Bhave, S., (2014). Birds: A bio-control agent in farmland ecosystem. *Indian J. of Agrofor.*, *16*, 15–20.
Marshall, E. J. P., West, T. M., & Kleijn, D., (2006). Impacts of an agri-environment field margin prescription on the flora and fauna of arable farmland in different landscapes. *Agric. Ecosyst. Environ.*, *113*, 36–44.
Masozera, M. K., & Alavalapati, J. R. R., (2004). Forest dependency and its implications for protected areas management: A case study from the Nyungwe Forest Reserve, Rwanda. *Scand. J. For. Res.*, *19*, 85–92.
Mass, B., Clough, Y., & Tscharntke, T., (2013). Bats and birds increase crop yield in tropical agroforestry landscapes. *Ecol. Lett.*, *16*, 1480–1487.
Mcdermott, M. E., Rodewald, A. D., & Matthews, S. N., (2015). Managing tropical agroforestry for conservation of flocking migratory birds. *Agrofor. Syst.*, *89*, 383–396.
McNaughton, K. G., (1988). Effects of windbreaks on turbulent transport and microclimate. *Agric. Ecosyst. Environ.*, *22*, 17–39.

McNeely, J. A., & Schroth, G., (2006). Agroforestry and biodiversity conservation-traditional practices, present dynamics and lessons for the future. *Biodivers. Conserv.*, *15*, 549–554.
McNeely, J. A., (2004). Nature vs nurture: Managing relationships between forests, agroforestry and wild biodiversity. *Agrofor. Syst.*, *61*, 155–165.
Mead, D. J., (2005). Forests for energy and the role of planted trees. *Crit. Rev. Plant Sci.*, *24*, 407–421.
Meiresonne, L., Schrijver, A. D., & Vos, B., (2007). Nutrient cycling in a poplar plantation (*Populus trichocarpa Populus deltoides* 'Beaupre'') on former agricultural land in northern Belgium. *Can. J. For. Res.*, *37*, 141–155.
Mendes, S. M., Santos, J., Freitas, H., & Sousa, J. P., (2011). Assessing the impact of understory vegetation cut on soil epigenic macrofauna from a cork-oak Montado in South Portugal. *Agrofor. Syst.*, *82*, 139–148.
Milder, J. C., DeClerck, F. A. J., Sanfiorenzo, A., Sanchez, D. M., Tobar, D. E., & Zuckerberg, B., (2010). Effects of farm and landscape management on bird and butterfly conservation in western Honduras. *Ecosphere*, *1*, 1–22.
Millennium Ecosystem Assessment, (2005). *Ecosystems and Human Well-Being: Biodiversity Synthesis*. World Resources Institute, Washington, DC.
Moguel, P., & Toledo, V. M., (1999). Biodiversity conservation in traditional coffee systems of Mexico. *Conserv. Biol.*, *13*, 11–21.
Montagnini, F., & Nair, P. K. R., (2004). Carbon sequestration: An underexploited environmental benefit of agroforestry systems. *Agrofor. Syst.*, *61*, 281–295.
Morandin, L. A., & Kremen, C., (2013). Bee preference for native versus exotic plants in restored agricultural hedgerows. *Restor. Ecol.*, *21*, 26–32.
Morandin, L. A., Long, R. F., & Kremen, C., (2016). *Pest Control and Pollination Cost-Benefit Analysis of Hedgerow Restoration in a Simplified Agricultural Landscape, 109*, 1020–1027.
Morgan, J. A., Follett, R. F., Allen, L. H., Grosso, S. D., Derner, J. D., Dijkstra, F., Franzluebbers, A., et al., (2010). Carbon sequestration in agricultural land of the United States. *J. Soil Water Conserv.*, *65*, 6A–13A.
Muliawati, E. S., Budiastuti, M. T. S., Suprayogo, D., & Sutrisno, J., (2018). Agrobiodiversity in rural homegardens as food reserves for climate change adaptation (case study: Samin Sub-Watershed, Central Java, Indonesia). *Bulg. J. Agric. Sci.*, *24*, 759–767.
Munyuli, M. B. T., (2012). Butterfly diversity from farmlands of central Uganda. *Psyche, 2012*, 1–23. Article 481509.
Murniati, M. D. A., Garrity, D. P. N., & Gintings, A. N., (2001). The contribution of agroforestry systems to reducing farmers' dependence on the resources of adjacent national parks: A case study from Sumatra, Indonesia. *Agrofor. Syst.*, *52*, 171–184.
Murrieta-Galindo, R., Gonzalez-Romero, A., Lopez-Barrera, F., & Parra-Olea, G., (2013). Coffee agrosystems: An important refuge for amphibians in central Veracruz, Mexico. *Agrofor. Syst.*, *87*, 767–779.
Nair, P. K. R., Kumar, B. M., & Nair, V. D., (2009). Agroforestry as a strategy for carbon sequestration. *J. Plant Nutr. Soil Sci.*, *172*, 10–23.
Nair, V., Haile, S. G., Michel, G. A., & Nair, P. K. R., (2007). Environmental quality improvement of agricultural lands through silvopasture in southeastern United States. *Sci. Agric.*, *64*, 513–519.
Natarajan, I., (1996). Trends in firewood consumption in rural India, National Council of Applied Economic Research (NCAER). *Margin*, *28*, 41–47.

Njurumana, G. N., & Octavia, D., (2020). Conservation of NTFP's species through agroforestry for community livelihoods in Sikka, east Nusa Tenggara. *J. Sylva Indonesiana, 3*, 1–16.

Numa, C., Verdu, J. R., & Sanchez-Palomino, P., (2005). Phyllostomid bat diversity in a variegated coffee landscape. *Biol. Conserv., 122*, 151–158.

Okello, B. D., O'Connor, T. G., & Young, T. P., (2001). Growth, biomass estimates and charcoal production of *Acacia drepanolobium* in Laikipia, Kenya. *For. Ecol. Manag., 142*, 143–153.

Ouinsavi, C., Sokpon, N., & Bada, S. O., (2005). Utilization and traditional strategies of in situ conservation of iroko (*Miliciaexcelsa* Welw. C.C. Berg) in Benin. *For. Ecol. Manag., 207*, 341–350.

Palacios, C. P., Aguero, B., & Simonetti, J. A., (2013). Agroforestry systems as habitat for herpetofauna: Is there supporting evidence? *Agrofor. Syst., 87*, 517–523.

Pandey, D. N., & Mohan, D., (1993). Nest site selection by cavity nesting birds on *Melia azedarach* L. & management of multiple use forests. *J. Bombay Nat. Hist. Soc., 90*, 58–61.

Parrish, J. D., & Petit, L. J., (1995). The value of shade coffee plantations to tropical birds: Landscape and vegetation effects. In: Lockeretz, W., (ed.), *Environmental Enhancement Through Agriculture* (pp. 113–124). Tufts University, School of Nutrition Science and Policy, Boston, Massachusetts, USA.

Paudel, B. R., Udawatta, R. P., & Anderson, S. H., (2011). Agroforestry and grass buffer effects on soil quality parameters for grazed pasture and row-crop systems. *App. Soil Ecol., 48*, 125–132.

Perfecto, I., & Castiñeiras, A., (1998). Deployment of the predaceous ants and their conservation in agroecosystems. In: Barbosa, P., (ed.), *Conservation Biological Control* (pp. 269–289). Academic Press, San, Diego, CA, USA.

Perfecto, I., & Vandermeer, J., (2002). Quality of agroecological matrix in a tropical montane landscape: Ants in coffee plantations in southern Mexico. *Conserv. Biol., 16*, 174–182.

Perfecto, I., Rice, R., Greenberg, R., & Van, D. V. M. E., (1996). Shade coffee: A disappearing refuge for biodiversity. *BioScience, 46*, 598–608.

Perry, C. H., Miller, R. C., & Brooks, K. N., (2001). Impacts of short rotation hybrid poplar plantations on regional water yield. *For. Ecol. Manag., 143*, 143–151.

Peters, V. E., & Carroll, C. R., (2012). Temporal variation in coffee flowering may influence the effects of bee species richness and abundance on coffee production. *Agrofor. Syst., 85*, 95–103.

Philpott, S. M., & Armbrecht, I., (2006). Biodiversity in tropical agroforests and the ecological role of ants and ant diversity in predatory function. *Ecol. Entomol., 31*, 369–377.

Philpott, S. M., & Foster, P. F., (2005). Nest-site limitation in coffee agroecosystems: Artificial nests maintain diversity of arboreal ants. *Ecol. Appl., 15*, 1478–1485.

Philpott, S. M., (2005). Changes in arboreal ant populations following pruning of coffee shade-trees in Chiapas, Mexico. *Agrofor. Syst., 64*, 219–224.

Philpott, S. M., Arendt, W. J., Armbrecht, I., Bicier, P., Diestch, T. V., Gordon, C., Greenberg, R., et al., (2008a). Biodiversity loss in Latin American coffee landscapes: Review of the evidence on ants, birds, and trees. *Conserv. Biol., 22*, 1093–1105.

Philpott, S. M., Perfecto, I., & Vandermeer, J., (2008b). Effects of predatory ants on lower trophic levels across a gradient of coffee management complexity. *J. Anim. Ecol., 77*, 505–511.

Philpott, S., Greenberg, R., Bichier, P., & Perfecto, I., (2004). Impacts of major predators on tropical agroforest arthropods: Comparisons within and across taxa. *Oecologia, 140*, 140–149.

Pineda, E., Moreno, C., Escobar, F., & Halffter, G., (2005). Frog, bat, and dung beetle diversity in the cloud forest and coffee agroecosystems of Veracruz, Mexico. *Conserv. Biol., 19*, 400–410.

Pinho, R. C., Miller, R. P., & Alfaia, S. S., (2012). Agroforestry and the improvement of soil fertility: A view from Amazonia. *App. Environ. Soil Sci.,* 1–11. ID 616383.

Ponisio, L. C., Gaiarsa, M. P., & Kremen, C., (2017). Opportunistic attachment assembles plant-pollinator networks. *Ecol. Lett., 20*, 1261–1272.

Pratiwi, A., & Suzuki, A., (2019). Reducing agricultural income vulnerabilities through agroforestry training: Evidence from a randomized field experiment in Indonesia. *Bull Indones. Econ. Stud., 55*, 83–116.

Prihatini, J., Iskandar, J., Partasasmita, R., & Nurjaman, D., (2018). The impacts of traditional homegarden conversion into the commercial one: A case study in Sukapura village of the Upstream Citarum Watershed, West Java, Indonesia. *Biodiversitas, 19*, 1926–1940.

Rahman, P. M., Varma, R. V., & Sileshi, G. W., (2012). Abundance and diversity of soil invertebrates in annual crops, agroforestry and forest ecosystems in the Nilgiri biosphere reserve of Western Ghats, India. *Agrofor. Syst., 85*, 165–177.

Rahman, S. A., Sunderland, T., Kshatriya, M., Roshetko, J. M., Pagella, T., & Healey, J. R., (2016). Towards productive landscapes: Trade-offs in tree-cover and income across a matrix of smallholder agricultural land-use systems. *Land Use Policy, 58*, 152–164.

Reitsma, R., Parrish, J. D., & McLarney, W., (2001). The role of cacao plantations in maintaining forest avian diversity in southeastern Costa Rica. *Agrofor. Syst., 53*, 185–193.

Rice, R. A., & Greenberg, R., (2000). Cacao cultivation and the conservation of biological diversity. *Ambio, 3*, 167–176.

Rivest, D., Cogliastro, A., Bradley, R. L., & Oliver, A., (2010). Intercropping hybrid poplar with soybean increases soil microbial biomass, mineral N supply and tree growth. *Agrofor. Syst., 80*, 33–40.

Roberts, D. L., Cooper, R. J., & Petit, L. J., (2000a). Use of premontane moist forest and shade coffee agroecosystems by army ants in western Panama. *Conserv. Biol., 14*, 192–199.

Roberts, D. L., Cooper, R. J., & Petit, L. J., (2000b). Flock characteristics of ant-following birds in premontane moist forest and coffee agroecosystems. *Ecol. Appl., 10*, 1414–1425.

Rosas-Ramos, N., Banos-Picon, L., Tormos, J., & Asis, J. D., (2019). The complementarity between ecological infrastructure types benefits natural enemies and pollinators in a Mediterranean vineyard agroecosystem. *Ann. Appl. Biol., 175*, 193–201.

Round, P. D., Gale, G., & Brockelman, W. Y., (2006). A comparison of bird communities in mixed fruit orchards and natural forest at Khao Luang, southern Thailand. *Biodivers. Conserv., 15*, 2873–2891.

Rumberger, M., Munzenberger, B., Bens, O., Ehrig, F., Lentzsch, P., & Huttl, R., (2004). Changes in diversity and storage function of ectomycorrhiza and soil organo profile dynamics after introduction of beech into Scots pine forests. *Plant Soil, 264*, 111–126.

Sabino, W., Costa, L., Andrade, T., Teixeira, J., Araujo, G., Acosta, A. L., Carvalheiro, L., & Giannini, T. C., (2022). Status and trends of pollination services in Amazon agroforestry systems. *Agric. Ecosyst. Environ., 355*, 108012.

Sakchoowong, W., Jaitrong, W., & Ogata, K., (2008). Ant diversity in forest and traditional hill-tribe agricultural types in northern Thailand. *Kasetsart J. (Nat. Sci.), 42*, 617–626.

Samyn, J. M., (1993). *Quelquesre'Flexionssur la Consommation de Combustibles Ligneuxdans les Campagnes Rwandaises*. Ministe`re de l'Agriculture, de l'Elevage et de l'Environnement, Kigali.

Sauer, T. J., & Hernandez-Ramirez, G., (2011). Agroforestry. In: Hatfield, J. L., & Sauer, T. J., (eds.), *Soil Management: Building A Stable Base for Agriculture* (pp. 351–370). American Society of Agronomy and Soil Science of America, Madison, WI.

Schirmel, J., Albrecht, M., Bauer, P. M., Sutter, L., Pfister, S. C., & Entling, M. H., (2018). Landscape complexity promotes hoverflies across different types of semi-natural habitats in farmland. *J. Appl. Ecol.*, *55*, 1747–1758.

Schmitz, O. J., Hamback, P. A., & Beckerman, A. P., (2000). Trophic cascades in terrestrial systems: A review of the effects of carnivore removals on plants. *Am. Nat.*, *155*, 141–153.

Scholten, H., (1988). Snow distribution on crop fields. *Agric. Ecosyst. Environ.*, *22, 23*, 363–380.

Schroth, G., Fonseca, G. A. B., Harvey, C. A., Gascon, C., Vasconcelos, H. L., & Izac, A. M. N., (2004). *Agroforestry and Biodiversity Conservation in Tropical Landscapes* (p. 523). Island Press, Washington.

Schulze, C. H., Waltert, M., Kessler, P. J. A., Pitopang, R., Shahabuddin, Veddeler, D., Muhlenberg, M., et al., (2004). Biodiversity indicator groups of tropical land-use systems: Comparing plants, birds, and insects. *Ecol. Appl.*, *14*, 1321–1333.

Seobi, T., Anderson, S. H., Udawatta, R. P., & Gantzer, C. J., (2005). Influences of grass and agroforestry buffer strips on soil hydraulic properties. *Soil Sci. Soc. Am. J.*, *69*, 893–901.

Sepp, K., Mikk, M., Mand, M., & Truu, J., (2004). Bumblebee communities as an indicator for landscape monitoring in the agri-environmental program. *Landsc. Urban Plan.*, *67*, 173–183.

Sharma, P. K., & Kapoor, K. K., (2005). Problems and prospective of nitrogen fixation in agroforestry systems. *Proc Indian Natl. Sci. Acad. B*, *71*, 145–161.

Sharrow, S. H., & Ismail, S., (2004). Carbon and nitrogen storage in agroforests, tree plantations, and pastures in western Oregon, USA. *Agrofor. Syst.*, *60*, 123–130.

Shastri, C. M., Bhat, D. M., Nagaraja, B. C., Murali, K. S., & Ravindranath, N. H., (2002). Tree species diversity in a village ecosystem in Uttara Kannada district in Western Ghats, Karnataka. *Curr. Sci.*, *82*, 1080–1084.

Sileshi, G., Akinnifesi, F. K., Ajayi, O. C., Chakeredza, S., Kaonga, M., & Matakala, P. W., (2007). Contributions of agroforestry to ecosystem services in the miombo eco-region of eastern and southern Africa. *Afr. J. Environ. Sci. Technol.*, *1*, 68–80.

Simonetti, J. A., Grez, A. A., & Bustamante, R. O., (2006). Interacciones y procesos en el bosquemaulinofragmentado. In: Grez, A. A., Simonetti, J. A., & Bustamante, R. O., (eds.), *Biodiversidad en Ambientesfragmentados de Chile: Patrones y Procesos a Diferentesescalas* (pp. 99–114). Editorial Universitaria, Santiago.

Singh, M., Babanna, S. K., Kumar, D., Dwivedi, R. P., Dev, I., Kumar, A., Tewari, R. K., et al., (2021). Valuation of fuelwood from agroforestry systems: A methodological perspective. *Agrofor. Syst.*, *95*, 977–993.

Siriri, D., & Raussen, T., (2003). The agronomic and economic potential of tree fallows on scoured benches in the humid highlands of south-western Uganda. *Agric. Ecosyst. Environ.*, *95*, 359–369.

Slazak, A., Freese, D., Matos, E. S., & Huttl, R. F., (2010). Soil organic phosphorus fraction in pine-oak forest stands in Northeastern Germany. *Geoderma*, *158*, 156–162.

Smith, B. D., & Lewis, T., (1972). The effects of windbreaks on the blossom-visiting fauna of apple orchards and on yield. *Ann. Appl. Biol.*, *72*, 229–238.
Soderstrom, B., Svensson, B., Vessby, K., & Glimskar, A., (2001). Plants, insects and birds in semi-natural pastures in relation to local habitat and landscape factors. *Biodivers. Conserv.*, *10*, 1839–1863.
Sodhi, N. S., Koh, L. P., Prawiradilaga, D. M., Darjono, Tinulele, I., Putra, D. D., & Tan, T. H. T., (2005). Land use and conservation value for forest birds in Central Sulawesi (Indonesia). *Biol. Conserv.*, *122*, 547–558.
Staley, J. T., Botham, M. S., Amy, S. R., Hulmes, S., & Pywell, R. F., (2018). Experimental evidence for optimal hedgerow cutting regimes for brown hairstreak butterflies. *Insect. Conserv. Divers.*, *11*, 213–218.
Staley, J. T., Botham, M. S., Chapman, R. E., Amy, S. R., Heard, M. S., Hulmes, L., Savage, J., & Pywell, R. F., (2016). Little and late: How reduced hedgerow cutting can benefit *Lepidoptera. Agric. Ecosyst. Environ.*, *224*, 22–28.
Stamps, W. T., Woods, T. W., Linit, M. J., & Garrett, H. E., (2002). Arthropod diversity in alley cropped black walnut (*Juglans nigra* L.) stands in Eastern Missouri, USA. *Agrofor. Syst.*, *56*, 167–175.
Stenchly, K., Clough, Y., & Tscharntke, T., (2012). Spider species richness in cocoa agroforestry systems, comparing vertical strata, local management and distance to forest. *Agric. Ecosyst. Environ.*, *149*, 189–194.
Sutter, L., Albrecht, M., & Jeanneret, P., (2018). Landscape greening and local creation of wildflower strips and hedgerows promote multiple ecosystem services. *J. Appl. Ecol.*, *55*, 612–620.
Thorlakson, T., & Neufeldt, H., (2012). Reducing subsistence farmer's vulnerability to climate change: Evaluating the potential contributions of agroforestry in Western Kenya. *Agric. Food Security*, *1*, 1–13.
Timberlake, T. P., Vaughan, I. P., & Memmott, J., (2019). Phenology of farmland floral resources reveals seasonal gaps in nectar availability for bumblebees. *J. Appl. Ecol.*, *56*, 1585–1596.
Tylianakis, J. M., Klein, A. M., & Tscharntke, T., (2005). Spatiotemporal variation in the diversity of Hymenoptera across a tropical habitat gradient. *Ecology*, *86*, 3296–3302.
Tyndall, J. C., & Colletti, J. P., (2007). Mitigating swine odor with strategically designed shelterbelt systems: A review. *Agrofor. Syst.*, *69*, 45–65.
Udawatta, R. P., & Jose, S., (2012). Agroforestry strategies to sequester carbon in temperate North America. *Agrofor. Syst.*, *86*, 225–242.
Udawatta, R. P., Anderson, S. H., Motavalli, P. P., & Garrett, H. E., (2011a). Clay and temperature influences on sensor measured volumetric soil water content. *Agrofor. Syst.*, *82*, 61–75.
Udawatta, R. P., Gantzer, C. J., Anderson, S. H., & Garrett, H. E., (2008). Agroforestry and grass buffer effects on pore characteristics measured by high-resolution X-ray computed tomography. *Soil Sci. Soc. Am J*, *72*, 295–304.
Udawatta, R. P., Garrett, H. E., & Kallenbach, R. L., (2011b). Agroforestry buffers for non-point source pollution reductions from agricultural watersheds. *J. Environ. Qual.*, *40*, 800–806.
Udawatta, R. P., Kremer, R. J., Adamson, B. W., & Anderson, S. H., (2008b). Variations in soil aggregate stability and enzyme activities in a temperate agroforestry practice. *Appl. Soil Ecol.*, *39*, 153–160.

Udawatta, R. P., Kremer, R. J., Garrett, H. E., & Anderson, S. H., (2009). Soil enzyme activities and physical properties in a watershed managed under agroforestry and row-crop system. *Agric. Ecosyst. Environ., 131*, 98–104.

Udawatta, R. P., Krstansky, J. J., Henderson, G. S., & Garrett, H. E., (2002). Agroforestry practices, runoff, and nutrient loss: A paired watershed comparison. *J. Environ. Qual., 31*, 1214–1225.

Uri, V., Tullus, H., & Lohmus, K., (2002). Biomass production and nutrient accumulation in short-rotation grey alder (*Alnusincana* (L.) Moench) plantation on abandoned agricultural land. *For. Ecol. Manag., 161*, 169–179.

Van, B. S. A., Philpott, S. M., Greenberg, R., Bichier, P., Barber, N. A., Mooney, K. A., & Gruner, D. S., (2008). Birds as predators in tropical agroforestry systems. *Ecology, 89*, 928–934.

Van, N. M., Lawson, G., Soumare, A., Groot, J. J. R., & Hairiah, K., (1996). Root distribution of trees and crops: Competition and/or complementarity. In: Ong, C. K., & Huxley, P., (eds.), *Tree–Crop Interactions: A Physiological Approach* (pp. 319–364). CAB International, Wallingford.

Vandermeer, J., Perfecto, I., Nunez, G. I., Phillpott, S., & Ballinas, A. G., (2002). Ants (*Azteca* sp.) as potential biological control agents in shade coffee production in Chiapas, Mexico. *Agrofor. Syst., 56*, 271–276.

Vaughan, C., Ramirez, O., Herrera, G., & Guries, R., (2007). Spatial ecology and conservation of two sloth species in a cacao landscape in limon, Costa Rica. *Biodivers. Conserv., 16*, 2293–2310.

Vickery, J. A., Feber, R. E., & Fuller, R. J., (2009). Arable field margins managed for biodiversity conservation: A review of food resource provision for farmland birds. *Agric. Ecosyst. Environ., 133*, 1–13.

Vickery, J., Carter, N., & Fuller, R. J., (2002). The potential value of managed cereal field margins as foraging habitats for farmland birds in the UK. *Agric. Ecosyst. Environ., 89*, 41–52.

Wal, H. V. D., Pena-Alvarez, B., Arriaga-Weiss, S. L., & Hernandez-Daumas, S., (2012). Species, functional groups and habitat preferences of birds in five agroforestry classes in Tabasco, Mexico. *Wilson J. Ornithol., 124*, 558–571.

Wanger, T. C., Iskandar, D. T., Motzke, I., Brook, B. W., Sodhi, N. S., Clough, Y., & Tscharntke, T., (2010). Composition of tropical amphibians and reptiles in Sulawesi, Indonesia. *Conserv. Biol., 24*, 795–802.

Wanger, T. C., Saro, A., Iskandar, D. T., Brook, B. W., Sodhi, N. S., Clough, Y., & Tscharntke, T., (2009). Conservation value of cacao agroforestry for amphibians and reptiles in South-East Asia: Combining correlative models with follow-up field experiments. *J. Appl. Ecol., 46*, 823–832.

Way, M. J., & Khoo, K. C., (1992). Role of ants in pest management. *Annu. Rev. Entomol., 37*, 479–503.

Weerasekara, C., Udawatta, R. P., Jose, S., Kremer, R. J., & Weerasekara, C., (2016). Soil quality differences in a row-crop watershed with agroforestry and grass buffers. *Agrofor. Syst., 90*, 829–838.

Williams-Guillen, K., & Perfecto, I., (2010). Effects of agricultural intensification on the assemblages of leaf-nosed bats (*Phyllostomidae*) in a coffee landscape in Chiapas, Mexico. *Biotropica, 42*, 605–613.

Williams-Guillen, K., McCann, C., Martinez, S. J. C., & Koontz, F., (2006). Resource availability and habitat use by mantled howling monkeys in a Nicaraguan coffee plantation: Can agroforests serve as core habitat for a forest mammal? *Anim. Conserv., 9*, 331–338.

Williams-Guillén, K., Perfecto, I., & Vandermeer, J., (2008). Bats limit insects in a neotropical agroforestry system. *Science, 320*, 70.

Wordley, C., Altringham, J., & Raman, T. R. S., (2014). Bats in Indian coffee plantations: Doing more good than harm? *Curr. Sci., 107*, 1958–1960.

Wunderle, J. M. Jr., & Latta, S. C., (1998). Avian resource use in Dominican shade coffee plantations. *Wilson Bull., 110*, 271–281.

Yang, C., Wang, M., Cai, W., & Li, J., (2012). Bensulfuron-methyl biodegradation and microbial parameters in a riparian soil as affected by simulated saltwater incursion. *Clean–Soil Air Water, 40*, 348–355.

Yashmita-Ulman, & Singh, M., (2021). Density, habitat associations and conservation status of *Gekko gecko* (Tokay gecko) in human-dominated landscapes around Nameri Tiger Reserve, Assam, India. *Acta Ecologica Sinica, 42*, 188–197.

Yashmita-Ulman, Kumar, A., & Sharma, M., (2017). Traditional home garden agroforestry systems: Habitat for conservation of Baya Weaver *Ploceus philippinus* (*Passeriformes*: *Ploceidae*) in Assam, India. *J. Threat. Taxa, 9*, 10076–10083.

Yashmita-Ulman, Kumar, A., & Sharma, M., (2018). Agroforestry systems as habitat for avian species: Assessing its role in conservation. *Proc. Zool. Soc., 71*, 127–145.

Yashmita-Ulman, Singh, M., Kumar, A., & Sharma, M., (2021). Conservation of plant diversity in agroforestry systems in a biodiversity hotspot region of northeast India. *Agric. Res., 10*, 569–581.

Yashmita-Ulman, Singh, M., Kumar, A., & Sharma, M., (2021a). Conservation of wildlife diversity in agroforestry systems in eastern Himalayan biodiversity hotspot. *Proc. Zool. Soc., 74*, 171–188.

Yukhnovskyi, V., Polishchuk, O., Lobchenko, G., Khryk, V., & Levandovska, S., (2021). Aerodynamic properties of windbreaks of various designs formed by thinning in central Ukraine. *Agrofor. Syst., 95*, 855–865.

Zamora, D. S., Jose, S., & Napolitano, K., (2009). Competition for 15N labelled nitrogen in a loblolly pine-cotton alley cropping system in the southeastern United States. *Agric. Ecosyst. Environ., 131*, 40–50.

Zuria, I., & Gates, J. E., (2013). Community composition, species richness, and abundance of birds in field margins of central Mexico: Local and landscape-scale effects. *Agrofor. Syst., 87*, 377–393.

CHAPTER 16

Tree Species Composition in Traditional Agroforestry Systems in Various Agroclimatic Regions of India

KAUSHAL SINGH,[1] MOHAMMED NASIR,[2] VIVEK VAISHNAV,[3] and ASHOK GEHLOT[4]

[1]*Department of Silviculture and Agroforestry, College of Forestry, Banda University of Agriculture and Technology (BUAT), Banda, Uttar Pradesh, India*

[2]*Department of Forest Products Utilization, College of Forestry, BUAT, Banda, Uttar Pradesh, India*

[3]*Laboratory for Conservation and Genetic Improvement of Forest Trees, Department of Forestry and Environmental Science, Manipur University, Imphal, Manipur, India*

[4]*Biotechnology Division, CSIR–IHBT, Palampur, Himachal Pradesh, India*

ABSTRACT

Agroforestry is a land use system that integrates woody perennials with crops and/or animals. It is the most effective tool to mitigate and adapt to adverse effects of climate change. The present communication focuses on the various tree species traditionally grown or retained in agricultural fields to obtain various products like fruit, fodder, fuel, timber, medicine, etc. A total of 155 plant species (93 tree species and 62 cultivated species) belonging to 50 families were recorded through a literature survey of secondary sources covering all agro-climatic regions of India. We found that traditional agroforestry systems play an important role in the conservation of native tree species.

Sustainable Forest Resources Management: Issues and Implications. Jahangeer A. Bhat, Gopal Shukla, Manmohan J. Dobriyal, Sumit Chakravarty, A. Arunachalam, and Rainer W. Bussmann (Eds.)

16.1 INTRODUCTION

Agriculture is the key occupation of the Indian population and is known as the backbone of the Indian economy (Chandra & Malaya, 2011). The rapid growth of the population has changed the scenario of the race of modern development, changing the land use system and causing environmental degradation in India and many developing countries (Punia, 2022). Population growth and land degradation lead to the loss of biodiversity (Bishaw & Asfaw, 2010). The international concern is to find an alternative farming system that will be ecologically and economically sound and acceptable by society (Molla & Kewessa, 2015). Agroforestry is a land use system that integrates trees, crops, and animals as a practically feasible, scientifically sound, ecologically desirable, and socially acceptable practice (Nair, 1979). It is a very old practice in which woody perennials are deliberately grown with agricultural crops or fodder crops in the form of a spatial mixture and/or a temporal sequence (Viswanath et al., 2018). Agroforestry is a land use system that integrates trees and shrubs on farmlands and rural landscapes to enhance productivity, profitability, diversity, and ecosystem sustainability (Policy, 2014). Traditional agroforestry is a set of age-old agroforestry systems, which are generally devoid of intentional intensified cultivation of agricultural or forage crops and have been practiced with varying structure, function, socioeconomic attributes, and ecological services across the globe (Viswanath et al., 2018).

In India, people have selected many multipurpose tree and shrub species to cope with adverse environmental conditions under different traditional agroforestry systems. These traditional agroforestry systems are location-specific according to the needs of rural communities and the available indigenous multipurpose plant species. A lot of research work has been carried out in different climatic zones (Table 16.1) of India, and a number of traditional agroforestry systems have been documented. In this chapter, a compilation of trees (Table 16.2) and crops (Table 16.3) has been done through a literature survey of well-known traditional agroforestry systems adopted in various agroclimatic regions of India.

Based on various scientific principles of long-term resource efficiency and sustainability, the geographical area of India is divided into 15 agroclimatic regions and 73 sub-regions by the Planning Commission in 1988 (Table 16.1).

TABLE 16.1 Agro-Climatic Zones of India

Agro-Climatic Zones	States Under Agro-Climatic Zone
Western Himalayan region	Jammu and Kashmir, Himachal Pradesh, and Uttarakhand
Eastern Himalayan region	Arunachal Pradesh, Assam Manipur, Meghalaya, Mizoram, Nagaland, Sikkim, Tripura, West Bengal
Lower Gangetic Plains region	West Bengal
Middle Gangetic Plains region	Uttar Pradesh, Bihar
Upper Gangetic Plains region	Uttar Pradesh
Trans-Gangetic Plains region	Punjab, Haryana, Delhi, and Rajasthan
Eastern Plateau and Hills region	Chhattisgarh, Jharkhand, Madhya Pradesh, Maharashtra, Odisha, West Bengal
Central Plateau and Hills region	Madhya Pradesh, Rajasthan, Uttar Pradesh
Western Plateau and Hills region	Maharashtra, Madhya Pradesh
Southern Plateau and Hills region	Andhra Pradesh, Karnataka, Tamil Nadu
East Coast Plains and Hills region	Orissa, Andhra Pradesh, Tamil Nadu, and Pondicherry
West Coast Plains and Ghats region	Tamil Nadu, Kerala, Goa, Karnataka, Maharashtra
Gujarat Plains and Hills region	Gujarat, Dadra & Nagar Haveli, Daman, and Diu
Western Dry region	Rajasthan
Island region	Andaman and Nicobar, Lakshadweep

Source: Adapted from: Planning Commission (Khanna, 1989).

TABLE 16.2 List of Tree Species of Traditional Agroclimatic Regions Various Agroclimatic Regions of India

Tree Species	Family	Tree Species	Family
Acacia auriculiformis A. Cunn. Ex. Benth.	Fabaceae	*Grevillea robusta* A. Cunn. Ex R. Br.	Proteaceae
Acacia leucophloea (Roxb.) Willd.	Fabaceae	*Grewia optiva* J.R. Drumm. Ex Burret	Malvaceae
Acacia nilotica var. cupressiformis (J.L. Stewart) Vajr. and Kamble	Fabaceae	*Hardwickia binnata* Roxb.	Fabaceae
Acacia nilotica var. indica (Benth.) Kyal. and Boatwr.	Fabaceae	*Juglans regia L.*	Juglandaceae
Acacia senegal (L.) Willd.	Fabaceae	*Leucaena leucocephala* (L.) de Wit	Fabaceae
Acacia tortilis (Forssk.) Hayne	Fabaceae	*Litsea spp.*	Lauraceae
Adina cordifolia (Roxb.) Brandis	Rubiaceae	*Madhuca latifolia* Var. latifolia (Roxb.) A. Chev.	Sapotaceae

TABLE 16.2 *(Continued)*

Tree Species	Family	Tree Species	Family
Aegle marmelos (L.) Correa	Rutaceae	*Malus domestica* (Suckow) Borkh.	Rosaceae
Ailanthus excels Roxb.	Simaroubaceae	*Mangifera indica L.*	Anacardiaceea
Albizia lebbeck (L.) Benth.	Fabaceae	*Melia azadirach L.*	Meliaceae
Albizia procera (Roxb.) Benth.	Fabaceae	*Magnolia champaca* L. Baill. Ex Pierre	Magnoliaceae
Albizia stipulate (DC.) Boivin	Fabaceae	*Morus alba L.*	Moraceae
Alnus nepalensis D. Don	Betulaceae	*Mytenus emarginata* (Wild.) Ding Hou	Celastraceae
Anacardium occidentale L.	Anacardiaceae	*Ougeinia oojeinensis* (Roxb.) Hochr.	Fabaceae
Annona squamosal L.	Annonaceae	*Parkia roxburghii* G. Don	Fabaceae
Anogeissus latifolia (Roxb. ex. DC.) Wall. ex Guill. and Perr.	Combretaceae	*Pongamia pinnata* (L.) Pierre	Fabaceae
Anogeissus pendula Edgew.	Combretaceae	*Populus deltoids* W. Bartram ex Marshall	Salicaceae
Anthocephalus cadamba (Roxb.) Miq.	Rubiaceae	*Populus nigra* L.	Salicaceae
Aquiliaria spp.	Thymelaeaceae	*Prosopis cineraria* (L.) Druce.	Fabaceae
Areca catechu L.	Arecaceae	*Prunus armeniaca* L.	Rosaceae
Artocarpus heterophyllus L.	Moraceae	*Prunus cerasoides* Buch.-Ham. Ex D.Don	Rosaceae
Azadirachta indica A. Juss.	Meliaceae	*Prunus domestica* L.	Rosaceae
Bombax ceiba L.	Malvaceae	*Prunus persica* L.	Rosaceae
Borassus flabellifer L.	Arecaceae	*Psidium guajava* L.	Myrtaceae
Bridelia retusa (L.) A.Juss.	Phyllanthaceae	*Pterocarpus santalinus* L.f	Fabaceae
Butea monosperma (L.) Kuntze	Fabaceae	*Pyrus communis* L.	Rosaceae
Cassia siamea L.	Fabaceae	*Quercus leucotrichophora A. Camus*	Fagaceae
Casuarina equisetifolia L.	Casuarinaceae	*Robinia pseudoacacia* L.	Fabaceae
Ceiba pentandra (L) Gaertn.	Malvaceae	*Salix alba* L.	Salicaceae
Celtis australis L.	Cannabaceae	*Salvadora oleoides* L.	Salvadoraceae
Citrus reticulate Blanco	Rutaceae	*Samanea saman* (Jacq.) Merr.	Fabaceae
Cocos nucifera L.	Arecaceae	*Santalum album* L.	Santalaceae

TABLE 16.2 *(Continued)*

Tree Species	Family	Tree Species	Family
Dalbergia latifolia Roxb.	Fabaceae	*Schima wallichii* (DC.) Korth.	Theaceae
Dalbergia sissoo Roxb. ex DC.	Fabaceae	*Semecarpus anacardium* L.f.	Anacardiaceae
Diospyros melanoxylon Roxb.	Ebenaceae	*Sesbania sesban* (L.) Merr.	Fabaceae
Dysoxylum binectiferum (Roxb.) Hook.f. ex Bedd.	Meliaceae	*Shorea robusta* C.F. Gaertn.	Dipterocarpaceae
Phyllanthus emblica L.	Phyllanthaceae	*Syzygium cumini* (L.) Skeels	Myrtaceae
Erythrina indica L.	Fabaceae	*Tamarindus indica* L.	Fabaceae
Eucalyptus hybrid Maiden	Myrtaceae	*Tecomella undullata* (Sm.) Seem.	Bignoniaceae
Eucalyptus tereticornis Sm.	Myrtaceae	*Tectona grandis L.* f.	Lamiaceae
Ficus amplissima Sm.	Moraceae	*Terminalia arjuna* (Roxb. Ex DC.) Wight and Arn.	Combretaceae
Ficus benghalensis L.	Moraceae	*Terminalia bellirica* (Gaertn.) Roxb.	Combretaceae
Ficus mysorensis var. pubescens Roem and Schult.	Moraceae	*Terminalia chebula* Retz.	Combretaceae
Ficus racemose L.	Moraceae	*Terminalia tomentosa* Wight and Arn.	Combretaceae
Ficus religiosa L.	Moraceae	*Ulmus wallichiana* Planch.	Ulmaceae
Ficus virens Aiton	Moraceae	*Zizyphus mauritiana* L.	Rhamnaceae
Gmelina arborea Roxb. ex Sm.	Lamiaceae	–	–

TABLE 16.3 List of Various Crop Species Cultivated in Traditional Agroforestry Systems in Various Agroclimatic Regions of India

Tree Species	Family	Tree Species	Family
Abelmoschus esculentus (L.) Moench	Malvaceae	*Lens esculenta Moench.*	Fabaceae
Amaranthus frumentaceus Buch.-Ham. ex Roxb.	Amaranthaceae	*Linum usitatissimum* L.	Linaceae
Amorphophallus peoniifolius (Dennst) Nicolson	Araceae	*Macrotyloma uniflorum* (L.) Verdc.	Poaceae
Ananas comosus var. cosmosus	Bromeliaceae	*Medicago falcate* L.	Fabaccac
Apluda mutica L.	Poaceae	*Medicago sativa L.*	Fabaceae

TABLE 16.3 *(Continued)*

Tree Species	Family	Tree Species	Family
Arachis hypogaea L.	Fabaceae	*Oplismenus compositus* (L.) P.Beauv.	Poaceae
Arundinella nepalensis Trin.	Poaceae	*Oryza sativa L.*	Poaceae
Brassica compestris L.	Brassicaceae	*Panicum maximum* Jacq.	Poaceae
Brassica oleracea var. botrytis L.	Brassicaceae	*Panicum miliaceum* L.	Poaceae
Brassica oleracea var. Capitata L.	Brassicaceae	*Pennisetum glaucum* (L.) R.Br.	Poaceae
Cajanus cajan (L.) Millsp.	Fabaceae	*Perilla frutescens* (L.) Britton	Lamiaceae
Capsicum annum L.	Solanaceae	*Phaseolus vulgaris* L.	Fabaceae
Chrysopogon montanus	Poaceae	*Piper betle* L.	Piperaceae
Cicer arietinum L.	Fabaceae	*Pisum sativa* L.	Fabaceae
Cleome viscosa L.	Cleomaceae	*Saccharum officinarum* (L.)	Poaceae
Colocasia esculenta L.	Araceae	*Sesamum indicum* L.	Pedaliaceae
Curcuma longa L.	Zingiberaceae	*Setaria italic* (L.) Beauv.	Poaceae
Cyamopsis tetragonoloba (L.) Taub.	Fabaceae	*Solanum lycopersicum* L.	Solanaceae
Dicanthium annulatum (Forssk.) Stapf.	Poaceae	*Solanum melongena* L.	Solanaceae
Echinochloa frumentacea Link	Poaceae	*Solanum tuberosum* L.	Solanaceae
Eleusine coracana (L.) Gaertn.	Poaceae	*Sorghum bicolor* (L.) Moench	Poaceae
Eruca sativa Mill.	Brassicaceae	*Themada anathera* (Nees ex steud.) Hack.	Poaceae
Fagopyrum esculentum Moench	Polygonaceae	*Trichosanthes dioica* Roxb.	Cucurbitaceae
Glycine max (L.) Merr.	Fabaceae	*Trigonella foenum-graceum* L.	Fabaceae
Gossypium hirsutum L.	Malvaceae	*Triticum aestivum* L.	Poaceae
Heteropogon contortus (L.) P.Beauv. ex Roem. and Schult.	Poaceae	*Vigna acontifolia* (Jacq.) Marechal	Fabaceae
Hordeum vulgare L.	Poaceae	*Vigna angularis* (Willd.) Ohwi and H. Ohashi	Fabaceae
Iris lacteal Pall.	Iridaceae	*Vigna mungo (*L.) Hepper	Fabaceae
Ischaemum aristatum L.	Poaceae	*Vigna unguiculata* (L.) Walp.	Fabaceae
Lagenaria siceraria (Molina) Standl.	Cucurbitaceae	*Zea mays* L.	Poaceae
Lathyrus odoratus L.	Fabaceae	*Zingiber officinale* Roscoe	Zingiberaceae

16.2 WESTERN HIMALAYAN REGION

Climate, geography, ecology, plant communities, and land use types are all diverse in the Western Himalayan region. In Jammu and Kashmir (Ladakh), the average annual precipitation is 80 mm, but in Himachal Pradesh and Uttarakhand, it is 200 cm. The people in this region, who practice both sedentary and nomadic types of livestock grazing, rely heavily on raising cattle as a source of income. Horticulture practices have significantly thrived in the western Himalayan agroclimatic zone. Saffron, maize, wheat, peaches, apricots, cherries, almonds, and litchis are some of the main crops grown here. The majority of the people here live in rural areas, where they have very low living standards and no access to marketing or storage—two essential obstacles to successful agroforestry practice. The tree species grown in the traditional agroforestry systems in the region can be divided into two categories: nitrogen-fixing species *viz. Celtis australis* L., *Prunus cerasoides* Buch.-Ham. ex D. Don, *Bauhinia variegata* L., *Grewia optiva* J.R. Drumm. ex Burret, *Quercus leucotrichophora A. camus*, *Melia azedarach* L.; and non-nitrogen-fixing trees viz. Ougeinia dalbergioides Benth., Albizia stipulata B. Boivin, *Dalbergia sissoo* DC, *Alnus nepalensis* D. Don. In the state of Jammu and Kashmir, agri-silvi-horticulture, and homestead forestry systems are the most practiced. The woody and fruit tree species most commonly adopted for agroforestry plantations are *Populus deltoides, Salix alba, Robinia pseudoacacia, Morus alba, P. nigra, Juglans regia, Ailanthus excelsa, Ulmus wallichiana, Pyrus communis, Malus domestica, Prunus persica, P. armeniaca,* etc. (Islam et al., 2015).

Rangelands play a major role in the livelihood of over 25% of Jammu and Kashmir's (J&K) nomads and semi-nomads, including the Gujjars, Bakerwals, Chopans, Changpas, and Gaddis. There are issues with pastoral ecosystem integrity and the sustainability of cattle grazing systems. The most important aspect of grazing management seems to be managing grazing timing, length, and intensity. Red clover, white clover, tall fescue, orchard grass, sain foin, lucerne, and tree fodder are examples of forages that are frequently used to feed animals. The resource basis of grazing and pasture areas has deteriorated as a result of recent land use changes, which has led to an imbalance between the supply and demand for different types of feed due to the high livestock population. To restore the damaged grasslands and increase the supply of fodder, agroforestry systems, such as silvopastoral, agri-silvicultural, agri-horticultural, and horti-pastoral systems, have recently been implemented (Mugloo et al., 2020).

The trans-Himalayan zone includes a frigid desert that includes Ladakh. It is a dry area devoid of vegetation and does not benefit from the southeast monsoon because it is located in the Himalayan Mountain system's rain shadow. A place where the climate exhibits high climatic extremes of heat (+35°C in summers) and cold (–40°C in winters) with considerable dryness is referred to as a cold desert environment. The soil in the Ladakh region is predominantly alkaline, has a low fertility status, poor water retention, and a sparse plant cover. The indigenous agroforestry systems are managed scientifically. In the agri-silviculture system, wheat (*Triticum aestivum*), barley *(Hordeum vulgare*), buckwheat (*Fagopyrum esculentum*), foxtail millet (*Setaria italica*), etc., are grown with poplar (*Populus nigra*), willow (*Salix alba*), etc. In Agri-Horti-Silviculture system, pea (*Pisum sativum*), turnip (*Brassica rapa*), rajma/red bean (*Phaseolus vulgaris*), cabbage (*Brassica oleracea var. botrytis*), phool gobi/cauliflower (*Brassica oleracea var. capitata*), etc., are grown with horticulture trees apricot (*Prunus armeniaca*), and apple (*Malus domestica*); and forest trees poplar (*Populus nigra*), and willow (*Salix alba*), etc. In silvopasture system, grasses like alfalfa (*Medicago sativa* and *Medicago falcata*), and iris (*Iris lactea*) are cultivated for fodder with fodder trees poplar (*Populus nigra*), and willow (*Salix alba*), etc. In Horti-Silvi-Pasture system these grasses are grown with fruit trees apricot (*Prunus armeniaca*), apple (*Malus domestica*), mulberry (*Morus alba*), and walnut (*Juglans regia*), and forest trees poplar (*Populus nigra*), willow (*Salix alba*), etc., and grasses alfalfa (*Medicago sativa* and *Medicago falcata*), and iris (*Iris lactea*) (Butola et al., 2012).

In Himachal Pradesh, forest tree species *Leucaena leucocephala L. Benth, Grewia optiva Drumm. ex Burr, Albizia lebbek Benth., Toona ciliata M. Roem., Acacia catechu Willd., Pinus roxburghii Sargent, Eucalyptus sp.*, and *Bauhinia variegata Linn.*; and horticultural tree species mango, lemon, and banana are dominant in different agroforestry systems. Among the systems, Agrisilviculture and pastoral silviculture systems are among the dominant agroforestry systems in the state (Dar et al., 2018). In these systems, the crops of pea, wheat, barley, and mustard, and vegetables of cauliflower, cabbage, maize, beans, colocassia, tomato, capsicum, turmeric, and ginger are grown with trees of *Celtis australis, Grewia optiva, Pinus roxburghii, Quercus leucotrichophora, Robinia pseudoacasia, Bauhinia variegata, Pistacia integerrima, Myrica esculanta, Toona ciliata, Morus alba, Pyrus pashia, Ficus roxburghii, Ficus palmata*, and *Prunus padus*. These trees help to retain soil on the bunds of agriculture fields, whereas fruit trees of *Prunus armeniaca, Prunus saliciana*, and *Pyrus communis* are planted in and around the agriculture fields. The dominant grass species are *Chrysopogon montanus, Panicum maximum, Heteropogon contortus, Dicanthium annulatum, Apluda mutica, Ischaemum aristatum, Themeda anathera, Arundinella nepalensis*, and *Oplismenus compositus* (Kaler et al., 2017).

In Uttarakhand, *Celtis australis, Bauhinia vahlii, Grewia optiva, Cinnamomum tamala, Prunus cerasoides, Melia azedarach, Toona serrata, Quercus leucotrichophora*, etc., are the most preferred tree species, which are grown in the indigenous agroforestry system. These agroforestry species serve for numerous uses including medicine, food, fodder, fuelwood, timber, fibers, and religious purposes. Fodder species are an essential source of animal feed during winter. These are planted on the bunds or regenerated naturally. As per the utilization by the local people, *Grewia optiva* is the dominant tree species grown with different crops, i.e., *Oryza sativa, Echinochloa frumentoacea, Amaranthus frumentaceus, Eleusine coracana, Setaria italica, Zea mays, Glysine max, Cajanus cajan, Macrotyloma uniflorum, Vigna mungo, Phaseolus vulgaris,* and *V. angularis* are cultivated. These 12 crop species have special importance in the daily life of the local community of the region; and thus have been grouped in a unique category called Baranaja (12 varieties of crops). On the other hand, five crop varieties, i.e., *Hordeum vulgare, Triticum aestivum, Pisum sativum, Lens esculenta, Brassica compestris,* are cultivated during the winter cropping season, and *Triticum aestivum* is the major cereal crop among these. However, during the last decade, some of the traditional crops such as *Panicum miliaceum, Setaria italica, Sesamum indicum, Cleome viscosa, Perilla frutescens,* and *Trigonella foenum* cultivated previously have become locally extinct from the area (Kala, 2010).

16.3 EASTERN HIMALAYAN REGION

The Eastern Himalayan Region exhibits very rugged, hilly, and unexplored terrain and receives annual rainfall ranging from 2000 mm to 4000 mm, which is the highest in the world. Due to an abundance of red soil, frequent forest fire events, and shifting cultivation preferred by most of the hill communities, the majority of arable lands are degraded and not suitable for agriculture. The plant and animal diversity in the various climatic zones of the area may be preserved using agroforestry model-based systems. Around 60% of the area (98,236 sq km) is suitable for agroforestry. In terms of the overall area with high potential for agroforestry, Assam, Mizoram, and

Manipur are the core states, with land areas of 22,641 sq km (36%), 6966 sq km (47%), and 4565 sq km (30%), respectively (Nath et al., 2021).

In these multistory agroforestry systems, many plant species are mixed together due to traditional knowledge and subsistence needs. Indigenous tribes are recognized for maintaining household gardens using diverse traditional agroforestry models that produce pineapples with paddy, bamboo, eri silk, fruits, and piper beetles. The three types of agroforestry systems that are most common in the area are horti-, agri-, and silvi-pastoral systems. In Agri-silviculture, tree species such as *Alnus nepalensis, Parkia roxburghii, Michelia champaca, Gmelina arborea,* and *Schima wallichii* are planted, whereas *Gmelina arborea, Taxus spp., Salix, Litsea spp., Michelia champaca, Parkia roxburghii, Aquilaria spp., Schima wallichii, Ceiba pentandra, Terminalia spp., and Bombax ceiba* are the favored species in Horti-silviculture, and *Alnus nepalensis* in Silvi-pastoral systems.

16.3.1 TECHNICAL INTERVENTIONS FOR AGROFORESTRY SYSTEMS IN HIMALAYAN REGION

Two very successful technical interventions have also been implemented and found suitable to promote mixed cropping with different agroforestry systems *viz.* Sloping Watershed Environmental Engineering Technology (SWEET) and Sloping Agricultural Land Technology (SALT).

16.3.1.1 SLOPING AGRICULTURAL LAND TECHNOLOGY (SALT)

The Sloping Agricultural Land Technology (SALT) is a model farming method that uses fast-growing perennial nitrogen-fixing tree or shrub species to establish dense hedgerows along contour lines. It was first initiated by the Mindanao Baptist Rural Life Centre in the southern Philippines to help control soil erosion and increase crop yields. This creates a living barrier that traps sediments and progressively converts the sloping land to terraced land. The hedgerows serve as a source of organic matter, which significantly reduces soil erosion and helps to improve and/or maintain soil fertility. SALT has become one of the most popular strategies for reducing soil erosion in the Himalayan region. The following features of SALT technology, taken as a whole, could be economically advantageous to the communities that it is intended to serve:

1. **Contour Farming:** Farming along contours will prevent soil erosion and allow rainwater to be retained, increasing soil productivity.
2. **Planting Nitrogen-Fixing Varieties:** As part of the general concept of SALT, trees and nitrogen-fixing crops suitable for current agro-climatic conditions are planted as part of the crop rotation. In areas subjected to heavy agriculture, this leads to improved soil fertility and recovery of depleted soil nutrients. In the long term, this contributes to increased output as well as a decrease in the use of artificial fertilizers.
3. **Planting Crops with Varying Maturity Periods:** Planting seasonal crops, perennial crops like horticulture varieties, and tree plantations helps farmers diversify their sources of income and provides protection from the risks posed by seasonal natural fluctuations that could negatively impact one of the sources of income.
4. **Labor-Intensive Activities:** SALT-related tasks like contour trenching, plantation, and tending operations, etc., demand a lot of manual labor in addition to certain specific knowledge. Therefore, SALT operations have the potential to significantly increase local employment and prevent seasonal migration with the right government backing.
5. The key to ensuring food security and fending off the whims associated with an increasingly unpredictable climate variability linked to the phenomenon of global climate change is improved agricultural techniques that incorporate soil amelioration and productivity-enhancing measures while also respecting the traditional practices of the local populace. Thus, with careful planning and implementation of SALT, agricultural productivity can increase significantly compared to the investment needed to implement such techniques.

The SALT was created with the intention of promoting soil and moisture conservation techniques, nutrient augmentation in nutrient-poor areas, and crop production enhancements, all of which are anticipated to have a favorable environmental impact.

16.3.1.2 SLOPING WATERSHED ENVIRONMENTAL ENGINEERING TECHNOLOGY (SWEET)

An R&D-based program called SWEET and invented and introduced by Govind Ballabh Pant Institute of Himalayan Environment and Development

(Uttarakhand, India) aims to improve stakeholders' livelihoods while also restoring the ecological integrity of damaged regions. This package integrates straightforward technologies including water collection, soil conservation, species selection for plantations on wasteland, nursery-raised plants for afforestation, bio-fencing, and plantation maintenance with input from stakeholders. For the ecological and socioeconomic suitability of plant species, community feedback is gathered. Additionally, plants are evaluated for factors related to their ecophysiology and vigor. Water is an essential input for plantations; hence rainwater runoff is collected in poly-ponds. Saplings are raised in poly-pits and poly-tunnels to protect them from the cold and frost. For plantings, pits (60 cm x 60 cm x 60 cm in size) are filled with gravel-free soil mixed with 3 kg of well-composted farmyard manure in each pit. To supplement farmyard manure, weeds and agricultural waste can be bio-composted.

16.4 LOWER GANGETIC PLAINS REGION

This zone covers 14 districts of West Bengal out of 18 districts, except Cooch Behar, Darjeeling, Jalpaiguri, and Purulia districts. Soils are highly fertile because of the Ganges and its several tributaries. It forms the world's largest deltaic zone and mangrove forest, known as Sundarbans. This zone is again subdivided into four subzones, namely Barind plains, Central alluvial plains, alluvial coastal saline plains, and Rarh plains (Sirohi, 2006). Soils of Barind plains are sandy loam to silty clay in texture, yellowish to reddish-brown in color with a pH of 5.5 to 7.5, whereas in the Rarh region, soils are lateritic and red with a pH value of 5.5 to 6.6. Central alluvial plains have loamy to heavy clay with good water-holding capacity and a pH of 6.0 to 7.8. The soils of alluvial coastal saline plains are of tidal origin, with a silty clay texture and a pH that varies from 6.4 to 7.6, but face problems of high salinity and poor drainage. This zone is a rice-rice or rice-vegetable cropping zone with well-established irrigation facilities. Dhara et al. (2016) reported six major agroforestry systems from this region, namely agri-silviculture (agricultural crops + timber trees), agri-horticulture (agricultural crops + fruit trees), agri-horti-silviculture (agricultural crops + timber trees + fruit trees), horti-agriculture (fruit trees + vegetables), silvi-apiculture (timber trees + honey-bees), horti-apiculture (fruit trees + honeybees), boundary plantation (trees on boundary + crops), aqua-forestry (fish + timber trees), and homestead agroforestry (combination of timber trees, fruit trees, and vegetables, etc.).

Tree species such as *Dysoxylum binectiferum, Anthocephalus cadamba, Gmelina arborea, Tectona grandis, Eucalyptus tereticornis*, etc., are potential multipurpose species, whereas mango, guava, ber, and litchi are fruit tree species in agroforestry systems. Rice, maize, gram, wheat, mustard, groundnut, lentil, jute, and tea are major agricultural crops. Similarly, among the vegetables, pointed gourd, pea, okra, colocasia, tomato, elephant foot, brinjal, potato, bottle gourd, etc., are important vegetable crops in this region.

16.5 MIDDLE GANGETIC PLAINS REGION

This region is extended in Bihar and eastern Uttar Pradesh and 10 districts of Uttar Pradesh namely Mau, Ballia, Azamgarh, Ayodhya, Jaunpur, Ghazipur, Sant Ravidas Nagar, Varanasi, Mirzapur, and Sonbhadra come under this region (Srivastava, 2006). The Ganga river and its tributaries drain this region resulting in a fertile alluvial plain with gently sloping terrain. Soils of the region are khaddar and bangar, derived from alluvium. The rainfall varies from 1050–1300 mm (Srivastava, 2006). Hot, wet summers and cool, dry winters characterize the climate of the region. *Artocarpus heterophyllus, Azadirachta indica, Psidium guajava, Dendrocalamus strictus, Musa paradisiaca,* and *Citrus species* are the common tree species in traditional agroforestry systems of the region (Rana et al., 2007). Rice, maize, pigeon pea, moong, wheat, lentil, Bengal gram, sesame, and peas are the main agricultural crops (Srivastava, 2006).

16.6 UPPER GANGETIC PLAINS REGION

This region is known for rich water resources and deep, fertile soils. The topography of this region is plain land with a gentle slope. More than 65% of the area is irrigated, and irrigation is done through canals and tube wells. This region covers the most developed 32 districts of Uttar Pradesh. The mean annual rainfall varies between 700 and 1000 mm. The main agricultural crops are wheat, rice, pearl millet, moong, sugarcane, potato, brinjal, cabbage, etc., along with animal husbandry. *Populus deltoids, Eucalyptus hybrid, Dalbergia sissoo, Anthocephalus cadamba, Leucaena leucocephala* are important wood and fodder species, while Mangifera indica, Psidium guajava are fruit-producing species that are generally retained inside or on the bunds of agricultural fields.

16.7 TRANS-GANGETIC PLAINS REGION

This region has good soil and water resources and covers parts of Punjab, Haryana, Rajasthan, Chandigarh, and Delhi. The topography of this region is plain with medium and light soils. The Yamuna, Ghaggar, Sutlej, Ravi, and Beas are important rivers for irrigation in this region. This region has the highest intensity of irrigation at 93% in Punjab, 78% in Haryana, and 95% in the Sriganganagar district of Rajasthan (Verma & Singh, 2006). The climate of this region is arid, semiarid, and sub-humid, with rainfall varying from 190–1150 mm. *Melia azadirach, Acacia nilotica, Butea monosperma, Albizia lebbeck, Dalbergia sissoo, Prosopis cineraria*, and *Mangifera indica* are common tree species grown on the bunds of fields as a traditional agroforestry practice along with wheat, rice, cotton, bajra, guar, etc., as agricultural crops in this region.

16.8 EASTERN PLATEAU AND HILLS REGION

This is the largest agroclimatic region which spreads over Madhya Pradesh, Chhattisgarh, Jharkhand, Maharashtra, West Bengal, and Orissa. The temperature in July ranges from 26°C to 34°C and from 10°C to 27°C in January. The average rainfall is about 1350 mm. Irrigation facilities are very poor and agricultural productivity is quite low. Shifting cultivation has been the traditional pattern of farming among the tribal people of this region (Pandey, 2006). The soils are shallow, red, and yellow with some patches of lateritic and alluvium. Crops like rice, maize, potato, pigeon pea, soybean, groundnut, etc., and the important tree species of traditional agroforestry systems are *Acacia nilotica, Madhuca latifolia, Borassus flabellifer, Terminalia arjuna* in Orissa, and *Artocarpus heterophyllus, Azadirachta indica, Dalbergia sissoo, Gmelina arborea, Leucaena leucocephala, Melia azadirach, Syzygium cumini, Tectona grandis* in Jharkhand (Singh et al., 2017). In Koraput, Orissa, Panigrahi et al., 2021 reported various species of palm, *Acacia nilotica, Shorea robusta, Tamarindus indica, Azadirachta indica, Syzygium cumini, Gmelina arborea, Albizia lebeck, Pongamia pinnata, Samanea saman, Erythrina indica, Terminalia spp.*, and *Dendrocalamus strictus* were grown along with agricultural crops such as groundnut, black gram, green gram, mustard, etc. Tree species like *Terminalia arjuna, T. tomentosa, Albizia procera, Mangifera indica, Butea monosperma, Ziziphus mauritiana, Azadirachta indica,* and *Gmelina arborea* have been reported from the farm bunds of the Balaghat district of Madhya Pradesh (Sarvanan, 2021).

16.9 CENTRAL PLATEAU AND HILLS REGION

This region is interspersed with plateaus and hilly areas in Madhya Pradesh, Rajasthan, and Uttar Pradesh states. Water runoff is very high, large areas are barren and uncultivated (Pandey, 2006). In Madhya Pradesh, it extends over Keymore plateau, Vindhyachal plateau, Satpura plateau, ravines of Gird, and plains of Narmada valley and Bundelkhand. It covers Bharatpur, Dhaulpur, Madhopur, Alwar, Jaipur, Tonk, Bundi, Bhilwara, Pali, Chittaurgarh, Banswara, Dungarpur, Udaipur, and Sirohi districts in Rajasthan. Soils are red and black (Keymore), deep black (Vindhya plateau), deep black and clayey (Narmada valley), mixed red, yellow alluvial (Gird area), Lithosols (Bhilwara, Udaipur), and light reddish-brown to black clay loam in Satpura plateau (Pandey, 2006). In Bundelkhand region of UP and MP during Kharif season, few crops are cultivated like pigeon pea, moong, urad, groundnut; otherwise, most of the agricultural land is left fallow. In Rabi season, green gram, sweet pea, soybean, mustard, linseed, and wheat crops are cultivated along with trees of *Madhuca longifolia, Acacia nilotica, A. leucophloea, Phyllanthus emblica, Azadirachta indica, Syzygium cumini, Mangifera indica, Tamarindus indica, Ziziphus mauritiana, Butea monosperma,* and *Psidium guajava*. In Sapura plateau of Madhya Pradesh, traditionally agriculture crops like wheat, paddy, jowar, bajra, moong, soybean, urad, and moong are cultivated with tree species such as *Tectona grandis, Dalbergia sissoo, Shorea robusta, Acacia nilotica, Madhuca longifolia, Mangifera indica, Annona squamosa*, etc. (Bijalwan et al., 2019). In the Rajasthan region, agriculture crops are cultivated with tree species such as *Prosopis cineraria, Tecomella undulata, Salvadora oleoides, Dalbergia sissoo*, and *Azadirachta indica* as a traditional agroforestry practice in this region (Prahlad, 2018).

16.10 WESTERN PLATEAU AND HILLS REGION

It covers major parts of Maharashtra, Madhya Pradesh, and one district from Rajasthan. This zone is further subdivided into four subzones: (i) Western hills and plains, which includes Kolhapur, Satara, Nashik, Pune, and Jhabua districts of Maharashtra. The climate is semiarid with less than 1000 mm rainfall and soils are shallow to deep black loamy texture (Pandey, 2006). (ii) Scarcity region, it includes Dhule, Sangli, Ahmadnagar, and Solapur districts of Maharashtra. Rainfall is low, about 600 mm in this region. (iii) Central plateau, this subregion includes 10 districts in Maharashtra, 11 districts in

Madhya Pradesh, and one district Jhalawar from Rajasthan. (iv) Central Vidarbha spread over four districts namely Yavatmal, Wardha, Nagpur, and Nanded in Maharashtra. Climate is semi-arid to sub-humid dry, and soils are deep black clayey. Tree species like *Azadirachta indica, Butea monosperma, Acacia catechu, Acacia nilotica, Dalbergia sissoo, Terminalia bellerica, Aegle marmelos,* and *Semecarpus anacardium, and Phyllanthus emblica, Tamarindus indica, Zizyphus mauritiana, Mangifera indica, Citrus reticulata, Punica granatum, Citrus aurantifolia, Annona squamosa, Psidium guajava,* and *Musa paradisiaca* have been reported in agroforestry systems (Deshmukh et al., 2017). They are also reported from the rainfed farms of Akola.

16.11 SOUTHERN PLATEAU AND HILLS REGION

This region comprises 35 districts of Andhra Pradesh, Karnataka, and Tamil Nadu states. This is a peninsular plateau locked land. This region is mostly rainfed, drought-prone, and livelihood depends on agriculture (Annamalai, 2006). Cropping intensity of this region is low (112%). In Mandya district of Karnataka, *Ficus* trees are integral components of rainfed traditional agroforestry systems. *Ficus benghalensis* was the major species followed by *F. religiosa, F. amplissima, F. virens, F. racemosa, and F. mysorensis var. pubescens* with green gram, red gram, horse gram, and cowpea (Bhaskar et al., 2013). In Bellary district, crops such as rice, maize, wheat, groundnut were practiced with agroforestry tree species viz. *Acacia nilotica, Azadirachta indica, Bambusa* spp., *Cocos nucifera*, and *Tectona grandis* (Ram Newaj et al., 2017).

16.12 EAST COAST PLAINS AND HILLS REGION

This agro-climatic zone in India consists of four traditional rice-growing states: Orissa, Andhra Pradesh, Tamil Nadu, and Pondicherry. The zone is further subdivided into six subzones: the Orissa coastal region, the North coastal Andhra and Ganjam region, the South coastal Andhra region, the North coastal Tamil Nadu region, Thanjavur region, and the South coastal Tamil Nadu region. This region covers 19.948 M Ha area and has 11.83% agroforestry area (Arunachalam et al., 2022). In addition, Tamil Nadu is also known for Kangayam, Pulpwood, and Timber-based agroforestry system (Chavan et al., 2015). Rainfall in this zone is received during both the South West and North East Monsoons. The area has a hot and dry summer due to the subtropical belt on the east coast of this zone. It is home to several important rivers, including the Mahanadi, Godavari, Krishna, Cauvery, and Vaigai. The land is extremely fertile due to the nutrient-rich historic sediment soils brought by the flowing rivers, and agriculture is the region's backbone. In this region, the most common soil types are red loam and black cotton in Andhra Pradesh, red noncalcareous, black soil, and sandy loam in Tamil Nadu, and alluvial sandy loam, loam clay loam, and Coastal alluvial in Orissa. Paddy, sugarcane, groundnut, mustard, sorghum, maize, other millets, pulses, vegetables, banana, coconut, mango, cashew nut, and orchard crops are the most important crops grown in the region. Important tree species cultivated in this region are *Ailanthus excelsa, Acacia leucophloea, Casuarina equisetifolia, Leucaena leucocephala, Pongamia pinnata, Artocarpus, Santalum album, Diospyros melanoxylon, Pterocarpus santalinus, Dalbergia latifolia, Tectona grandis, Grevillea robusta, Eucalyptus spp., Casuarina equisetifolia, Leucaena leucocephala, Mangifera indica, Casuarina equisetifolia, Acacia nilotica, Syzygium cumini, Psidium guajava, Azadirachta indica, Anacardium occidentale, Cocus nucifera, Ficus religiosa, eucalyptus, subabul, Pterocarpus santalinus, Bambusa nutan, B. vulgaris, B. tulda, Samania saman, Eucalyptus hybrid, Acacia auriculiformis, Cassia siamea, Casuarina equisetifolia, silver oak*, and *Tectona grandis* (Dhyani et al., 2009; Chavan et al., 2015).

16.13 WEST COAST PLAINS AND GHAT REGION

West Coast Plains and Ghat region is an important part of India's peninsular plateau and it covers the Malabar and Konkan coasts and the Sahyadris region. The region represents Goa, Karnataka, Kerala, Maharashtra, and Tamil Nadu states of India. This zone is well-known for the growing of plantation crops and spices. This is a humid region with annual rainfall above 200 cm and soils are laterite and coastal alluvial type. This region has an area of 11.69 million hectares and has 13.96% agroforestry area (Arunachalam et al., 2022). Maharashtra is known for Timber-based agroforestry (Chavan et al., 2015). The major crops grown in this agro-climatic zone are rice, coconut, oilseeds, sugarcane, millets, pulses, and cotton. This region has an important tree component, including *Acacia auriculiformis, Artocarpus heterophyllus, Cocos nucifera, Areca catechu, Casuarina equisetifolia, Hardwickia binnata, Acacia nilotica, Leucaena leucocephala, Prosopis, Eucalyptus, Moringa, Phoenix, sapota, Terminalia chebula, Tectona grandis, Madhuca latifolia, Tamarindus indica, Mangifera indica, Artocarpus integrifolia, Acacia catechu, Santalum album, Pterocarpus marsupium, Adina cordifolia, Terminalia tomentosa, Hardwickia binata, Syzygium cumini,* and *Albizia lebbeck* (Dhyani et al., 2009; Chavan et al., 2015).

16.14 GUJARAT PLAINS AND HILLS REGION

This is a well-known oilseed-producing region, which includes hills and plains of Kathiawar, and the fertile valleys of Mahi and Sabarmati rivers. The region extends in Gujarat, Dadra, and Nagar Haveli, Daman, and Diu. This region has a 18.673 M Ha area and 13.76% agroforestry area (Arunachalam et al., 2022). The main crops grown in this region are groundnut, cotton, rice, millets, oilseeds, wheat, and tobacco. Soils are regur in the plateau region, alluvium in the coastal plains, and red and yellow soils in Jamnagar. This region is well known for Pulpwood agroforestry system (Chavan et al., 2015). The major trees growing in this agro-climatic region are *Azadirachta indica, Prosopis juliflora, Acacia nilotica, Ailanthus excelsa, Eucalyptus, Casuarina equisetifolia, Tectona grandis, Mangifera indica, Madhuca latifolia, Acacia catechu, Dalbergia latifolia, Santalum album, Gmelina arborea, Anogeissus pendula,* and *Leucaena leucocephala* (Dhyani et al., 2009). The union territory Dadra and Nagar Haveli spreads over the foothills of Western Ghats between Gujarat and Maharashtra. The tribal population of this union territory depends on forests for their fuel, fodder, and small timber requirements (ISFR, 2019). In rural areas, the population cultivates various agricultural crops such as rice, ragi, jwar, maize, black gram while retaining some trees of *Tectona grandis, Butea monosperma, Terminalia crenulata, Bridelia retusa, Acacia catechu,* and *Anogeissus latifolia* for agroforestry, and Mango, cashew, guava, custard apple, sapota, etc., on the farms, along the ponds, and streams.

16.15 WESTERN DRY REGION

This agro-climatic zone comprises 12 districts in the west of the Aravallis of Rajasthan. The region is characterized by high evaporation, contrasting temperatures, absence of perennial rivers, and scanty vegetation. This region has a total area of 17.587 million hectares and 2.45% agroforestry area (Arunachalam et al., 2022). Rajasthan is known for its Khejri-based agro-forestry system (Chavan et al., 2015). Bajra, jowar, and moth are the main crops grown in this region. The region includes *Prosopis cineraria, Azadirachta indica,* and *Albizia lebbeck (*Dhyani et al., 2009). Tree species such as *Zizyphus mauritiana, Prosopis cineraria, Acacia senegal*, and *Capparis decidua* are combined with agricultural crops like bajra, moong, and gwar in Jaisalmer as the best traditional agroforestry practice. In Sri Ganganagar, agricultural crops such as bajra, moong, gwar, wheat, cotton, and rice are cultivated with *Acacia nilotica, Prosopis cineraria*, and *Acacia tortilis* trees. Meanwhile, in Bikaner, Barmer, and Jodhpur, crops like moong, bajra, gwar, and moth are traditionally practiced with tree species *Zizyphus mauritiana, Ziziphus nummularia, Prosopis cineraria, Capparis decidua,* and *Acacia senegal* (Tiwari & Singh, 2006). In Churu, Jhunjhunu, Sikar, Nagaur, and Jalor, agricultural crops such as bajra, moong, gwar, mustard, and wheat in irrigated areas (Nagaur) are cultivated with *Prosopis cineraria, Capparis decidua, Acacia nilotica, Salvadora oleoides*, and *Mytenus emarginata.* Pali district has slightly different climatic conditions compared to other western Rajasthan districts, with 92 dams for irrigation and drinking water. Agricultural crops such as jowar, bajra, moong, guar, and taramira are cultivated with remarkable tree species like *Acacia nilotica* var. *cupressiformis, Acacia nilotica* var. *indica, Acacia leucophloea, Acacia catechu, Salvadora oleoides*, etc.

16.16 ISLAND REGION

The island region includes Andaman and Nicobar Islands and Lakshadweep. The soils vary from sandy along the coast to clayey loam in valleys and lower slopes. This region has an area of 0.785 million hectares and 2.42% agroforestry area (Arunachalam et al., 2022). The main crops grown are rice, maize, millets, pulses, areca nut, turmeric, and cassava. Important trees found in this region include *Cocos nucifera, Bauhinia, Erythrina,* and *Leucaena* (Dhyani et al., 2009). In the coastal region, especially in Andaman and Nicobar Islands and Lakshadweep, most cropping systems are plantation crop-based, such as coconut, arecanut, cashew, red oil palm, clove, cardamom, and coffee as middle story crops, and betel nut and black pepper, while pulses, pineapple, tuber crops, grasses, and green legumes are grown as ground story crops (Dagar, 2020). In some areas of Andaman on the bunds of rice fields, tree species like coconut, *Acacia auriculiformis, Casuarina equisetifolia, Borassus flabellifer*, and *Sesbania sesban* are cultivated (Dagar, 2020).

16.17 CONCLUSION

The results of the present review reveal that traditional agroforestry systems play an important role in the conservation of tree diversity. Based on results obtained from this review study, the following suggestions are recommended: thorough documentation of various land-use systems is important to study the diversity of tree species. The appropriate utilization of tree species in various land uses and raising awareness at the grassroots level is essential to conserve the loss of valuable tree resources. The various agencies should promote agroforestry by adopting the plantation of native tree species in agroforestry for the conservation of native tree species.

KEYWORDS

- **agro-climatic zone**
- **agroforestry systems**
- **Himalayas**
- **native tree species**
- **plantation**
- **traditional agroforestry systems**
- **traditional farming**

REFERENCES

Annamalai, S. J. K., (2006). *Long-term Strategies and Programs for Mechanization of Agriculture in Agro Climatic Zone–X: Southern Plateau and Hills Region.* https://farmech.dac.gov.in/06035-04-ACZ10-15052006.pdf (accessed on 6 May 2024)

Arunachalam, A., Rizvi, R. H., Handa, A. K., & Ramanan, S. S., (2022). Agroforestry in India: Area estimates and methods. *Current Science, 123*(6), 743, 744.

Bhaskar, D., Vishwanath, S., & Purushothaman, S., (2013). Decomposition and nutrient release dynamics of *Ficus benghalensis* L. litter in traditional agroforestry systems of Karnataka, Southern India. *ISRN Forestry.* doi:10.1155/2013/524679.

Bijalwan, A., Verma, P., Dobriyal, M. J. R., Patil, A. K., Thakur, T. K., & Sharma, C. M., (2019). Trends and insight of agroforestry practices in Madhya Pradesh, India. *Current Science*, (117), 579–605.

Butola, J. S., Malik, A. R., & Baba, J. A., (2012). Livelihood diversification in cold arid desert of Indian Himalaya: Urgent need of transforming traditional agroforestry system with emphasis on adoption of herbal farming. *International Journal of Medicinal and Aromatic Plants, 2*, 544–548.

Chandra, D. G., & Malaya, D. B., (2011). Role of e-agriculture in rural development in Indian context. In: *2011 International Conference on Emerging Trends in Networks and Computer Communications (ETNCC)* (pp. 320–323). Udaipur, India. doi: 10.1109/ETNCC.2011.6255913.

Chavan, S. B., Keerthika, A., Dhyani, S. K., Handa, A. K., Ram, N., & Rajarajan, K., (2015). National agroforestry policy in India: A low hanging fruit. *Current Science, 108*(10), 1826–1834.

Dagar, J. C., (2020). Agroforestry systems for coastal and island regions. *Indian Journal of Agroforestry, 2*, 59–74.

Dar, M., Qaisar, K. N., Ahmad, S., & Wani, A. A., (2018). Inventory and composition of prevalent agroforestry systems of Kashmir Himalaya. *Advances in Research, 14*(1), 1–9.

Deshmukh, A. P., Deshmukh, H. K., Taide, Y. B., Cinchamalatpure, U. R., Walke, R. D., & Mane, V. P., (2017). Agroforestry systems practiced in southern part of Akola district of Maharashtra. *Journal of Tree Sciences, 36*, 92–100.

Dhara, P. K., Panda, S., Sarkar, S., & Das, N. C., (2016). Agroforestry systems practiced in new alluvial zone of West Bengal. In: *Proc Intl. Conf. Agriculture, Food Sci., Nat. Resource Management and Environmental Dynamics: The Technology, People and Sustainable Development* (pp. 162–164). Kalyani, India.

Dhyani, S. K., Kareemulla, K., Ajit, & Handa, A. K., (2009). Agroforestry potential and scope for development across agro-climatic zones in India. *Indian Journal of Forestry, 32*(2), 181–190.

Hussain, M., (2015). Agro-climatic zones and economic development of Rajasthan. *International Journal of Humanities and Social Science Invention*, 50–57.

ISFR, (2019). *Indian State of Forest Report, 2019.*

Islam, M. A., Masoodi, T. H., Gangoo, S. A., Sofi, P. A., Bhat, G. M., Wani, A. A., Gatoo, A. A., et al., (2015). Perceptions, attitudes and preferences in agroforestry among rural societies of Kashmir, India. *Journal of Applied and Natural Science, 7*(2), 976–983.

Kala, C. P., (2010). Status of an indigenous agro-forestry system in changing climate: A case study of the middle Himalayan region of Tehri Garhwal, India. *Journal of Forest Science, 56*(8), 373–380.

Khanna, S. S., (1989). *Farm Planning: The Agro-climatic Approach in The Hindu 1989.* Survey of Indian Agriculture, 1989, Madras.

Kumari, A., Sehgal, A. N., & Kumar, S., (2008). Traditional agroforestry systems practiced in Lahul (Lahul & Spiti) and Kinnaur district of Himachal Pradesh. *Indian Forester*, 1003–1010.

Molla, A., & Kewessa, G., (2015). Woody species diversity in traditional agroforestry practices of Dellomenna District, Southeastern Ethiopia: Implication for maintaining native woody species. *International Journal of Biodiversity*. http://dx.doi.org/10.1155/2015/643031.

Mugloo, J. A., Masoodi, T. H., Khan, P. A., Dar, M., Wani, A. A., & Raja, R., (2020). Management practices vis-a-vis agroforestry for the improvement of rangelands of Jammu and Kashmir in Northwestern Himalaya, India. In: Dagar, J. C., Gupta, S. R., & Teketay, D., (eds.), *Agroforestry for Degraded Landscapes*. Springer, Singapore. https://doi.org/10.1007/978-981-15-6807-7_2.

Nair, P. K. R., (1979). Intensive multiple cropping with coconuts in India. *Advances in Agro. Crop. Science, 6,* 1–12.

Nath, A. J., Kumar, R., Devi, N. B., Rocky, P., Giri, K., Sahoo, U. K., & Pandey, R., (2021). Agroforestry land suitability analysis in the Eastern Indian Himalayan region. *Environmental Challenges, 4,* 100199.

NS, K., & KS, P., (2017). Economic analysis of prevailing agroforestry systems among different categories of farmers in Kangra Valley of Himachal Pradesh, India. *International Journal of Bio-Resource & Stress Management, 8*(4).

Pandey, M. M., (2006). *Long-Term Strategies and Programs for Mechanization of Agriculture in Agro Climatic Zone–VIII: Central Plateau and Hills Region.* https://farmech.dac.gov.in/06035-04-ACZ8-15052006.pdf (accessed on 6 May 2024).

Panigrahi, S., Siri, P., Rout, S., & Sahoo, R. K., (2021). Fuel and fodder species used for agro forestry practices in Koraput, Odisha. *Akshar Wangmay' Special Issue, Sustainable Development and Environmental Issue* (Vol. II, pp. 94–100).

Prahlad, V. C., (2018). Climate resilient agro-forestry interventions to reduce farming risks in Rajasthan. *International Journal of Science, Environment and Technology, 7*(4), 1374–1386.

Punia, V., (2022). Agriculture and population growth. *ECONSPEAK, 12*(6), 32–43.

Ram, N., Chaturvedi, O. P., Kumar, D., Prasad, R., Rizvi, R. H., Alam, B., Handa, A. K., et al., (2017). Soil organic carbon stock in agroforestry systems in western and southern plateau and hill regions of India. *Current Science, 112*(11), 2191–2193.

Rana, B. S., Saxena, A. K., & Parihar, K. S., (2007). Status and scope of agroforestry development in eastern part of Uttar Pradesh. In: Puri, S., & Panwar, P., (eds.), *Agroforestry Systems and Practices* (pp. 191–212). New India Publishing Agency, New Delhi.

Rizvi, R. H., Ram, N., Handa, A. K., Sridhar, K. B., & Anil, K., (2019). *Agroforestry Mapping in India Through Geospatial Technology: Present Status and Way Forward* (pp. 1–35). Technical Bulletin-01/2019, ICAR-CAFRI, Jhansi.

Sarvanan, S., (2021). Role of trees in farm bunds – a case study in Balagahat district, Madhya Pradesh, India. International *Journal of Environment, Agriculture and Biotechnology, 6*(6), 382–388.

Singh, B. K., Oraon, P. R., Kumar, A., & Malik, M. S., (2017). Traditional agroforestry practices existing at Lohardaga district of Jharkhand. *Trends in Biosciences, 10*(23), 4651–4654.

Sirohi, N. P. S., (2006). *Long-Term Strategies and Programs for Mechanization of Agriculture in Agro Climatic Zone–III: Lower Gangetic Plains Region.* https://farmech.dac.gov.in/06035-04-ACZ3-15052006.pdf (accessed on 22 April 2024).

Srivastava, N. S. L., (2006). *Long-Term Strategies and Programs for Mechanization of Agriculture in Agro Climatic Zone–IV: Middle Gangetic Plains Region.* https://farmech.gov.in/06035-04-ACZ4-15052006.pdf (accessed on 22 April 2024).

Tiwari, V. P., & Singh, M., (2006). Tree crop interactions in the Thar Desert of Rajsthan (India). *Secheresse, 17*(1, 2), 1–7.

Verma, S. R., & Singh, S., (2006). *Long-term Strategies and Programs for Mechanization of Agriculture in Agro Climatic Zone–VI: Trans Gangetic Plains Region.* https://farmech.gov.in/06035-04-ACZ6-15052006.pdf

Viswanath, S., Lubina, P. A., Subhanna, S., & Sandhya, M. C., (2018). Traditional agroforestry systems and practices: A review. *Advanced Agricultural Research & Technology Journal, II*(1), 18–29.

CHAPTER 17

Fodder Trees: Potential Feed for Livestock in a Resource-Constrained World

ARCHIBOLD G. BAKARE,[1] TITUS J. ZINDOVE,[1] and NATION CHIKUMBA[2]

[1]*College of Agriculture, Fisheries and Forestry, Fiji National University, Nausori, Fiji*

[2]*Marondera University of Agricultural Sciences and Technology, Marondera, Zimbabwe*

ABSTRACT

Resource-constrained farmers rely on livestock for sustenance. However, their livestock do not perform to expectations because of scarce feed resources. If resources are available, they will be of poor quality. Non-conventional supplements for the livestock are expensive beyond the reach of many farmers. Hence, fodder trees are the only cheap alternative supplements for the livestock. The fodder trees can easily be propagated and require less capital and land. They have the potential to supply feed to livestock in times of extreme climatic conditions such as droughts. Also, they are nutritious, allowing livestock to acquire all nutrients for growth, maintenance, and reproduction. Farmers need to know all the challenges impeding the use of fodder trees, including anti-nutritional factors in each tree species, species appropriateness to different agro-ecological zones, and xeromorphic traits of each plant species. Technologies such as the use of tannin-binding compounds and exogenous enzymes are essential to understand how livestock can efficiently use nutrients from the fodder trees. The products (tannin-binding compounds and exogenous enzymes) are readily

Sustainable Forest Resources Management: Issues and Implications. Jahangeer A. Bhat, Gopal Shukla, Manmohan J. Dobriyal, Sumit Chakravarty, A. Arunachalam, and Rainer W. Bussmann (Eds.)

available on the market. The information also needs to be disseminated to the resource-constrained farmers.

17.1 INTRODUCTION

Rearing of livestock plays a vital role in alleviating poverty and improving food security and sustenance of livelihood for humans. They are used as a source of protein, providing meat, milk, and eggs. Also, when livestock is sold, it provides cash for buying other household utensils. However, to achieve all of this, animals should be fed with diets that meet growth, maintenance, and reproduction requirements. Ingredients for the diets mainly come from cereal grains, pulses, and agricultural by-products (Bakare & Chimonyo, 2017). Production of the crops is, however, thwarted by vagaries of nature, including droughts and floods. Furthermore, developing countries, for example, do not have the capital to produce enough grain/crops to feed human and livestock populations. This results in poor quality feed being offered to the livestock species.

For years, farmers have fed fodder trees growing naturally on their farms to their livestock (Figure 17.1). The fodder trees are distributed in different ecological systems in the world where they provide firewood to the human populace, support environmental services such as control of erosion, improve soil fertility, and control agricultural pests (Kuyah et al., 2016). For this reason, different methods of cultivating the fodder trees have been devised. These methods include using vegetative propagation and breaking seed dormancy. Therefore, the fodder trees can be propagated in different ecological regions and thus provide feed for livestock. Unlike herbaceous plant material (grasses), the nutrient composition of the fodder trees tends to vary little across seasons (Abebe et al., 2012). Also, the root system of the trees extracts minerals from deep soil layers. Hence, when fed to livestock, they are a good source of minerals.

The presence of anti-nutritional factors/plant secondary compounds limits the utilization of nutrients (Jerónimo et al., 2016). Anti-nutritional factors are a defense mechanism plants use to defend themselves against herbivory. It is believed that plants contain thousands of these secondary compounds. Depending on the situation, the secondary compounds can have beneficial or harmful effects on livestock. For example, if an animal eats foliage with high concentrations of the compound, they get sick, leading to a high aversion from the plant species (Burritt & Provenza, 1996). These behaviors can be transferred to the next generation, and the herbivorous animals avert from feeding

(a)

(b)

FIGURE 17.1 Cut and carry feeding system using fodder trees (a) cattle; and (b) goats.

on the plant species. Animal nutrition research has now made it possible to use the fodder trees' foliage despite anti-nutritional factors. Farmers worldwide have come up with different methods such as wilting, soaking, and adding chemical and biological compounds, just to mention a few.

The supply of foliage from fodder trees is also affected by the season. For example, deciduous trees can shed their leaves depending on the availability of moisture in the soil. The trees shed leaves to conserve water and to help them pollinate during spring. When trees shed their leaves, foliage biomass is reduced, which means animals will not have enough feed during periods of scarcity. In southern Africa, for example, dry periods occur during the period of May-September. Most of the trees shed their leaves during these periods. Farmers are, therefore, always encouraged to conserve the foliage to supply nutritious feed to their livestock in periods of feed scarcity. Vegetation is conserved as either hay or silage.

17.2 INVENTORY OF FODDER TREES BY ECOLOGICAL ZONE

There are many trees reported in literature that are potential feed for livestock. The tree species are classified as either leguminous or non-leguminous. Different trees, leguminous, and non-leguminous, are distributed in different ecological zones; namely, arid, semi-arid, subtropics, and tropics. Despite a large number of trees recorded, very few have real feed value. When farmers select trees to feed their animals, they have to look for several characteristics like nutritional value, anti-nutrients, plant parts (leaves and pods), and the tree's ability to regenerate leaves after pruning. Leguminous plant species have been the focus of most feeding in livestock. The reason is that they have high protein and low fiber content compared to non-leguminous fodder trees. According to Khan (2017), legumes are divided into three subfamilies: Caesalpinioideae, Faboideae, and Mimosoideae. Caesalpinioideae comprises more than 2200 species. Examples of plant species in this family include *Bauhinia variegata, Cassia fistula, Caesalpinia pulcherrima, Delonix regia, Saraca indica, Tamarindus indica*, and *Parkinsonia spp.* Faboideae includes genera *like Cicer, Dalbergia, Erythrina, Lathyrus, Lupinus, Robinia, Sophora,* and *Wisteria* spp. Mimosoideae includes over 3000 plant species, including *Acacia spp., Albizia spp.*, and *Mimosa* spp. The leguminous trees have the potential to replace expensive protein supplements for poor resource farmers.

Trees are also classified as either deciduous or evergreen. Deciduous trees shed leaves during dry or drought conditions. Before they shed off their leaves, nitrogen, and carbon from the foliage is channeled to parenchyma cells in the roots and bark. The nitrogen and carbon are re-used during spring or when water is available to grow new leaves/shoots (Srivastava, 2002). Evergreen tree species, on the other hand, do not lose leaves. The information is important for resource-constrained farmers to know the seasonal availability of leaves from fodder trees. Table 17.1 shows examples of different tree species from different ecological zones.

17.3 FACTORS AFFECTING FOLIAGE BIOMASS

Sustainable use of fodder trees for livestock depends on the amount of biomass produced by the plants. In this case, the consumable biomass consists of leaves, stems, flowers, pods/fruits, and bark. Leaves are the most preferred, and to some extent, pods or fruits. The leaves and pods/fruits are more nutritious compared to stems and barks. The availability of the plants' foliage biomass is affected by several factors, namely: the quantity of light available, temperature, plant age/size, plant species, nutrient availability, and water availability (Foroughbakhch et al., 2009; Irving, 2015). Nutrient supply, water availability, and temperature tend to have large effects on foliage biomass production.

17.4 NUTRIENT SUPPLY

A nutrient deficit in the soil negatively affects biomass production. For example, nitrogen deficiency strongly depresses vegetation flush. Nitrogen is a major component of chlorophyll, a compound which traps sunlight energy to produce carbohydrates from water and carbon dioxide (Evans, 1989). More photosynthesis will occur with more nitrogen and subsequent uptake of the soil's nutrients, leading to more leaf biomass production. The opposite is true when there is a deficit of nitrogen in the soil. Macro- and micro-minerals are equally important for plant growth and leaf biomass production. Controlling the optimum nutrient levels in the soil will ensure more biomass for use as feed is produced. Most nutrients are available in the soil at a pH range of 6.0–7.5. For example, nitrogen, phosphorus, and potassium are found in the soil at a pH of 5.5–9.0 (Brady & Weil, 1999). Hence, in these conditions, more minerals are absorbed from the soil and used for biomass production. Availability of nutrients is also affected by leaching. For example, most mobile nutrients like potassium, nitrogen, sulfur, and even magnesium, held more tightly than potassium (Goldy, 2013), can be leached

TABLE 17.1 Examples of Trees Species from Different Ecological Zones

Tree Species (Scientific Name)	Deciduous/Evergreen	Ecological Zone
Bassia latifolia	Deciduous	Semi-arid
Calliandra calothyrsus	Semi-deciduous/ evergreen	Humid and sub-humid
Sesbania grandiflora	Deciduous/evergreen	Arid and semi-arid
Ailanthus excels	Deciduous	Arid and semi-arid
Celtus australis	Deciduous	Sub-tropical and temperate
Acacia Nilotica	Deciduous	Dry and moist tropics
Bauhinia variegate	Semi-evergreen/ deciduous	Sub-tropics, moist, and dry tropics
Moringa oleifera	Deciduous	Moist tropics, sub-humid, humid
Prosopis cineraria	Deciduous	Arid and semi-arid
Robinia pseudoacacia	Deciduous	Arid and semi-arid
Azadirachta indica	Deciduous/evergreen	Dry and moist tropics
Albizia amara	Deciduous	Dry tropics
Quercus incana	Deciduous	Temperate
Dalbergia sissoo	Deciduous	Moist tropics
Morus alba	Deciduous	Moist tropics/temperate
Anogeissus leiocarpus	Deciduous	Tropics
Colophospermum mopane	Deciduous	Sub-tropics
Grewia bicolor	Deciduous	Sub-tropics
Aesculus indica	Deciduous/evergreen	Temperate
Barleria eranthmoides	Deciduous	Semi-arid
lpomea hardwickii	Deciduous	Sub-tropics, tropics
Feretia apodanthera	Deciduous	Tropics
Olea ferruginea	evergreen	Sub-tropics
Senegalia caffra	Deciduous	Sub-tropics
Leucaena leucocephala	Deciduous	Sub-humid, semi-arid
Ziziphus mauritiana	Deciduous	Dry and moist tropics
Gliricidia sepium	Deciduous	Humid, sub-humid
Morus laevigata	Deciduous	Sub-tropics
Prosopis chilensis	Deciduous	Dry tropics
Ficus gnaphalocarpa	Deciduous	Tropics
Celtis integrifolia	Deciduous	Tropics
Diospyros mespiliformis	Deciduous	Sub-tropics, tropics
Steganotaenia araliacea	Deciduous	Sub-tropics

to greater depths in the soil due to excessive rainfall. Only tree species with a deeper root system will absorb the nutrients, which will be used for metabolic processes in the plant.

17.5 WATER AVAILABILITY

The amount of foliage biomass varies significantly across ecological systems in the world. Ecological systems receiving a considerable amount of rainfall have plants producing large quantities of foliage biomass. When rainfall is high and evenly distributed throughout the year, this is favorable for vegetative shoot growth due to increased soil moisture (Figure 17.2; Li et al., 2018).

Amount of soil moisture tends to vary with seasons. For example, in subtropical ecological systems, foliage biomass is produced in large amounts during the rainy seasons when soil moisture is high. With time as rainfall decreases into the dry season, plant aboveground biomass gradually decreases as well. Low moisture content in the soil causes dehydration in plants. In response to dehydration, abscisic acid (ABA) is synthesized by root cells and later transported to the leaf apoplast (Seo & Koshiba, 2011). Water-stressed leaf cells also produce ABA. Increased ABA concentration induces leaf stomatal closure and leads to leaf drop. Hence, during the cool and dry season in the subtropics, there will be

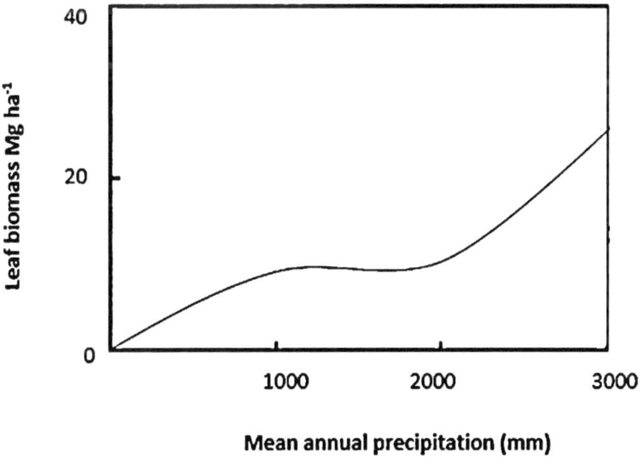

FIGURE 17.2 Relationship between precipitation and leaf biomass (Mg/ha).

less foliage biomass. Therefore, foliage biomass is harvested when in abundance during the rainy season and conserved as either hay or silage.

17.6 TEMPERATURE

Of equal importance as a factor affecting foliage production is temperature. Dendrochronology studies have often found reduced growth during cooler seasons (Lavergne et al., 2018; Fontana et al., 2019). Low temperatures limit many biological processes such as root and microbial activity, mineralization of organic matter, and photosynthesis. Reduced root and microbial activity result in a reduced supply of nitrogen and phosphorus, consequently affecting growth and foliage biomass accumulation (Zhou et al., 2002). On another note, high temperatures accelerate trees' metabolic rates, enhancing biologically driven processes. This improves plant growth by initiating plant photosynthesis. Studies have shown that leaf growth rate increases with temperature until the optimum temperatures are reached, but a continual increase in temperature will reduce growth and cause senescence of the leaves; hence, leaf biomass is reduced (Tamaki et al., 2002). This means fodder in the tropics and sub-tropics will be more compared to foliage in arid and semi-arid ecosystems. However, it should be noted that temperature alone cannot contribute to foliage growth and quantity on fodder trees. Other factors like water and nutrients available in the soil play an important role.

17.7 ADAPTIVE STRATEGIES OF FODDER TREES TO THE SURROUNDING ENVIRONMENT AND METHOD USED TO FEED THE ANIMALS

Trees also adapt to the surrounding environment. Studies have demonstrated that trees can withstand scarce water through the small size of their leaves. Figure 17.3 shows an illustration of evapotranspiration in different types of leaves. Smaller leaves increase the surface area through which a plant loses water. Hence, fodder trees most likely to thrive in harsh dry conditions should have smaller leaves. Farmers introducing fodder trees on their farms should consider such xeromorphic traits. Narrow leaves, if not fed fresh, can easily be conserved by sun drying. The smaller surface allows the leaves to dry within a short period, thus reducing chances of molding. When it comes to ensiling fodder trees, fodder trees in some subtropical and tropical ecosystems are not ensilage material compared to fodder trees in temperate regions (Tainton, 2000). They have low water-soluble carbohydrates (WSC), which leads to a higher buffering capacity during ensiling. Proteolysis in the leaves will occur and this reduces the efficiency of nitrogen utilization by the ungulates. Some measures can be used to prevent this from happening. These include the use of silage additives like urea, ammonia, and molasses, and exogenous enzymes. Other studies have shown that the mixing of legume tree leaves and crops can reduce the rate of proteolysis. A study by Mugweni (2000) successfully ensiled a mixture of maize with *Acacia boliviana, Calliandra calothyrsus, Gliricidia sepium,* and *Leucaena leucocephala.*

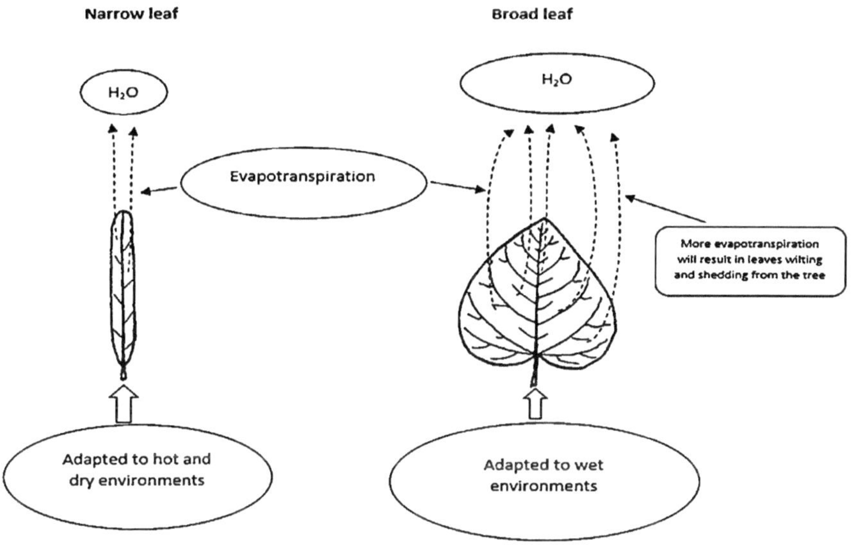

FIGURE 17.3 Illustration of evapotranspiration in different types of leaves.

17.8 NUTRIENT COMPOSITION

Herbivores can consume herbaceous plants (grasses) and foliage from trees and shrubs (Wood et al., 2015). Foliage from trees and shrubs is widely recognized for its nutrients relative to herbaceous plants (grasses). In addition, the nutrient composition of foliage from trees does not fluctuate across seasons. Hence, foliage from trees is a potential source of nutrients for livestock throughout the year. In the cool and dry season, trees shed their leaves. The remaining herbaceous plant material (grasses) is of low quality. It does not have enough nutrients to meet the requirements for growth, maintenance, and reproduction in animals. Inadequate protein in the diet may lead to stunted growth, decreased feed efficiency, reduced conception rate, and decreased milk production (Minson, 2012). Therefore, the use of foliage biomass from trees is the only cheap alternative to provide a source of protein to the animals. Examples of the tree species include *Calliandra calothyrsus* (CP 20–25%), *Sesbania sesban* (CP 20%), *Leucaena leucocephala* (CP 22%), *Albizia chinensis* (CP 22%), *Albizia lebbeck* (CP 24%), *Albizia saman* (CP 25%), *Erythrina poeppigiana* (CP 25%), just to mention a few.

Forages vary extensively in their fiber content. The fiber content in plants plays a vital role in determining the palatability of plant material. Neutral detergent fiber (NDF) and acid detergent fiber (ADF) are the common fiber types measured in forages. The variation in NDF and ADF of plant material is associated with plant part maturity (Hameed et al., 2008). New shoots have less fiber content compared to mature leaves. Also, fiber content varies according to the plant part. Leaves have less fiber content compared to stems and bark. The high fiber content in animals' diets increases chewing activity, consequently decreasing dry matter intake.

Of equal importance are minerals in leaves of fodder trees. Minerals are inorganic nutrients that play a vital role in animals' health, but they do not provide any energy. Minerals are categorized into macro- and micro-minerals. Examples of macro-minerals are calcium, phosphorus, magnesium, potassium, sodium, and sulfur. On the other hand, micro-minerals include copper, iron, manganese, cobalt, zinc, iodine, molybdenum, fluorine, chromium, selenium, and chlorine. To meet the mineral requirements, animals should ingest sufficient amounts of forages with the minerals. According to Juknevičius and Sabienė (2007) tree leaves contain more of these nutrients than other feedstuff like grasses and whole grain. This is because they have deep root systems, which enable them to extract minerals from greater soil depths. These minerals will end up in different plant parts such as leaves, fruits/pod, stems, and barks.

17.9 NUTRITIONAL VALUE OF FODDER TREES FOR LIVESTOCK

Fodder trees are widely recognized for their nutrients. They contain high crude protein levels and mineral matter and are highly digestible compared to herbaceous plant material (grasses). The fodder trees' nutritive value regulates the livestock's optimal body size; hence, they are a valuable feed resource fed as dry, fresh, or as silage. Digestibility is one parameter that can be used to estimate the nutritive value of feed ingredients. It is used to determine the amount of nutrients in fodder assimilated by the animal. It has been noted that fodder leaves are readily digested by livestock (cattle, sheep, and goats) and have led to improvements in their performance. Different tree species have different digestibility values, and this varies with the maturity of the plant part. Young leaves are more digestible compared to mature leaves. Additionally, leaves and fruits/pods are more digestible than stems and bark. *Grewia oppositifolia* has an *in vitro* dry matter digestibility of (62.2%), *Morus alba* (52.3%), *Betula celtis* (54.4%), *Celtis australis* (49.3%), *Diospyros lotus* (57.5%), Aesculus indica (60.6%), *Celtis caucasica* (46.3%), *Robinia pseudoacacia* (52.7%), *Olea ferruginea* (46.4%), and *Quercus incana* (42.4%) (Yaqoob & Javed, 2008). Regarding livestock performance, supplementation to Washera lambs with *Ficus thonningii* significantly increased total dry matter, organic matter, and feed intake (Mekuriaw & Asmare, 2018). Feeding *Acacia karroo* leaf meal to goats improved live weight gain (Brown et al., 2018). Similarly, Nguni cattle fed *A. karroo* leaf-meal produced beef with higher protein content (Mapiye et al., 2010). In dairy production feeding trials, 1 kg of *Calliandra* leaves increased milk production by 0.6–0.8 kg. Attempts have also been made to feed fodder tree leaves to non-ruminants. A review of literature showed that *Moringa oleifera, Azadirachta indica, Manihot esculenta* improved productivity of chickens (Odunsi et al., 2009; Gadzirayi et al., 2012; Bakare et al., 2020). Improvement in performance was also noted in pigs fed fodder tree leaves.

17.10 ANTI-NUTRITIONAL FACTORS

Utilization of nutrients from the different fodder tree leaves and fruits/pods is limited because of anti-nutritional factors/plant secondary compounds. Anti-nutritional factors are a defense mechanism plants employ to protect themselves against herbivory. The concentration of anti-nutritional factors in plants may vary depending on genotype, phenology, and season (Soetan & Oyewole, 2009; Huang et al., 2018).

Some examples of secondary compounds in plants are tannins, oxalates, saponins, cyanogenic glycosides, and foliar phytic acid (Table 17.2; Tadele, 2015). Foliar phytic acid levels are generally low; hence, most literature does not mention the occurrence of phytic acid in leaves (Hadi & Zotz, 2014). Nevertheless, anti-nutritional factors affect the digestion of plant material or have toxic effects on livestock after being absorbed after ingestion. Different

TABLE 17.2 Anti-Nutritional Factors and Effects on Animal

Anti-Nutritional Factor	Examples of Plant Species	Effects on Animals	Animals Most Affected	Remedial Measures
Tannins (condensed and hydrolysable)	All vascular plants	Reduced feed intake, negative N digestibility, loss in weight.	Poultry, swine	Soaking, boiling
Calcium oxalates	*Acacia berlandieri, Sesbania vesicaria, S. punicea, S. drummondii, Cyclanthus subpalmata, Pandanus leram, Crinum amabile, Heliconia longiflora* and *Guzmania zahnii*	Form a complex with dietary calcium rendering the complex unavailable for absorption and assimilation. Results in calcium deficiencies.	Horses	Soaking, boiling
Saponins	*Azadirachta indica, Commiphora glandulosa, Sesbania sesban*	Affects absorption of vitamin A and E, and lipids. This results in stunted growth in poultry and swine. They also cause bloating in ruminants.	Poultry, swine, ruminants	Soaking
Cyanogenic glycosides	*Acacia cunninghamii, Acacia sieberiana, Bambusa bambos, Albizia stipulate, Bassia latifolia, Sesbania sesban, Diospyros lycioides*	Inhibit ATPase formation. Tissues suffer from energy deprivation, which will consequently lead to death.	Poultry, swine, ruminants	Sun drying, fermentation

methods can be used to reduce the anti-nutritional factors to innocuous levels. These include fermentation, sun drying, soaking, and boiling. For example, in a study by Ngwa et al. (2004), fermentation (ensiling) reduced cyanogenic glycosides in pods of *Acacia sieberiana*. The use of these methods will enable many trees to be made available for livestock feed.

17.11 POTENTIAL TECHNOLOGIES TO IMPROVE UTILIZATION OF FODDER TRESS

Tannin-binding agents such as Polyethylene Glycol (PEG) and Polyvinylpyrrolidone have been used to study the potential adverse effects of tannins on rumen metabolism. The compounds are used to break tannin-protein complexes. Various researchers have utilized PEG of molecular weight 3500 or 4000 and alleviated the effects of tannins. Consumption of tannin-rich species (*Pistacia lentiscus* L.) increased in goats supplemented with PEG (Landau et al., 2002). It is believed that PEG binds to tannins and prevents astringency effects. Other effects of PEG supplementation include increasing fatty acids in the meat of various ruminant animals. The compound can be directly administered to animals through drinking water, mixing with concentrate feed, and spraying with tannin-rich feedstuffs like leaves from tannin-rich plants/trees (Besharati & Taghizadeh, 2011).

Recent advancements in the field of biotechnology have resulted in the production of exogenous microbial enzymes. The enzymes are produced using bacteria and fungi species. However, more emphasis is on the use of fungi because they produce large amounts of exogenous enzymes (Subramaniyam & Vimala, 2012). The enzymes digest fibrous feed material and make other nutrients available for animals. The enzymes are classified depending on the class of nutrients to be digested. Examples of these enzyme categories include non-starch polysaccharidases, amylases, proteases, phytases, and enzyme cocktails. Many studies have incorporated the microbial exogenous enzymes in feed for different livestock species (Salem et al., 2013; Cowieson & Kluenter, 2019; Oh et al., 2019; Recharla et al., 2019; Sallam et al., 2020). Most of these microbial exogenous enzymes are available commercially. If farmers get access to the enzymes, more of the nutrients from the fodder trees will be made available for use by the animals.

17.12 CONCLUSION

Fodder trees are an important feed source for livestock for resource-constrained farmers. They provide valuable nutrients for the livestock species. Anti-nutritional factors limiting the amount of nutrients for use by animals can easily be lowered to innocuous levels by simple heating/sun drying, boiling, fermentation, and soaking. Several technologies, such as the use of tannin-binding compounds (PEG and Polyvinylpyrrolidone), and microbial exogenous enzymes, are essential to understand how livestock can efficiently utilize nutrients from different fodder tree species.

KEYWORDS

- **anti-nutritional factors**
- **fodder trees**
- **foliage biomass**
- **livestock**
- **nutrients**
- **nutritional value**
- **xeromorphic traits**

REFERENCES

Abebe, A., Tolera, A., Holand, Ø., Ådnøy, T., & Eik, L. O., (2012). Seasonal variation in nutritive value of some browse and grass species in Borana rangeland, Southern Ethiopia. *Trop. Subtrop. Agroecosystems., 15*(2), 261–271.

Bakare, A. G., & Chimonyo, M., (2017). Relationship between time spent eating and nutritionally related blood metabolites of growing pigs fed on diets containing graded levels of fiber. *Anim. Prod. Sci., 57*(6), 1106–1111.

Bakare, A. G., Cawaki, P., Ledua, I., Kour, G., Jimenez, V., Sharma, A., & Tamani, E., (2020). Acceptability, growth performance and nutritional status of chickens fed cassava leaf meal (CLM)–based diets. *Trop. Anim. Health Prod., 4*, 1–9.

Besharati, M., & Taghizadeh, A., (2011). Effect of tannin-binding agents (polyethylene glycol and polyvinylpyrrolidone) supplementation on in vitro gas production kinetics of some grape yield by-products. *ISRN Vet. Sci.*

Brady, N. C., & Weil, R. R., (1999). *The Nature and Properties of Soils* (12th edn., p. 881). Prentice Hall. Upper Saddle River, NJ.

Brown, D., Ng'ambi, J. W., & Norris, D., (2018). Effect of tanniniferous *Acacia karroo* leaf meal inclusion level on feed intake, digestibility and live weight gain of goats fed a *Setaria verticillata* grass hay-based diet. *J. Appl. Anim. Res., 46*(1), 248–253.

Burritt, E. A., & Provenza, F. D., (1996). Amount of experience and prior illness affect the acquisition and persistence of conditioned food aversions in lambs. *Appl. Anim. Behav. Sci., 48*, 73–80.

Cowieson, A. J., & Kluenter, A. M., (2019). Contribution of exogenous enzymes to potentiate the removal of antibiotic growth promoters in poultry production. *Anim. Feed Sci. Technol., 250*, 81–92.

Evans, J. R., (1989). Photosynthesis and nitrogen relationships in leaves of C3 plants. *Oecol., 78*(1), 9–19.

Fontana, C., Santini-Junior, L., Olmedo, G. M., Botosso, P. C., Tomazello-Filho, M., & Oliveira, J. M., (2019). Assessment of the dendrochronological potential of *Licaria bahiana* Kurz, an endemic laurel of lowland Atlantic forests in Brazil. *Acta Bot. Bras., 33*(3), 454–464.

Foroughbakhch, R., Hernández-Piñero, J. L., Alvarado-Vázquez, M. A., Céspedes-Cabriales, E., Rocha-Estrada, A., & Cárdenas-Ávila, M. L., (2009). Leaf biomass determination on woody shrub species in semiarid zones. *Agrofor. Syst., 77*(3), 181–192.

Gadzirayi, C. T., Masamha, B., Mupangwa, J. F., & Washaya, S., (2012). Performance of broiler chickens fed on mature *Moringa oleifera* leaf meal as a protein supplement to soyabean meal. *Int. J. Poult. Sci., 11*(1), 5–10.

Goldy, R., (2013). *Knowing Nutrient Mobility is Helpful in Diagnosing Plant Nutrient Deficiencies*. Michigan State University Extension.

Hadi, A. H., & Zotz, G., (2014). Phytic acid in green leaves. *Plant Biol., 16*(4), 697–701.

Hameed, M., Naz, N., Ahmad, M. S., Shazad, I., & Riaz, A., (2008). Morphological adaptations of some grasses from the salt range, Pakistan. *Pak. J. Bot., 40*(4), 1571–1578.

Huang, Q., Liu, X., Zhao, G., Hu, T., & Wang, Y., (2018). Potential and challenges of tannins as an alternative to in-feed antibiotics for farm animal production. *Anim. Nutr., 4*(2), 137–150.

Irving, L. J., (2015). Carbon assimilation, biomass partitioning and productivity in grasses. *Agriculture, 5*(4), 1116–1134.

Jerónimo, E., Pinheiro, C., Lamy, E., Dentinho, M. T., Sales-Baptista, E., Lopes, O., & Silva, F., (2016). Tannins in ruminant nutrition: Impact on animal performance and quality of edible products. In: Combs, C. A., (ed.), *Tannins: Biochemistry, Food Sources and Nutritional Properties* (pp. 121–168). Nova Science Publishers, Inc., New York.

Juknevičius, S., & Sabienė, N., (2007). The content of mineral elements in some grasses and legumes. *Liet Moks Akad., 53*(1), 44–52.

Khan, A. S., (2017). *Medicinally Important Trees*. Springer.

Kuyah, S., Öborn, I., Jonsson, M., Dahlin, A. S., Barrios, E., Muthuri, C., Malmer, A., et al., (2016). Trees in agricultural landscapes enhance provision of ecosystem services in Sub-Saharan Africa. *Int. J. Biodivers. Sci. Ecosyst. Serv. Manag., 12*(4), 255–273.

Landau, S. Y., Perevolotsky, A., Kababya, D., Silanikove, N., Nitzan, R., Baram, H., & Provenza, F. D., (2002). Polyethylene glycol affects goats' feeding behavior in a tannin-rich environment. *Rangel Ecol. Manag., 55*(6), 598–603.

Lavergne, A., Daux, V., Pierre, M., Stievenard, M., Srur, A. M., & Villalba, R., (2018). Past summer temperatures inferred from dendrochronological records of *Fitzroya cupressoides* on the eastern slope of the Northern Patagonian Andes. *J. Geophys. Res. Biogeosci., 123*(1), 32–45.

Lie, Z., Xue, L., & Jacobs, D. F., (2018). Allocation of forest biomass across broad precipitation gradients in 'China's forests. *Sci. Rep., 8*, 10536.

Mapiye, C., Chimonyo, M., Dzama, K., Muchenje, V., & Strydom, P. E., (2010). Meat quality of Nguni steers supplemented with *Acacia karroo* leaf-meal. *Meat Sci., 84*(4), 621–627.

Mekuriaw, Y., & Asmare, B., (2018). Nutrient intake, digestibility and growth performance of Washera lambs fed natural pasture hay supplemented with graded levels of *Ficus thonningii* (Chibha) leaves as replacement for concentrate mixture. *Agric & Food Secur., 7*, 30.

Minson, D., (2012). *Forage in Ruminant Nutrition*. Elsevier.

Mugweni, B. Z., (2000). *The Effect of Feeding Mixed Maize-Forage Tree Legume Silages on Milk Production from Lactating Holstein Dairy Cows*. M. Sc. Animal Science. Thesis. University of Zimbabwe, Harare.

Ngwa, T. A., Nsahlai, I. V., & Iji, P. A., (2004). Ensilage as a means of reducing the concentration of cyanogenic glycosides in the pods of *Acacia sieberiana* and the effect of additives on silage quality. *J. Sci. Food Agric., 84*, 521–529.

Odunsi, A. A., Adegbile, S. A., Akande, T. O., & Olayeni, T. B., (2009). Neem (*Azadirachta indica*) seed cake in the diets of cockerel chickens. *Int. J. Poult. Sci., 8*(1), 47–51.

Oh, J., Harper, M., Melgar, A., Compart, D. P., & Hristov, A. N., (2019). Effects of *Saccharomyces cerevisiae*-based direct-fed microbial and exogenous enzyme products on enteric methane emission and productivity in lactating dairy cows. *Int. J. Dairy Sci., 102*(7), 6065–6075.

Recharla, N., Kim, D., Ramani, S., Song, M., Park, J., Balasubramanian, B., Puligundla, P., & Park, S., (2019). Dietary multi-enzyme complex improves in vitro nutrient digestibility and hind gut microbial fermentation of pigs. *PloS One, 14*(5), e0217459.

Salem, A. Z. M., Gado, H. M., Colombatto, D., & Elghandour, M. M. Y., (2013). Effects of exogenous enzymes on nutrient digestibility, ruminal fermentation and growth performance in beef steers. *Livest. Sci., 154*(1–3), 69–73.

Sallam, S. M., Kholif, A. E., Amin, K. A., El-Din, A. N. N., Attia, M. F., Matloup, O. H., & Anele, U. Y., (2020). Effects of microbial feed additives on feed utilization and growth performance in growing Barki lambs fed diet based on peanut hay. *Anim. Biotechnol., 31*(5), 447–454.

Seo, M., & Koshiba, T., (2011). Transport of ABA from the site of biosynthesis to the site of action. *J. Plant Res., 124*(4), 501–507.

Soetan, K. O., & Oyewole, O. E., (2009). The need for adequate processing to reduce the anti-nutritional factors in plants used as human foods and animal feeds: A review. *Afr. J. Food Sci., 3*(9), 223–232.

Srivastava, L. M., (2002). *Plant Growth and Development: Hormones and the Environment*. Oxford: Academic Press.

Subramaniyam, R., & Vimala, R., (2012). Solid state and submerged fermentation for the production of bioactive substances: A comparative study. *Int. J. Sci. Nat., 3*, 480–486.

Tainton, N., (2000). *Pasture Management in South Africa*. University of natal press. South Africa.

Tamaki, M., Kondo, S., Itani, T., & Goto, Y., (2002). Temperature responses of leaf emergence and leaf growth in barley. *J. Agric. Sci., 138*, 17–20.

Wood, S. A., Karp, D. S., DeClerck, F., Kremen, C., Naeem, S., & Palm, C. A., (2015). Functional traits in agriculture: Agrobiodiversity and ecosystem services. *Trends Ecol. Evol., 30*(9), 531–539.

Yaqoob, M., & Javed, I., (2008). Nutritional evaluation of fodder tree leaves of northern grasslands of Pakistan. *Pak. J. Bot., 40*(6), 2503–2512.

Zhou, G., Wang, Y., Jiang, Y., & Yang, Z., (2002). Estimating biomass and net primary production from forest inventory data, a case study of China's Larix forests. *For. Ecol. Manag., 169*, 149–157.

CHAPTER 18

Re-Carbonation of Terrestrial Production Landscapes Through Regenerative Agriculture

AJAY KUMAR MISHRA,[1] RAJESWARI DAS,[2] DIPTI GROVER,[3] MITA MEHER,[2] BISWORANJITA BISWAL,[4] SHEETAL SHARMA,[1] and VIVEK MANYAPU[5]

[1]*International Rice Research Institute South Asia Regional Center, Varanasi, Uttar Pradesh, India*

[2]*Department of Soil Science, School of Agriculture, Gandhi Institute of Agriculture and Technology University, Gunupur, Rayagada, Odisha, India*

[3]*Kurukshetra University, Kurukshetra, Haryana, India*

[4]*Agronomy Section, National Dairy Research Institute, Karnal, Haryana, India*

[5]*Department of Biotechnology, National Pingtung University of Science and Technology, Pingtung, Taiwan*

ABSTRACT

The sustainability of contrasting agro-ecologies of landscapes is threatened by multifaceted challenges offered by climate change and food insecurity. The global food production system is struggling with a wide range of ill effects, from hunger, poverty, and obesity; to industrialized farming, exaggerated application of chemical fertilizers and pesticides, low-grade (adulterated) food, environmental deprivation, extinction of biodiversity, over-exploitation of manpower and animals; to the dominance of the corporate sector and a dearth of resilience. In addition, the

Sustainable Forest Resources Management: Issues and Implications. Jahangeer A. Bhat, Gopal Shukla, Manmohan J. Dobriyal, Sumit Chakravarty, A. Arunachalam, and Rainer W. Bussmann (Eds.)

release of greenhouse gases and soil degradation compound the adversity of climate change on livelihood security. This chapter highlights promising solutions to revert degenerative agriculture to regenerative agriculture through various good agricultural practices such as carbon farming through agroforestry, including livestock in farming and using bio-inputs. It also upscales our understanding of diverse approaches and mechanisms of terrestrial carbon sequestration under different landscapes and improvement in soil health and carbon sequestration through multiple re-carbonation technologies and practices.

Moreover, Nature-based solutions (NbS) are designed to improve the ecosystem services nature provides. They can provide numerous benefits in agriculture, such as restoring degraded land, improving carbon sequestration, and increasing crop resilience and productivity. This chapter also provides insights on carbon farming, crediting, and trading as viable and scalable options for improving the economics and environmental security of the farming community.

18.1 INTRODUCTION

Terrestrial ecosystems, composed of plants and soil in uplands and wetlands, have a significant impact on the global carbon (C) cycle and, in their natural state, serve as a sink for atmospheric carbon dioxide (CO_2) and methane (CH_4). The anthropogenic greenhouse effect is primarily driven by the emission of CO_2, CH_4, and nitrous oxides (N_2O), which are produced by burning fossil and biomass fuels and the breakdown of above- and below-ground organic materials. Around 8000 BC, natural ecosystems were presumably converted to agroecosystems, urban lands, and mined lands (collectively known as managed ecosystems), turning them from sinks to sources of greenhouse gases (GHGs), primarily CO_2, CH_4, and N_2O, and depleting the terrestrial C. Since the beginning of agriculture, Ruddiman (2005) estimated that the terrestrial C stock (soil and plant) had been reduced by 456 Pg (502.65 × 109 tons). The cumulative soil organic carbon (SOC) pool depletion is approximately between 130 and 135 Pg (Lal, 2018). Because of the correlation between SOC stock and atmospheric CO_2 concentration, re-carbonization of some terrestrial biosphere (soil and vegetation) is crucial to reduce climate change due to anthropogenic activities and improving other ecosystem services (Lal, 2018; Mishra et al., 2013).

Furthermore, the production of biomass (from agriculture and forestry), storage of nutrients and water, filtering of waste and pollutants from water, provision of habitat for soil biodiversity, provision of soil carbon storage that interacts with the Earth's climate and prevents soil erosion by enhancing soil aggregate stability (Kumar et al., 2022; Mishra et al., 2022a; Mishra et al., 2022b; Mishra et al., 2022c; Mishra et al., 2022d; Mishra et al., 2022e). To understand the necessity of re-carbonization, FAO launched a program called RECSOIL that aims to improve soil health by preserving and enhancing soil organic C stores. It enables soil organic C to fulfill crucial ecological functions that have the potential to offer several advantages. Healthy soils resist climate change, increase soil resilience to extreme weather events and pandemics, improve food security and farm revenue, reduce poverty and hunger, provide critical ecosystem services, and contribute to the SDGs (Mishra et al., 2022f; Mishra et al., 2022g). From 2021, carbon emissions must stay below 250 Pg (918 $PgCO_2$) to meet the Paris Agreement's mandate of keeping global temperature increases below 2°C. At current rates, that much carbon dioxide will be released by 2045. There is a great deal of ambiguity regarding the scope (potential) and limitations of CO_2 fertilization, which has been discussed in the following chapter.

18.2 TERRESTRIAL PRODUCTION LANDSCAPES: CONSTRAINTS AND OPPORTUNITIES

Ecosystems are the primary source and sink of biogenic greenhouse gases, *viz.* CO_2, N_2O, CH_4. Even in mature, old-growth forest ecosystems, the C balance tends to be positive in undisturbed ecosystems because C intake from photosynthesis outweighs losses through respiration (Gough et al., 2008). Although respiration is the primary source of carbon dioxide flow from ecosystems, dissolved organic and inorganic C, and C lost through erosive processes, disturbance, such as fire accidents, the occurrence of drought, spread of diseases, or overgrazing, can result in significant losses of carbon from both soils and plants. Finding high-C sink storage priority biomes is the first step. Peatland, degraded lands, grasslands, and agricultural areas are key high-C-density biomes. Permafrost (underground earth material *viz.* soil, silt, and rocks frozen for at least two years in a row) soils may melt in the case of anticipated global warming, significantly increasing CO_2 and CH_4 emissions with positive feedback on climate change. The opportunities to re-carbonize soil lie in agricultural lands, forest lands, wetland/peat soils, degraded soils, and desertified lands. In agricultural lands: (i) crop lands can be re-carbonized by conservation tillage, mulch farming, cover cropping, integrated nutrient management, biochar, manuring, and agroforestry; and (ii) grazing lands can be improved for pastures and deferred rotational grazing. In forest lands – (i) afforestation on marginal and degraded lands; and (ii) plantation of fast-growing trees. Wetlands can be restored and regulated for a water table for tolerant vegetation. Restoration of eroded, salinized soils, polluted lands, desertified and wasteland can significantly tap carbon into soils. Efficient re-carbonization may be constrained by population increase, rise in CO_2 levels, tropical deforestation, urban encroachment, political issues, and policies. There are few publications on variables like soil organic matter quality or precise GHG flux measurements because of the difficulty in acquiring comprehensive information on each practice and the dearth of long-term trials with specific monitoring techniques. Any method intended to store carbon as organic matter that does not have the proper C/N ratio in the soil or the material supplied to the soil might increase CO_2 emissions and deplete the soil's availability of N. Furthermore, it is unclear how C-sequestering methods affect local climate through water balance and albedo (Conant, 2012). Local warming may be facilitated by actions that lower GHG concentrations (Chapin et al., 2008).

Additionally, Schlesinger (2000) reported that actions that improve soil C sequestration may result in higher levels of N_2O or CH_4 (fertilization to increase C inputs) or both (flooding to conserve organic soils). Uncertainties include the destiny of lost C and the influence of erosion on the loss of soil C reserves (Berhe et al., 2007). Disturbances are random and frequently unforeseen phenomena that might result in loss of C (Smith, 2005). Several obstacles must be overcome to realize the significant potential drawdown of atmospheric CO_2 through C sequestration in the biosphere, which is anticipated to be between 50 and 70 ppm CO_2 by 2150 (Hansen et al., 2008). The terrestrial C pool in soils and vegetation can be improved (Lenton, 2010). However, under site-specific conditions, soil, plant, and animal systems, methods must be recognized, confirmed, and rectified. After being discovered, these techniques must be promoted to locals for acceptance while taking into consideration human factors (i.e., economic, social, political, gender, and cultural issues). Attempting to maintain global warming within the 2°C limitations outlined in the Copenhagen Accord (Macintosh, 2010) calls for further investigation into the possibilities and viability.

18.3 AGROFORESTRY SYSTEMS (AFSS) AS A POTENTIAL SOURCE OF CARBON SEQUESTRATION

Carbon sequestration is a natural process in which atmospheric CO_2 is captured or fixed and stored in natural forms like soils, vegetation, rocks, etc., though it can be artificially carried out (Manyapu, 2021). Among all the natural processes, agroforestry is the pioneering approach to realize the purpose of CO_2 fixation and climate change mitigation and adaptation (Kumar et al., 2020; Giri et al., 2019). Many recent studies also suggest adopting AFSs because of their significant role in fixing CO_2 in above-ground biomass, even though AFSs are not mainly intended for carbon sequestration (Arya et al., 2021; Arya et al., 2019; Kumar et al., 2021) as well as in belowground biomass (Kumar et al., 2019). Agroforestry is frequently regarded as a financially advantageous method of combating climate change, and the different systems decrease soil greenhouse gas emissions and store carbon in woody biomass and soils.

18.4 STUDIES OF VARIOUS AFSS IN DIVERSE ECOLOGICAL CONDITIONS

Following the discovery of these systems' capacity to trap atmospheric CO_2 and store the C in plant parts and soil, land-use systems like agroforestry are significant in mitigating and adapting to climate change and have recently gained significant relevance. In this regard, AFSs have received significant attention. In sub-tropical highlands like Ziro valley, Arunachal Pradesh, India, two crops of rice per year (March to July and July to October) are grown and fish are also reared in the paddy fields, especially during the first crop of rice. In addition to pigs and chickens, other agricultural products include millets grown on berms between rice plots and domesticated animals like the bison (*Bos frontalis*) called "*Mithun*." In tropical humid and subhumid zones, coffee under the shade of innate trees like *Lagerstroemia microcarpa, Acrocarpus fraxinifolius, Syzygium cumini, Dalbergia latifolia,* silver oak (*G. robusta*), etc., are accompanied with robusta coffee (*Coffea robusta*) in Coorg, Karnataka. In the arid and semi-arid zones, scattered trees are planted in pasture lands like *Acacia leucophloea.* The silvopasture system of Tamil Nadu is where multipurpose tree species and shrubs integrate with crops and/or livestock.

Additionally, *Acacia leucophloea* is planted along with other crops like millets and horse gram. In the tropical wet humid gardens of Kerala, woody perennial crops like *Areca catechu*, fruit crops like *Mangifera indica* and *Tamarindus indica*, and Timber species like *Tectona grandis*, Ailanthus triphysa can be found easily. Pepper vines can be seen climbing on these tree species.

18.5 TREE-BASED AGRICULTURAL SYSTEM VS. TREELESS AGRICULTURAL SYSTEM

Although trees fix more carbon in agroforestry systems, crops also trap and store a considerable quantity of carbon (Table 18.1). Crop improvement increases soil organic matter, a vital element of the terrestrial carbon pool. The silvopastoral system could sequester 104–198 MgC/ha. The data also shows the Rice-Rice system can sequester 0.26 Mg/ha/year (Nayak et al., 2009). The potential for sequestering carbon in and out of the soil is higher in agroforestry systems with better management. Thus, by increasing Indian agriculture's capacity to sequester carbon, plants in agroforestry not only help small and marginal farmers improve their quality of life but also contribute to efforts to combat global warming.

18.6 C3 PLANTS VS. C4 PLANTS

Plants can play a significant role as carbon sinks by capturing atmospheric CO_2 through photosynthesis. The C4 plants are generally broad-leaf plants (e.g., maize, sugarcane) and have a greater capacity for CO_2 assimilation and binding compared to the C3 plants, which are narrow-leaf plants (e.g., wheat, oats, rice). C4 plants are more efficient in carbon accumulation and therefore produce more biomass than C3 plants. The biomass of C4 plants, which make up 5% of all known plant species on Earth, is relatively small. Despite their rarity, they are responsible for around 23% of terrestrial carbon fixation. An effective way to prevent climate change is to increase the amount of C4 plants on Earth. This would aid in the biosequestration of CO_2. The CO_2 flux varies with time, agronomic management, and plantation type (Kumar et al., 2019; Kumar et al., 2015).

18.7 TERRESTRIAL CARBON STOCKS IN DIFFERENT LANDSCAPES

All around the world, carbon is stocked up in various forms. The amount of carbon that is kept in a given system is referred to as a "pool" or "stock." The Earth's upper mantle and crust, where the majority of the planet's carbon is stocked, is formed over millions of years from sedimentary rock (Mackenzie & Lerman, 2006). The second-largest carbon stock is in the oceans and atmosphere, having relatively smaller amounts of carbon stocks, primarily in the form of carbon dioxide, a greenhouse gas (Kayler et al., 2017). Around 1,325 GtC is stored in the top few feet of soil, and when deeper layers are taken into account, that amount might reach 3,000 GtC (Köchy et al., 2015).

TABLE 18.1 Effect of Carbon Sequestration on Tree-based Agricultural System vs. Treeless Agricultural System

Sl. No.	System	Country	Years of Practice	Effect on C-Sequestration	References
		Tree-based Agricultural System			
1.	*Gmelina arborea* based agri-silvicultural system	India	6 years	31.37 tC/ha	Swamy, Puri, & Singh (2003)
2.	*Dalbergia sissoo* based agri-silvicultural system	India	11 years	48 to 52 t C/ha	Newaj & Dhyani (2008)
3.	Humid tropical high – agrosilvicultural	Africa	3 years	29 to 53 MgC/ha	Oelbermann, Voroney, & Gordon (2004)
4.	Agri-silvicultural system in humid tropical lands	Southeast Asia		12 to 228 MgC/ha	Murthy et al. (2013)
5.	Humid tropical; high humid tropical; low dry lowlands – Silvopastoral	North America	3	104 to 198 MgC/ha	Murthy et al. (2013)
6.	Agrisilviculture – with *Leucaena leucocephala*	Arid, semi-arid region – India	4	14.42 Mg C/ha	Prasad et al. (2012)
		Treeless Agricultural System			
7.	Rice-wheat	India	1	0.22 Mg C/ha	Ghosh et al. (2009)
8.	Rice-wheat-jute	India	1	0.07 Mg C/ha	Manna et al. (2005)
9.	Soybean-wheat-maize	India	1	0.02 Mg C/ha	Hati et al. (2007)
10.	Rice-rice	India	1	0.26 Mg C/ha	Nayak et al. (2009)
11.	Maize-chickpea	India	1	0.02 Mg C/ha	Vineela et al. (2008)
12.	Soybean-wheat	India	1	0.14 Mg C/ha	Behra et al. (2007)

This theme of potential carbon sequestration in soils is also referred to as the "recarbonization" of soils (FAO, 2019).

Carbon stock refers to the quantity of carbon sequestered through the ecosystem in the form of five different pools: (i) aboveground biomass; (ii) belowground biomass; (iii) soil; (iv) deadwood; and (v) litter. An increase in carbon stock means an enhanced ecosystem capacity to absorb and sequester carbon dioxide (CO_2) from the atmosphere through photosynthesis. Terrestrial ecosystems biologically remove CO_2 from the atmosphere by the mechanism of photosynthesis. Terrestrial carbon stocks include soil and biomass carbon stocks, which act as the biological "scrubber" for CO_2.

Either the net removal of CO_2 from the atmosphere or the prevention of CO_2 emissions from terrestrial ecosystems are considered as terrestrial sequestration. It is estimated that the terrestrial biosphere sequesters around 2 billion tons (2 Gt) of carbon per year. About 2,000 Gt ± 500 of carbon is estimated to be stored in soils and vegetation globally (Singh, 2015).

The most critical terrestrial carbon sinks include soils, landscapes such as forests, grasslands, mangroves, and peatlands. Croplands and agroforestry absorb around 25% of global carbon emissions. CO_2 is effectively locked in plant tissues during biomass growth and soil complexes. Climate change mitigation is possible through carbon sequestration if atmospheric CO_2 can be converted to terrestrial carbon (vegetation or soil). This can be accomplished in two ways: (i) by increasing net photosynthesis, such as by adding new grass or tree plantings to existing areas; or (ii) by decreasing the rate of soil organic carbon decomposition through changes in land management, such as reducing the intensity of tillage or changing how water is managed (Lemus & Lal, 2005). Terrestrial carbon can be lost through wildfires and human activities like deforestation. Terrestrial ecosystems can mitigate climate change by creating plantations, restoring forests, and avoiding the loss of carbon stocks by reducing deforestation and degradation.

18.8 CARBON STOCK IN SOIL ECOSYSTEM

Soil is the primary stock of terrestrial organic carbon. More carbon is stored in soil than is present in plants and the atmosphere combined (Post & Kwon, 2000), and soil is a relatively stable reservoir of different organic and inorganic carbon fractions. Total carbon in the terrestrial ecosystem is approximately 3170 Pg. Nearly 80% (2500 Pg) of this C is found in soil (Lal, 2008). The soil ecosystem has a significant role in the world carbon budget and regulation of greenhouse gases. The natural transformation of CO_2 into soil-based inorganic compounds like calcium and magnesium carbonates is known as direct sequestration. Indirect sequestration is the process by which plants build biomass through photosynthesis. Eventually, this biomass is moved into the soil, decomposed there, and stored as soil organic carbon. The top 1 meter of soil contains 1500 to 2400 Pg of organic carbon (Ciais et al., 2013, Stockman et al., 2013). This is about 3–4 times the carbon stored in vegetation. A similar study by (Lal, 2004) found that the total amount present is approximately 1500 Pg of organic C in soils, which is about double the atmospheric C content. The carbon stocks of soils depend on various parameters such as soil texture, moisture content in the soil, mineral composition, landscape position, soil aggregates stability, and temperature regimes (Jimenez et al., 2007). Various management strategies, including afforestation, land use patterns, climate, conservation practices, ideal cultural practices, crop management practices, and activities of soil organisms, including microflora and fauna, all affect the sequestration capability of soil (Lal et al., 2007).

18.9 CARBON STOCK IN FOREST ECOSYSTEM

Globally, forests store ~861 GtC (gigatons of carbon), with 44% in soil (0–1 m depth), 42% in above and belowground biomass, 8% in dead wood, and 5% in litter (Pan et al., 2011). About 92% of the world's terrestrial biomass is in forest ecosystems, but this is not homogeneously distributed across the globe. The quantity of carbon stored by various forest types varies, and most of this variance is influenced by the climate in a certain region of the world (Figure 18.1). Tropical rainforests account for only 30% of global tree cover but contain 50% of the world's carbon stored in trees. Boreal forests store enormous amounts of carbon in soils, while tropical forests store the majority of their carbon in vegetation (Hengl et al., 2017).

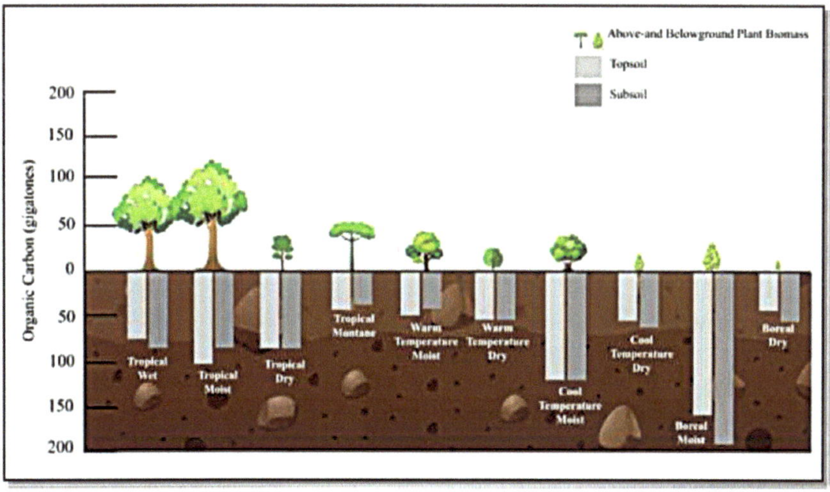

FIGURE 18.1 Different forest ecosystems influencing soil organic carbon stock (gigatons).

Source: Adapted from: Scharlemann et al. (2014).

It is estimated that in the forest ecosystem, biomass is approximately 536 Pg C, whereas the soil comprises around 704 Pg C. The world's forests store around 70% of soil organic carbon (SOC). Forests play a significant role in carbon reduction from the atmosphere. Carbon is stored in forests with the growth of woody tissues. Young forests have their highest rate of carbon uptake, which steadily decreases with time. Although old forests store carbon for a more extended period, their carbon uptake potential is small. Currently, approximately 20–25% of the organic pool of soil is being depleted due to human activities, such as the conversion of forest land for agricultural purposes (Lal, 2005). At the same time, it is estimated that approximately 1.6–1.7 Pg C is depleted due to deforestation per year (Watson et al., 2000). Forest land should be set aside for trees and not be converted for other land uses; this anthropogenic disturbance needs to be reduced, and other activities like afforestation, reforestation, and other social forestry should be promoted. Reforestation with long-lived trees (>100 years) will trap carbon for a longer duration, release it gradually, and offset carbon's climate impact. On our Earth, we can plant about 1.2 trillion trees. Planting so many trees and conserving them may lead to a carbon sequestration potential of approximately 205 billion tons and offset 10 years of CO_2 emissions (Bastin et al., 2019). Around two-thirds of all carbon emissions might be captured by restoring damaged forests worldwide, which would trap up to 205 billion tons of carbon (Chazdon & Brancalion, 2019).

18.10 CARBON STOCK IN MANGROVE FORESTS AND PEATLANDS

Despite having a relatively smaller area of coverage, mangroves, and peatlands are disproportionately more important for climate regulation and adaptation. Tropical mangroves accumulate over 1000 tons of carbon per hectare (Donato et al., 2011), which is approximately 5 times larger amount compared to other forest types and protects shorelines from storms. Peat bogs serve as a carbon sink by accumulating partially degraded material that would otherwise be destroyed. Peatlands operate as carbon sinks or springs, depending on the weather in different regions of the world and at varying periods of the year (Strack, 2008). The creation of new wetlands or the improvement of existing ones can enhance the amount of bulk carbon stocks (Lovett, 2008). Mangroves are not only remunerative to the local communities but also play an essential role in the form of biomass carbon and soil carbon stocks by assimilating atmospheric CO_2 (Mishra et al., 2014).

18.11 CARBON STOCK UNDER URBAN FORESTRY

Urban forestry means introducing trees to new locations and planting along avenues. By reducing atmospheric carbon dioxide, urban forests can be crucial in mitigating the effects of climate change (McHale et al., 2007). Urban trees and shrubs absorb CO_2 and convert it into biomass through photosynthesis (aboveground and belowground biomass), storing carbon in stems, branches, or roots (Nowak & Crane, 2002). Urban forest soils also store significant amounts of organic carbon before releasing it into the atmosphere (Pouyat et al., 2002, 2006). The impacts of urban forestry can vary depending on the type of vegetation used, serving as both a carbon sink and a source of emissions. Furthermore, by reducing carbon emissions from fossil fuel combustion, urban forests decrease the need for heating in winter and cooling in summer (McPherson & Simpson, 2003).

18.12 CARBON STOCKS IN AGROFORESTRY SYSTEMS

Agroforestry systems (AFS) are now known as sustainable land-use systems that can improve food security by enhancing carbon sequestration. AFS can be a successful strategy for climate change mitigation and providing environmental services (Nair et al., 2012). Trees from various systems, such as agrosilviculture, silvopasture, backyard gardens, hedgerows, and shelterbelts, are typically included in agroforestry (Nair et al., 2009). Perennial plants like trees and shrubs, as well as other vegetal and animal species, are included in these systems. Factors responsible for carbon stocks in agroforestry systems can be types of trees, tree age, species diversity, and density. However, due to their spatial distribution, trees on other land systems, such as farmlands, have a greater potential for emission and sequestration. More than 45% of agricultural fields have more than 10% tree cover, according to a recent global assessment (Zomer et al., 2009). Additionally, it has been determined that the biomass carbon reserves in agricultural fields range from 3 to 18 t C ha^{-1} (Nair, 2012; Nair & Nair, 2014). Agroforestry practices have more significant potential to increase C sequestration, especially in mixed/intercropping agricultural landscapes, than monocrop agriculture (Morgan et al., 2010). Agroforestry approaches in the tropics helped to restore 35% of the original carbon stock of the destroyed forest, compared to only 12% by crops and pastures (Palm et al., 2010). Agroforestry systems in tropical regions have higher carbon densities than field crops or pastures (Albrecht & Kandji, 2003; Nair et al., 2009; Nair, 2012). Worldwide, agroforestry is practiced on an estimated 1000 billion hectares of land (Zoomer et al., 2009).

Agroforestry systems store 30 to 322 Pg C yr^{-1} (Jose & Bardhan, 2012). If tree management practices are improved, more carbon (12,000 t C yr^{-1} to 17,000 Mg C yr^{-1}) can be stored by 2040.

18.13 CARBON STOCKS IN GRASSLANDS

About 25% of the Earth's total grasslands are covered (3.4 billion ha), and 12% of its terrestrial carbon stores are found there (Adams et al., 1990; Ojima et al., 1993). Grasslands have significantly less biomass in the aboveground biomass carbon pool because dominated species are herbaceous (nonwoody) vegetation. Also, because of harvest, grazing, fire, and senescence; aboveground biomass carbon is comparatively short-lived. In contrast, because of extensive fibrous root systems and sequestration of abundant carbon to soils, resulting in deep, fertile soils with huge organic carbon content; belowground biomass is higher. Soil carbon and belowground biomass are affected by variations in annual precipitation and temperatures. So, carbon stocks are higher in regions where rainfall is higher, such as in prairie, a humid temperate region of the tall grasses found in the United States.

18.13.1 IMPROVEMENT IN SOIL HEALTH AND CARBON SEQUESTRATION THROUGH MULTIPLE RE-CARBONATION APPROACHES

The rising CO_2 concentration in the atmosphere results from human activity, causing severe environmental degradation. Thus, techniques for sequestering and reducing atmospheric carbon are required. Environmentally friendly approaches like agroforestry, conservation agriculture, biochar application, organic farming, permaculture, regenerative agriculture, and integrated nutrient management (INM) have been performed worldwide to restore soil organic carbon (SOC) and to reduce its emission as harmful atmospheric CO_2 (Kumar et al., 2022; Mishra et al., 2022c). Organic matter is an essential component of soil, influencing its physical, chemical, and biological properties and thus significantly contributing to the proper functioning on which human societies rely. Soil organic matter (SOM) benefits include improved soil quality due to increased water and nutrient retention, resulting in higher plant productivity in natural and agricultural settings (Pu et al., 2019). SOM improves soil structure and reduces erosion, which leads to improved water quality in groundwater and surface waters, increased food security, and reduced negative impacts on ecosystems (Bai et al., 2022).

Nowadays, changing land use patterns for conserving soil organic carbon has been widely adopted worldwide. Agroforestry systems in various countries have been reported to hugely influence soil organic carbon stock with different types of cropping systems and land covers (Ramos et al., 2018; Noiha et al., 2018). A series of experiments carried out on conservation agriculture practices, which is a farming system that can prevent arable land loss while also regenerating degraded lands, have resulted in an improved SOC in the upper layer (0–30 cm) of soil as compared to the conventional agriculture practices involving plowing operations (Pu et al., 2019). There are several different CA practices with crop residue retention, CA with zero tillage, CA with planting bed and many more, which directly or indirectly impact various soil quality parameters. A few of the CA practices and their impact on various soil quality parameters have been described in Table 18.2. Biochar, the lightweight black residue of carbon and ashes left over from biomass pyrolysis, has immense benefits on soil carbon sequestration when applied for a long time. According to the report of Zheng et al.

TABLE 18.2 Soil Quality Parameters as Influenced by Various CA Practices

Sl No.	Type of CA Practiced	Influenced Soil Quality Parameters	Remark	References
1.	Permanent Broad bed (PBB)+ residue retention	Infiltration rate (IR) and hydraulic conductivity (HC).	A significant increase in IR and HC over a depth of 30 cm has been observed due to PBB and residue retention.	Govaerts et al. (2007); Bai et al. (2022)
2.	CA based on Zero tillage (ZT) and planting bed (PB.)	Soil compaction	A reduction in soil compaction was observed along with BD compared to conventional tillage (CT) practices.	Bai et al. (2022)
3.	ZT + residue retention	Bulk density and soil organic carbon	There was a significant reduction in BD and an enhancement in SOC. ZT influenced the increase in SOC upto 30 cm, but residue retention influences over 60 cm depth of soil.	Sommer et al. (2014); Szostek et al. (2022)
4.	ZT with bed planting (ZTB) ZT with flat planting (ZTF)	Total soil nitrogen	Both practices influence enhancing total soil N as compared to CT practices. The influence of ZTB was more as compared to ZTF.	Hafif (2014)

TABLE 18.3 Evidence-based Worldwide Influence of Multiple Re-Carbonation Approaches on Soil Carbon Stock

Sl. No.	Re-Carbonation Approaches	Country	Duration	Experiment	Influence on Soil Carbon Sequestration	References
1.	Agroforestry	Brazil	8 years	Agroforestry framework with oil palm (AFS_p) and oil palm + cacao (AFS_{P+C})	Soil carbon stock (Mg C ha^{-1}) increased with depth in both the systems and AFS_{P+C} was recorded as highest in soil carbon stock (116.7 ± 1.5 Mg C ha^{-1}) and biomass C.	Ramos et al. (2018)
		Cameroon	–	Three kinds of land cover as indicated by the time of stands, i.e., low (0–10 yr), medium (10–20 yr) and high (>20 yr) of Eucalyptus, accepting Savannah as control.	The most seasoned stands (108.51 ± 26.46 t C/ha) against 13.62 ± 3.03 t C/ha in Savannah was the most broad load of carbon. Eucalyptus saligna put away 39.66 t C/ha (4 t C ha^{-1} $year^{-1}$) in youthful stands; 57.28 t C/ha (6 t C ha^{-1} $year^{-1}$) in medium stands and 85.46 t C/ha (9 t C ha^{-1} $year^{-1}$) in old stands. The sequestration potential was higher in eucalyptus stands (398.25 t CO_2eq/ha) than in savannah (50.05 t CO_2eq/ha).	Noiha et al., 2018
2.	Conservation agriculture	North China	4 years	A bifactorial experiment was conducted with three levels for tillage practices (no-tillage, rotary tillage, and conventional tillage) and two levels for residue management (retention and removal).	Irrespective of residue retention or removal, no-tillage resulted in increased soil organic carbon stock (SOC) in the upper layer of soil. Residue retention resulted in the significantly increased value of SOC as compared to residue removal in all the tillage practices.	Pu et al. (2019)
		North China	11 years	Four tillage systems for winter wheat established in 2001 were: moldboardplow tillage with maize residues removed (PT0), moldboard-plow tillage with maize residues incorporated (PT), rotary tillage with maize residues incorporated (RT), and no-till with maize residues retained on the soil surface (NT) in Wheat-Maize cropping system.	Compared with PT0 and PT, significantly higher SOC and N concentrations were observed in the surface layer (0–10 cm depth) under NT and RT in 2001. After 11 years, in 2012, the net profile (0–30 cm) SOC sequestration rate was 10.60, 13.95, 13.65, and 14.92 Mg ha^{-1} 2012 under PT0, PT, RT, and NT, respectively.	Dikgwatlhe et al. (2014)

TABLE 18.3 *(Continued)*

Sl. No.	Re-Carbonation Approaches	Country	Duration	Experiment	Influence on Soil Carbon Sequestration	References
3.	Biochar	China	4 years	Continuous application of biochar for 4 years at different rates, i.e., 3, 6, and 12 t ha^{-1}.	Biochar application increased the contents of SOC and recalcitrant organic C fraction by 11.02–22.13% and 18.41–32.31% at 6 and 12 t ha^{-1}, respectively, and the increased proportion for recalcitrant organic C fraction was more significant than that for SOC.	Zheng et al. (2022)
		Egypt	2 years	Biochar was added to a saline-sodic soil at five rates: 0, 5 (BC_5), 10 (BC_{10}), 15 (BC_{15}), and 20 (BC_{20}) ton ha^{-1}.	The total amount of organic-C sequestered by the pearl millet was 6.00–9.45 ton of C ha^{-1} per season, while the total CO_2 emissions were 1.62–1.82 ton of organic-C ha^{-1} per season. Applying biochar enhanced the nutrient availability and uptake; moreover, it mitigated the salinity effects in the leaf tissues. Carbon sequestration by pearl millet amended with biochar is an effective strategy to mitigate gas emissions from saline-sodic soils.	Ali et al. (2021)
4.	Organic farming	Switzerland	15 years	Reduced tillage vs. ploughing	Reduced tillage enhanced SOC (25%), microbial biomass (32%) and microbial activity (34%) of topsoil as compared to plowing.	Krauss et al. (2020)
		France, Germany, the Netherlands, and Switzerland	8–21 years	Reduced tillage and mold board plowing under organic farming practices.	SOC increased with increased organic inputs and increased clay-to-silt ratio. Reduced tillage enhanced the SOC in the surface layer (0–30 cm) by 20% as compared to plowing practice.	Krauss et al. (2022)
5.	Permaculture	France	7 years	Intensive cultivation under gardening practices with high organic inputs vs. local pastures.	The concentrations of total OC and nitrogen (N) in bulk soils were higher under permaculture practices due to significant inputs of manure and compost, resulting in higher concentrations of the bioavailable nutrients Ca, Mg, K, and P.	de Tombeur et al. (2018)

TABLE 18.3 *(Continued)*

Sl. No.	Re-Carbonation Approaches	Country	Duration	Experiment	Influence on Soil Carbon Sequestration	References
6.	Regenerative agriculture	13 different ecoregions	10 years	Empirical soil laboratory data under different land use practices.	Land conversion to a rotational grazing system offers the highest carbon sequestration potential over 10 years.	Wiltshire & Beckage (2022)
7.	Integrated nutrient management	India	2 years	Nine treatments with three replications in a randomized block design, including various doses of organic manure and recommended dose of fertilizers (RDF) of N, P, K, and Zinc.	The highest organic carbon was recorded with treatment receiving 4 t ha^{-1} and RDF of N, P, K, and Zinc compared to the sole application of organic manure and RDF.	Sur et al. (2010)
			2 years	Principal component analysis (PCA) of 13 treatments encompassing different combinations of chemical fertilizers, organic manure, and biofertilizers	The treatments receiving the INM package exhibited the highest soil quality index (SQI) and SOC.	Shah et al. (2022)

(2022), expanding the pace of biochar application throughout some time in China has expanded the items in SOC and hard-headed organic C part by 11.02–22.13% and 18.41–32.31% at 6 and 12 t ha^{-1}, respectively, and the expanded extent for stubborn organic C portion was more significant than that for SOC. Regenerative Agriculture, which refers to farming and grazing practices that, among other things, reverse climate change by rebuilding soil organic matter and restoring degraded soil biodiversity, resulting in carbon drawdown, has a significant effect on C sequestration (Mishra et al., 2022a). Studies have reported that continuous practice of RA has resulted in significant development of soil organic carbon pool and has reduced CO_2 emission (Wiltshire & Beckage, 2022). Adopting different re-carbonation approaches has resulted in the enhancement of soil health due to the buildup of SOC stock in the upper layer of soil. Few worldwide evidence-based re-carbonation approaches and their effect on soil carbon sequestration have been listed and explained in Table 18.3 of this section.

18.13.1.1 CARBON FARMING, CREDITING, AND TRADING: A VIABLE AND SCALABLE OPTION FOR ECONOMIC AND ENVIRONMENTAL SECURITY OF THE FARMING COMMUNITY

Climate change is one of the most serious threats we face. Driving a car or riding a motorcycle, using air conditioning and heating, and lighting homes all use energy and emit greenhouse gases (GHG), which contribute to climate change. When GHG emissions rise, the Earth's climate, average weather, and average temperatures change. Various sources and sinks in agribusiness and ranger service produce, retain, and store three sorts of GHGs: carbon dioxide (CO_2), methane (CH_4), and nitrous oxide (N_2O). Many rural and ranger service practices emit ozone-depleting substances into the atmosphere. Carbon sequestration occurs when a tree or plant absorbs atmospheric carbon dioxide and stores it as cellulose in its trunk, branches, twigs, leaves, and natural product while releasing oxygen back into the atmosphere. Carbon dioxide is also taken up by the roots of trees and plants. District-specific Nature-based arrangements (NbS) that should be investigated to determine how much carbon is stored in the soil are greater than the total sum in vegetation and the climate.

Carbon Farming is an entire homestead way to deal with further developing carbon catch on working scenes by carrying out practices that have been displayed to expand the rate at which CO_2 is eliminated from the environment and put away in plant material or potentially soil organic matter

(Jansson et al., 2022). Carbon Farming is a philosophy for cooperating with the agroecosystem processes that drive framework change. Carbon farming unequivocally recognizes that solar energy drives farm environment elements and that carbon is the energy transporter inside the farm system. The growing global population has increased food demand, resulting in a carbon footprint from agricultural activities that continues to harm the environment (Sharma et al., 2021).

While part of the problem, agriculture can also be part of the solution. CO_2 emissions from agricultural production currently account for 11% of global greenhouse gas emissions, with the livestock sector contributing the most. As a result, farmers, the private sector, and governments are increasingly interested in carbon markets tailored to farming and agricultural activities (Sharma et al., 2021). Farmers and ranchers, as land stewards, are sequestering carbon through sustainable soil, crop, livestock, and agroforestry management practices. Carbon markets are online financial marketplaces where people can buy and sell carbon credits. Numerous initiatives and efforts are underway to reduce carbon emissions and promote carbon storage and removal activities. Carbon has thus become a valuable economic commodity (Conant et al., 2011). All GHGs are converted to CO_2 equivalents to find a common unit for this commodity. CO_2 equivalents are traded on carbon markets. Markets function similarly to financial markets. Carbon credits are the currency used in these markets. Carbon credits are the money for exchanging fossil fuel byproducts. The unit for one carbon credit is comparable to one ton of CO_2 emissions (Funk et al., 2014). There are two types of carbon markets: regulatory compliance and voluntary. Organizations and governments that are legally required to report their GHG emissions use the compliance market. Mandatory carbon reduction schemes govern it at the national, regional, and global levels. The three Kyoto Protocol mechanisms for the regulatory market are essential: the Clean Development Mechanism (CDM), Joint Implementation (JI), and the EU Emissions Trading System (ETS). While some countries have not formally ratified the Kyoto Protocol, they have other legally binding state and regional GHG reduction initiatives such as methane mitigation (Pakistan), biogas (Armenia), biomass production (India), reforestation programs (India), and afforestation of grasslands (Tanzania). Only developing countries are eligible to participate in the CDM. Some examples of CDM projects are provided below, and their distribution is illustrated in Figure 18.2.

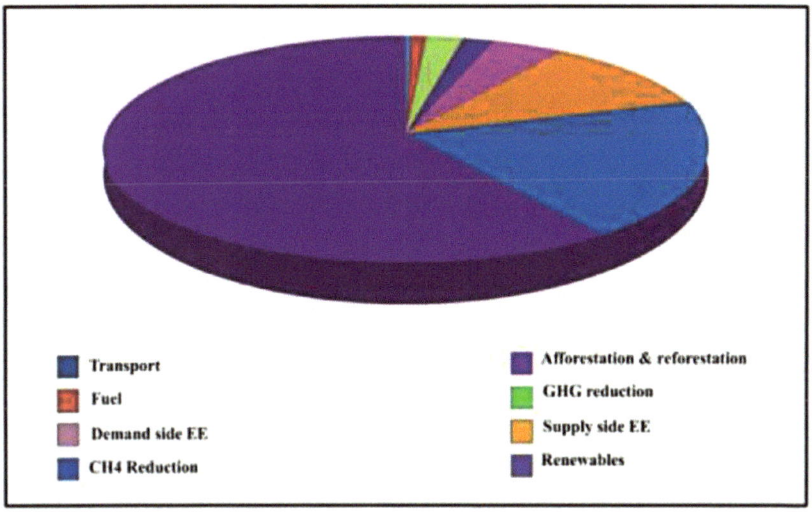

FIGURE 18.2 Percentage of CDM projects in each category [EE: Energy efficiency].
Source: Adapted from UNEP, 2009

The deliberate market has filled in significance for agribusiness and ranger service projects. The majority of willful carbon credits are purchased by the confidential area (VER). The most well-known reasons for purchasing carbon credits are corporate social responsibility (CSR) and advertising. Other factors include confirmation, notoriety, and environmental and social benefits. Carbon credits on purposeful business sectors are known as Variable Emission Reduction (VER).

18.13.1.1.1 Cost-Benefit Ratio Involved in Carbon Projects

There are numerous expenses associated with fostering a venture, which is why it is critical to locate a task engineer who can assist with the development and obtain financing for the venture. It will take two to five years (depending on how quickly the venture is created and the payment arrangement) before the assets are obtained from the sale of carbon credits. This is because carbon credits would be paid only after carbon sequestration occurred and could be estimated in all agriculture, forestry, and other land use (AFOLU) projects. The cost of the projects can be segregated into the cost of project implementation, the cost of raising awareness among local participants, and costs that are directly associated with project documentation, registration, and verification (Baalmann & Schlamadinger, 2008).

18.13.1.1.2 Challenges in Carbon Farming

Carbon farming initiatives (CFI) call for agro-ecological strategies that encourage ranchers to use the best homestead management practices. However, ranchers are frequently hesitant to participate in such projects due to the complexity of the plan and its execution, as well as policymakers' and ranchers' competing objectives (Salas Castelo et al., 2017). Some of the impediments to carbon farming are directly related to landowner interests and insufficient abilities or management capacities. Political insecurity has a significant impact on the acceptance and execution of such practices (Liu et al., 2018). Furthermore, vulnerability to environmental effects and a lack of awareness of such plans and strategies may hinder their implementation. Many ranchers are unfamiliar with the term "carbon cultivation" and require specific information about the benefits and drawbacks of carbon cultivation. The situation is further exacerbated by high information costs and concerns about the impact of carbon cultivation on yield and ranch efficiency (Valdivia et al., 2012; Ingram et al., 2016). Other significant barriers to CFI include a lack of supported techniques and systems, higher managerial costs, and difficulty in becoming certified as a carbon offset supplier.

18.14 INCLUSIVE APPROACH OF REGENERATIVE AGRICULTURE TO ACHIEVE SDGS

RA focuses on the natural component of manageability, which includes subjects such as improving and further developing soil wellbeing, advancing asset management, easing environmental change, working on nutrient cycling and water quality and availability, as articulated by the two targets (e.g., improve soil quality) and activities (e.g., use perennials). These topics improve food security by increasing provisioning (e.g., food, feed, and fiber), regulating (e.g., environmental guidance, soil erosion, and water purification), and supporting (e.g., nutrient cycling and soil formation) ecosystem services (Figure 18.3). The primary tenets of RA include system and landscape-based conservation agriculture, which involves no-till farming combined with residue mulching, cover cropping, integrated nutrient and pest management, diverse rotations, and crop integration with trees and livestock. RA is inclusive, and its site-specific package(s) should be tailored to biophysical and human factors. RA is soil-driven rather than seed-driven, and it is based on the principle that "the well-being of the soil, plants, animals, and people is one and inseparable." In the present context of climate change and environmental issues, it is appropriate to extend the concept by stating that the health of the soil, plants, animals, people, and the environment is one and indivisible.

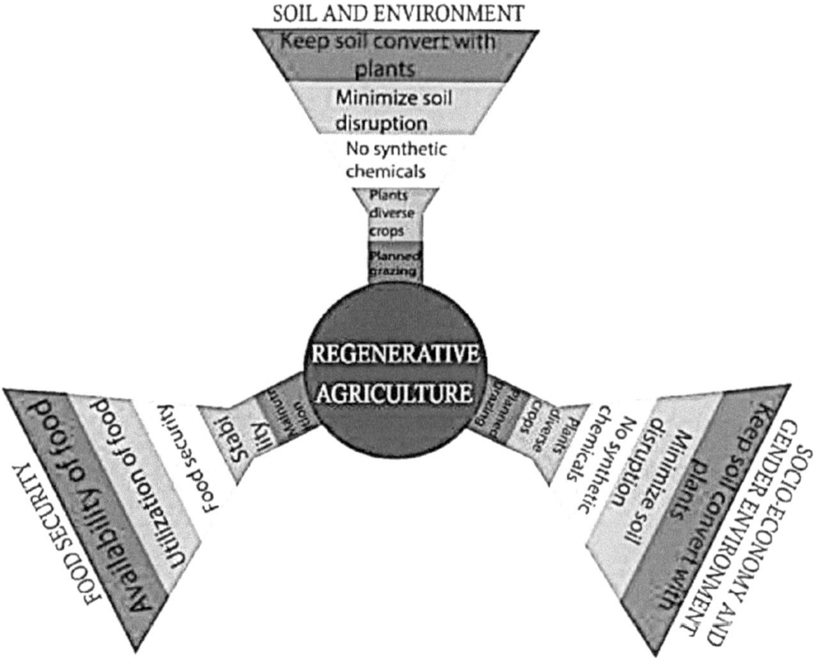

FIGURE 18.3 Multiple dimensions of regenerative agriculture practices.

The socio-economic dimension in RA includes economic prosperity, social upliftment, rural employment, and gender inclusion which relates to aspects of cultural ecosystem services. RA can be the entry point to regenerate and contribute to multiple provisioning, regulating, and supporting services to enhance the environment and the social and economic dimensions of sustainable food production.

18.15 CONCLUSION AND WAY FORWARD

Multiple actors and stakeholders must come together on one platform (policy makers, farmers, researchers, funding agencies, marketers, and consumers) to refine and propose better RA solutions. Regenerative systems have the potential to enhance productivity, profitability, and sequester terrestrial carbon. Additionally, these systems offer various ecosystem services. Designing a worldwide roadmap and strategic framework for the holistic adaptability of RA systems is the need of the hour to reduce environmental impacts through recarbonization and an inclusive approach.

KEYWORDS

- **agroforestry systems**
- **carbon farming**
- **carbon sequestration**
- **climate change**
- **GHG emissions**
- **GHG mitigations**
- **nature-based solutions**

REFERENCES

Adams, J. M., Faure, H. F., Faure-Denard, L., McGlade, J. M., & Woodward, F. I., (1990). Increases in terrestrial carbon storage from the last glacial maximum to the present. *Nature, 348*(6303), 711–714.

Ajit, Dhyani, S. K., Handa, A. K., Sridhar, K. B., Jain, A. K., Uma, Sasindran, P., et al., (2014). Carbon sequestration assessment of block plantations at JSW steels limited. In: *Compendium of Abstracts*, 3rd *World Agroforestry Congress* (pp. 354, 355). Organized by ICAR, WAC and ISAF at Delhi.

Ali, E. F., Al-Yasi, H. M., Kheir, A. M. S., et al., (2021). Effect of biochar on CO_2 sequestration and productivity of pearl millet plants grown in saline sodic soils. *J. Soil Sci. Plant Nutr., 21*, 897–907. https://doi.org/10.1007/s42729-021-00409-z.

Arvind, K., (2022). Coping with climate change. In: Arvind, K., (ed.), *Ecosystem-Based Adaptation* (pp. 143–233). Elsevier.

Arya, R., Mishra, A. K., & Chaudhry, S., (2018). Variation in soil properties and carbon stocks under roadside plantation and rice-wheat cropping system in North Western Haryana, India. *Int. J. Curr. Microbiol. App. Sci., 7*(4), 1939–1949. https://doi.org/10.20546/ijcmas.2018.704.222.

Bai, L., Kong, X., Li, H., Zhu, H., Wang, C., & Ma, S., (2022). Effects of conservation tillage on soil properties and maize yield in Karst Regions, Southwest China. *Agriculture, 12*, 1449. https://doi.org/10.3390/ agriculture12091449.

Bastin, J. F., Finegold, Y., Garcia, C., Mollicone, D., Rezende, M., Routh, D., Zohner, C. M., & Crowther, T. W., (2019). The global tree restoration potential. *Science, 365*(6448), 76–79.

Behra, U. K., Sharma, A. R., & Pandey, H. N., (2007). Sustaining productivity of wheat–soybean cropping system through integrated nutrient management practices on the vertisols of central India. *Plant and Soil, 297*, 185–199.

Berhe, A. A., Harte, J., Harden, J. W., & Torn, M. S., (2007). The significance of the erosion-induced terrestrial carbon sink. *BioScience*, *57*(4), 337–346.

Chapin, III. F. S., Randerson, J. T., McGuire, A. D., Foley, J. A., & Field, C. B., (2008). Changing feedbacks in the climate–biosphere system. *Frontiers in Ecology and the Environment*, *6*(6), 313–320.

Chaudhari, S. K., Bardhan, G., Kumar, P., Singh, R., Mishra, A. K., Rai, P., Singh, K., & Sharma, D. K., (2015a). Short-term tillage and residue management impact on physical properties of a reclaimed sodic soil. *J. Indian Soc. Soil Sci., 63*(1), 30–38.

Chaudhari, S. K., Kumar, P., Mishra, A. K., Singh, K., Rai, P., Singh, R., & Sharma, D. K., (2015b). Labile carbon fractions buildup and dynamics under vertical stratification of *Populus deltoides* and *Eucalyptus tereticornis* based agroforestry systems in trans-Gangetic plains of India. *Ann. Agric. Res., 36*(1), 1–9.

Chaudhari, S. K., Kumar, P., Singh, K., Mishra, A. K., Rai, P., Singh, R., & Sharma, D. K., (2014). Aggregate fractions and organic carbon dynamics in partially reclaimed sodic soils growing *Eucalyptus tereticornis*. *Journal of Soil Salinity and Water Quality, 6*(2), 96–100.

Chaudhary, A., Venkatramanan, V., Kumar, M. A., & Sharma, S., (2022). Agronomic and environmental determinants of direct seeded rice in South Asia. *Circular Economy and Sustainability*, 1–38.

Chazdon, R., & Brancalion, P., (2019). Restoring forests as a means to many ends. *Science, 365*(6448), 24, 25.

Ciais, P., Sabine, C., Bala, G., Bopp, L., Brovkin, V., Canadell, J., et al., (2013). In: Stocker, T. F., Qin, D., Plattner, G. K., Tignor, M., Allen, S. K., Boschung, J., Nauels, A., et al., (eds.), *Carbon and Other Biogeochemical Cycles in Climate Change 2013: The Physical Science Basis* (pp. 465–570). Contribution of Working Group I to the Fifth Assessment Report of the Intergovernmental Panel on Climate Change. Cambridge, UK; New York, NY: Cambridge University Press.

Conant, R. T., (2012). Grassland soil organic carbon stocks: Status, opportunities, vulnerability. In: *Recarbonization of the Biosphere* (pp. 275–302). Springer, Dordrecht.

Conant, R. T., Ogle, S. M., Paul, E. A., & Paustian, K., (2011). Measuring and monitoring soil organic carbon stocks in agricultural lands for climate mitigation. *Front. Ecol. Environ., 9*, 169–173.

de Tombeur, F., Sohy, V., Chenu, C., Colinet, G., & Cornelis, J. T., (2018). Effects of permaculture practices on soil physicochemical properties and organic matter distribution in aggregates: A case study of the Bec-Hellouin farm (France). *Front. Environ. Sci., 6*, 116. doi: 10.3389/fenvs.2018.00116.

Dikgwatlhe, S. B., Chen, Z. D., Lal, R., Zhang, H. L., & Chen, F., (2014). Changes in soil organic carbon and nitrogen as affected by tillage and residue management under wheat-maize cropping system in the North China plain. *Soil & Tillage Research, 144*, 110–118.

Donato, D. C., Kauffman, J. B., Murdiyarso, D., Kurnianto, S., Stidham, M., & Kanninen, M., (2011). Mangroves among the most carbon-rich forests in the tropics. *Nature Geoscience, 4*(5), 293–297.

Funk, J. M., Field, C. B., Kerr, S., & Daigneault, A., (2014). Modeling the impact of carbon farming on land use in a New Zealand landscape. *Environ. Sci. Policy, 37*, 1–10.

Ghosh, P. K., Saha, R., Gupta, J. J., Ramesh, T., Das, A., Lama, T. D., Munda, G. C., et al., (2009). Long-term effect of pastures on soil quality in acid soil of North-East India. *Australian Journal of Soil Research, 47*(4), 372–379.

Giri, A., Kumar, G., Arya, R., Mishra, S., & Mishra, A. K., (2019). Carbon sequestration in *Populus deltoides* based agroforestry system in Northern India. *Int. J. Chem. Stud., 7*(1), 2184–2188.

Gough, C. M., Vogel, C. S., Schmid, H. P., & Curtis, P. S., (2008). Controls on annual forest carbon storage: Lessons from the past and predictions for the future. *Bioscience, 58*(7), 609–622.

Govaerts, B., Sayre, K. D., Lichter, K., et al., (2007). Influence of permanent raised bed planting and residue management on physical and chemical soil quality in rain fed maize/wheat systems. *Plant Soil, 291*, 39–54. https://doi.org/10.1007/s11104-006-9172-6.

Hafif, B., (2014). Effect of tillage on soil nitrogen; a review. *International Journal on Advanced Science, Engineering and Information Technology, 4*(4), 16–19.

Hansen, J., Sato, M., Kharecha, P., et al., (2008). Target atmospheric CO_2: Where should humanity aim? *Open Atmos. Sci., 2*, 217–231.

Hati, K. K., Swarup, A., Dwivedi, A. K., Mishra, A. K., & Bandyopadhyay, K. K., (2007). Changes in soil physical properties and organic carbon status at the topsoil horizon of a vertisol of central India after 28 years of continuous cropping, fertilization and manuring. *Agriculture, Ecosystems and Environment, 119*, 127–134.

Hengl, T., Mendes De, J. J., Heuvelink, G. B., Ruiperez, G. M., Kilibarda, M., Blagotić, A., Shangguan, W., et al., (2017). SoilGrids250m: Global gridded soil information based on machine learning. *PLoS One, 12*(2), e0169748. https://doi.org/10.1080/10549811.2020.1749856.

Ingram, J., Mills, J., Dibari, C., Ferrise, R., Ghaley, B. B., Hansen, J. G., Iglesias, A., et al., (2016). Communicating soil carbon science to farmers: Incorporating credibility, salience and legitimacy. *J. Rural Stud., 48*, 115–128.

Jansson, C., Faiola, C., Wingler, A., Zhu, X. G., Kravchenko, A., De Graaff, M. A., Ogden, A. J., et al., (2021). Crops for carbon farming. *Front. Plant Sci., 12*, 636709. doi: 10.3389/fpls.2021.636709.

Jimenez, J. J., Lal, R., Leblanc, H. A., & Russo, R. O., (2007). Soil organic carbon pool under native tree plantations in the Caribbean lowlands of cost Rica. *Forest Ecology and Management, 241*, 134–144. doi: 10.1016/j.foreco.2007.01.022.

Kayler, Z., Janowiak, M., & Swanston, C., (2017). *Global Carbon.* Climate Change Resource Center. US Department of Agriculture, Forest Service, Washington, DC.

Köchy, M., Hiederer, R., & Freibauer, A., (2015). Global distribution of soil organic carbon–Part 1: Masses and frequency distributions of SOC stocks for the tropics, permafrost regions, wetlands, and the world. *Soil, 1*(1), 351–365.

Krauss, M., Berner, A., Perrochet, F., et al., (2020). Enhanced soil quality with reduced tillage and solid manures in organic farming – a synthesis of 15 years. *Sci. Rep., 10*, 4403. https://doi.org/10.1038/s41598-020-61320-8.

Krauss, M., Wiesmeier, M., Don, A., Cuperus, F., Gattinger, A., Gruber, S., Haagsma, W. K., et al., (2022). Reduced tillage in organic farming affects soil organic carbon stocks in temperate Europe. *Soil and Tillage Research, 216*, 105262, https://doi.org/10.1016/j.still.2021.105262.

Kumar, P., Chaudhari, S. K., Mishra, A. K., Singh, K., Rai, P., Singh, R., & Sharma, D. K., (2014). Labile carbon dynamics and soil amelioration in six-year-old *Eucalyptus tereticornis* plantation in sodic soils. *J. Soil Salinity Water Qua., 6*(2), 91–95.
Kumar, P., Fagodiya, R. K., Chaudhari, S. K., Singh, R., Mishra, A. K., Singh, K., & Sharma, D. K., (2022). Effect of different nitrogen management options on nutrient uptake, biomass carbon sequestration and grain yield of maize-wheat system in reclaimed sodic soil. *Journal of Plant Nutrition, 45*(8), 1240–1252.
Kumar, P., Mishra, A. K., Chaudhari, S. K., Basak, N., Rai, P., Singh, K., Singh, R., et al., (2018). Carbon pools and nutrient dynamics under eucalyptus-based agroforestry system in the semi-arid region of North-west India. *J. Indian Soc. Soil Sci., 66*(2), 188–199.
Kumar, P., Mishra, A. K., Chaudhari, S. K., et al., (2021). Different Prosopis species influence sodic soil ecology by favoring carbon build-up and reclamation in North-West India. *Trop. Ecol., 62*, 71–81. https://doi.org/10.1007/s42965-020-00126-1.
Kumar, P., Mishra, A. K., Chaudhari, S. K., Sharma, D. K., Rai, A. K., Singh, K., Rai, P., & Singh, R., (2020). Carbon sequestration and soil carbon buildup under eucalyptus plantation in semi-arid regions of North-West India. *J Sustain. For.*
Kumar, P., Mishra, A. K., Chaudhari, S. K., Singh, R., Singh, K., Rai, P., Pandey, C. B., & Sharma, D. K., (2016). Biomass estimation and carbon sequestration in *Populus deltoides* plantations in India. *J. Soil Salinity Water Qual., 8*(1), 25–29.
Kumar, P., Mishra, A. K., Chaudhari, S. K., Singh, R., Yadav, K., Rai, P., & Sharma, D. K., (2022). Conservation agriculture influences crop yield, soil carbon content and nutrient availability in the rice–wheat system of north-west India. *Soil Research, 60*(6), 624–635.
Kumar, P., Mishra, A. K., Kumar, M., Chaudhari, S. K., Singh, R., Singh, K., Rai, P., & Sharma, D. K., (2019). Biomass production and carbon sequestration of *Eucalyptus tereticornis* plantation in reclaimed sodic soils of North-West India. *Indian J. Agric. Sci., 89*(7), 1091–1095.
Lal, R., (2004). Soil carbon sequestration to mitigate climate change. *Geoderma, 123*(1–22).
Lal, R., (2005). Forest soils and carbon sequestration. *Forest Ecology and Management, 220*(1–3), 242–258.
Lal, R., (2008). Carbon sequestration. *Philosophical transactions of the Royal Society of London; Series B. Biological Sciences, 362*, 815–830.
Lal, R., Follett, R., Stewart, B. A., & Kimble, J. M., (2007). Soil carbon sequestration to mitigate climate change and advance food security. *Soil Science, 172*(12), 943–956.
Lal, R., Smith, P., Jungkunst, H. F., Mitsch, W. J., Lehmann, J., Ramachandran, N. P. K., McBratney, A. B., et al., (2018). The carbon sequestration potential of terrestrial ecosystems. *Journal of Soil and Water Conservation, 73*(6), 145A–152A.
Lemus, R., & Lal, R., (2005). Bioenergy crops and carbon sequestration. *Critical Reviews in Plant Sciences, 24*(1), 1–21.
Lenton, T. M., (2010). The potential for land-based biological CO_2 removal to lower future atmospheric CO_2 concentration. *Carbon Management, 1*(1), 145–160.
Liu, T., Bruins, R. J., & Heberling, M. T., (2018). Factors influencing farmers' adoption of best management practices: A review and synthesis. *Sustainability, 10*, 432.
Lovett, R., (2008). Burying biomass to fight climate change. *New Scientist,* 2654.
Macintosh, A., (2010). Keeping warming within the 2 Climit after Copenhagen. *Energy Policy, 38*(6), 2964–2975.
Mackenzie, F. T., & Lerman, A., (2006). *Carbon in the Geobiosphere: Earth's Outer Shell.* Springer Science & Business Media.

Manna, M. C., Swarup, A., Wanjari, R. H., Ravankar, H. N., Mishra, B., Saha, M. N., Singh, Y. V., et al., (2005). Long-term effect of fertilizer and manure application on soil organic carbon storage, soil quality and yield sustainability under subhumid and semi-arid tropical India. *Field Crops Research, 93*, 264–280.

Manyapu, V., (2021). 16. carbon sequestration. In: *Modern Agriculture – Approach and Applications* (p. 225). Brillion Publishing, New Delhi, India.

McHale, M. R., McPherson, E. G., & Burke, I. C., (2007). The potential of urban tree plantings to be cost effective in carbon credit markets. *Urban Forestry & Urban Greening, 6*(1), 49–60.

McPherson, E. G., & Simpson, J. R., (2003). Potential energy savings in buildings by an urban tree planting program in California. *Urban Forestry & Urban Greening, 2*(2), 73–86.

Mishra, A. K., & Sharma, S., (2022). Rice-wheat crop manager: An effective crop management tool for rice-based cropping in Eastern Uttar Pradesh, Indian. *Farmer, 9*(5), 161–164.

Mishra, A. K., Das, R., Sahoo, S., & Biswal, B., (2018a). Global regulations and legislations on nanoparticles usage and application in diverse horizons. In: Turan, N. B., Engine, G. O., & Bilgili, M. S., (eds.), *Environmental Nanotechnology: Implications and Impact, Comprehensive Analytical Chemistry* (pp. 261–290). Elsevier.

Mishra, A. K., Grover, D., Das, R., Mishra, J., & Biswal, B., (2022). Carbon sequestration potential in agricultural systems. In: *Agro-biodiversity and Agri-ecosystem Management* (pp. 87–111). Springer, Singapore.

Mishra, A. K., Sharma, S., Pinaki, D., Grover, C., Pallavi, C., & Jyotiprakash, M., (2022). *Natural Farming Systems as Hotspots of Eco-Tourism and Their Potential for Sustainable Developmen*t. Nova Science Publishers, Inc. New York.

Mishra, A. K., Sheoran, H. S., Mishra, S., Pandey, A., Sah, D., Bhat, M. A., & Sharma, S., (2022). Effect of conservation agriculture on energy consumption and carbon emission. In: *Agriculture, Livestock Production and Aquaculture* (pp. 75–96). Springer, Cham.

Mishra, A. K., Shinjo, H., Jat, H. S., Jat, M. L., Jat, R. K., Funakawa, S., & Sutaliya, J. M., (2022). Farmers' perspectives as determinants for adoption of conservation agriculture practices in Indo-Gangetic plains of India. *Resources, Conservation & Recycling Advances, 15*, 200105.

Mishra, A. K., Singh, J., Kumar, V., Srivastava, R., & Srivastava, S., (2013). Standing carbon stock estimation in different tree species grown in dry tropical forests of Vindhyan highland, Mirzapur, India. *Ecol. Environ. Conserv., 19*(2), 401–407.

Mishra, A. K., Singh, S. P., Das, R., Kumari, V., Mishra, S., Sharma, S., & Sahoo, S., (2022). Innovative pathways to increase resource conservation and nutrient use efficiency in rice-wheat cropping systems for food security and decreased environmental footprints. In: *New Horizons in Wheat and Barley Research* (pp. 511–543). Springer, Singapore.

Mishra, A. K., Sinha, D. D., Grover, D., Mishra, S., Tyagi, R., Sheoran, H. S., & Sharma, S., (2022). Regenerative agriculture as climate smart solution to improve soil health and crop productivity thereby catalyzing farmers' livelihood and sustainability. In: *Towards Sustainable Natural Resources* (pp. 295–309). Springer, Cham.

Mishra, A. K., Yeggina, P. K., Sinha, D. D., Srivastava, A. K., & Sharma, S., (2022). Scope for production of pulses in rice fallow lands in South Asia. In: *Advances in Legumes for Sustainable Intensification* (pp. 9–26). Academic Press.

Morgan, J. A., Follett, R. F., Allen, L. H., Grosso, S. D., Derner, J. D., Dijkstra, F., Franzluebbers, A., et al., (2010). Carbon sequestration in agricultural land of the United States. *J. Soil Water Conserv., 65*, 6A–13A.

Murthy, I. K., Gupta, M., Tomar, S., Munsi, M., Tiwari, R., Hegde, G. T., & Ravindranath, N. H., (2013). Carbon sequestration potential of agroforestry systems in India. *J. Earth Sci. Clim. Change*, *4*(1), 1–7.

Nair, P. K. R., (2012). Carbon sequestration studies in agroforestry systems: A reality-check. *Agroforestry Systems, 86*, 243–253.

Nair, P. K. R., Kumar, B. M., & Nair, V. D., (2010). Agroforestry as a strategy for carbon sequestration. *J. Plant Nutr. SoilSci., 172*, 10–23.

Nair, P. R., & Nair, V. D., (2014). 'Solid–fluid–gas': The state of knowledge on carbon-sequestration potential of agroforestry systems in Africa. *Current Opinion in Environmental Sustainability, 6*, 22–27.

Nair, P. R., Nair, V. D., Kumar, B. M., & Haile, S. G., (2009). Soil carbon sequestration in tropical agroforestry systems: A feasibility appraisal. *Environmental Science & Policy, 12*(8), 1099–1111.

Nayak, P., Patel, D., Ramakrishnan, B., Mishra, A. K., & Samantaray, R. N., (2009). Long-term application effects of chemical fertilizer and compost on soil carbon under intensive rice–rice cultivation. *Nutrient Cycling in Agroecosystems, 83*, 259–269.

Newaj, R., & Dhyani, S. K., (2008). Agroforestry for carbon sequestration: Scope and present status. *Indian Journal of Agroforestry, 10*, 1–9.

Noiha, N. V., Zapfack, L., Hamadou, M. R., et al., (2018). Floristic diversity, carbon storage and ecological services of eucalyptus agrosystems in Cameroon. *Agroforest Syst., 92*, 239–250. https://doi.org/10.1007/s10457-017-0130-5.

Nowak, D. J., & Crane, D. E., (2002). Carbon storage and sequestration by urban trees in the USA. *Environmental Pollution, 116*(3), 381–389.

Oelbermann, M., Voroney, R. P., & Gordon, A. M., (2004). Carbon sequestration in tropical and temperate agroforestry systems: A review with examples from Costa Rica and Southern Canada Agriculture. *Ecosystems and Environment, 104*, 359–377.

Ojima, D. S., Dirks, B. O., Glenn, E. P., Owensby, C. E., & Scurlock, J. O., (1993). Assessment of C budget for grasslands and drylands of the world. *Water, Air, and Soil Pollution, 70*(1), 95–109.

Osborne, C. P., & Beerling, D. J., (2006). Nature's green revolution: The remarkable evolutionary rise of C4 plants. *Philosophical Transactions of the Royal Society of London; Serie B Biological Sciences, 361*(1465), 173–194. doi:10.1098/rstb.2005.1737.

Pan, Y., Birdsey, R. A., Fang, J., Houghton, R., Kauppi, P. E., Kurz, W. A., & Hayes, D., (2011). A large and persistent carbon sink in the world's forests. *Science, 333*(6045), 988–993.

Post, W. M., & Kwon, K. C., (2000). Soil carbon sequestration and land-use change: Processes and potential. *Global Change Biology, 6*, 317–327.

Pouyat, R. V., Yesilonis, I. D., & Nowak, D. J., (2006). Carbon storage by urban soils in the United States. *Journal of Environmental Quality, 35*(4), 1566–1575.

Prasad, J. V. N. S., Srinivas, K., Srinivasarao, C., Ramesh, C., Venkatravamma, K., & Venkateswarlu, B., (2012). Biomass productivity and carbon stocks of farm forestry and agroforestry systems of leucaena and Eucalyptus in Andhra Pradesh, India. *Current Science, 103*(5), 536–540.

Pu, C., Kan, Z., Liu, P., Ma, S., Qi, J., Zhao, X., & Zhang, H., (2019). Residue management induced changes in soil organic carbon and total nitrogen under different tillage practices in the North China plain. *Journal of Integrative Agriculture, 18*(6), 1337–1347.

Ramos, H. M. N., Vasconcelos, S. S., Kato, O. R., et al., (2018). Above- and belowground carbon stocks of two organic, agroforestry-based oil palm production systems in eastern Amazonia. *Agroforest Syst., 92*, 221–237. https://doi.org/10.1007/s10457-017-0131-4.

Ruddiman, W. F., (2005). *Plows, Plagues and Petroleum: How Humans Took Care of Climate*. Princeton: Princeton University Press.
Salas, C. E. M., (2017). *The Role of Factors That Influence the Adoption of the Australian Carbon Farming Initiative-Emissions Reduction Fund: A Mixed Methods Study*. Ph.D. Thesis, James Cook University, Douglas, Australia.
Scharlemann, J. P., Tanner, E. V., Hiederer, R., & Kapos, V., (2014). Global soil carbon: Understanding and managing the largest terrestrial carbon pool. *Carbon Management, 5*(1), 81–91.
Schlesinger, W. H., (2000). Carbon sequestration in soils: Some cautions amidst optimism. *Agriculture, Ecosystems & Environment, 82*(1–3), 121–127.
Shah, T. I., Shah, A. M., Bangroo, S. A., Sharma, M. P., Aezum, A. M., Kirmani, N. A., Lone, A. H., Jeelani, M. I., Rai, A. P., Wani, F. J., et al., (2022). Soil quality index as affected by integrated nutrient management in the Himalayan foothills. *Agronomy, 12*, 1870. https:// doi.org/ 10.3390/agronomy12081870.
Sharma, M., Kaushal, R., Kaushik, P., & Ramakrishna, S., (2021). Carbon farming: Prospects and challenges. *Sustainability, 13*, 11122. https://doi.org/10.3390/ su131911122.
Singh, S. K., Thawale, P. R., Sharma, J. K., Gautam, R. K., Kundargi, G. P., & Juwarkar, A. A., (2015). Carbon sequestration in terrestrial ecosystems. In: *Hydrogen Production and Remediation of Carbon and Pollutants* (pp. 99–131). Springer, Cham.
Smith, P., (2005). An overview of the permanence of soil organic carbon stocks: Influence of direct human-induced, indirect and natural effects. *Eur. J. Soil Sci., 56*(5), 673–680. doi: 10.1111/j. 1365-2389.2005.00708.x.
Sommer, R., Piggin, C., Feindel, D., Ansar, M., Delden, L., Shimonaka, K., Abdalla, J., et al., (2014). Effects of zero tillage and residue retention on soil quality in the mediterranean region of Northern Syria. *Open Journal of Soil Science, 4*, 109–125. doi: 10.4236/ ojss.2014.43015.
Srivastava, K., Jat, H., Meena, M., Choudhary, M., Mishra, A., & Chaudhari, S. K., (2016). Long-term impact of different cropping systems on soil quality under silty loam soils of Indo-Gangetic plains of India. *J. Nat. Appl. Sci., 8*(2), 584–587. https://doi.org/10.31018/ jans.v8i2.841.
Stockmann, U., Adams, M. A., Crawford, J. W., Field, D. J., Henakaarchchi, N., Jenkins, M., Minasny, B., et al., (2013). The knowns, known unknowns and unknowns of sequestration of soil organic carbon. *Agriculture, Ecosystems & Environment, 164*, 80–99.
Strack, M., (2008). *Peatlands and Climate Change* (pp. 13–23). Calgary: University of Calgary. ISBN 978-952-99401-1-0.
Sur, P., Mandal, M., & Das, D. K., (2010). Effect of integrated nutrient management on soil fertility and organic carbon in cabbage (*Brassica oleracea* var. capitata) growing soils. *Indian Journal of Agricultural Sciences, 80*(8), 38–41.
Swamy, S. L., Puri, S., & Singh, A. K., (2003). Growth, biomass, carbon storage and nutrient distribution in Gmelina arborea Roxb. stands on red lateritic soils in Central India. *Bioresource Technology, 90*, 109–126.
Szostek, M., Szpunar-Krok, E., Pawlak, R., Stanek-Tarkowska, J., & Ilek, A., (2022). Effect of different tillage systems on soil organic carbon and enzymatic activity. *Agronomy, 12*, 208. https://doi.org/10.3390/ agronomy12010208.
UNEP Risoe. (2009). *The CDM/JI Pipeline Analysis and Database*. http://cdmpipeline.org (accessed on 22 April 2024).

Valdivia, C., Barbieri, C., & Gold, M. A., (2012). Between forestry and farming: Policy and environmental implications of the barriers to agroforestry adoption. *Can. J. Agric. Econ. Rev. Can. Agroecon., 60*, 155–175.

Velasco, E., Roth, M., Norford, L., & Molina, L. T., (2016). Does urban vegetation enhance carbon sequestration? *Landscape and Urban Planning, 148*, 99–107.

Vineela, C., Wani, S. P., Srinivasarao, C., Padmaja, B., & Vittal, K. P. R., (2008). Microbial properties of soils as affected by cropping and nutrient management practices in several long term manurial experiments in the semi-arid tropics of India. *Applied Soil Ecology, 40*, 165–173.

Viswanath, S., Lubina, P. A., Subbana, S., & Sandhya, M. C., (2018). Traditional agroforestry systems and practices: A review. *Advanced Agricultural Research & Technology Journal, II*(1).

Watson, R. T., Noble, I. R., Bolin, B., Ravindranath, N. H., Verardo, D. J., & Dokken, D. J., (2000). *Land Use, Land-Use Change and Forestry* (p. 375). Cambridge, UK: Cambridge University Press. ISBN-13: 9780521800839.

Wiltshire, S., & Beckage, B., (2022). Soil carbon sequestration through regenerative agriculture in the US state of Vermont. *PLOS Clim., 1*(4), e0000021. https://doi.org/10.1371/journal.pclm.0000021.

Zheng, H., Liu, D., Liao, X., Miao, Y., Li, Y., Li, J., Yuan, J., Chen, Z., & Ding, W., (2022). Field-aged biochar enhances soil organic carbon by increasing recalcitrant organic carbon fractions and making microbial communities more conducive to carbon sequestration. *Agriculture, Ecosystems & Environment, 340,* 108177.

Zomer, R. J., Trabucco, A., Coe, R., & Place, F., (2009). *Trees on Farm: Analysis of Global Extent and Geographical Patterns of Agroforestry*. ICRAF Working Paper no. 89. Nairobi, Kenya: World Agroforestry Center.

Index

A

D

G

Q

R

S

T

U

V